SYNTHETIC FUELS

SYNTHETIC FUELS

RONALD F. PROBSTEIN
Ford Professor of Engineering, Emeritus
Massachusetts Institute of Technology

R. EDWIN HICKS
Consulting Chemical Engineer

DOVER PUBLICATIONS, INC.
Mineola, New York

CONTENTS

PREFACE TO THE DOVER EDITION

When this book was first published in 1982, it appeared that the production of synthetic fuels was on the verge of becoming a reality. At that time oil supplies appeared to be dwindling in relation to demand, and the production of clean gaseous and liquid hydrocarbon fuels from coal, oil shale, tar sands, and biomass was envisaged as being capable of supplying society's needs until a transition to more abundant clean energy sources such as solar energy could be accomplished. Available coal supplies were and still are estimated to be in the range of several hundred years. But predictions of the early demise of plentiful and cheap oil and natural gas were overstated, at least on the scale of two decades. Recently oil and gas prices have increased dramatically as estimates once again emphasize the increasing world demand and limited lifetime of these remaining resources, with fifty years a common figure used to express economic depletion.

In the more than two decades since the original publication of *Synthetic Fuels*, the fundamental concepts of fuel conversion presented in the book remain unchanged, because emphasis was placed on the fundamental chemistry and physics of the conversion processes with detailed technologies utilized to illustrate the concepts and to describe the state of the art. With the world seemingly awash with cheap fuel, regrettably there has been little technology development since the original publication of this text. This is not to say that nothing has been done, but rather that any efforts have been of small scale in contrast to the major industrial development that is needed for effective large-scale production of synthetic fuels.

Many thanks are due John Grafton, who understood the basic fundamental character of the book and the need to make it available again through publication in an affordable edition. The original edition is unaltered except for the correction of a number of small errors.

Ronald F. Probstein
R. Edwin Hicks
September 2005

PREFACE

The production of synthetic fuels to replace dwindling supplies of natural fuels is already a reality in some countries, and in the United States is an urgent national priority on the verge of becoming a major industry. The scale of this development will probably be larger than any previous technological undertaking. To carry it out will require large numbers of engineers and scientists from many backgrounds to be trained in, and have an understanding of, the new field of synthetic fuels. The seeds of this field are to be found in the fact that the conversion of carbonaceous materials to synthetic fuels can be viewed as a process of hydrogenation. This can be taken to imply that the subject is built up from a limited number of fundamental chemical, physical, and engineering principles. In this book we endeavor to present synthetic fuels as a unified engineering subject, while at the same time recognizing that many of its principles are well understood aspects of various engineering fields. Although the subject is a rapidly developing one, we feel that a book which attempts its definition is needed now. We recognize that advances in the science and technology of synthetic fuels mandate future revision; however, by concentrating on what we believe to be the fundamentals, we hope that we have written a text which will be both timely and lasting.

The book is an outgrowth of a graduate course taught at M.I.T. on synthetic fuel production. The course, like the book, is directed toward beginning graduate students and advanced undergraduates, mainly, but not exclusively, in chemical and mechanical engineering. By including material on the necessary chemical and physical fundamentals, we have tried to make the text sufficiently self-contained so that it will prove of value to students from a wide variety of engineering and scientific disciplines, including environmental, mining, petroleum, and industrial engineering, as well as chemistry. It is our intention that the book also serve as a reference and guide for professionals desirous of understanding or entering this new and exciting field. In presenting the review of chemical and physical fundamentals, we have utilized examples of relevance to synthetic fuels which are applied later in the text. For those familiar with the particular subject the examples themselves will be of interest, while for those who have had only limited or no exposure, the examples provide a vehicle for understanding or reviewing the topic covered.

Synthetic fuels are generally understood to include liquid and gaseous fuels, as well as clean solid fuels, produced by the conversion of coal, oil shale, or tar sands, and various forms of biomass. The aim of the book is to provide an understanding of the methods and processes for carrying this out. Although every effort has been made to be reasonably comprehensive, the book is not a design manual, and it makes no attempt to examine every possible process and product fuel. Emphasis has been placed on the conversion of coal, oil shale, and tar sands, for which a unified picture can be developed, although biomass conversion is covered. The discussions on resources and technology development have been biased toward developed and developing technologies as seen from an American perspective. In view of the many areas encompassed by the field, the choice of material is in part dictated by the authors' preference and background. We have been helped in this regard by the fact that one of us is a chemical engineer and the other a mechanical engineer.

Wherever possible we have tried to present facts sufficient to enable the reader to draw general conclusions regarding the suitability of a process or product from rational considerations, without reliance on developers' or manufacturers' detailed design and economic studies, or empirical results. Extended or lengthy calculational details and approaches have been avoided. Where we give unit or plant mass and energy balances, they are almost always simplified and are intended to illustrate a method, a concept, or a principle, and not to replace detailed engineering calculations. Apart from pedagogical reasons, this approach is often necessary because manufacturers or developers have limited their published data for proprietary reasons. As a result we have had to fill in gaps, so that our estimates may not always conform to the latest practice or design but should be sufficient for the illustrative purposes intended.

Many times questions have arisen due to conflicting data reported in the literature. In these cases, we have never tried to sweep the problem under the rug, but have attempted to provide an answer based on the principles discussed. Occasionally we have not been successful and have been able only to ask the question. In this regard, we would note that although much of the material is drawn from published sources, much is not and represents unpublished work of the authors. Every effort has been made to acknowledge the work of others. However, because the book is intended as a text and because we do not wish to burden the reader with an overwhelming bibliography, we have often chosen not to reference an original source directly, but instead to reference books or compilations in which the original work is described and cited. Our directly cited references have been drawn mainly from the English language literature.

SI units are used consistently throughout the text. In the United States, where the transition to these units is still in progress, this may cause some moments of hesitation, particularly for professional readers in industry where British units are still prevalent in design and engineering estimations. In some cases nominal units on which plant designs are based, such as "barrels per day," are given in parentheses next to the appropriate SI unit.

Many of the disciplines on which the subject of synthetic fuels is built are applied in a number of fields, each with its own preferred terminology and nomenclature. This is most evident with thermodynamics. In view of the overall dominance of chemical processing, we have generally adopted systems used in chemistry, and they are described in context.

Problem solving is an important part of any student's learning experience, and we apologize for not having included problems in the text. This was done solely in the interest of not delaying publication of the book. However, the student will find a sufficient number of "problems for the reader" throughout the text to partly serve this purpose. It is our intention to prepare a problem manual within as short a time after publication of the book as is practicable.

This book is dedicated to our wives whose assistance and forbearance made it possible.

ACKNOWLEDGMENTS

More than a formal acknowledgment is due Water Purification Associates and the Massachusetts Institute of Technology for their backing. WPA provided support for the authors as well as aid and services for the project. M.I.T. on its part relieved one of us from teaching duties so that the book might be completed rapidly. In this regard, Herbert H. Richardson, chairman of the Mechanical Engineering Department at M.I.T., deserves special thanks for his encouragement.

The book owes its existence to the course on Synthetic Fuel Production taught at M.I.T. The first time the course was given, a number of individuals served as guest lecturers and contributed their ideas on the subject matter. We wish particularly to thank Lawrence E. Swabb, Jr. of Exxon Research and Engineering Co., Bernard S. Lee of the Institute of Gas Technology, Steven R. Reznek of the U.S. Environmental Protection Agency, and Sidney Katell of West Virginia University. We especially want to acknowledge Lawrence Swabb, who prepared a written set of notes on the topic of coal liquefaction from which we have benefited greatly.

Many individuals from WPA provided us with technical assistance and written material which was invaluable in our presentation of the various topics. We wish in particular to acknowledge the contributions of Olu Aiyegbusi on biomass conversion, David J. Goldstein on chemistry fundamentals and gas purification, Mark R. S. Manton on catalysis, Yong C. Hong on chemical equilibrium, and John G. Casana on environmental aspects. Other individuals to whom we are indebted for assistance are Eric H. Reichl, consultant, who clarified some historical aspects of synthetic fuel development; R. Tracy Eddinger of COGAS Development Co., who reviewed and commented on our section on the COGAS and COED processes; Graham C. Taylor of the Denver Research Institute who reviewed and commented on our discussion of economic considerations; and Jack B. Howard, John P. Longwell, Adel F. Sarofim, and James Wei of M.I.T., who kindly supplied us with unpublished manuscripts.

An especial debt of gratitude is due Nancy D. Flaherty of WPA for her skillful processing of the manuscript onto magnetic disks for direct production of the book by electronic means. We express our grateful appreciation to Robert H. Dano of WPA who expertly prepared all of the figures in a form that enabled their direct reproduction. We also thank Margaret Conlin of M.I.T. for her capable handling of many of the secretarial details, and Derith Glover of WPA for her fine editorial assistance.

Photographs for the book were furnished by many groups and individuals to whom collectively we extend our appreciation here, and individually where each photograph appears.

Finally, we sincerely thank our editors, Diane D. Heiberg and Madelaine Eichberg, and our publisher for their many efforts and cooperation in producing a text not only of high quality, but one printed by the newest methods available in order to shorten the publication time.

Ronald F. Probstein
R. Edwin Hicks

SYNTHETIC FUELS

ONE

INTRODUCTION

1.1 SYNTHETIC FUELS AND THEIR MANUFACTURE

Gaseous or liquid synthetic fuels are obtained by converting a carbonaceous material to another form. In the United States the most abundant naturally occurring materials suitable for this purpose are coal and oil shale. Tar sands are also suitable, and large deposits are located in Canada. The conversion of these raw materials is carried out to produce synthetic fuels to replace depleted, unavailable, or costly supplies of natural fuels. However, the conversion may also be undertaken to remove sulfur or nitrogen that would otherwise be burned, giving rise to undesirable air pollutants. Another reason for conversion is to increase the calorific value of the original raw fuel by removing unwanted constituents such as ash, and thereby to produce a fuel which is cheaper to transport and handle.

Biomass can also be converted to synthetic fuels and the fermentation of grain to produce alcohol is a well known example. In the United States, grain is an expensive product which is generally thought to be more useful for its food value. Wood is an abundant and accessible source of bio-energy but it is not known whether its use to produce synthetic fuels is economical. The procedures for the gasification of cellulosic materials have much in common with the conversion of coal to gas. We consider the conversion of biomass in the book, but primary emphasis is placed on the manufacture of synthetic fuels from coal, oil shale, and tar sands. Most of the conversion principles to be discussed are, however, applicable to the spectrum of carbonaceous or cellulosic materials which occur naturally, are grown, or are waste.

For our purposes we regard the manufacture of synthetic fuels as a process of hydrogenation, since common fuels such as gasoline and natural gas have a higher hydrogen content than the raw materials considered. The source of the hydrogen which is added is water. The mass ratio of carbon to hydrogen for a variety of fuels is shown in Table 1.1. Generally, the more hydrogen that is added to the raw material, the lower is the boiling point of the synthesized product. Also, the more hydrogen that must be added, or alternatively the more carbon which must be removed, the lower is the overall conversion efficiency in the manufacture of the synthetic fuel.

Table 1.1 Carbon-to-hydrogen ratio for various fuels

Fuel	C/H Mass Ratio	Molar Representation
Bituminous coal	~15	$CH_{0.8}$
Benzene	12	$CH_{1.0}$
Crude oil	~ 9	$CH_{1.33}$
Gasoline	6	CH_2
Methane	3	CH_4

The organic material in both tar sands and in high-grade oil shale has a carbon-to-hydrogen mass ratio of about 8, which is close to that of crude oil and about half that of coal. For this reason, processing oil shale and tar sands to produce liquid fuels is considerably simpler than making liquid fuels from coal. However, the mineral content of rich tar sands in the form of sand or sandstone is about 85 mass percent, and the mineral content of high-grade oil shale, which is a fine-grained sedimentary rock, is about the same. Therefore, very large volumes of solids must be handled to recover relatively small quantities of organic matter from oil shale and tar sands. On the other hand, the mineral content of coal in the United States averages about 10 percent.

In any conversion to produce a fuel of lower carbon-to-hydrogen ratio, the hydrogenation of the raw fossil fuel may be direct, indirect, or by pyrolysis, either alone or in combination. Direct hydrogenation involves exposing the raw material to hydrogen at high pressure. Indirect hydrogenation involves reacting the raw material with steam, with the hydrogen generated within the system. In pyrolysis the carbon content is reduced by heating the raw hydrocarbon until it thermally decomposes to yield solid carbon, together with gases and liquids having higher fractions of hydrogen than the original material.

To obtain fuels that will burn cleanly, sulfur and nitrogen compounds must be removed from the gaseous, liquid, and solid products. As a result of the hydrogenation process, the sulfur and nitrogen originally present in the raw fuel are reduced to hydrogen sulfide and ammonia, respectively. Hydrogen sulfide and ammonia are present in the gas made from coal or released during the pyrolysis of oil shale and tar sands, and are also present in the gas generated in the hydrotreating of pyrolysis oils and synthetic crudes.

Synthetic fuels include low-, medium-, and high-calorific value gas; liquid fuels such as fuel oil, diesel oil, gasoline; and clean solid fuels. Consistent with SI units, we use the shorthand terms low-, medium-, and high-CV gas, where CV denotes calorific value, in place of the terms low-, medium-, and high-Btu gas which are appropriate to British units. Low-CV gas, often called producer or power gas, has a calorific value of about 3.5 to 10 million joules per cubic meter (MJ/m^3). This gas is an ideal turbine fuel whose greatest utility will probably be in a gas-steam combined power cycle for the generation of electricity at the location where it is produced. Medium-CV gas is loosely defined as having a calorific value of about 10–20 MJ/m^3, although the upper limit is somewhat arbitrary, with existing gasifiers yielding somewhat lower values. This gas is also termed power gas or sometimes industrial gas, as well as synthesis gas. It may be used as a fuel gas, as a source of hydrogen for the direct liquefaction of coal to liquid fuels, or for the synthesis of methanol and other liquid fuels. Medium-CV gas may also be used for the production of high-CV gas, which has a calorific value in the range of about 35 to 38 MJ/m^3, and is normally composed of more than 90 percent methane. Because of its high calorific value, this gas is a substitute for natural gas and is suitable for economic pipeline transport. For these reasons it is referred to as substitute natural gas (SNG) or pipeline gas. Lom and Williams[1] have pointed out that originally SNG stood for synthetic natural gas, but it was observed that what was natural could not very well be synthetic.

In Figure 1.1 are shown the principal methods by which the synthetic gases can be produced from coal. Gas can be manufactured by indirect hydrogenation by reacting steam with coal either in the presence of air or oxygen. When air is used, the product gas will be diluted with nitrogen and its calorific value will be low in comparison with the gas manufactured using oxygen. The dilution of the product gas with nitrogen can be avoided by supplying the heat needed for the gasification from a hot material that has been heated with air, in a separate furnace, or in the gasifier itself before gasification. In all of the cases, the gas must be cleaned prior to using it as a fuel. This purification step involves the removal of the hydrogen sulfide, ammonia, and carbon dioxide, which are products of the gasification. Medium-CV gas, consisting mainly of carbon monoxide and hydrogen, can be further upgraded by altering the carbon monoxide-to-hydrogen ratio catalytically and then, in another catalytic step, converting the resulting synthesis gas mixture to methane. A high-CV gas can be produced by direct hydrogenation, termed hydrogasification, in which hydrogen is contacted with the coal. A procedure still under development, which allows the direct production of methane, is catalytic gasification. In this method the catalyst accelerates the steam gasification of coal at relatively low temperatures and also catalyzes the upgrading and methanation reactions at the same low temperature in the same unit.

Gas can also be produced by pyrolysis, that is, by the distillation of the volatile components. Oil shale or tar sands are generally not thought of as primary raw materials for gas production, although the use of oil shale has been discussed.

Clean synthetic liquid fuels can be produced by several routes, as shown in Figures 1.2 to 1.4. For example, in indirect liquefaction (Figure 1.2), coal is first gasified and then the liquid fuel is synthesized from the gas. This procedure is not thermally efficient, relating to the fact that the carbon bonds in the coal must first be broken, as in gasification, and then in a further step some of them must be put back together again. Another procedure, illustrated in Figure 1.3, is pyrolysis, the distillation of the natural oil out of the coal, shale, or tar sands. The oil vapors are condensed, the resulting pyrolysis oil is treated with hydrogen, and the sulfur and nitrogen in it is reduced. This is similar to the procedure used in upgrading crude oil in a refinery to produce a variety of liquid fuels. Pyrolysis may also be carried out in a hydrogen atmosphere, a process termed hydropyrolysis, in order to increase the liquid and gas yield. In direct liquefaction (Figure 1.4) there are two basic procedures, hydroliquefaction and solvent extraction. In hydroliquefaction the coal is mixed with recycled coal oil and, together with hydrogen, fed to a high pressure catalytic reactor where

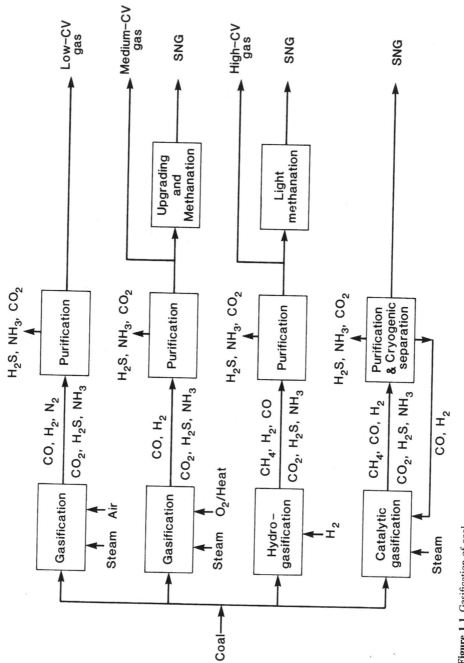

Figure 1.1 Gasification of coal.

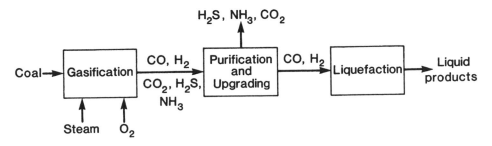

Figure 1.2 Indirect liquefaction of coal.

the hydrogenation of the coal takes place. In solvent extraction, also termed "solvent refining," the coal and the hydrogen are dissolved at high pressure in a recycled coal-derived solvent which transfers the hydrogen to the coal. After phase separation, the coal liquid is cleaned and upgraded by refinery procedures to produce liquid fuels. In solvent refining, with a low level of hydrogen transfer, a solid, relatively clean fuel termed "solvent refined coal" is obtained.

In order to compare the size and output of synthetic fuel plants producing different products, a nominal reference standard is desirable and should incorporate both the calorific value of the product and the rate of production. In the United States the nominal accepted calorific values of various synthetic fuel products are shown in Table 1.2. These values are all defined in British units and we have shown their SI equivalents. It should be emphasized that the calorific value per barrel of liquid fuel is a nominal one that holds approximately for synthetic crudes, fuel oils, and some transportation fuels.

Although there are not now "standard-size" plants, nor are there likely to be in the future, there has nevertheless grown up in the U.S. literature certain nominal plant sizes which have come to be accepted as "standard." These sizes are shown in Table 1.3 together with their SI equivalents. The nominal standard sizes shown were not necessarily accepted because they are economically optimum. Indeed, the accepted nominal liquid fuel plant size has changed from 100 thousand barrels per day, a value which was adopted because it was believed that the synthetic fuel products would be upgraded in existing refineries which no longer would have a source of natural petroleum crude.

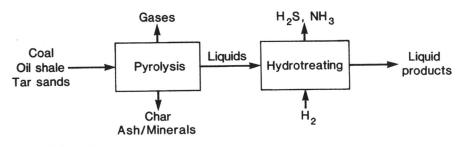

Figure 1.3 Pyrolysis.

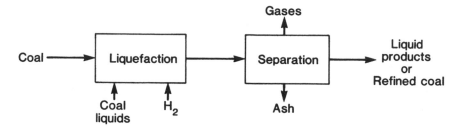

Figure 1.4 Direct liquefaction of coal.

For the purposes of this book, where needed we have chosen to adopt nominal standard plant sizes that are convenient in SI units and not at great variance with the commonly accepted sizes in British units. We are justified in this to the extent that the sizes in Table 1.3 are arbitrary and to the extent that they do not necessarily represent the actual sizes of plants to be built. Our nominal standards are shown in Table 1.4, where it can be seen that the calorific value outputs of the plants are more nearly comparable than for the sizes in Table 1.3. Moreover, the "ten million" unit size may ease the transition to SI units for those accustomed to "barrels per day" and "standard cubic feet per day." In any case, we emphasize the nominal character of the selection, the main purpose of which is to provide a basis for comparison amongst the different types of plants.

1.2 HISTORY

Contrary to widespread opinion, the manufacture of synthetic fuels is neither new nor has it been limited in the past to small-scale development. This is perhaps most true of the production of gas from coal.[1,2] As early as 1792, Murdoch, a Scottish engineer, distilled coal in an iron retort and lighted his home with the coal gas produced.[3] By the early part of the nineteenth century, gas manufactured by the distillation of coal was introduced for street lighting, first in London in 1812, following which its use for this purpose spread rapidly throughout the major cities of the world. The early coal gas contained noxious and poisonous impurities, but techniques were soon developed for cleaning and purification. Gas distribution networks were rapidly built up in most large- and medium-size cities, particularly in the industrialized European countries,

Table 1.2 Nominal calorific values of synthetic fuels defined in British units

Fuel	British	SI Equivalent
Synthetic liquid fuels	6×10^6 Btu/bbl*	40 MJ/L
Substitute natural gas	1×10^3 Btu/scf†	37 MJ/m³
Refined coal	16×10^3 Btu/lb	37 MJ/kg

* bbl = barrel = 42 U.S. gallons.
† scf = standard cubic foot.

Table 1.3 Nominal standard-size synthetic fuel plants defined in British units

	British		SI Equivalent	
Product	Output	10^{11} Btu/day	Output	10^8 MJ/d
Synthetic liquid fuels	50×10^3 bbl/day	3.0	8.0×10^6 L/d	3.2
Substitute natural gas	250×10^6 scf/day	2.5	7.1×10^6 m³/d	2.7
Refined coal	10×10^3 ton/day	3.2	9.1×10^6 kg/d	3.4

along with gasworks for the manufacture of the coal gas. By the last quarter of the nineteenth century, coal gas came into widespread use for home lighting. Toward the end of the nineteenth century its use as a source of heat for domestic and industrial applications took on even greater importance.

Coal gas as distributed contained about 50 percent hydrogen, from 20 to 30 percent methane, with carbon monoxide the remaining principal constituent. Its calorific value was about 19 MJ/m³ and this value served as the benchmark for the "town gas" industry. A solid coke byproduct of limited utility was produced in coal processing by "carbonization." However, in the latter half of the nineteenth century, complete gasification of coke was achieved commercially by means of a cyclic gas generator in which the coke was alternately blasted with air to provide heat and steam to generate "blue water gas," a name given to the gas because it burned with a blue flame. The discovery of blue gas is attributed to Fontana in 1780, who proposed making it by passing steam over incandescent carbon.[4] The blue gas was composed of about 50 percent hydrogen and 40 percent carbon monoxide, with the remainder about equal parts carbon dioxide and nitrogen. It had a calorific value of about 11 MJ/m³, and to increase this value to that of town gas it was enriched with cracked oil gas.

At about the same time as the cyclic gas generator was commercialized, the complete gasification of coal was carried out commercially in continuous fixed bed "gas producers," which manufactured a low-CV gas with a calorific value of 5–6.5 MJ/m³ by reaction of air and steam with coal. By the turn of the century, gas producers had come into widespread use for on-site generation of gas for industrial heating.

In the first half of the twentieth century the availability of natural gas with a calorific value of 37 MJ/m³ began to displace the use of manufactured gas. In the United States following the end of World War II, with discoveries of large quantities of natural gas in Texas and because of the growth of the natural gas pipeline transportation network, the town gas industry virtually disappeared. The same was also

Table 1.4 Nominal standard-size synthetic fuel plants defined in SI units

	SI		British Equivalent	
Product	Output	10^8 MJ/d	Output	10^{11} Btu/day
Synthetic liquid fuels	10^7 L/d	4.0	63×10^3 bbl/day	3.8
Substitute natural gas	10^7 m³/d	3.7	350×10^6 scf/day	3.5
Refined coal	10^7 kg/d	3.7	11×10^3 ton/day	3.5

true in Europe where low-cost natural gas became available while the price of coal rose.

Since the Arab oil embargo of 1973, in coal-producing countries such as the United States, attention has once again turned to the alternatives for manufacturing gas from coal. Development work is now directed toward improving gasifier technology and unit output for the large-scale production of industrial gas, gas to be used in the manufacture of SNG, and synthesis gas to be used for the manufacture of liquid fuels and chemicals. Development of processes to produce SNG in single units is also being carried forward. A number of commercial-scale coal conversion projects, each with an output of 3.5 million cubic meters per day or greater, are also underway, including ones for the production of medium-CV industrial fuel gas as well as SNG.

The history of coal liquefaction is considerably more recent than that of coal gasification. Direct coal liquefaction[5,6] can be traced back to the work of Bergius in Germany, recorded in his publications from 1912 to 1926. In his studies and developments, he produced coal liquids by reacting pulverized coal or coal-oil slurries with hydrogen at high pressures and temperatures. In 1931 he received the Nobel Prize in Chemistry for his work. The development of commercial-size coal hydrogenation units for the production of motor fuels was begun in Germany in 1926. By 1939, production was estimated for the year to be 1.1 million tons, which is roughly equivalent to a production of 4 million liters per day of gasoline. In 1935, a commercial coal hydrogenation plant was started up in England and smaller-scale experimental plants were put in operation elsewhere throughout the world. During World War II, production of liquid fuels, mostly aviation gasoline, from coal and coal tars by direct liquefaction, was greatly expanded in Germany. Over a one year period it is estimated to have peaked at an average production of 4.2 million tons annually,[7] which is equivalent to a rate of about 16 million liters of gasoline per day. This output came from 12 plants, the largest of which produced some 2.7 million liters per day of gasoline equivalent.

Following the end of World War II, the plants for the production of liquid fuels from coal were phased out as they became uneconomical in an era of cheap natural petroleum crudes. Work in the United States on direct coal liquefaction by the Bergius process continued under the sponsorship of the Bureau of Mines, which from 1949 to 1953 operated a pilot plant with a nominal coal feed rate of 45 tons per day and a nominal synthetic fuel oil production rate of 32 thousand liters per day. Further efforts waned until 1973 when interest in coal hydrogenation, particularly in the United States, increased markedly. Most developments since that time have been in the direction of designing processes to operate at milder conditions than those prevailing in the original Bergius plants. The German units were run at pressures from 25 to 70 MPa and temperatures from 450 to 500°C. Second generation units now being developed in the United States run at pressures from 10 to 20 MPa at temperatures of about 450°C. These processes have not operated at commercial scale, although medium-size pilot plants feeding up to 45 tons per day of coal have been run, while pilot plants of 5 times and 12 times this capacity are now starting up. Demonstration plants moving to 5500-ton-per-day coal feeds, or about 2.5 million liters per day of synthetic crude output, are underway for completion in the mid-1980s.

The principal method of indirect liquefaction is to react carbon monoxide and hydrogen in the presence of a catalyst to form hydrocarbon vapors, which are then condensed to liquid fuels. This procedure for synthesizing hydrocarbons is based on the work of Fischer and Tropsch in Germany in the 1920s.[6,8] In the mid-1930s a number of Fischer-Tropsch plants were constructed and by 1938 over 590 thousand tons of oil and gasoline were being produced annually in Germany from synthesis gas manufactured from coal,[9] an amount equivalent to some 2.3 million liters per day of gasoline. During the 1930s plants were also constructed in Japan, Britain, and France. In 1938, Germany slowed development of Fischer-Tropsch plants in favor of direct liquefaction. The reason for this was that the synthetic fuels industry was to be rapidly expanded in preparation for war, and the direct liquefaction plants were technologically far more developed and of much larger capacity. Reactor trains in the direct hydrogenation plants were eventually built that processed up to about 350 tons per day of coal, to yield 250 thousand liters per day of gasoline. On the other hand, the individual reactors in the early Fischer-Tropsch plants yielded only some 5 thousand liters per day of gasoline, so in the first plant there were literally 100 such reactors.[10] As a result of the German decision, the maximum nominal capacity during World War II from 9 Fischer-Tropsch plants went up to only 660 thousand tons, though little more than half this amount was actually produced during 1943.[9] As with the direct liquefaction facilities, those that were not destroyed during World War II were shut down subsequently as cheap petroleum crude and natural gas became available.

One country which has continued to pursue the development of indirect coal liquefaction by Fischer-Tropsch synthesis is South Africa. Since 1955 the South African Coal, Oil, and Gas Corporation has operated the Sasol plant for the production of motor fuels both for commercial and development purposes. Through 1980 the plant was the only fully operational commercial-size facility for the production of motor fuels from coal, producing about 1.3 million liters per day of gasoline and diesel fuel. A second generation plant, Sasol II, designed for about 8 million liters per day of motor fuels, is scheduled for initial production in 1981, while a third plant of the same size is also under construction with startup scheduled for 1984. The individual reactor units in these plants have capacities about 100 times greater than the original units commercialized in Germany.

Compounds of carbon, hydrogen, and oxygen—such as methanol—can also be catalytically synthesized from a mixture of carbon monoxide and hydrogen.[6] In fact, the synthesis of methanol predates Fischer-Tropsch synthesis. The procedure was described in a German patent issued to Badische Anilin und Soda Fabrik in 1913. The production of methanol from synthesis gas, or water gas as it was then called, in the original BASF process took place at quite high pressures of from 100 to 150 MPa. In modern procedures these pressures have been reduced considerably. Methanol synthesis quickly became a well-established process and by 1940 in the United States alone production had almost reached 500 thousand liters per day. After World War II, low-cost natural gas and light petroleum distillates replaced coal as the feedstock. Although little methanol is today being manufactured from coal, it is likely that this will change. Moreover, a development by Mobil, wherein methanol is converted to high octane gasoline by means of specially engineered catalysts, is a promising indirect liquefaction route. A pilot plant to manufacture 16 thousand liters per day of gasoline from coal using this process is underway in West Germany, as is a 2-million-liter-per-day commercial plant in New Zealand where natural gas will be the feedstock.

Oil shale deposits occur in many countries of the world, although the United States has particularly large reserves. In the United States, as a result of the volatile economics of production, oil shale development has been turned on and off for more than a century. The earliest shale oil industry was started in France in 1838 where oil shale was distilled to make lamp fuel. In 1862, production of oil from shale was begun in Scotland. The French industry operated intermittently up to the late 1950s when it was shut down, although oil shale exploitation is again being considered. On the other hand, the Scottish oil shale industry ran continuously for about a hundred years, being shut down only when all the high-grade reserves had been depleted. At its peak in 1913 about 3 million tons of oil shale were processed, corresponding to a production of about 1 million liters per day of shale oil. Oil shale operations were started in many other countries, and one still in continuous operation since it was begun in 1921 is that in Estonia, where gas, oil, and electric power are produced from relatively rich shale. Oil shale retorting was started in Fushun, Manchuria, in 1929 and while under the control of the Japanese during World War II, production reached a rate of 575 thousand liters per day of crude shale oil. It has been estimated that production of the Chinese oil shale industry in the mid-1970s was expanded to between 6.5 and 9.5 million liters per day.[11]

In the United States many pilot retorting processes have been tested for short periods. Among the largest were a semi-commercial-size retort operated by Union Oil in the late 1950s, which processed 1100 tons per day of high-grade shale; and one operated by The Oil Shale Corp. (now Tosco Corp.), which in the early 1970s processed 900 tons per day of high-grade shale. For a shale grade of 150 liters per ton, these feed rates correspond, respectively, to productions of 165 thousand liters per day and 135 thousand liters per day of crude shale oil. Again today a number of demonstration retorts are being started up and there is a rush of commercial developers to Colorado, including Union Oil and Tosco, to undertake the construction of commercial-size oil shale plants to produce shale oil at the rate of 8 million liters per day or more. It seems possible that this time around the industry may develop to commercial scale.

Tar sands, also called oil sands, are found in large deposits on every continent except Australia, but the most sizeable measured ones are in the Athabasca area of the province of Alberta in Canada, and in Venezuela. The large Athabasca deposits were confirmed at the turn of the century and attempts to commercially exploit them date from that time. Between 1930 and 1960, ''commercial enterprises were formed, failed, and were reformed with some regularity.''[12] Behind many of these efforts were Sidney Ells and Karl Clark. It was Clark who in the late 1940s developed the hot water extraction process named after him for the recovery of the bitumen from the sand. Beginning in the mid-1950s, extensive pilot plant operations were undertaken in the Athabasca region, and in 1965 Great Canadian Oil Sands, Ltd., undertook construction of a commercial plant to produce synthetic crude oil from tar sands. The plant went into operation in 1968 and is presently producing 8 million liters per day of synthetic crude oil. Syncrude Canada, Ltd., began operation of a second large commercial facility in 1978, with production designed to reach 20 million liters per day. Two other plants are also going forward, each with roughly 20 million liter per day projected output. Tar sand deposits in the United States are only a fraction of those in Canada, and little beyond small-scale development work has been done to exploit this resource for synthetic fuel production.

1.3 PROPERTIES OF COAL, OIL SHALE, AND TAR SANDS

In synthetic fuel conversion processes, a number of chemical and physical properties of the raw fuel are important for specifying the most appropriate conversion method and characterizing the products. A brief review of some of the most relevant properties for the purposes of the book is given below for each of the fuels considered.

Coal

Coals are classified by their "type" and "rank." Coal type is determined by the nature of the plant material of which it is composed or from which it originated. The microscopic study of coal and the relation of its different visible features to the original plant material forms the basis of coal petrography. It has been found that different coal types are composed of the same petrographic entities but in differing amounts. The petrographic components are termed "macerals," by analogy with minerals in inorganic rocks. Macerals are combined into three main groups called vitrinite, exinite, and inertinite. The different macerals making up the groups and their composition or derivation are shown in Table 1.5. The designations are generally indicative of the source materials or appearance, except for inertinite, whose name derives from the fact that the macerals of this group behave as inert infusible diluents when a coal containing them is thermally pyrolyzed. In this context, vitrinite and exinite are termed "reactive" components.[6] Coal originated in peat deposits and wood was the dominant plant material, so typical coals consist mainly of vitrinite. The average vitrinite content

Table 1.5 Coal macerals and maceral groups[6]

Maceral Group	Maceral	Composed of or Derived from
Vitrinite	Collinite	Humic gels
	Tellinite	Wood, bark, and cortical tissue
Exinite	Sporinite	Fungal and other spores
	Cutinite	Leaf cuticles
	Resinite	Resin bodies and waxes
	Alginite	Algal remains
Inertinite	Micrinite	Unspecified detrital matter, <10 μm
	Macrinite	Similar, but 10-100 μm grains
	Semifusinite ⎫	
	Fusinite ⎭	"Carbonized" woody tissue
	Sclerotinite	Fungal sclerotia and mycelia

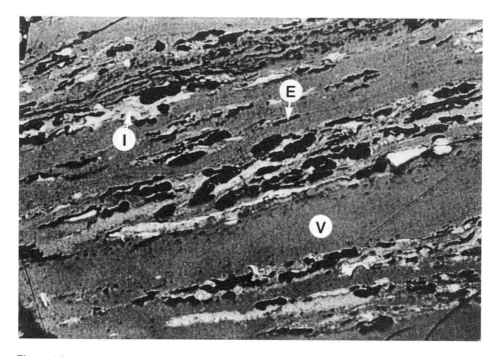

Figure 1.5 Photomicrograph of polished section of coal: V = vitrinite, E = exinite, I = inertinite.[13] *(Courtesy of G. J. Pitt, National Coal Board, England, and Academic Press Inc. (London) Ltd.)*

of American coals is in excess of 70 percent and their inertinite fraction is on the order of 10 to 15 percent. Exinite usually constitutes only a few percent, and the remaining material is inorganic mineral matter. Figure 1.5 is a photomicrograph of a polished section of coal exposing the maceral groups.

Rank is the most important property of coal and signifies the degree of maturation in the process of coal formation or "coalification." It indicates the degree of chemical change that occurred in the coal. The coal of lowest rank is lignite, followed in increasing rank by subbituminous coal, bituminous coal, and anthracite. The character of the chemical change with rank is illustrated in Table 1.6. It can be seen that the lower the coal rank, the lower its fixed carbon content and the higher its oxygen and hydrogen content, although the hydrogen content may drop somewhat from the bituminous to lignite coals. Generally, the lower the coal rank, the lower also is the calorific value and the higher the fraction of moisture and volatile matter.

The calorific value and the "proximate analysis" defining the fraction of moisture, volatile matter, ash, and fixed carbon are important properties of coal which must be known to characterize the conversion process. In Figure 1.6 these properties are shown for coals of different rank. The proximate analyses are presented on an ash-free basis. The ash content of coals in the United States ranges from about 2.5 to 33 mass percent and averages about 9 percent.[14]

To obtain a proximate analysis,[6,15] coal is first air-dried at 10–15°C above room temperature if it is too wet to crush without losing moisture. The coal is then crushed

Table 1.6 Chemical change with coal rank

| | Mass Percent† | | |
Material	Carbon	Hydrogen	Oxygen
Wood (cellulose)*	44	6	50
Peat*	59	6	35
Lignite	71	5	24
Subbituminous coal	74	5	21
Bituminous coal	84	5	11
Anthracite	94	3	3
Graphite*	100	–	–

* Not a coal.

† Values are representative averages.

and further dried in a forced-air circulation oven at 105 to 110°C. The sum of the mass losses from the air-drying and oven-drying is recorded as total moisture. The total moisture of coal that is fully saturated with water but does not contain visible water on its surface is termed the "bed moisture." The bed moisture reflects the pore volume of the coal accessible to moisture. The "as-received" moisture content may be smaller if the coal has been allowed to dry, or greater if there is an excess quantity of surface water. After the moisture determination, the coal is heated in a covered crucible for about 7 minutes to a higher temperature of about 950°C and the further loss in mass

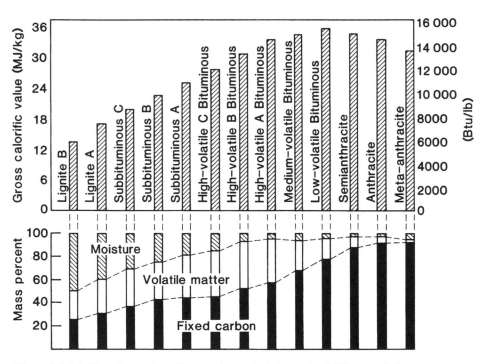

Figure 1.6 Calorific values and proximate analyses of ash-free coals of different rank.[14]

Table 1.7 Format of proximate and ultimate analyses

Proximate (mass %)	Ultimate (mass %)
Moisture	Moisture
Ash	Carbon (C)
Volatile Matter (VM)	Hydrogen (H)
Fixed Carbon	Nitrogen (N)
	Sulfur (S)
	Oxygen (O)
	Ash
100% (total)	100% (total)
Gross Calorific Value, GCV (MJ/kg)	Gross Calorific Value, GCV (MJ/kg)
Sulfur (mass %)	

is recorded as volatile matter. The volatile matter consists principally of tar, lighter oils, hydrocarbon gases, hydrogen, oxides of carbon, and water as decomposition products of the coal. Finally, the coal is oxidized (burned) and the loss in mass is the fixed carbon. The solid residue is ash, the inorganic mineral matter, and any contamination from mining.

More detailed properties of the coal are given by an ultimate analysis, a breakdown according to the five most important elements: carbon, hydrogen, oxygen, nitrogen, and sulfur. There are several ways of presenting an ultimate analysis. In this book, the analyses are given on an "as-received" (not dried) basis. Also, the moisture is shown separately, which is different from some presentations where the moisture is shown by adding its hydrogen and oxygen contents to the hydrogen and oxygen in the dry coal. Table 1.7 shows how proximate and ultimate analyses are normally presented. Typical ultimate analyses are shown in Table 1.8 for coals from the major fields in the United States. An analysis for anthracite coal is not shown, since this is not normally considered a coal which would be used in the manufacture of synthetic fuels.

Table 1.8 Ultimate analyses in mass percent of representative coals of the United States

Component	Fort Union Lignite	Powder River Subbituminous	Four Corners Subbituminous	Illinois C Bituminous	Appalachia Bituminous
Moisture	36.2	30.4	12.4	16.1	2.3
Carbon	39.9	45.8	47.5	60.1	73.6
Hydrogen	2.8	3.4	3.6	4.1	4.9
Nitrogen	0.6	0.6	0.9	1.1	1.4
Sulfur	0.9	0.7	0.7	2.9	2.8
Oxygen	11.0	11.3	9.3	8.3	5.3
Ash	8.6	7.8	25.6	7.4	9.7
Total	100.0	100.0	100.0	100.0	100.0
GCV (MJ/kg)	15.6	18.4	19.6	24.9	31.2

The gross calorific value (GCV) of coal, also called the higher heating value, is the amount of heat liberated by complete combustion at constant volume under specified conditions. The gross calorific value of the coal can be estimated from the ultimate analysis using the Dulong formula:[15]

$$Q_{GCV} \text{ (MJ/kg)} = 33.83C + 144.3\left(H - \frac{O}{8}\right) + 9.42S \qquad (1.1)$$

where C, H, O, and S are, respectively, the mass fractions of carbon, hydrogen, oxygen, and sulfur. The calorific value may be either on an as-received basis, dry, or dry and ash-free (daf) basis according to the mass fractions used in the formula. The gross calorific value Q_{GCV} includes heat released by condensation of steam formed during combustion. However, this heat of condensation is normally not recovered in combustion, and thus the net calorific value (NCV) is applicable. The net calorific value, also called lower heating value, may be estimated from the relation

$$Q_{NCV} = Q_{GCV} - 21.6H \qquad (1.2)$$

The gross calorific values shown in Table 1.8 are close to those given by Eq. (1.1). They also agree with the calorific values in Figure 1.6 when those values are appropriately reduced to account for the ash content.

For purposes of estimation in this book, where the coal calorific value is not specified, the following average calorific values are used for the three major coal ranks: bituminous, 30 MJ/kg; subbituminous, 23 MJ/kg; and lignite, 16 MJ/kg.

The sulfur content of coal in the United States ranges from about 0.3 to 8 percent.[14] Sulfur in coal is principally found in the form of either pyritic or organic sulfur. In the low-sulfur coals, mainly the western coals, most of the sulfur is in the organic form, and in the high-sulfur eastern coals most is in the pyritic form. Pyritic sulfur is sulfur combined with iron, mainly as pyrite but also as marcasite. These two minerals have the same chemical composition but different crystalline forms. Organic sulfur is sulfur that is chemically linked to coal. It cannot be separated out physically, and chemical means are needed. On the other hand, pyritic sulfur has a high specific gravity and can be separated from the coal by various physical means, of which the most important is washing with water.

The specific gravity of clean coal ranges from about 1.2 for lignite to about 1.7 for anthracite, with an average value of about 1.3 for subbituminous and 1.4 for bituminous.[16] Within any rank the specific gravity increases with increasing ash content.

There are properties of coal associated with its behavior on heating which are of considerable importance in coal conversion. One of these is the characteristic wherein certain coals on heating soften and fuse together, swelling and resolidifying into a porous char or coke which is greater in volume than the original coal. The term coke is applied to char that has fused into lumps of a marketable size and quality, a property that depends on the coal type. Coals that behave in the manner described are termed "caking" or "coking" coals, the distinction not being of importance for our purposes. It is probable that the gas which is formed while the coal is in the plastic state is responsible for the volume increase. The agglomerating characteristic, or tendency of the coal to fuse together, is important in coal conversion since coals that agglomerate can cause blockages in the reactor. In this regard, high-volatile bituminous coals are not necessarily ideal for conversion processes, as they tend to agglomerate at high pressure and temperature.

Another property manifested on heating is that the carbon in coal appears to be of two forms which behave differently.[17,18] One is called by Johnson[17] "base carbon" and the other is carbon in volatile matter. Volatile carbon is evolved by thermal pyrolysis, while base carbon remains in the residual char after devolatilization is complete. The carbon associated with the volatile content of the coal is highly reactive, while the base carbon is less reactive.

Reactivity is an important kinetic property of coal and the char formed from it. It may be loosely defined as the ability of the coal or char to react with oxygen, steam, carbon dioxide, and hydrogen. The reactivity of coal tends to increase with decreasing coal rank. Most criteria for reactivity are empirical and defined by a variety of measurements. From semi-empirical correlations of the rate of base carbon conversion in steam and hydrogen gasification, Johnson[17] has defined a relative reactivity factor. The value of this dimensionless factor is dependent on the particular carbonaceous solid and is normalized with respect to its value for a specific batch of bituminous coal char. Johnson has shown the relative reactivity factor for base carbon gasification to range from a value of about 10 for a North Dakota lignite to 0.3 for a low-volatile bituminous char. For many coals, the reactivity factor may be estimated from an empirical relation given by Johnson[19]

$$f_0 = 6.2\,C(1 - C) \tag{1.3}$$

where f_0 is the reactivity factor and C is the mass fraction of total carbon in the original coal on a dry ash-free basis. The equation is not generally applicable to lignites, which show a wide variability in reactivity and for which it has been suggested that the catalytic effect of the inorganic mineral matter may be largely responsible for the observed high reactivities.

Other specific properties of coal that are of interest are discussed in context.

Oil Shale

In the United States there are two principal oil shale types, that from the Green River Formation in Colorado, Utah, and Wyoming, and the Devonian-Mississippian black shale of the East and Midwest. The Green River shales are considerably richer, occur in thicker seams, and therefore are the more likely to be exploited for synthetic fuel manufacture. For this reason we shall emphasize their properties.

The shared property of all oil shales is the presence of "kerogen," a high-molar-mass organic material which is almost totally insoluble in all common organic solvents and is not a member of the petroleum family. The organic material in oil shale also contains a fraction of soluble organic matter called bitumen, typically around 10 percent for Green River shale, which is a highly viscous crude hydrocarbon that is the principal organic component of tar sands. However, the major part of the oil is derived from the pyrolysis of kerogen. In the Green River oil shale, the kerogen is not bound to a particular type of rock such as shale, so that the name oil shale is somewhat misleading. Instead, the largest concentrations of kerogen are found in sedimentary non-reservoir rocks such as "marlstone," a mix of carbonates, silicates, and clays. In contrast the black shales are true shales, composed predominantly of the clay illite.[16] Figure 1.7 is a magnified photograph of a polished section of Green River shale in which the layered, sedimentary structure of the material is evident.

By definition,[20] oil shale yields a minimum of 42 liters of oil per ton of shale (written 42 L/t) and up to 420 L/t. We note that these values are conversions from British units of 10 U.S. gal/ton and 100 U.S. gal/ton, respectively. Lower-grade shales are those averaging below about 100 L/t. From data presented in Ref. 16, we may derive the following empirical relation between the mass percent of total organic matter in Green River oil shale and the modified Fischer assay oil yield in L/t:

$$\text{Yield (L/t)} = 8.22 \times \text{Organic Matter (mass \%)} - 10.8 \qquad (1.4)$$

This corresponds to about 100 L/t yield for shale with 13.5 percent organic matter, and 300 L/t for shale with 38 percent organic matter. The "modified Fischer assay" refers to the pyrolysis oil yield from a 100 gram sample of crushed and dried shale,

Figure 1.7 Polished section of Green River oil shale showing sedimentary structure. *(Courtesy of E. H. Leland, Colorado School of Mines Research Institute.)*

slowly heated for 40 minutes to a temperature of 500°C in a special aluminum retort with air excluded. The temperature is held at 500°C until oil formation ceases, usually 20 minutes, but up to 40 minutes for richer shales. The distillate from the retort is condensed and the liquids collected. It is important to note that the yield from modified Fischer assay retorting is not necessarily optimum, and that another pyrolysis procedure could give a different value. The assay does, however, provide a reproducible basis for comparing different shales.

Table 1.9 shows typical ultimate analyses of organic matter in Green River oil shale from the Mahogany Zone in the Piceance Creek Basin of Colorado, and New Albany, Devonian shale from Kentucky. A calorific value may be assigned to shale using the Dulong formula (1.1), although oil shale is normally not considered a fuel. The gross calorific values of the organic matter calculated for the analyses of Table 1.9 are 41.1 MJ/kg for the Mahogany Zone shale and 37.5 MJ/kg for the New Albany shale. The gross calorific value for a given shale is specified approximately once the grade, that is, the fraction of organic matter is known. Some heat goes into decomposing the inorganic mineral matter but the amount is small. From data reproduced in Ref. 16, we may represent the gross calorific value of Green River oil shale by the empirical relation

$$Q_{GCV} \text{ (MJ/kg)} = 0.0533 \times \text{Shale Grade (L/t)} - 0.343 \qquad (1.5)$$

For 100 L/t shale this value is about 5 MJ/kg.

About half of the inorganic mineral portion of Green River oil shale consists of carbonates, of which about two-thirds is dolomite and the remainder calcite. Clays make up about 35 percent of the inorganic material and quartz about 15 percent. On the other hand, the inorganic matter in Devonian oil shales, which are true shales, is about half clay with the remainder principally quartz and pyrite.[16]

Of importance in oil shale processing is the characteristic that oil shales are relatively hard impermeable rocks through which fluids will not flow. Also of interest is the specific gravity of the oil shale since much of the inorganic material must be disposed. The approximate range of specific gravities for Green River shales lies

Table 1.9 Typical composition in mass percent of organic matter in Mahogany Zone and New Albany shale[16]

Component	Green River Mahogany Zone	Devonian New Albany
Carbon	80.5	82.0
Hydrogen	10.3	7.4
Nitrogen	2.4	2.3
Sulfur	1.0	2.0
Oxygen	5.8	6.3
Total	100.0	100.0
H/C atom ratio	1.54	1.08

between about 1.7 and 2.5. The organic component has a specific gravity of about 1.05 and the mineral fraction has an average value of about 2.7, so an increase in organic content of an oil shale causes a decrease in its specific gravity. An estimate of the oil shale specific gravity is readily obtained if the modified Fischer assay yield is known. Using Eq. (1.4), the organic fraction can be determined, which together with the average specific gravity values given for the organic and mineral components specifies the oil shale specific gravity. More detailed data on this and other properties of oil shale can be found in Ref. 16.

Tar Sands

Tar sands are normally a mixture of sand grains, water, and a high-viscosity crude hydrocarbon called bitumen. Unlike the kerogen in oil shale, bitumen is a member of the petroleum family in that it dissolves in organic solvents and decreases in viscosity when heated. At room temperature the bitumen is semi-solid and cannot be pumped, but at temperatures of about 150°C it will convert to a thick fluid which will become runny at still higher temperatures. Camp[12] has observed that tar sand might more properly be called bituminous sand, tar being a term applied to the liquid residue of the thermal distillation of organic materials. Tar sand is also called oil sand, but Camp points out that this may be an allusion to the synthetic crude oil which can be manufactured from bitumen.

In the Athabasca deposits of Alberta, Canada, the bitumen is present in a porous sand matrix, with porosities of typically 25 to 35 percent. The sand consists predominantly of fine quartz grains about 75 to 250 μm in size, with small amounts of clay minerals attached to the grain surfaces. The sand grains are all in direct contact with one another, and wet with a thin layer of water a few micrometers thick that forms a continuous sheath. Filling the void volume among them is the bitumen, which forms a continuous phase through the pores. The bitumen is not in direct contact with the sand grains, being separated from them by the water layer. Most of the deposits in the United States do not have the water layer around the sand grain so that the bitumen adheres directly to the grains, a characteristic of importance in defining the extraction method. Figure 1.8 shows a photomicrograph of bitumen-free sands from the Athabasca deposit typical of the bitumen-saturated sands. Because the sands constitute a stable framework, the absence or mobilization of the bitumen essentially does not alter the structure. The schematic representation next to the photograph shows a typical *in situ* arrangement of the material making up the tar sands. Normally tar sands are unconsolidated, that is, not cemented together as in the photograph, with their mechanical properties governed by the slightly interlocking nature of the grain contacts. However, as with some U.S. deposits, the tar sands may be consolidated, that is, cemented together, forming a sandstone or limestone. An important feature of most tar sand deposits is that fluids will flow through the mineral matrix, in contrast to the relatively impermeable character of oil shale.

The amount of bitumen in Alberta tar sands covers a range up to 18 mass percent. However, the sum of the bitumen and water generally totals about 17 mass percent. The remaining 83 mass percent is the sand grains. The largest and richest tar sand

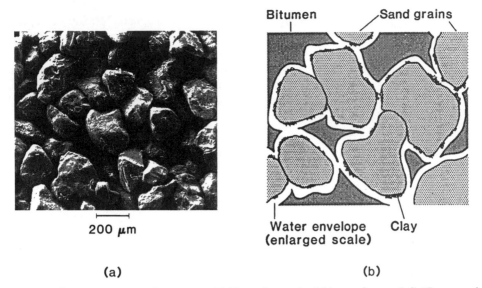

(a)

(b)

Figure 1.8 Athabasca tar sands structure: (*a*) Photomicrograph of bitumen-free sands.[21] *(Courtesy of D. J. Barnes, Hardy Associates (1978) Ltd., Edmonton.)* (*b*) Schematic of typical *in situ* arrangement.

deposits in the United States are in Utah. In these deposits, the bitumen saturation goes up to almost 17 mass percent. A deposit with about 14 percent bitumen is considered rich.

The chemical composition of the bitumen from the Athabasca deposit is relatively constant over a large geographical area. A typical ultimate analysis is shown in Table 1.10. When compared with kerogen (Table 1.9), the sulfur content is seen to be considerably higher and the oxygen and nitrogen contents considerably lower. The sulfur levels range from about 4.5 to 5 mass percent and the nitrogen levels from about 0.4 to 0.5 mass percent. The compositions of the United States tar sands are considerably more variable, although some of the tar sands in southeastern Utah are comparable, except for the sulfur levels, which are somewhat lower. The bitumens in many of the Utah deposits have very low sulfur levels with values typically less

Table 1.10 Typical composition of bitumen in Athabasca tar sands[16]

Component	Mass %
Carbon	83.1
Hydrogen	10.3
Nitrogen	0.4
Sulfur	4.6
Oxygen	1.6
Total	100.0
H/C atom ratio	1.49

than 0.5 percent. On the other hand, Utah bitumen is considerably more viscous than Alberta bitumen and therefore more difficult to extract. The average specific gravity of the mineral matter in Athabasca tar sands is about 2.7, and that of the tar sands about 1.9 to 2.0. The Utah tar sands have roughly the same density characteristics.

1.4 RESOURCES

To assess a region's viability for manufacturing synthetic fuels from coal, oil shale, and tar sands it is necessary to know the availability of the raw fuel resource. Many synthetic fuel plants will be located where the raw fuel is mined, and plants will be designed for lives of from 20 to 30 years. A reserve must therefore be sufficient to support the operation of the plant or industry over its expected lifetime. Although quantities of a particular fuel may be spoken of on a nation-wide basis, locally the picture can be quite different. We therefore wish to know not only the disposition of the resource but also the economics of mining it and the amount which can be recovered.

In defining a "resource" of a mineable material, geologists include in their estimates for a given region only those materials for which economic extraction is currently or potentially feasible.[14,22] Those parts of the resource assumed to be ready for extraction and consumption are considered "reserves."

In the U.S. Geological Survey classification of mineable resources,[14,22] estimates of "total resources" are comprised of two main classes, "identified" and "undiscovered." The identified resources include those that have been located and measured or have been estimated by inference from geographical continuity of the resource. Undiscovered resources comprise "hypothetical resources" inferred by extrapolation of geological data to unexplored parts of areas known to contain the mineable material, plus "speculative resources" for areas not yet discovered. The use of the term "total resource" is somewhat ambiguous, even taking into account the rate at which the resource is being mined, since it is subject to upward or downward revision as new geological information is obtained.

We have already noted that one variable in defining a resource is economic recoverability. Following McKelvey[22] and Averitt[14] we show in Figure 1.9 a two-dimensional map of the various categories of resources. One coordinate represents the degree of certainty about the existence or magnitude of the resource and the other coordinate the economic feasibility of recovery. As can be seen from this map, it is only to the category of identified resources that the geologist applies the word reserve, and then only to those resources considered to be economically mineable. The reserves that can be counted upon are measured or "proved reserves," which have been defined by detailed exploration. They are usually only a small fraction of the total resource. Another closely linked category is "indicated reserves," which are based partly on measurement and partly on reasonable geologic projection. The proved and indicated reserves together make up what is known as the "demonstrated reserve base." In the case of coal, it is this figure which has most often been cited to indicate the vast amount available in the United States.

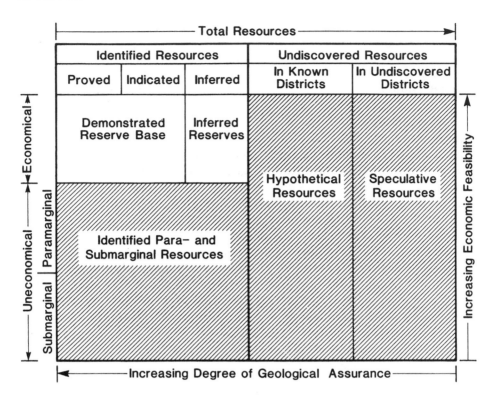

Figure 1.9 U.S. Geological Survey classification of mineable resources (after McKelvey[22] and Averitt.[14])

Coal

By far the largest reserves of the world's non-renewable energy resources are to be found in coal. Coal represents about 80 percent of the world's proved energy reserves.[12] Table 1.11 shows the distribution of the world's proved coal reserves in order of size. The United States possesses about one-quarter of these reserves.

Table 1.11 World's proved coal reserves in billions of tons[6]

Country	Reserves
Soviet Union	273
United States	182
China, People's Republic	101
German Federal Republic	40
Australia	24
India	12
Republic of South Africa	11
Other	55
Total	698

The locations of the principal coal deposits in the conterminous United States are shown in Figure 1.10. The largest coal deposits are in the Northern Great Plains and Rocky Mountain area encompassing the Powder River, Fort Union, and Four Corners regions, and in the Appalachian and Illinois Basins.

The demonstrated coal reserve base distributed according to region is shown in Table 1.12, compiled from the data of Averitt.[14] This reserve refers to identified resources (see Figure 1.9) suitable for mining by present methods, where at least 50 percent is recoverable. The coal in this category lies less than 300 meters below the surface. Table 1.12 also shows the potential methods by which the coal can be mined. In the Northern Great Plains and Rocky Mountain region, where almost half of the nation's coal is to be found, more than 40 percent of the coal can be surface mined. Surface or strip mining can be done more economically and in most cases with a much higher proportion of the coal recovered.

The reserve base represents a resource in the ground, although not all of it can be recovered. The U.S. Bureau of Mines has defined a "recoverability factor" for coal as the percentage of the reserve base that can be recovered by established mining practices. The "recoverable reserve" is the recoverability factor multiplied by the reserve base—it is the amount of coal actually mined as distinguished from the amount present in the ground. Based on past records for underground mining, the Bureau has used a recoverability factor of 50 percent. For surface mining, the recovery factor has

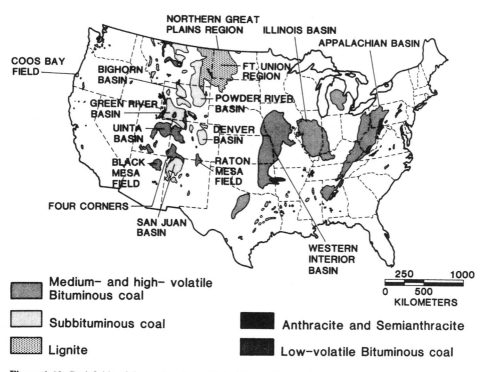

Figure 1.10 Coal fields of the conterminous United States.[14]

Table 1.12 Demonstrated coal reserve base of the United States in billions of tons by region and potential method of mining[14]

Region	Underground	Surface	Total	Percent of Grand Total
Northern Great Plains and Rocky Mountain	103	78	181	46
Appalachian Basin	88	15	103	26
Illinois Basin	64	16	80	20
Other	15	15	30	8
Grand Total	270 ×6.3 = 81	124 ×0.45=55.8	394	100

136.8 ×10⁹ Tons

been variously estimated between 80 and 90 percent or more. Schmidt[23] has pointed out, however, that the recoverability factor in any given mine may be characterized by these values, but that there are substantial areas between mines, under cities, in pockets too small for economical recovery, badly faulted, etc., for which mining is not possible or is otherwise precluded. The effect is to lower the recoverable reserves. Based on data Schmidt has assembled, he suggests use of the recoverability factors shown in Table 1.13 for defining recoverable reserves.

Two of the most important factors in determining the recoverability of coal are the bed or seam thickness and the bed depth or thickness of overburden. Overburden is generally any material that overlies the coal deposits and is of little utility. Table 1.14 shows the categories of bed thickness used by the U.S. Geological Survey in estimating resources for the three ranks of coal most important for producing synthetic fuels. These categories may differ from the recommended standards in a few states. Coal resource data is also categorized according to the thickness of overburden, approximately: 0 to 300 m, 300 to 600 m, and 600 to 900 m. In some states, where the overburden is thin, other categories are used.

In most of the major coal regions of the United States, except for the Appalachian Basin, a large fraction of the proved reserves can be surface mined (see Table 1.12). In the Appalachian Basin the coal seams have an average thickness of about 1.2 m. The average seam thicknesses of surface mineable coals are generally larger, as can be seen from Table 1.15, which presents seam thickness, overburden, and yield for surface mineable coal in the major U.S. coal regions.

Table 1.13 Recoverability factors for defining recoverable reserves of coal by mining method[23]

Mining Method	Commonly Estimated Recoverable Reserves, % of Reserve Base	Indicated Recovery, % of Reserve Base
Underground	50	30
Surface	90	45

Table 1.14 Categories of bed thickness in meters,* used in calculating resources of coal of different ranks[14]

Coal	Thin	Intermediate	Thick
Bituminous	0.35–0.70	0.70–1.05	>1.05
Subbituminous and Lignite	0.75–1.5	1.5–3.0	>3.0

*Data converted from British units and rounded.

For a given thermal efficiency of a synthetic fuel plant and a fixed plant size (determined by the calorific value of the product), the rate of coal mined is set by its calorific value. Table 1.16 gives an approximate range of overall process thermal efficiencies, defined by the sum of the calorific values of the product fuel and byproducts compared to the calorific value of the raw coal. Using average values of the efficiency ranges, we show in Table 1.17 the mining rates required for our nominal standard-size plants. From these rates, we also show in Table 1.17 the recoverable reserves that would be required for mine-plant complexes based on a 30-year mine life.

From a knowledge of recoverable reserves, the required coal reserve can be calculated. This has been done using the indicated recoverability factors shown in Table 1.13 and the recoverable reserves shown in Table 1.17, averaged over the different plants. The results are presented in Table 1.18. The coal reserve required for any one plant is seen to be of the order of a billion tons compared to the hundreds of billions of tons of demonstrated U.S. coal reserves shown in Table 1.12. However, were, say, 100 to 200 synthetic fuel plants converting coal to be constructed, it is evident that the coal reserve needed would represent a significant fraction of the demonstrated coal reserve.

Oil Shale

Oil shale resources[20,25] are usually expressed in terms of crude-shale-oil equivalence in the oil shale. The world's identified oil shale resources, assaying at greater than 42 L/t, are estimated to be 500 trillion liters. The United States possesses about two-thirds of these resources and Brazil about one-quarter. Smaller deposits are found in the Soviet Union, Zaire, Canada, Italy, China, and elsewhere. A survey compiled in

Table 1.15 Average seam thickness, overburden, and yield for surface mineable coal in the United States[24]

Region	Principal Rank	Seam Thickness, m	Overburden, m	Coal Yield, t/m²
Powder River	Subbituminous	20	<60	4.5–27
Fort Union	Lignite	5–6	<60	3.1–9.0
Four Corners	Subbituminous	3.5	<75	8.3
Illinois Basin	C Bituminous	0.9–1.2	6–60	4.3

Table 1.16 Thermal efficiencies for coal conversion processes

Process	Efficiency, %
Indirect liquefaction	40–50
Gasification	60–65
Direct liquefaction	65–70
Solvent refining	70–75

1980 by the Federal Institute for Geosciences and Natural Resources in Hanover, Germany, indicates that the world resources may be considerably larger than previously estimated.

Only about 10 percent of the United States' identified resources of 320 trillion liters are from the Devonian black shales. These shales are all of lower grade, assaying at between 42 and 105 L/t, and are generally found in the same regions as the coal basins of the East and Midwest. One estimate indicates a considerably higher resource base of Devonian shales and oil yields significantly above Fischer assay attainable by special retorting procedures.[26] These shales are discussed in Section 7.1, but not considered further here. The greatest promise for commercial production lies where the high-grade oil shale is found, in areas in Colorado, Utah, and Wyoming underlain by what is called the Green River Formation. Figure 1.11 outlines the oil shale areas in this formation.[20] They are seen to comprise a relatively small region. The identified high-grade shales with yields greater than 105 L/t have an oil equivalence of about 100 trillion liters.[16] The richest, most accessible zones are estimated to yield about half this amount. About 80 percent of the high-grade material is located in Colorado in the Piceance Creek Basin. These deposits are in strata varying from 3 to 600 meters

Table 1.17 Coal mining rates and total recoverable reserve required for standard-size plants producing synthetic fuels from three major coals

	Liquid Fuel (40 MJ/L) Ind. Liquefaction, 10^7 L/d	Pipeline Gas (37 MJ/m³) Gasification, 10^7 m³/d	Liquid Fuel (40 MJ/L) Dir. Liquefaction, 10^7 L/d	Refined Coal (37 MJ/kg) Solvent Refining, 10^7 kg/d
	Daily production rate (10^6 kg/d)			
Bituminous*	30	20	20	17
Subbituminous*	39	26	26	22
Lignite*	56	37	37	31
	Total recoverable reserve required (10^6 t)†			
Bituminous	290	190	190	170
Subbituminous	380	250	250	220
Lignite	540	360	360	300

* Calorific values: bituminous, 30 MJ/kg; subbituminous, 23 MJ/kg; lignite, 16 MJ/kg.

† Based on 325-days-per-year production and a 30-year mine life.

Table 1.18 Total coal reserves required in millions of tons for an average single standard-size synthetic fuel mine-plant complex for underground and surface mining

| | | Total Coal Reserve | |
| | Total Recoverable | Underground | Surface |
Coal	Reserve*	Mining	Mining
Bituminous	200	700	450
Subbituminous	300	1000	700
Lignite	400	1300	900

* Based on 325-days-per-year production and a 30-year mine life.

in thickness, at depths up to 500 meters below the surface and in outcroppings at the surface.[24]

So far we have spoken only of "identified resources" and not "reserves," which would imply economic recoverability. The reason for this is evident, since large-scale commercial production is only now being undertaken. Estimates have been made at various times of the U.S. resources which are thought to be economically recoverable, and they span a wide range. Since the richest and most accessible zones yield about half of the identified resources, a value half again of that would seem to be a

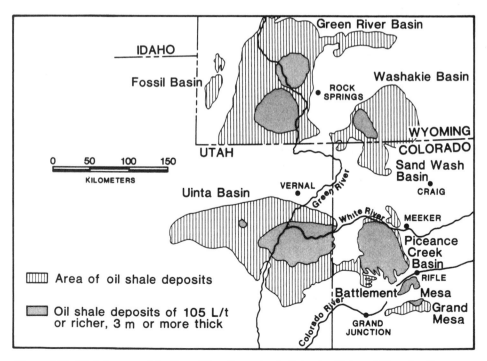

Figure 1.11 Oil shale areas of the Green River Formation in Colorado, Utah, and Wyoming.[20]

conservative estimate for "identified reserves" and would amount to 25 trillion liters of shale oil. This value is about the middle of the range of published estimates.[27] Even this figure represents a very large potential for synthetic fuel development.

The mining of oil shale for surface processing is roughly the same as for coal. Underground and surface mining is envisaged, although the amount of shale that can be economically and readily open-pit mined is estimated to be only about 10 to 15 percent of the total.[28]

Large quantities of shale must be mined and large quantities of material disposed of to produce relatively small quantities of synthetic fuels. For a surface retorting plant producing a synthetic crude oil, an overall conversion efficiency of 70 percent is typical for present shale processes. This would mean that about 114 thousand tons of 125 L/t shale would have to be mined daily and about 95 thousand tons of spent shale disposed of daily, for a self-sufficient integrated mine and plant designed to produce 10 million liters per day of synthetic crude. For different shale grades, the mined and disposed tonnage would be correspondingly higher or lower. For the example given, a total recoverable reserve of 140 billion liters of shale oil resources is needed, assuming 325-days-per-year production and a 30-year mine life. For an underground mine, about 30 percent of the shale remains in place with conventional room-and-pillar mining techniques, so that a reserve base of 200 billion liters of shale oil is required. This may be compared to our conservative estimate of identified reserves of about 25 trillion liters. These resources are more than sufficient to support a large-scale oil shale industry.

Tar Sands

Tar sands resources are expressed in terms of a crude bitumen equivalence. The world's identified resources[29] are estimated at about 360 trillion liters, of which about 150 trillion liters are attributable to Canada, 110 trillion liters to Venezuela, and 96 trillion liters to the U.S.S.R. The Soviet estimates, which may prove to be too low, are for tar sand deposits in Siberia. The United States has only relatively small deposits with an identified resource of about 4.5 trillion liters, almost all of which is in the eastern half of the state of Utah. We would emphasize that these are identified resources and not reserves, which are but a small fraction of the resource. However, the 1980 survey by the Federal Institute for Geosciences and Natural Resources, mentioned in connection with shale, has surveyed tar sand reserves, and their data indicate that world reserves may be larger than indicated by earlier estimates.

Three major factors define whether the resource falls into the category of a reserve: the overburden thickness, the thickness of the tar sand zone, and bitumen saturation.[16] The Athabasca tar sands of Canada have zone thicknesses averaging about 45 meters. Current mining methods are applicable to shallow deposits characterized by a ratio of thickness of overburden to thickness of the tar sand zone on the order of one or less. On this basis alone, only about 10 percent of the Canadian deposits are mineable within current technology. Newer mining technologies are under development to make deep deposits and deposits with large overburden ratios economical. Included is underground or, "*in situ,*" recovery. Bitumen saturations of less than 10 percent are also not considered economical to mine at present because of the large amount of material which must be processed and disposed of to recover a small volume of bitumen. When these two factors, together with mining efficiency, are taken into account, it is estimated that only a little more than 4 percent of the Canadian resources, or about 6 trillion liters, can be classified as proved reserves. This is nevertheless an exceedingly large resource, one capable of supporting a great number of 10-million-liter-per-day plants, each of which would produce 100 billion liters over a 30-year lifetime.

The location of many of the U.S. deposits in Utah presents formidable recovery problems because of inaccessibility. This may be compounded by a dearth of water resources in the region. Considering only shallow tar sand deposits, probably no more than 10 percent of the U.S. resource can be classed as recoverable. At plant conversion efficiencies of 70 percent, this resource would only support about three 10-million-liter-per-day plants with 30-year lifetimes. Newer technologies, including *in situ* recovery, will have to be developed to more fully exploit the U.S. tar sand resources.

REFERENCES

1. Lom, W. L., and Williams, A. F., *Substitute Natural Gas*. Wiley-Halsted, New York, 1973.
2. Elliott, M. A., and Linden, H. R., "Manufactured Gas," in *Kirk-Othmer, Encyclopedia of Chemical Technology*, Vol. 10, pp. 353-442. 2nd Edition, Interscience Publishers, New York, 1969.
3. Shnidman, L., "Utilization of Coal Gas," in *Chemistry of Coal Utilization*, Vol. II (H. H. Lowry, ed.), pp. 1252-1286. Wiley, New York, 1945.
4. Morgan, J. J., "Water Gas," in *Chemistry of Coal Utilization*, Vol. II (H. H. Lowry, ed.), pp. 1673-1749. Wiley, New York, 1945.
5. Storch, H. H., "Hydrogenation of Coal and Tar," in *Chemistry of Coal Utilization*, Vol. II (H. H. Lowry, ed.), pp. 1750-1796. Wiley, New York, 1945.
6. Berkowitz, N., *An Introduction to Coal Technology*. Academic Press, New York, 1979.
7. Donath, E. E., "Hydrogenation of Coal and Tar," in *Chemistry of Coal Utilization*, Supplementary Volume (H. H. Lowry, ed.), pp. 1041-1080. Wiley, New York, 1963.
8. Storch, H. H., "Synthesis of Hydrocarbons from Water Gas," in *Chemistry of Coal Utilization*, Vol. II (H. H. Lowry, ed.), pp. 1797-1845. Wiley, New York, 1945.
9. Storch, H. H., *et al.*, "Synthetic Liquid Fuels from Hydrogenation of Carbon Monoxide," U.S. Bureau of Mines Technical Paper 709, Government Printing Office, Washington, D.C., 1948.
10. Reichl, E., "Synthetic Fuels: A Brief History," *Coal Technology '80*, 3rd Intn'l. Coal Utilization Conf., Houston, Texas, Nov. 1980.
11. Prien, C. H., "Survey of Oil Shale Research in the Last Three Decades," in *Oil Shale* (T. F. Yen and G. V. Chilingarian, eds.), pp. 235-267. Elsevier, New York, 1976.
12. Camp, F. W., "Tar Sands," in *Kirk-Othmer, Encyclopedia of Chemical Technology*, Vol. 19, pp. 682-732. 2nd Edition, Interscience Publishers, New York, 1969.
13. Pitt, G. J., and Milward, G. R., *Coal and Modern Coal Processing: An Introduction*. Academic Press, New York, 1979.
14. Averitt, P., "Coal Resources of the United States, January 1, 1974," U.S. Geological Survey Bulletin 1412, Government Printing Office, Washington, D.C., 1975.
15. Ergun, S., "Coal Classification and Characterization," in *Coal Conversion Technology* (C. Y. Wen and E. S. Lee, eds.), pp. 1-56. Addison-Wesley, Reading, Mass., 1979.
16. Baughman, G. L., *Synthetic Fuels Data Handbook*. 2nd Edition, Cameron Engineers, Inc., Denver, Colorado, 1978.

17. Johnson, J. L., "Kinetics of Bituminous Coal Char Gasification with Gases Containing Steam and Hydrogen," in *Coal Gasification* (L. G. Massey, ed.), pp. 145-178. Advances in Chemistry Series No. 131, American Chemical Society, Washington, D.C., 1974.
18. Wen, C. Y., and Huebler, J., "Kinetic Study of Coal-Char Hydrogasification," *Ind. Eng. Chem. Process Des. & Dev.* **4**, 142-154, 1965.
19. Johnson, J. L., "Relationship between the Gasification Reactivities of Coal Char and the Physical and Chemical Properties of Coal and Coal Char," *Am. Chem. Soc., Div. Fuel Chem., Preprints* **20**(4), 85-101, August 1975. Also in *Kinetics of Coal Gasification* (collected papers of J. L. Johnson), pp. 237-260. Wiley, New York, 1979.
20. Duncan, D. C., and Swanson, V. E., "Organic-Rich Shale of the United States and World Land Areas," U.S. Geological Survey Circular 523, Government Printing Office, Washington, D.C., 1965.
21. Barnes, D. J., "Micro-fabric and Strength Studies of Oil Sands," M.Sc. Thesis, Dept. of Civil Engineering, University of Alberta, Edmonton, Alberta, 1980.
22. McKelvey, V.E., "Mineral Potential of the United States," in *The Mineral Position of the United States 1975-2000* (E. N. Cameron, ed.), pp. 67-82. Univ. of Wisconsin Press, Madison, Wisconsin, 1973.
23. Schmidt, R.A., *Coal in America*. McGraw-Hill, New York, 1979.
24. Probstein, R. F., and Gold, H., *Water in Synthetic Fuel Production*. MIT Press, Cambridge, Mass., 1978.
25. Culbertson, W. C., and Pitman, J. K., "Oil Shale," in *United States Mineral Resources*, U.S. Geological Survey Paper 820, Government Printing Office, Washington, D.C., 1973.
26. Janka, J. C., and Dennison, J. M., "Devonian Oil Shale," in *Symposium Papers: Synthetic Fuels from Oil Shale, Atlanta, Georgia, December 3-6, 1979*, pp. 21-116. Institute of Gas Technology, Chicago, Ill., 1980.
27. Rattien, S., and Eaton, D., "Oil Shale: The Prospects and Problems of an Emerging Energy Industry," in *Annual Review of Energy*, Vol. 1 (J. M. Hollander and M. K. Simmons, eds.), pp. 183-212. Annual Reviews, Inc., Palo Alto, Calif., 1976.
28. Schmidt-Collerus, J. J., "The Disposal and Environmental Effects of Carbonaceous Solid Wastes from Commercial Oil Shale Operations," Report No. NSF-RA-E-74-0004 (NTIS Catalog No. PB 231 796), Denver Research Institute, Denver, Colo., Jan. 1974.
29. Bowman, C. W., and Carrigy, M. A., "World-Wide Oil Sand Reserves," in *The Future Supply of Nature-Made Petroleum and Gas, The First UNITAR Conference on Energy and the Future* (R. F. Meyer, ed.), pp. 732-738. Pergamon Press, New York, 1977.

TWO

CHEMICAL AND PHYSICAL FUNDAMENTALS

This chapter serves to introduce those fundamental concepts required later in the book when assessing synthetic fuel technologies. We do not pretend, nor is it necessary, to provide a complete and self-contained review of any of the disciplines covered. Each subject has been presented in innumerable texts, some of which have been used extensively in preparing this chapter and are referenced for further reading.[1-10] The material presented here should be sufficient to serve as a guide to those unfamiliar with the subjects and as a quick review for others. The examples are usually specific to synthetic fuel production and should be of interest even to those familiar with the fundamental disciplines.

Some of the subjects such as thermodynamics are applied in many fields, each with its own preferred terminology and nomenclature. In view of the overall dominance of chemical processing we have tended to adopt systems used in chemistry; in particular the nomenclature recommended by the International Union of Pure and Applied Chemistry (IUPAC) for physiochemical quantities and units. This notation is adopted by Denbigh,[4] whom we in turn have followed.

2.1 CHEMISTRY FOR SYNTHETIC FUELS

Carbonaceous or carbon-containing materials are the key substances used for synthetic fuel production. By definition, all compounds containing carbon are classified as organic, a name originally selected to differentiate between substances produced by living organisms and inorganic substances that are not associated with living organisms. While organic chemistry is properly the chemistry of all carbon compounds, substances such as carbon monoxide, carbon dioxide, carbon disulfide, and metal carbonate and cyanide salts are traditionally considered to be inorganic. A clear distinction between organic and inorganic materials is not necessary for an understanding of the chemistry of synthetic fuel production. In fact, during the processing and eventual combustion of synthetic fuels most of the organically bound carbon ends up as "inorganic" carbon in the form of carbon oxides, while organically bound sulfur and nitrogen may be released as hydrogen sulfide and ammonia during processing, or sulfur and nitrogen oxides during combustion. These are inorganic materials. However, as synthetic fuel production is predominantly characterized by organic chemicals and their reactions, much of this introduction is necessarily devoted to organic chemistry concepts.

All substances are composed of the approximately 100 known *elements*. The smallest part of an element which retains the chemical character of the element is the *atom*. The smallest stable entity of some elements at ambient temperatures is the *molecule*, which consists of two or more atoms joined together. Chemical *compounds* are molecules made up from different atoms. The atoms join in fixed proportions determined by their *valency*, which is defined as the number of hydrogen atoms the element can combine with or replace in a compound. Although an element may be associated with a common valency that occurs in the majority of its compounds, it can have different valencies as determined by the characteristics of the atoms to which it is joined. Elements present in the substances discussed in the book are listed in Table 2.1 together with their common valency.

The formation of new compounds from elements or other compounds is represented by a chemical equation

$$C + 2H_2 \longrightarrow CH_4$$
$$12 \qquad 4 \qquad\qquad 16$$

$$(2.1)$$

Here an atom of carbon reacts with two molecules of hydrogen to form methane. Atoms are conserved in a chemical reaction so that the number of carbon atoms and

Table 2.1 Some elements, their masses, and common valences

Name	Chemical Symbol or Formula	Relative Atomic Mass	Molar Mass, kg/kmol	Common Valency
Aluminum	Al	26.98	26.98	3
Calcium	Ca	40.08	40.08	2
Carbon	C	12.01	12.01	4
Chlorine	Cl_2	35.45	70.90	1
Chromium	Cr	52.00	52.00	2,3,6
Cobalt	Co	58.93	58.93	2,3
Copper	Cu	63.55	63.55	1,2
Gold	Au	196.97	196.97	1,3
Hydrogen	H_2	1.01	2.02	1
Iron	Fe	55.85	55.85	3
Molybdenum	Mo	95.94	95.94	—
Nickel	Ni	58.71	58.71	2
Nitrogen	N_2	14.01	28.02	3,5
Oxygen	O_2	16.00	32.00	2
Platinum	Pt	195.09	195.09	2,4
Potassium	K	39.10	39.10	1
Silicon	Si	28.09	28.09	4
Silver	Ag	107.87	107.87	1
Sodium	Na	22.99	22.99	1
Sulfur	S	32.06	32.06	—
Zinc	Zn	65.38	65.38	2

the number of hydrogen atoms must be the same on each side of the equation. This allows the *stoichiometric coefficient*, which is the number of atoms or molecules of each species participating in the reaction, to be determined. The stoichiometric coefficients in the above equation are 2 for hydrogen and 1 for carbon and methane. As atoms are conserved, the sum of the mass of the atoms on each side of the equation must be the same. The relative mass of the atoms is included in Table 2.1 and can be used to determine the mass of each species in the reaction. For example, the mass of methane is made up of the mass of carbon plus the mass of 4 hydrogen atoms, that is, $12 + (4 \times 1) = 16$. This relative mass expressed in grams represents an amount of methane called a *mole*. The mass of a mole of substance is called the *molar mass* M and in units of g/mol or kg/kmol has the same magnitude as the formerly used molecular weight. The molar mass of methane is thus 16 kg/kmol. In general m kg of a substance is equal to m/M kmol of that substance. The molar masses multiplied by the stoichiometric coefficient are written underneath their respective species in Eq. (2.1) and show that 12 kg of carbon and 4 kg of hydrogen are required to form 16 kg of methane. This is the *mass balance* for the reaction. Actual masses participating in a chemical reaction will be in proportion to these quantities.

Carbon is normally tetravalent and so will combine with four univalent elements such as hydrogen to form methane as shown above, or two divalent elements such as oxygen to form carbon dioxide, or sulfur to form carbon disulfide. Carbon further has the most important property of combining with itself to form chains and rings:

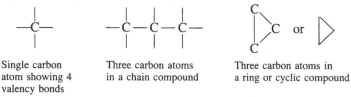

Single carbon atom showing 4 valency bonds   Three carbon atoms in a chain compound   Three carbon atoms in a ring or cyclic compound

When all the bonds between adjacent carbon atoms are single bonds, then the compound is said to be *saturated* in that it must contain a full complement of hydrogen or other elements to satisfy the remaining valency requirements. However, double or triple bonds between carbon atoms can occur, in which case the compound is said to be unsaturated as the extra bonds are readily available for joining with other substances to form a new compound. Reactions in which other substances are introduced into the molecule by reducing the number of bonds between a carbon pair are termed addition reactions. Saturated compounds can also undergo substitution reactions in which one element or group of elements is replaced by another, but these occur less readily than addition reactions. Saturated compounds are consequently relatively unreactive and unsaturated compounds are relatively reactive.

Compounds that contain hydrogen and carbon only are called hydrocarbons. The hydrocarbons are classified into *aliphatic* and *aromatic* compounds. The aliphatics are chain compounds and these hydrocarbons are further subdivided into paraffins, olefins, and acetylenes as shown in Table 2.2. Paraffins are saturated aliphatics and are a major constituent of most petroleums. Olefins and acetylenes are unsaturated aliphatics containing a double or a triple bond respectively. Some cyclic hydrocarbons have properties which are similar to the aliphatics. These are called alicyclics and are

Table 2.2 Classification and examples of aliphatic hydrocarbon compounds

Common Name	Chemical Formula	Structural Representation	Characteristics
Paraffins or Alkanes, General Formula C_nH_{2n+2}			
Methane	CH_4	H–C–H (with H above and below) or –C–	The paraffins occur naturally in petroleum. They are saturated compounds and as the name paraffin (little affinity) implies they are generally not reactive at ambient conditions. With the exception of the first four members—methane, ethane, propane, and butane—they are named according to their carbon number using the suffix -ane.
Ethane	C_2H_6	CH_3–CH_3 or –C–C–	
Octane	C_8H_{18}	–C–C–C–C–C–C–C–C–	
Olefins (Oil-forming) or Alkenes, General Formula C_nH_{2n}			
Ethylene	C_2H_4	CH_2=CH_2 or C=C	Olefins are obtained in the cracking of petroleum. They are reactive due to the presence of the double bond. Nomenclature is similar to the paraffins but with the suffix -ylene.

Acetylenes or Alkynes, General Formula C_nH_{2n-2}

Acetylene	C_2H_2	CH≡CH or —C≡C—	Acetylenes contain a triple bond and are highly reactive. Acetylene was the major starting material for organic synthesis until displaced by large quantities of cheap ethylene produced in petroleum refineries.

Alicyclics

Trimethylene	C_3H_6	or	May be saturated or unsaturated. Saturated members are naphthenes, which occur in petroleum.
Cyclohexane	C_6H_{12}		

considered to be a subclass of aliphatics, and have therefore been included in Table 2.2. Naphthenes are an important group of saturated alicyclic compounds which occur naturally in petroleum and oil shale.

Aromatic compounds are those cyclic compounds containing at least one benzene ring as shown in Table 2.3. Benzene, C_6H_6, may be represented by three different structural formulae

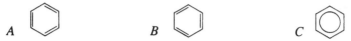

It is not possible to distinguish between the formulae A and B, as the molecule exhibits properties of both. In consequence the formula C is sometimes used and represents both A and B. Benzene and the aromatics are partly unsaturated and relatively reactive. If hydrogenated to the saturated cyclohexane compound, they are called "hydroaromatics." Hydrogenation of one of the rings in naphthalene, for example, results in the formation of the hydroaromatic compound tetrahydronaphthalene, often called tetralin:

$$(2.2)$$

Naphthalene   Tetralin

Tetralin reverts to naphthalene under certain conditions making the hydrogen available for other hydrogenation reactions. It is typical of the hydrogen donor solvents used in direct coal liquefaction processes as discussed in Section 3.4.

In addition to carbon and hydrogen, organic compounds can contain many other elements of which oxygen, nitrogen, and sulfur are the most important in fuel chemistry. These elements occur in the so-called "functional groups," which determine the characteristics of the compound and are used to classify organic chemicals as shown in Table 2.4.

The symbol R in Table 2.4 represents a hydrocarbon *radical*. When combined with a hydroxyl functional group it is an alcohol. For example, the methyl radical which is denoted by CH_3— is derived from methane. It can combine with the hydroxyl group to form methyl alcohol or, more correctly, methanol, CH_3OH. Methanol used to be produced by the thermal decomposition or destructive distillation of wood and was commonly known as wood alcohol. The alcohol made by fermentation of sugars is ethanol, C_2H_5OH, which contains the ethyl radical, C_2H_5—, derived from ethane. Table 2.5 contains a list of common radicals and their related paraffins.

As the properties of aromatics can be significantly different from those of aliphatics, the aromatic or *arryl radical* is often distinguished by the symbol Ar. Thus monohydroxy benzene, which is phenol, is represented by the structural formula

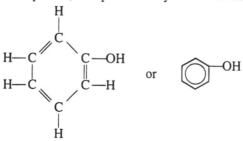

or

Table 2.3 Examples of aromatic hydrocarbon compounds

Common Name	Chemical Formula	Structural Representation	Characteristics
Benzene	C_6H_6	CH=CH, CH=CH, CH=CH (ring) or (hexagon) or (hexagon)	Named for their pleasant odor. Many occur naturally in coal tar, the liquid product obtained on pyrolysis of coal. Naphthalene is the largest single constituent of coal tar.
Toluene	C_7H_8	C_6H_5—CH_3 and CH_3 (ring)	
Xylene	C_8H_{10}	CH_3, CH_3 ortho- ; CH_3, CH_3 meta- ; CH_3, CH_3 para-	
Naphthalene	$C_{10}H_8$	(two fused rings)	
Diphenyl	$C_{12}H_{10}$	C_6H_5—C_6H_5 or (two connected rings)	
Anthracene	$C_{14}H_{10}$	(three fused rings)	

Table 2.4 Examples of functional groups of organic compounds

Type of Compound	Functional Group		Example		General Structural Formula
	Formula	Name			
Hydrocarbons (Table 2.2)					
Paraffins	$-\overset{\vert}{\underset{\vert}{C}}-\overset{\vert}{\underset{\vert}{C}}-$	(-ane)	Propane	C_3H_8	R—H
Olefins	C=C	(-ene)	Propylene or Propene	C_3H_6	R—C=C—R'
Acetylenes	C≡C	(-yne)	Methylacetylene or Propyne	C_3H_4	R—C≡C—R'
Oxygen-containing (Table 2.11)					
Alcohols	—OH	Hydroxyl	Methanol	CH_3OH	R—O—H
Phenols	—OH	Hydroxyl	Phenol	C_6H_5OH	Ar—O—H
Acids	—COOH	Carboxyl	Acetic acid	CH_3COOH	$R-\overset{O}{\overset{\Vert}{C}}-O-H$
Aldehydes	—CHO		Acetaldehyde	CH_3CHO	$R-\overset{O}{\overset{\Vert}{C}}-H$
Ethers	—O—	(oxy)	Ethyl ether	$(C_2H_5)_2O$	R—O—R'
Ketones	C=O	Carbonyl	Acetone	$(CH_3)_2CO$	$R-\overset{O}{\overset{\Vert}{C}}-R'$
Nitrogen-containing (Table 2.10)					
Primary amines	—NH₂	Amino	Methylamine	CH_3NH_2	$R-N\begin{smallmatrix}H\\\\H\end{smallmatrix}$
Secondary amines	=NH	—	Dimethylamine	$(CH_3)_2NH$	$R-N\begin{smallmatrix}H\\\\R'\end{smallmatrix}$
Tertiary amines	≡N	—	Trimethylamine	$(CH_3)_3N$	$R-N\begin{smallmatrix}R'\\\\R''\end{smallmatrix}$
Sulfur-containing (Table 2.9)					
Thiols	—SH	Mercapto	Methylmercaptan	CH_3SH	R—S—H
Sulfides	=S	—	Methyl sulfide	$(CH_3)_2S$	R—S—R'

and can be written ArOH. In phenol the arryl radical is the phenyl radical C_6H_5—. Another arryl radical is the toluyl group, $CH_3.C_6H_4$—, which occurs in toluol

Phenol, toluol, and other compounds with the hydroxyl group attached to the aromatic ring are slightly acidic and are not true alcohols. On the other hand, benzyl alcohol

$$\text{⟨◯⟩—CH}_2\text{OH}$$

is an alcohol similar to methanol or ethanol.

Phenolic compounds such as phenol and toluol are formed in the thermal decomposition of coal tar. The simple term "phenols" (written in the plural) is usually used to mean a wide variety of phenolic compounds. When phenols are produced during coal gasification they end up in process wastewaters from which they have to be removed.

The aliphatic acids, R—COOH, are true acids and form crystalline salts with alkalis such as caustic soda, NaOH. Thus acetic acid, $CH_3.COOH$ (which if diluted is vinegar) forms sodium acetate, $CH_3.COONa$. The term "fatty acid" is sometimes used because long chain aliphatic acids can be prepared from fats. Bases also occur in organic chemistry. Amines, which may be regarded as derived from ammonia, NH_3, are a most important class of bases.

Oxygen, nitrogen, sulfur, and other elements can also be present within the ring structure, in which case the compounds fall into the *heterocyclic* group of organics.

Table 2.5 Some members of the paraffin series and their radicals

Carbon Number, n	Name	Boiling Point, °C	Formula C_nH_{2n+2}	Radical Name C_nH_{2n+1}
1	Methane	−161.4	CH_4	Methyl
2	Ethane	−88.6	$CH_3.CH_3$	Ethyl
3	Propane	−42.2	$CH_3.CH_2.CH_3$	Propyl
4	n-Butane	−0.6	$CH_3.CH_2.CH_2.CH_3$	n-Butyl
	Isobutane	−10	$\begin{matrix} CH_3 \\ \diagdown \\ CH.CH_3 \\ \diagup \\ CH_3 \end{matrix}$	Isobutyl
5	n-Pentane	36.3	$CH_3.(CH_2)_3.CH_3$	n-Amyl
	Isopentane	28.0	$(CH_3)_2CH.CH_2.CH_3$	Isoamyl
	Neopentane	9.5	$(CH_3)_2.C.(CH_3)_2$	Neoamyl
6	n-Hexane	69.0	$CH_3.(CH_2)_4.CH_3$	n-Hexyl
7	n-Heptane	98.4	$CH_3.(CH_2)_5.CH_3$	n-Heptyl
8	n-Octane	125.7	$CH_3.(CH_2)_6.CH_3$	n-Octyl
9	n-Nonane	150.5	$CH_3.(CH_2)_7.CH_3$	n-Nonyl
10	n-Decane	174.0	$CH_3.(CH_2)_8.CH_3$	n-Decyl

These compounds typically contain five- or six-membered rings, contain double bonds, and exhibit aromatic character. Furan, thiophen, and pyrrole are examples of five membered ring heterocyclics

| Furan, C_4H_4O | Thiophen, C_4H_4S | Pyrrole, C_4H_4NH |

Much of the organic nitrogen and sulfur in petroleum, oil shale, tar sands, and coal is present in heterocyclic groups.

Coal can be considered as being made up of aromatic macromolecules, which may include some saturated rings and some heterocyclic rings. Macromolecules containing more than one ring are called polycyclic, polynuclear, or polycondensed rings. The term "condensed" is used because often two rings join together with the exclusion of a smaller molecule such as water. The rings may be fused together as in naphthalene or they may be held together by a bond as in diphenyl, see Table 2.3. In coal, the ring macromolecules are typically linked by small aliphatic chains of methylene (CH_2) which may contain oxygenated functional groups. These linkages are broken in direct liquefaction processes, releasing gases and leaving liquids of a highly aromatic character.

While radicals have been used above as a convenient shorthand method of presenting an organic compound, it is possible for radicals to exist independently in the free state. These so-called "free radicals" have one or more unpaired electrons and so are unstable, highly reactive, and can normally exist independently for a short time only. They arise as intermediate species in many organic reactions, including the complex series of reactions occurring in the thermal decomposition of coal and oil shale.

A further property of organic molecules is their shape or structure. For example the saturated hydrocarbon having eight carbon atoms, C_8H_{18}, is known as normal octane or n-octane if the carbon atoms form a single chain. However, some of the carbon atoms may be present in a branch chain, as in 2:2:4 trimethylpentane, commonly but incorrectly called "isooctane":

$$CH_3(CH_2)_6CH_3$$

n-Octane

$$CH_3—\overset{\overset{\textstyle CH_3}{|}}{\underset{\underset{\textstyle CH_3}{|}}{C}}—CH_2—\overset{\overset{\textstyle CH_3}{|}}{CH}—CH_3$$

2:2:4 Trimethylpentane or "Isooctane"

In fact the prefix iso- is reserved for the grouping $(CH_3)_2C—H$ as in the true isooctane and the prefix neo- for the grouping $(CH_3)_3C$ as in neooctane:

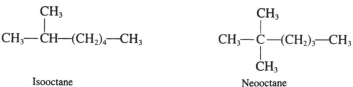

Isooctane   Neooctane

It can be seen that "isooctane" contains both an iso- and a neo-grouping and that a more specific nomenclature such as 2:2:4 trimethylpentane is required. Here the compound is named as a derivative of the longest chain, and the position of the attached radicals is stated.

The existence of more than one organic compound having the same chemical formula, as discussed above for octane, is known as *structural isomerism*. Examples of structural isomerism for aromatic compounds are shown in Table 2.3 for the xylenes. Here the prefixes *ortho-, meta-,* and *para-* are used to indicate the relative positions of the radicals attached to the benzene ring. The non-specificity of molecular formulae for many organic molecules is the reason the graphical or structural representation is used; this representation is not necessary for inorganic compounds.

The structure of a hydrocarbon has a strong bearing on its characteristic as a fuel. One important fuel property is its tendency to knock or explode on compression in internal combustion engines. This tendency to knock is quantified by the octane number which is the percent "isooctane" in a mixture of "isooctane" and *n*-heptane which will knock under the same conditions as the fuel being tested. Because "isooctane" has better antiknock properties than most commercial gasolines, it is arbitrarily assigned an octane number of 100. A poor fuel, *n*-heptane, is assigned an octane rating of zero. Generally, *n*-paraffins have lower octane numbers than branched paraffins, while olefins and aromatics have high octane numbers. Olefins do not occur in significant concentrations in most petroleums, but are formed during refining operations and are important in increasing the octane number of fuels now that the use of lead-containing additives is being phased out.

Refining[11] of petroleums involves separation of the crude feedstock into various fractions by distillation, chemical processing to change the size and structure of molecules to desirable products, and purification steps for reduction of nitrogen and sulfur species. The fractions separated from petroleum and coal tar are normally named according to their boiling point, or the boiling range at which they distill off from the mixture, rather than according to their composition. However, the boiling point of related organic compounds increases essentially as the size of the molecule increases, as can be seen in Table 2.5 for the paraffins, so the number of carbon atoms in the compounds falling in each fraction is fairly well defined. Typical fractions based on nominal boiling ranges are listed in Table 2.6 for petroleum and in Table 2.7 for coal tar. The overlap in the carbon number for some of the boiling ranges is related to whether or not the carbon atoms are in a chain or ring, and the effect of unsaturated bonds and branch chains if present.

The words tar and pitch which appear in Tables 2.6 and 2.7 are broad terms referring to materials obtained on the distillation of coal or petroleum. *Tar* is the viscous liquid remaining after pyrolysis of wood, coal, or shale; while *pitch* is the solid residue left on distillation of a tar. *Bitumen* on the other hand is a class of material, not a boiling fraction, and refers specifically to substances that are soluble in carbon disulfide. Petroleum is a bituminous liquid and asphalt is a bituminous solid. Organic substances that are insoluble in carbon disulfide are *pyrobitumens*. Kerogen, the main organic constituent in oil shale, and the organics in coals are pyrobitumens and by definition are not petroleums.

Table 2.6 Typical petroleum fractions

Fraction	Carbon Number	Boiling Range, °C	Uses
Gases	C_1–C_4	<0	Fuel
Light naphtha	C_5–C_7	26–93	Solvent
Heavy naphtha	C_6–C_{10}	93–177	Motor fuel
Kerosene	C_9–C_{15}	149–260	Jet fuel
Light gas oil	C_{13}–C_{18}	204–343	Fuel oil
Heavy gas oil	>C_{22}	385–565	Heavy fuel oil
Lube oils/waxes	>C_{22}	385–538	Lubricants
Residuum	>C_{40}	>565	Asphalt tar
Gasoline boiling range	C_5–C_{12}	30–210	
Diesel fuel boiling range	C_{11}–C_{21}	200–370	

Chemical processing to alter molecular structure may be by decomposition or cracking of large molecules, combination or polymerization of smaller molecules, and reforming of molecules. *Cracking* of heavy or high boiling oils to yield products in the gasoline boiling range is induced thermally by heating to above about 360°C, and may be assisted by the presence of catalysts. Reactions occurring during cracking may be represented by the following decomposition of a paraffin to a smaller paraffin and an olefin[11]

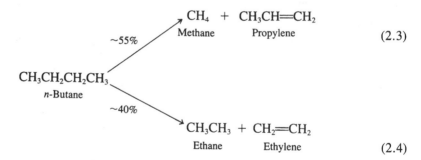

$$CH_4 \; + \; CH_3CH{=}CH_2$$
Methane   Propylene   (2.3)

$$CH_3CH_2CH_2CH_3$$
n-Butane

~55%

~40%

$$CH_3CH_3 \; + \; CH_2{=}CH_2$$
Ethane   Ethylene   (2.4)

Although paraffins are readily cracked, polycyclic aromatic compounds do not crack significantly unless one or more of the rings are first hydrogenated, as is done in hydrocracking. Hydrogen may be produced within the cracker by dehydrogenation of

Table 2.7 Typical coal tar fractions

Fraction	Boiling Range, °C	Typical HC Constituents and Their Carbon Numbers
Ammoniacal liquor	~100	—
Light oil	<170	Benzene, C_6; Toluene, C_7; Xylene, C_8
Middle oil or carbolic oil	170–230	Naphthalene, C_{10}
Heavy oil or creosote oil	230–270	
Green oil or anthracene oil	270–360	Anthracene, C_{14}
Residue or pitch	>360	—

paraffins to olefins, or may have to be provided from an external source. The degree of hydrogenation must be carefully controlled, not only to minimize costly hydrogen consumption, but to prevent hydrogenation of olefins and single-ring aromatics to compounds with lower octane numbers.

The thermal decomposition or *pyrolysis* of coal and oil shale may be likened to cracking in that their polycyclic structure is decomposed to smaller, lighter components. *Coking* is a competing reaction in which highly reactive free radicals produced during bond fission may combine or *polymerize* to form a residue of larger molecules sometimes called coke. Coking may be enhanced by increasing residence times in crackers to allow these polymerization reactions to proceed. Removal of carbon by coking effectively increases the hydrogen-to-carbon ratio of the remaining hydrocarbons. In refineries, gasoline products are produced by coking of heavy oils and by polymerization of gases formed in cracking operations. Liquid phase coking is normally not desired in the pyrolysis of oil shale and coal as it represents a carbon loss, and may be controlled by pyrolyzing in a hydrogen atmosphere, for example. The free radicals then combine with hydrogen to form paraffins rather than with other free radicals to form coke.

Reforming is a further chemical process used at refineries and important in some synthetic fuel processes. Unlike cracking and coking, the molar mass is not significantly altered, but the molecule structure is manipulated to improve product quality. Reforming often involves dehydrogenation to decrease the saturation as in the formation of aromatics from naphthenes, and olefins from paraffins. In isomerization reforming a normal paraffin is converted to the iso- compound without any change in saturation. All these reactions serve to improve the octane number of gasoline.

A chemical process generally not practiced in refining, but that is the basis of the indirect liquefaction of coal, is *synthesis*. Synthesis, like polymerization, is the building up of larger molecules from smaller ones. It is distinguished from polymerization in that the reacting molecules need not be similar. *Fischer-Tropsch* synthesis, for example, produces gasoline from synthesis gas:

$$nCO + 2nH_2 \xrightarrow{\text{catalyst}} C_nH_{2n} + nH_2O \qquad (2.5)$$

$$nCO + (2n + 1)H_2 \xrightarrow{\text{catalyst}} C_nH_{2n+2} + nH_2O \qquad (2.6)$$

About 75 to 80 percent of the useful product is olefinic, with the remainder being paraffinic and aromatic. Thus, by a process of complete thermal decomposition followed by a reconstruction process, an aliphatic liquid fuel having good antiknock properties can be produced. As coal is essentially aromatic in character, direct liquefaction produces aromatic fuels having properties different from petroleum gasolines. In this respect it is noted that aromatic and naphthenic hydrocarbons are a detriment in diesel fuels. In addition, some aromatic compounds are carcinogenic. However, direct liquefaction in theory has the advantage of relatively high conversion efficiencies.

Sulfur compounds in fuel generally have poor antiknock characteristics, but are probably better known for their foul odors. They can also cause corrosion problems in engines and in refineries, and poison some hydrocracking and reforming catalysts. The extent of the occurrence of sulfur compounds in fuels and fuel precursors is summarized in Table 2.8, while the types of sulfur compounds of relevance are listed in Table 2.9.

The term "sour" was originally applied to crude oils containing hydrogen sulfide, but is now generally used for liquids containing any odiferous substance. The mercaptans or thiols are responsible for the objectionable odor in many petroleums, and are removed in "sweetening" processes which include washing with alkalis, as well as extraction, adsorption, and oxidation processes. Hydrodesulfurization[8] is an important sulfur removal process in refineries that releases the sulfur as hydrogen sulfide on mild hydrogenation and sometimes cracking

$$RSH + H_2 \longrightarrow RH + H_2S \tag{2.7}$$

$$RSR' + 2H_2 \longrightarrow RH + R'H + H_2S \tag{2.8}$$

$$\boxed{}_S + 3H_2 \longrightarrow C_4H_8 + H_2S \tag{2.9}$$

Similar reactions probably account for the hydrogen sulfide released in oil shale retorting and coal conversion processes. The hydrogen sulfide, itself an obnoxious and highly toxic gas, is generally separated from the main gas stream and recovered as elemental sulfur. Any sulfur remaining in the fuel will on combustion be oxidized to the sulfur oxides SO_x, principally sulfur dioxide, SO_2, but also some sulfur trioxide, SO_3. These too are toxic, corrosive gases and generally have to be controlled. Gas purification processes are discussed in Section 5.1.

Apart from organically bound sulfur, coal and oil shale contain inorganic sulfur, usually as iron disulfide, FeS_2, in the crystalline forms of iron pyrites and marcasite.

Table 2.8 Representative elemental concentrations in mass percent of several fuels*

Fuel	C	H	O	N	S†
Petroleum crude					
Pennsylvania, low sulfur	86.1	13.9	0.0	0.0	0.06
Texas, medium sulfur	85.1	12.3	0.0	0.7	1.8
Mexico, high sulfur	83.7	10.2	—	—	4.2
No. 1 fuel oil	86.4	13.6	0.01	0.003	0.09
Coal‡ bituminous	83.6	5.6	6.0	1.6	3.2
subbituminous	76.6	5.8	15.0	1.5	1.1
lignite	72.3	5.1	19.9	1.1	1.6
Coal liquids§	87.5	11.5	0.7	0.3	0.01
Kerogen, Mahogany Zone	80.5	10.3	5.8	2.4	1.0
Shale oil	84.8	11.6	0.8	1.9	0.9
Shale oil syncrude§	86.1	13.8	0.1	0.01	0.02
Tar sand bitumen, Athabasca	83.1	10.3	1.6	0.4	4.6
Tar sand syncrude§	87.1	12.7	0.04	0.07	0.1

* Based on data in Refs. 11, 12, or given elsewhere in this book.
† Inorganic plus organic sulfur.
‡ Dry and ash-free basis.
§ After upgrading by hydrogenation.

Table 2.9 Some sulfur compounds representative of fuel processing

Compound	Group or Class	Typical Occurrence
Hydrogen sulfide, H_2S	Sulfide	Released on hydrotreating petroleum and in coal processing.
Iron pyrites and marcasite, FeS_2	Sulfide	Inorganic sulfur in oil shale, coal.
Carbon disulfide, CS_2	Disulfide	Coal tar; used as solvent.
Sulfur dioxide, SO_2	Oxide	Flue gases, exhausts.
Sulfuric acid, H_2SO_4	Acid	Used in petroleum processing.
Methyl mercaptan, CH_3SH	Thiol	Petroleum.
Thiophen, C_4H_4S	Heterocyclic	Coal tar and shale oils.
Methyl sulfide, $(CH_3)_2S$	Sulfide	Petroleum, cracked gasoline.
Methyl disulfide, $(CH_3)_2S_2$	Disulfide	On oxidation of thiols.

Iron disulfide decomposes at about 400°C and will form sulfur dioxide on combustion,

$$4FeS_2 + 11O_2 \longrightarrow 2Fe_2O_3 + 8SO_2 \qquad (2.10)$$

Nitrogen that occurs in petroleum, coal, oil shale, and tar sands is normally all organic nitrogen. The extent to which nitrogen occurs in these substances is shown in Table 2.8, while the nitrogen compounds of interest are shown in Table 2.10. Although ammonia does not occur naturally in coal and oil shale, it is released during thermal decomposition of these substances, and is produced during hydrodenitrogenation[8] of petroleum. It is obnoxious and toxic if concentrated. As it is highly soluble in water, it may be removed from gas mixtures by washing or "scrubbing" with water. Hydrogen sulfide and carbon dioxide are only slightly soluble in pure water, but will dissolve in ammoniacal solutions to form ammonium bisulfide, NH_4HS, and ammonium bicarbonate, NH_4HCO_3. In partial oxidation of coal, when a lot of carbon dioxide is formed in the same reactor as ammonia, scrubbing solutions and water condensates usually contain ammonium bicarbonate with very little ammonium bisulfide. This is because CO_2, which is much more concentrated than H_2S in the gas phase, uses up

Table 2.10 Some nitrogen compounds representative of fuel processing

Compound	Group or Class	Typical Occurrence
Ammonia, NH_3	Alkali	Released in hydrotreating, retorting, and gasification processes.
Methylamine, CH_3NH_2	Aliphatic amine	Petroleum
Aniline, $C_6H_5NH_2$	Aromatic amine	Coal tar
Pyrrole, C_4H_4NH	Heterocyclic	Coal tar
Pyridine, C_5H_5N	Heterocyclic	Coal tar
Quinoline, C_9H_7N	Heterocyclic	Coal tar
Nitric oxide, NO	Oxide	Exhaust flue gases

the dissolved ammonia and prevents significant quantities of H_2S from dissolving. In liquefaction of coal when hydrogen is produced externally and reacted with coal, little CO_2 is formed and so the watery condensates usually contain a lot of NH_4HS. Solutions of ammonium bisulfide which are formed during scrubbing operations in refineries are known as "sour waters." The dissolved gases are stripped from these waters by heating with steam and may be separated and the ammonia recovered in the anhydrous form, NH_3. Alternatively, ammonia may be recovered as fertilizer grade ammonium sulfate by dissolving the stripped ammonia in sulfuric acid

$$2NH_3 + H_2SO_4 \longrightarrow (NH_4)_2SO_4 \tag{2.11}$$

This latter route is no longer favored, as anhydrous ammonia and elemental sulfur generally have a higher market value and are cheaper to transport and store.

Many nitrogen compounds are basic and are removed in petroleum refining as they can deactivate catalysts. Nitrogen compounds may also contribute to gum formation in fuel oils, and the primary aromatic amines arising in some coal liquids are carcinogenic. On combustion, nitrogen compounds form oxides of nitrogen, NO_x, which may combine with hydrocarbon emissions in the presence of sunlight to form peroxyacetyl nitrate (PAN), an important constituent in photochemical smog:

$$NO_2 \xrightarrow{\text{light}} NO + O \xrightarrow[\text{oxygen}]{\text{hydrocarbons}} CH_3COO_2NO_2 \tag{2.12}$$

| Nitrogen dioxide | Nitric oxide | PAN |

It should be noted, however, that atmospheric nitrogen is the major source of nitrogen for NO_x formation in spark ignition gasoline engines, and that the photochemical smog problem may be controlled by reducing hydrocarbon emissions, for example, by oxidation in catalytic converters to form carbon dioxide and water:

$$CH_x + \left(1 + \frac{x}{4}\right)O_2 \xrightarrow{\text{catalyst}} CO_2 + \frac{x}{2}H_2O \tag{2.13}$$

The final group of compounds to be considered are those containing *oxygen*, some of which are listed in Table 2.11. Although oxygen is essential to combustion, it in itself has no calorific value and its presence reduces the energy available from a given mass of fuel. For example, methane has a calorific value of 55.5 MJ/kg, while the values for methanol and ethanol are 22.7 MJ/kg and 29.7 MJ/kg respectively. Although oxygen constitutes half the molecular mass of methanol, it represents only a quarter of the total oxygen demand for complete combustion:

$$CH_3OH + \tfrac{3}{2}O_2 \longrightarrow CO_2 + 2H_2O \tag{2.14}$$

Nevertheless, it is contended that the presence of some oxygen in the fuel molecule improves combustion, and alcohols are sometimes added to gasolines for this reason. The primary reason for the promotion of gasohol, a blend of approximately 90 percent gasoline and 10 percent methanol or ethanol, is the extension of fuel supplies using a non-petroleum source. Methanol may be produced from synthesis gas, while ethanol is a product of fermentation of the starch or sugars in grain and biomass:

Table 2.11 Some oxygen compounds arising in synthetic fuel processes

Compound		Group or Class	Typical Occurrence
Formula	Name		
H_2O	Water	—	Free or combined in coals, etc. In exhausts, flue, and retort off gases.
CO	Carbon monoxide	Oxides ⎫	In exhausts, flue, gasifier, and retort off gases. CO_2 is a byproduct of fermentation.
CO_2	Carbon dioxide	Oxides ⎭	
CH_3OH	Methanol	Alcohols	Produced from synthesis gas.
C_2H_5OH	Ethanol	Alcohols	Fermentation of biomass.
CH_3COOH	Acetic acid	Carboxylic acids	Oil shale wastewaters; the carboxylic group COOH occurs in coals, especially lignites, and releases CO_2 on heating.
$(C_6H_{10}O_5)_n$	Cellulose	Carbohydrates*	Wood, coal precursor.
$C_nH_{2n}O_n$	Sugars	Carbohydrates*	Biomass.
C_6H_5OH	Phenol	Phenols ⎫	
$C_6H_4(OH)_2$	Dihydroxybenzene†	Phenols ⎬	Coal tar, coal gasification off gases.
$C_6H_3(OH)_3$	Trihydroxybenzene‡	Phenols ⎭	
$C_{10}H_7OH$	Naphthol	Phenols	Coal tar.

* Hydrates of carbon, $C_x(H_2O)_y$.
† Three possible isomers: catechol, resorcinol, quinol.
‡ Common isomer: pyrogallol.

$$(C_6H_{10}O_5)_n + \frac{n}{2}H_2O \longrightarrow \frac{n}{2}C_{12}H_{22}O_{11} \qquad (2.15)$$

Starch   Sugar (maltose)

$$C_{12}H_{22}O_{11} + H_2O \longrightarrow 2C_6H_{12}O_6 \longrightarrow 4C_2H_5OH + 4CO_2 \qquad (2.16)$$

Sugar   Glucose   Ethanol

The fermented liquor contains up to 14 percent ethanol and is fractionated to produce "rectified spirit," a liquid mixture containing 93 to 95 percent ethanol, to which up to 10 percent methanol and small amounts of other substances may be added to produce the methylated spirits or "denatured" alcohol available commercially. The carbon dioxide, which according to the above chemical equation is produced at a molar rate equivalent to that of the ethanol, on a small scale may be sold as a byproduct. However, biomass processing on a large scale will inevitably result in release of CO_2 to the atmosphere.

It is interesting that the stoichiometry of coal hydrogenation to methane via the char-steam reaction

$$2C + 2H_2O \longrightarrow CH_4 + CO_2 \qquad (2.17)$$

also shows a CO_2 production equivalent to the product methane rate. Although carbon dioxide is not toxic, its release is of interest as, according to the "greenhouse model," small increases of carbon dioxide concentration in the atmosphere can result in increases in ambient temperature and consequent climatic changes.

In assessing carbon dioxide "pollution," the total amount released must be considered in relation to the final energy recovered from the fuel. Carbon dioxide is formed on the combustion of carbon to provide the heat needed for synthetic fuel production. Additionally, in oil shale production carbon dioxide is released by the decomposition of carbonates in the shale according to the reaction

$$CaCO_3 \xrightarrow{\text{heat}} CaO + CO_2 \qquad (2.18)$$

Calcite or | Lime or
Calcium | Calcium
carbonate | oxide

Generally, however, the major source of carbon dioxide release is in the combustion of the product fuel.

The phenols, which are highly toxic, and carboxylic or fatty acids are not considered fuels, but are of interest as they are found in the wastewater streams of several synthetic fuel processes. Biological treatment is one method that is practiced at synthetic fuel plants for controlling these and other dissolved organics in wastewaters. When the biological decomposition takes place in the presence of free oxygen, carbon dioxide and water are produced, together with a settleable bacterial sludge. In this way, part of the organic matter is oxidized and part is converted to cell material. This type of breakdown, caused by aerobic bacteria, is termed *aerobic decomposition*. An example is the decomposition of acetic acid

$$CH_3COOH \xrightarrow[\text{microorganisms}]{\text{aeration}} 2CO_2 + 2H_2O + \text{sludge} \qquad (2.19)$$

Acetic acid

If the breakdown occurs in the absence of free oxygen and the bacteria use combined oxygen, then carbon dioxide, methane, and some inert material are the end products. This is termed *anaerobic decomposition* and may be represented by the reaction

$$(CH_2O)_n \xrightarrow[\text{formers}]{\text{acid}} RCOOH \xrightarrow[\text{formers}]{\text{methane}} CO_2 + CH_4 \qquad (2.20)$$

2.2 THERMODYNAMICS FOR SYNTHETIC FUELS

Synthetic fuel processes require an energy input to accomplish the conversion. In other words, only part of the energy entering a synfuel plant is recovered in the products and byproducts, the rest is lost in the process. As indicated in Table 1.16, the amount of energy not recovered can be substantial and, consequently, considerable attention is directed towards improving process efficiencies. One way of increasing the efficiency of a process is to increase plant capital investment, for example by providing larger heat exchange surfaces; but this is, of course, constrained by economic considerations.

In addition, a portion of the energy loss is inherent to the process and the associated chemical changes and so cannot be reduced. The laws of thermodynamics provide the means for determining this inherent energy loss and for establishing upper efficiency bounds for specific conversions. While these calculated upper bounds are in general applicable to ideal processes, that is for systems operating without losses due to friction, uncontrolled heat loss to the atmosphere, etc., they nevertheless provide an important basis both for assessing the performance of real plants, and for indicating those process steps most likely to benefit from process modifications. Much of the discussion here is directed toward the conversion process itself; in later chapters the end use of the product fuel will be considered in gauging the overall energy effectiveness of specific synthetic fuels.

Mass and Energy Balances

Mass and energy conservation is basic to process evaluation. The balances expressing this conservation may be applied either to the overall process or to a specific unit in the process. Boundaries are drawn around the process step of interest, and rates of mass or energy are equated according to

$$\text{IN} + \text{GENERATED} = \text{OUT} + \text{ACCUMULATED} \qquad (2.21)$$

The accumulation term is zero for processes operating at steady state. The generation term is also zero when balancing conserved properties such as chemical elements, total mass, and total energy.

A specific type of mass or energy may be generated or consumed in the process. Consider, for example, the production of methane from synthesis gas following the reaction

$$CO + 3H_2 \longrightarrow CH_4 + H_2O \qquad (2.22)$$

According to the chemical reaction, one mole of water will be formed for each mole of methane produced. If it is assumed that the reaction goes to completion as written, then one mole of carbon monoxide and three moles of hydrogen will form one mole each of methane and water. Mass balance data calculated on this basis in Figure 2.1 can be seen to conform with Eq. (2.21). The *material conversion* factor we define by

$$\text{Material conversion} = \frac{\text{Mass of chemical species in desired product}}{\text{Total mass of this species into the process}} \qquad (2.23)$$

For the process shown in Figure 2.1 the material conversion for hydrogen, an expensive feedstock, is two-thirds, with the remaining one-third lost as water. This loss of hydrogen is inherent in the process due to the prevailing stoichiometry and would occur even if the process operated ideally and there were no other restrictions. It is called the *stoichiometric constraint*. We shall see later that there are other limitations on material conversions apart from stoichiometric constraints.

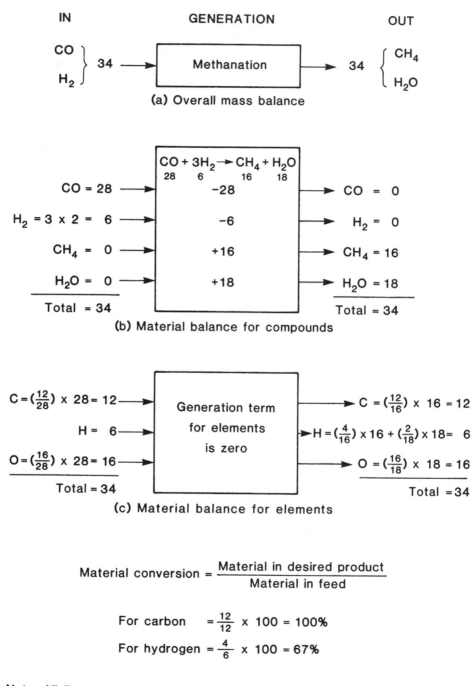

Material conversion = $\dfrac{\text{Material in desired product}}{\text{Material in feed}}$

For carbon $= \dfrac{12}{12} \times 100 = 100\%$

For hydrogen $= \dfrac{4}{6} \times 100 = 67\%$

Note: All flows in kg

Figure 2.1 Illustrative mass balance calculations for complete methanation of 1 mole of carbon monoxide and 3 moles of hydrogen.

Thermal Efficiency

One measure of the energy efficiency of the process is the thermal efficiency

$$\eta_{l} = \frac{\text{Heat in desired product}}{\text{Total heat into process}} \qquad (2.24)$$

Heat is taken to be the total energy content of the fuel as measured by its calorific value plus any sensible heat above some datum temperature. It includes also the thermal equivalent of electricity, or pressure energy above a datum pressure, or any other energy form that is imported or exported across the process boundaries. The thermal efficiency is sometimes called a "first law" efficiency as, like the first law of thermodynamics, it concerns total energy. According to the first law, energy is conserved, that is it can neither be created nor can it be destroyed. According to Eq. (2.21), therefore, all the energy entering a steadily operating process must leave it in one form or another. Some of this unrecovered energy is dissipated as low-grade heat to the atmosphere in flue gases or cooling tower plumes. However, byproducts formed may have a calorific content, and can account for a significant quantity of energy in some processes. If the byproduct represents a useable fuel, its energy content may legitimately be included in the numerator of Eq. (2.24).

Total energy, which is called the *internal energy U* in thermodynamics, is related to heat and work by the first law

$$\Delta U = q + w \qquad \text{(closed system)} \qquad (2.25)$$

where ΔU is the change in internal energy which occurs when a quantity of heat q flows into the system and the total work done on the system is w. The internal energy is a *function of state*, that is it depends only on conditions such as temperature, pressure, and composition of the system and is independent of its previous history.

Another function of state called the *entropy S* also changes when heat flows into the system

$$\Delta S = \frac{q_{\text{rev}}}{T} \qquad (2.26)$$

where ΔS is the change in entropy that occurs when the heat flows into the system at temperature T. The subscript "rev" means that all changes in the intrinsic states of the system occur *reversibly*, which requires that the system is only infinitesimally displaced from equilibrium as the process proceeds. If the process under consideration were to proceed irreversibly, the net heat flow would be smaller for the same entropy change, that is,

$$\Delta S > \frac{q}{T} \qquad \text{(irreversible process)} \qquad (2.27)$$

The internal energy change ΔU also remains the same and so it follows from Eq. (2.25) that the work done on the system for a given change is a minimum when the change occurs reversibly. The additional work required for an irreversible change is, according to Eq. (2.25), dissipated as heat. Equations (2.26) and (2.27) form a basis for the second law of thermodynamics which states that all spontaneous processes are to some extent irreversible and are accompanied by a "degradation of energy."

It follows also that the change in total or internal energy does not alone provide complete information on the amount of work obtainable from the system. For this reason a "second law" efficiency based on "available" energy has been introduced

$$\eta_{II} = \frac{\text{Available energy in desired product}}{\text{Available energy into process}} \tag{2.28}$$

Available energy is that portion of the total energy that is available for conversion to useful work. Unlike total energy, available energy is not conserved and is generally consumed in real processes to effect the conversion.

Available energy can be introduced in simplistic terms as follows.[13] Consider an electric capacitance at voltage E' and containing a charge C'. The electrical energy of the system is given by

$$\text{Electrical energy} = E'C'$$

In general, the energy theoretically obtainable from the system may be expressed as the product of a potential or driving force and the quantity of the energy commodity in the system. For an incompressible fluid at a pressure p and having a volume V:

$$\text{Hydraulic energy} = pV$$

Similarly for a system at a temperature T and with a heat capacity C:

$$\text{Thermal energy} = TC$$

In practice only some fraction of the energy, the available energy, can be obtained from the system:

$$\text{Available electrical energy} = (E' - E_0')C'$$
$$\text{Available hydraulic energy} = (p - p_0)V$$
$$\text{Available thermal energy} = (T - T_0)C$$

This limitation arises because the process can proceed only until the system is in equilibrium with some ground condition. In the case of electrical energy, the ground voltage can be arbitrarily defined as zero, but this cannot be done for pressure and temperature as atmospheric pressure is the standard ground pressure and 25°C or 298 K is generally taken to be the reference temperature. Temperature and pressure lower than the ground values can be achieved, but only by expending energy.

Consider the conversion of a quantity of thermal energy q,

$$q = TC \tag{2.29}$$

into work w. The maximum amount of work that can be produced, say, by conversion to hydraulic energy in an ideal process is equivalent to the available thermal energy:

$$w = (p - p_0)V = (T - T_0)C \tag{2.30}$$

Combining Eqs. (2.29) and (2.30) gives

$$w = \left(1 - \frac{T_0}{T}\right)q \qquad \text{(reversible process)} \tag{2.31}$$

which is the Carnot expression defining the maximum amount of heat available for conversion into work using a heat engine operating between T and T_0 without losses. As with the loss of hydrogen discussed in relation to Eq. (2.22), the unavailable heat in Eq. (2.31) is inherent to the process; actual heat losses are usually significantly larger.

While available energy analyses are important when considering heat-work conversions such as in steam-electric power generation, they are of less significance in fuel conversion processes in which the fraction of available energy in the feed and product fuel is similar.

Heat of Reaction

Another energy term that is not conserved but which is useful in analyzing closed systems at constant pressure, or flow processes such as flow through a turbine, is the *enthalpy H*,

$$H = U + pV \tag{2.32}$$

It can be expressed in an analogous way to the internal energy as given in Eq. (2.25)

$$\Delta H = q + w_u \quad \text{(constant } p \text{ or flow process)} \tag{2.33}$$

where w_u is the *useful work* done on the system. In a flow system it is the shaft work of a compressor driving the fluid, $+ w_u$, or of a turbine driven by the fluid, $- w_u$. In a closed system at constant pressure the useful work is work other than that due to expansion against the constant pressure, $p\Delta V$, and could arise from chemical changes. For a reversible chemical reaction

$$w_u = \Sigma \mu_i \, dn_i \quad \text{(constant } p, \text{ reversible)} \tag{2.34}$$

where μ_i is the *chemical potential* and n_i is the number of moles of substance i in the system. The useful work could, for example, be electrical work in a galvanic cell. The type of reactions of interest here, however, normally proceed irreversibly at constant pressure with no useful work involved. If in addition the initial and final temperatures are the same so that sensible heat effects are eliminated, then according to Eq. (2.33) the enthalpy change is simply the heat flow due to the reaction itself. It is termed the *heat of reaction* and is a measure of the energy absorbed by a stated reaction *proceeding to completion*.

For example, consider the combustion of methane in oxygen

$$CH_4 + 2O_2 \longrightarrow CO_2 + 2H_2O(l) \qquad \Delta H^{\circ}_{298} = -890.3 \text{ kJ/mol} \tag{2.35}$$

Here ΔH°_{298} is the heat of reaction at 298 K or 25°C per mole of methane burnt. The subscript denotes the reference temperature in kelvins, which is normally 298 K, although other values may be chosen. The superscript zero is used to denote the standard heat of reaction, and implies that the constant pressure prevailing is the standard pressure, which is conveniently selected to be one atmosphere. It also generally implies that the reactants are in the normal state of the substance, either gaseous, liquid, or solid, at the stated temperature and standard pressure. In cases where the normal state is not obvious or if a different state is intended, this should be specified. For example water, $H_2O(l)$, which is normally a liquid at 25°C and one atmosphere often enters into reactions as steam and is then shown as $H_2O(g)$ in the chemical equation. A positive value of the heat of reaction means that the system must gain heat from the surroundings to proceed isothermally, and the reaction is said to be endothermic. Combustion reactions transfer heat to the surroundings and are exothermic. Heats of combustion are normally reported as positive values, that is as $-\Delta H°$. The calorific value of a fuel is the heat released in a constant volume calorimeter and so will generally differ from the heat of combustion, which is for constant pressure conditions.

When the reaction involved is that of a compound being formed from its elements the enthalpy change is termed the *heat of formation*, and is here denoted by $\Delta_f H$. For example, the combustion of carbon in oxygen to CO_2 is a formation reaction since both C and O_2 are elements:

$$C + O_2 \longrightarrow CO_2 \qquad \Delta_f H°_{298} = -393.5 \text{ kJ/mol} \qquad (2.36)$$

The superscript zero and the temperature subscript have the same meaning as before; in particular note that the normal state of oxygen is the O_2 molecule and not atomic oxygen. Tabulated heats of formation such as given in Table 2.12 may be used to calculate the heat of reaction or combustion for any system. For example, to calculate the heat of combustion of methane according to Eq. (2.35) we need to add the equations for the formation of H_2O to that for CO_2 and subtract the equation for the formation of CH_4:

$$2H_2 + O_2 \longrightarrow 2H_2O(l) \qquad \Delta H°_{298} = 2\Delta_f H°_{298} = 2(-285.8) \text{ kJ/mol} \qquad (2.37)$$

$$(+) \quad C + O_2 \longrightarrow CO_2 \qquad\qquad \Delta_f H°_{298} = -393.5 \text{ kJ/mol} \qquad (2.36)$$

$$(-) \, C + 2H_2 \longrightarrow CH_4 \qquad\qquad \Delta_f H°_{298} = -74.8 \text{ kJ/mol} \qquad (2.38)$$

$$(=) \, CH_4 + 2O_2 \longrightarrow 2H_2O(l) + CO_2 \qquad \Delta H°_{298} = -890.3 \text{ kJ/mol} \qquad (2.35)$$

The calculated heat of reaction agrees with that given previously.

The demonstrated additive property of enthalpies can also be applied to entropies and other energies and is known as *Hess's law*. Note that the heat of combustion of methane estimated from the combustion of its components, which are one mole of carbon and two moles of hydrogen, is $393.5 + 2(285.8) = 965.1$ kJ/mol. This value differs from the correct value of 890.3 kJ/mol given in Eq. (2.35) by more than 8 percent. The discrepancy can be thought of simply as the energy associated with the carbon-hydrogen bonds in methane. All substances have energies associated with them relative to their constituent elements. A decrease in energy content normally indicates that the compound is more stable than the elements at the prevailing conditions. For example, the formation of methane from carbon and hydrogen is accompanied by a rejection of heat at 25°C, so methane is more stable than a mixture of carbon and hydrogen at this temperature. The heat of formation of methane given in Eq. (2.38) corresponds to the difference in the heat of reaction in Eq. (2.35) and that calculated above for the combustion of the constituent elements. Although the energy rejected represents an inherent loss in the formation of methane, part of this energy can usually be recovered and used elsewhere in a synthetic fuel process.

Tabulated values of $\Delta_f H°$ are for reactions proceeding at constant pressure and in which the initial and the final temperature of the reactants and products is the same.

Table 2.12 Heats and free energies of formation, and heats of combustion at 1 atmosphere and 298K*
Values in kJ/mol

Substance		Molar Mass	State	Heat of Formation, $\Delta_f H°_{298}$	Free Energy of Formation, $\Delta_f G°_{298}$	Heat of Combustion, $-\Delta H°_{298}$	
						GCV†	NCV‡
C	Carbon	12.01	Graphite	0	0	393.5	393.5
CH$_4$	Methane	16.04	Gas	− 74.8	− 50.8	890.3	802.3
C$_2$H$_6$	Ethane	30.07	Gas	− 84.7	− 32.9	1560	1428
C$_8$H$_{18}$	"Isooctane"	114.22	Gas	−224.1	13.1	5497	5100
C$_2$H$_4$	Ethylene	28.05	Gas	52.3	68.1	1411	1323
C$_8$H$_{16}$	1-Octene	112.20	Gas	− 82.9	+104.2	5351	4999
(—CH$_2$—)§	—	14.03	Gas	− 10.4	+ 13.0	668.9	624.9
			Liquid	− 16.1	—	663.2	619.2
C$_6$H$_6$	Benzene	78.11	Gas	82.9	129.7	3302	3170
			Liquid	49.2	—	3268	3136
C$_2$H$_2$	Acetylene	26.04	Gas	226.7	209.2	1300	1256
CH$_3$OH	Methanol	32.04	Gas	−201.2	−161.6	763.9	675.9
			Liquid	−238.7	−166.5	726.4	638.4
C$_2$H$_5$OH	Ethanol	46.07	Gas	−218.5	−168.3	1410	1278
			Liquid	−277.6	−174.7	1367	1235
C$_6$H$_5$OH	Phenol	94.11	Liquid	−158.2	− 46.1	3060	2928
CH$_3$COOH	Acetic acid	60.05	Liquid	−486.2	−391.5	872.4	784.4
CO	Carbon monoxide	28.01	Gas	−110.5	−137.3	283.0	283.0
CO$_2$	Carbon dioxide	44.01	Gas	−393.5	−394.4	0	0
H$_2$	Hydrogen	2.016	Gas	0	0	285.8	241.8
H$_2$O	Water	18.016	Steam	−241.8	−228.6	0	0
			Liquid	−285.8	−237.2	0	0
N$_2$	Nitrogen	28.02	Gas	0	0	—	—
NH$_3$	Ammonia	17.03	Gas	− 45.9	− 16.3	—	—
NO	Nitric oxide	30.01	Gas	90.4	86.7	—	—
NO$_2$	Nitrogen dioxide	46.01	Gas	33.3	51.3	—	—
O$_2$	Oxygen	32.00	Gas	0	0	0	0
S	Sulfur	32.06	Rhombic	0	0	296.8	296.8
H$_2$S	Hydrogen sulfide	34.08	Gas	− 20.0	− 32.8	562.6	518.6
SO$_2$	Sulfur dioxide	64.06	Gas	−296.8	−299.9	—	—
SO$_3$	Sulfur trioxide	80.06	Gas	−394.9	−370.7	—	—

* Data based on values in Ref. 12.
† To CO$_2$(g) and H$_2$O(l).
‡ To CO$_2$(g) and H$_2$O(g).
§ Taken to be one-eighth of 1-Octene.

Although the reference temperature is normally 298 K, a value at any other temperature, here indicated by a subscript T, may be estimated from

$$\Delta H_T^\circ = \Delta H_{298}^\circ + \sum \left(\int_{298}^T C_{p,i} \, dT \right)_{products} - \sum \left(\int_{298}^T C_{p,i} \, dT \right)_{reactants} \qquad (2.39)$$

where $C_{p,i}$ is the heat capacity at constant pressure of substance i and T is the absolute temperature.

Chemical Equilibrium

It has been implied that the stoichiometric constraint is not the only factor limiting the conversion of a substance into a desired product. Another limitation arises if the chemical reaction does not proceed completely to the desired products but instead is constrained on reaching an equilibrium condition. This limitation is superficially analogous to the available energy limitation imposed by a ground pressure or temperature. A mixture of reacting substances is said to be in chemical equilibrium when there is no net change in the quantities of reactants and products present. At equilibrium the reacting mixture is at its most stable composition, a condition which is met when the entropy of the system is a maximum and the Gibbs *free energy G* of the system is at a minimum. The equilibrium condition is met when the change in free energy dG for a small change in composition dn_i is zero for all i components present, that is,

$$\sum dG/dn_i = 0 \quad \text{(at equilibrium)} \qquad (2.40)$$

The free energy is related to the entropy and enthalpy by the relation

$$G = H - TS \qquad (2.41)$$

For an isothermal change,

$$\Delta G = \Delta H - T\Delta S \quad \text{(constant } T) \qquad (2.42)$$

and using Eqs. (2.33) and (2.26) the useful work becomes

$$w_u = \Delta G + (q_{rev} - q) \quad \begin{pmatrix} \text{constant } T, p \text{ in closed system} \\ \text{constant } T \text{ in flow process} \end{pmatrix} \qquad (2.43)$$

For a reversible process, therefore, the useful work done on the system is a minimum. Alternatively, the useful work done by the system is a maximum and is equivalent to the decrease in the free energy:

$$-w_{u,max} = -\Delta G \quad \begin{pmatrix} \text{reversible, constant } T, p \text{ in closed system} \\ \text{reversible, constant } T \text{ in flow process} \end{pmatrix} \qquad (2.44)$$

The constant temperature restriction in the above equations applies only to that part of the process in which heat is transferred.

    If a chemical reaction could be made to proceed reversibly, it would deliver $-\Delta G$ amount of useful work; if it proceeds irreversibly and if no useful work is done, it releases $-\Delta H$ amount of heat.

By applying the previously stated conditions for equilibrium it can be shown[4] that

$$K = \exp\left(\frac{-\Delta G_T^\circ}{RT}\right) = \frac{\Pi(a_i^{\nu_i})_{\text{products}}}{\Pi(a_i^{\nu_i})_{\text{reactants}}} \tag{2.45}$$

where K is the equilibrium constant. It is a dimensionless number that characterizes the reaction, and whose value is dependent on the temperature only. The quantity ΔG_T° is the standard free energy change for the reaction proceeding to completion. It can be determined from the standard free energies of formation of the reacting species tabulated in Table 2.12, in the same way as was done for the heats of reaction. The right-hand side of the equation is the ratio of the continued product of the "activities," a_i, of the products to that of the reactants, each activity term being raised to the power of the stoichiometric coefficient ν_i for the specific species.

The activity can be thought of as a thermodynamic concentration or pressure normalized with respect to its value at some reference state. The activity is dimensionless. In mixtures of ideal gases at the standard pressure of one atmosphere and in dilute or ideal solutions, it is simply the mole fraction of the substance in the reaction mixture. *Mole fraction* is a concentration term calculated as the ratio of the number of moles of the substance to the total number of moles present. An ideal gas is one which obeys the relation

$$pV = nRT \tag{2.46}$$

Here V is the volume of n kilomoles of the gas, and R is the universal gas constant. Partial pressures may be used in place of mole fractions for gases at all pressures for which the gas mixture behaves ideally. Useful formulae for calculating partial pressures and other gas concentration terms are summarized in Table 2.13. While all pressure terms in the table are in pascals, partial pressures used in place of activities in Eq. (2.45) must be dimensionless and should be normalized with respect to the standard atmospheric pressure of 101.3 kPa.

Assume by way of example that a mixture of carbon and steam can be made to react such that the only products formed are methane and carbon dioxide. It is found that the only way the reaction can be written for these conditions is if equal quantities of methane and carbon dioxide are produced

$$C + H_2O(g) \longrightarrow \tfrac{1}{2}CH_4 + \tfrac{1}{2}CO_2 \tag{2.47}$$

This hypothetical reaction is of interest as it represents the overall reaction for methane production by steam gasification of coal. It shows that at best only 50 percent of the carbon is recovered in the desired product. This is the *stoichiometric constraint*.

Assuming further that the reaction proceeds at the reference temperature of 298 K and the standard pressure of one atmosphere, the data in Table 2.12 can be used to show

$$\Delta H_{298}^\circ = 7.7 \text{ kJ/mol C}$$
$$\Delta G_{298}^\circ = 6.0 \text{ kJ/mol C}$$

and from Eq. (2.45),

$$K = \frac{(p_{CO_2})^{1/2}(p_{CH_4})^{1/2}}{p_{H_2O}} = \exp(-\Delta G_{298}^\circ/RT) \approx 0.09 \tag{2.48}$$

Table 2.13 Useful formulae for gas concentrations

p = pressure, Pa; T = temperature, K; V = volume, m³; m = mass, kg; M = molar mass, kg/kmol; R = gas constant = 8314 J/(kmol·K)

Component i		
Number of kilomoles	$n_i = m_i/M_i$	(1)
Total number of kilomoles in gas mixture	$n = \sum n_i$	(2)
Mole or volume fraction of i in gas mixture	$y_i = n_i/n$	(3)
Mass fraction of i in gas mixture	$Y_i = m_i/m$	(4)
	$\sum y_i = \sum Y_i = 1$	(5)
Partial pressure of i in ideal gas mixture	$p_i = y_i\, p$	(6)
	$\sum p_i = p$	(7)
Mean molar mass of gas mixture	$\overline{M} = \sum M_i\, y_i$	(8)
	$= 1/\sum (Y_i/M_i)$	(9)
Mean molar mass of air	$\overline{M}_{air} = 28.97$ kg/kmol	
Mass to mole (volume) conversion	$y_i = Y_i\, \overline{M}/M_i$	(10)
Wet Gas		
Mole (volume) fraction of water vapor in gas mixure, y_w		
Mole fraction	y_i (dry) $= y_i$ (wet)/$(1 - y_w)$	(11)
Equivalent equation in Y for mass fraction		
Vapor-Liquid Equilibrium		
Mole fraction in liquid, x_i; vapor pressure of pure i, $p_{v,i}$		
Partial pressure in ideal gas mixture	$p_i = x_i p_{v,i}$	(12)
Eq. (12) is Raoult's law for $x_i \rightarrow 1$, and Henry's law for $x_i \rightarrow 0$		
Henry's law; Henry's law constant, h_i	$p_i = h_i x_i$	(13)
Single Gas or Mixture		
Ideal gas law	$pV = nRT$	(2.46)
Volume of gas at NTP*	$V_0 = 22.41\, n$	(14)

* Normal temperature and pressure, $T_0 = 273.15$ K and $p_0 = 101.3$ kPa.

where the p_i are the partial pressures. An ideal gas mixture has been assumed and as discussed below, the activity of pure carbon is unity. Because equal quantities of carbon dioxide and methane are formed according to the hypothetical reaction, their partial pressures in the equilibrium gas mixture are equal and Eq. (2.48) becomes

$$p_{H_2O} \approx 11 p_{CH_4}$$

Three important bits of information have now been learned. First of all the heat of reaction is small but positive so that some heat must be provided to the system. It can be shown by application of Hess's law that if this heat is provided by combustion of carbon according to Eq. (2.36), then the overall heat of reaction is zero when 51 percent of the carbon is oxidized to CO_2, leaving 49 percent for conversion to methane. This is the *thermal constraint*, which we define as the maximum amount of carbon that can be converted to useful product for the assumed reaction scheme proceeding at 100 percent thermal efficiency. In the above example it has been assumed that steam is freely available; in reality the carbon burned to raise the steam must be accounted for as well. The difference between the stoichiometric and thermal constraints calculated above represents the carbon oxidized to provide the energy requirements for the formation of the methane and carbon dioxide products from their elements, and for the breakdown of steam into its elements. This, too, can be checked by application of Hess's law.

Secondly, the calculated free energy change for the reaction is positive and it therefore cannot be expected to proceed spontaneously. The calculated value is sufficiently small, however, to suggest that the reaction can be made to proceed by selecting alternative conditions, for example by selecting a higher reaction temperature.

Thirdly, it is seen that if the reacting system were to reach equilibrium, 11 moles of steam would leave the reaction vessel for every mole of methane produced. This is the *equilibrium constraint* at the temperature under consideration.

The prospect of approaching the ideal Carnot conversion factor, Eq. (2.31), is by now a bleak one, particularly as real systems operate far below the thermodynamically estimated efficiencies that are based on ideal operation.

In many synthetic fuel processes a liquid or solid participates in a gaseous reaction, and care has to be exercised in selecting the correct concentration term in calculating the equilibrium constant. Consider the case of water rather than steam reacting with carbon in the foregoing example. As there is no net reaction between water and water vapor when the two phases are in equilibrium, the vapor pressure of water at the reaction temperature may be used for the water concentration. In this case the standard free energy change must be based on the free energy of formation of steam, even though it is water that is participating in the reaction. This procedure is not suitable for solids, which typically have immeasurably small vapor pressures. It can, however, be shown[4] that when the free energy change for a reaction involving gases is calculated from the standard free energy change for the species as they occur in the reaction, then the concentration terms of the liquids and solids may be omitted from the equilibrium constant expression. This in effect means that the activities of the liquids and solids are unity. A necessary condition is that the non-gaseous substances are pure and that dissolution and heat of mixing effects are zero. It is stressed again that activities, concentrations, and partial pressures in Eq. (2.45) must be dimensionless.

Equilibrium Dependence on Temperature and Pressure

It is instructive to examine how changes in temperature and pressure affect chemical equilibrium. The effect of temperature is related to changes in the heat of reaction according to the *van't Hoff equation*

$$\frac{d \ln K}{dT} = \frac{\Delta H_T^\circ}{RT^2} \qquad (2.49)$$

The equation shows that the equilibrium constant of endothermic reactions can be increased by increasing the temperature. Evaluation of equilibrium constants by integration of this equation requires information on the temperature dependence of ΔH_T°; see for example Table 2.14 or Eq. (2.39). Fortunately, equilibrium constant data is often directly available for a range of temperatures.[14] Some equilibrium constant data of relevance to synthetic fuel processing are given in Table 2.15 and Figure 2.2. The scales in Figure 2.2 are the logarithm of the equilibrium constant versus $1/T$. This type of plot is frequently called a "van't Hoff plot." For reaction heats that are independent of temperature it can be seen from integrating Eq. (2.49) that $\ln K$ is linear in $1/T$. Figure 2.2b includes a plot of the equilibrium constant for the hypothetical gasification reaction, Eq. (2.47), and shows that as the reaction temperature is increased above 1000 K the equilibrium constant approaches unity. At high temperatures, therefore, the moles of steam leaving in the equilibrium mixture will be about equal to the moles of methane produced. The equilibrium constant data can be calculated either from Eq. (2.49) and the ΔH_T° data in Table 2.14, or directly from the appropriate combination of equilibrium constants in Table 2.15.

As stated previously, the equilibrium constant is dependent on temperature only and consequently is not affected by changes in pressure. Even so, the composition of the equilibrium mixture may be pressure dependent. Consider the reaction of carbon monoxide and hydrogen to form methanol

$$CO + 2H_2 \longrightarrow CH_3OH(g) \qquad \Delta H_{298}^\circ = -90.7 \text{ kJ/mol} \tag{2.50}$$

$$\Delta G_{298}^\circ = -24.3 \text{ kJ/mol}$$

$$K = 18 \times 10^3 = \frac{p_{CH_3OH}}{(p_{CO})(p_{H_2})^2} \tag{2.51}$$

If stoichiometric quantities of carbon monoxide and hydrogen are fed to the reactor, then Eq. (2.51) can be solved to show that the hydrogen in the equilibrium mixture at a total pressure of one-tenth of one atmosphere would be nearly 28 percent by volume of the methanol. This figure becomes 5 percent for operation at atmospheric pressure and 1 percent for 10 atmospheres, assuming the gas mixture continues to behave ideally. Increasing the pressure therefore increases the amount of hydrogen converted to product; we say that the *yield* of the reaction is increased. In general, whenever a gas reaction takes place with a decrease in the number of molecules, a rise in pressure increases the fraction of reactants converted. In the above example, three molecules combine to form one molecule. As the volume of a gas is proportional to the number of molecules present (see Eq. (14) in Table 2.13), the reaction results in a decrease in volume of 67 percent. The shift in the equilibrium composition with change in pressure can be thought of as a compensation effect, as stated in *Le Chatelier's principle*: "If a change occurs in one of the factors, such as temperature or pressure, under which the system is in equilibrium, the system will tend to adjust itself to annul, as far as possible, the effect of that change." The increase with temperature of the equilibrium constant for an endothermic reaction is in accordance with this principle.

Temperature and pressure are consequently important parameters in controlling

Table 2.14 Evaluation of thermodynamic functions for formation reactions of gases*

$\Delta_f H_T^o = \Delta H_{298}^o + aT + bT^2 + cT^3 + dT^4 + eT + f$ kJ/mol

$\Delta_f G_T^o = \Delta H_{298}^o - aT \ln T - bT^2 - (c/2)T^3 - (d/3)T^4 - (e/2T) + f + gT$ kJ/mol

$\Delta_f S_T^o = (\Delta_f G_T^o - \Delta_f H_T^o)/T$ kJ/(mol·K)

Formula	Name	$\Delta_f H_{298}^o$	a	b	c	d	e	f	g
CH_4	Methane	−74.8	−4.620E−2	+1.130E−5	+1.319E−8	−6.647E−12	−4.891E+2	+1.411E+1	−2.234E−1
C_2H_6	Ethane	−84.7	−9.834E−2	+6.414E−5	−9.311E−9	−3.553E−12	−9.782E+2	+2.717E+1	−4.535E−1
C_2H_4	Ethylene	+52.3	−7.281E−2	+5.802E−5	−1.861E−8	+5.648E−13	−9.782E+2	+2.032E+1	−4.076E−1
C_2H_2	Acetylene	+226.7	−2.269E−2	+2.228E−5	−1.208E−8	+1.618E−12	−9.782E+2	+8.373E+0	−2.044E−1
C_6H_6	Benzene	+82.9	−1.824E−1	+1.903E−4	−8.670E−8	+1.208E−11	−2.935E+3	+4.950E+1	−9.787E−1
CH_3OH	Methanol	−201.2	−5.834E−2	+2.070E−5	+1.491E−8	−9.614E−12	−4.891E+2	+1.688E+1	−2.467E−1
C_2H_5OH	Ethanol	−218.5	−1.088E−1	+8.252E−5	−1.706E−8	−4.056E−12	−9.782E+2	+2.886E+1	−5.189E−1
C_6H_5OH	Phenol	−96.4	−1.983E−1	+2.522E−5	−1.499E−7	+3.375E−11	−2.935E+3	+5.025E+1	−1.000E+0
CO	Carbon monoxide	−110.5	+5.619E−3	−1.190E−5	+6.383E−9	−1.846E−12	−4.891E+2	+8.684E−1	−6.131E−2
CO_2	Carbon dioxide	−393.5	−1.949E−2	+3.122E−5	−2.448E−8	+6.946E−12	−4.891E+2	+5.270E+0	−1.207E−1
H_2O	Water	−241.8	−8.950E−3	−3.672E−6	+5.209E−9	−1.478E−12	0	+2.868E+0	−1.722E−2
NH_3	Ammonia	−45.9	−2.896E−2	+8.345E−6	+8.124E−9	−4.366E−12	0	+7.711E+0	−8.876E−2
H_2S	Hydrogen sulfide	−84.9	−1.315E−2	−4.225E−6	+1.270E−8	−4.849E−12	0	+3.999E+0	−5.023E−2
NO_2	Nitrogen dioxide	+33.3	−1.944E−2	+2.755E−5	−1.721E−8	+4.193E−12	0	+3.769E+0	−5.551E−2
NO	Nitric oxide	+90.4	−2.850E−4	+2.921E−6	−4.124E−9	+1.744E−12	0	−7.917E−2	−1.307E−2
SO_2	Sulfur dioxide	−361.7	−2.220E−2	+3.316E−5	−2.234E−8	+5.979E−12	0	+4.216E+0	−6.026E−2

* All products and reactants are gases except carbon. Constants calculated on basis of Eqs. (2.39) and (2.49) and data for specific heats of gases from Ref. 15. Specific heat of carbon (graphite) given by $11.184 + 1.095 \times 10^{-2} T - 4.891 \times 10^5/T^2$ J/(mol · K) and for sulfur gas by $17.95 - 6.276 \times 10^{-3} T$ J/(mol · K).[12]

Table 2.15a Calculated heats of reaction in kJ/mol for formation reactions of gases*

Reaction	298 K	700 K	1000 K	1500 K
$C + \frac{1}{2}O_2 \rightarrow CO$	-110.5	-110.5	-111.9	-116.1
$C + O_2 \rightarrow CO_2$	-393.5	-394.0	-394.5	-395.0
$C + 2H_2 \rightarrow CH_4$	-74.8	-85.3	-89.5	-94.0
$H_2 + \frac{1}{2}O_2 \rightarrow H_2O$	-241.8	-245.6	-247.8	-250.5
$C + 2H_2 + \frac{1}{2}O_2 \rightarrow CH_3OH(g)$	-201.2	-212.9	-217.2	-223.9
$2C + 3H_2 + \frac{1}{2}O_2 \rightarrow C_2H_5OH(g)$	-218.5	-233.6	-238.0	-245.9
$2C + 3H_2 \rightarrow C_2H_6$	-84.7	-100.4	-105.6	-110.8
$2C + 2H_2 \rightarrow C_2H_4$	$+52.2$	$+42.3$	$+38.7$	$+33.2$
$2C + H_2 \rightarrow C_2H_2$	$+226.7$	$+225.0$	$+223.2$	$+217.9$
$6C + 3H_2 \rightarrow C_6H_6(g)$	$+82.9$	$+66.9$	$+62.7$	$+53.6$
$6C + 3H_2 + \frac{1}{2}O_2 \rightarrow C_6H_5OH(g)$	-96.4	-108.9	-111.3	-113.2
$\frac{1}{2}N_2 + \frac{3}{2}H_2 \rightarrow NH_3$	-45.9	-52.6	-55.0	-57.5
$\frac{1}{2}N_2 + \frac{1}{2}O_2 \rightarrow NO$	$+90.4$	$+90.6$	$+90.6$	$+91.4$
$S(g) + O_2 \rightarrow SO_2$	-361.7	-363.0	-362.9	-361.3
$S(g) + H_2 \rightarrow H_2S$	-84.9	-89.0	-90.4	-91.8

* For any other reaction R which can be written in terms of the formation reactions R_i such that $R = \sum a_i R_i$, then $\Delta H° = \sum a_i \Delta_f (H°)_i$.

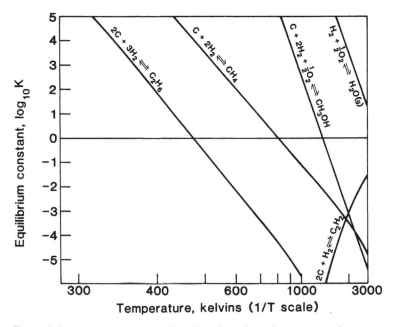

Figure 2.2a Equilibrium constants for selected reactions. Formation reactions.

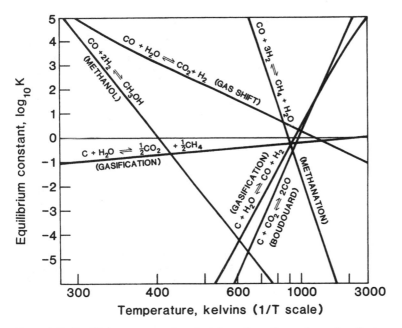

Figure 2.2b Equilibrium constants for selected reactions. Conversion and synthesis reactions.

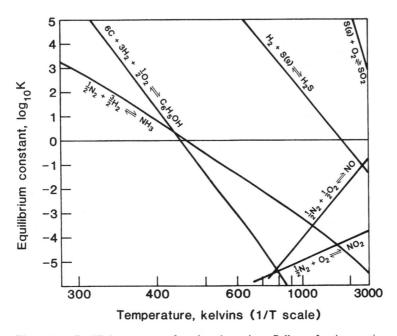

Figure 2.2c Equilibrium constants for selected reactions. Pollutant-forming reactions.

Table 2.15b Calculated equilibrium constants for formation reactions of gases*
Values are $\log_{10} K$.

Reaction	298 K	700 K	1000 K	1500 K
$C + \frac{1}{2}O_2 \rightarrow CO$	$+24.065$	$+12.968$	$+10.483$	$+ 8.507$
$C + O_2 \rightarrow CO_2$	$+69.134$	$+29.502$	$+20.677$	$+13.801$
$C + 2H_2 \rightarrow CH_4$	$+ 8.906$	$+ 0.958$	$- 0.999$	$- 2.590$
$H_2 + \frac{1}{2}O_2 \rightarrow H_2O$	$+40.073$	$+15.590$	$+10.070$	$+ 5.733$
$C + 2H_2 + \frac{1}{2}O_2 \rightarrow CH_3OH(g)$	$+28.331$	$+ 7.593$	$+ 2.780$	$- 1.042$
$2C + 3H_2 + \frac{1}{2}O_2 \rightarrow C_2H_5OH(g)$	$+29.500$	$+ 6.852$	$+ 1.572$	$- 2.618$
$2C + 3H_2 \rightarrow C_2H_6$	$+ 5.771$	$- 3.413$	$- 5.719$	$- 7.594$
$2C + 2H_2 \rightarrow C_2H_4$	$- 11.940$	$- 7.097$	$- 6.189$	$- 5.551$
$2C + H_2 \rightarrow C_2H_2$	$- 36.674$	$- 13.908$	$- 8.888$	$- 5.035$
$6C + 3H_2 \rightarrow C_6H_6(g)$	$- 22.736$	$- 15.097$	$- 13.646$	$- 12.607$
$6C + 3H_2 + \frac{1}{2}O_2 \rightarrow C_6H_5OH(g)$	$+ 5.770$	$- 4.518$	$- 6.984$	$- 8.941$
$\frac{1}{2}N_2 + \frac{3}{2}H_2 \rightarrow NH_3$	$+ 2.859$	$- 2.041$	$- 3.246$	$- 4.222$
$\frac{1}{2}N_2 + \frac{1}{2}O_2 \rightarrow NO$	$- 15.198$	$- 6.090$	$- 4.063$	$- 2.482$
$S(g) + O_2 \rightarrow SO_2$	$+59.672$	$+23.190$	$+15.064$	$+ 8.753$
$S(g) + H_2 \rightarrow H_2S$	$+12.853$	$+ 4.140$	$+ 2.131$	$+ 0.548$

* For any other reaction R which can be written in terms of the formation reactions R_i such that $R = \sum a_i R_i$, then $\log K = \sum a_i \log K_i$.

the yield of a reaction. Another method is to use an excess of one of the reactants to increase the yield relative to the other reactants. Unconverted reactants are not necessarily wasted; they can be recovered and reused. The cost of separation relative to the value of the increased yield will determine the extent to which any reactants are used in excess. It may be possible to continuously remove the desired product from the reaction mixture, as in the production of a gas from liquid and solid reactants. In this case the reactants can theoretically continue to react to completion. Finally, if competing parallel reactions occur, it may be possible to select operating conditions or, as discussed in the next section, to use a catalyst to increase the rate of formation of the desired product relative to unwanted products.

One important question that arises when considering synthetic fuel processes concerns the optimum operating conditions for a required conversion. Theoretically, a mixture of carbon and hydrogen can be used to synthesize any hydrocarbon. According to the foregoing chemical equilibrium analysis, the composition of an equilibrium mixture of hydrocarbons will be such that the free energy of the system is a minimum. The composition will therefore depend on the system temperature and pressure as well as on the carbon-to-hydrogen ratio. Methods of applying the minimum free energy constraint to a set of equilibrium equations to determine the equilibrium composition of a mixture undergoing a number of independent reactions are described in the literature.[4,12]

Some calculated compositions for carbon, hydrogen, and oxygen systems are shown in Figure 2.3. The lower alcohols including phenol, the lower molar mass aliphatic and aromatic hydrocarbons, and CO, CO_2, H_2, H_2O, as well as solid and gaseous carbon were included in the calculations. From Figure 2.3a for the carbon and hydrocarbon species, it can be seen that methane is the dominant hydrocarbon at low temperatures while acetylene dominates at high temperatures. Ethane, ethylene, and

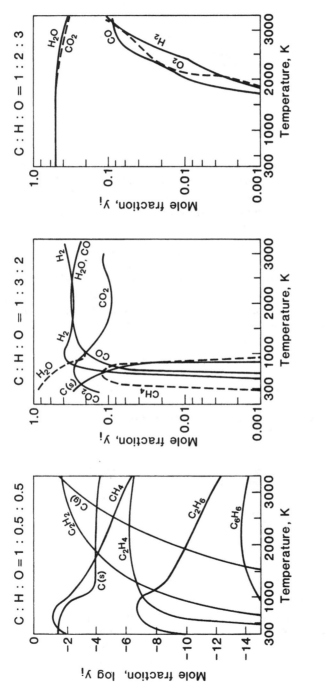

Figure 2.3 Equilibrium compositions at atmospheric pressure for three carbon, hydrogen, and oxygen mixtures. (*a*) Equal parts of hydrogen and oxygen showing the extremely low level of benzene. (*b*) Important components representative of coal gasification with steam and oxygen. (*c*) Effect of reaction with excessive oxygen.

benzene occur at relatively low concentrations, and higher hydrocarbon members can be expected to be even less significant. The equilibrium mole fractions of methanol and phenol which are not included in the figure are of the order of 10^{-12} and 10^{-26}, respectively. The other plots in Figure 2.3 show the change in important constituents in equilibrium coal gasification off gases as the concentration of oxygen in the system is increased relative to that of hydrogen.

Availability

It was shown by Eq. (2.44) that the maximum useful work occurring in certain processes is given by the free energy change. If the process proceeds to equilibrium with its surroundings at the ground state of atmospheric pressure and temperature, then in accordance with the earlier description of available energy, the amount of work done is the maximum useful work available from the system. This available work is often referred to as the availability ϕ. Equation (2.42) may be used to give a general definition of availability:

$$\phi = -(\Delta H_a - T_a \Delta S_a) \quad \text{(flow process)} \quad (2.52)$$

where T_a is the reference atmospheric temperature and ΔH_a and ΔS_a are the changes arising as the system proceeds from its original state to the atmospheric ground state. The change in availability that occurs for a system proceeding from state 1 to state 2 can be obtained by applying Eq. (2.52) to each of the two states and taking the difference

$$\Delta\phi = \Delta H - T_a \Delta S \quad (2.53)$$

where ΔH and ΔS represent any change and not necessarily to atmospheric conditions.

Table 2.16a Useful energy formulae: No work done

For a flow of heat dq into a system of n kilomoles and volume V, at temperature T and pressure p.

Heat capacity	$C = dq/dT$		(1)
J/K	$C_V = dq/dT$	Constant V	(2)
	$C_p = dq/dT$	Constant p	(3)
	$C_p - C_V = nR$	(Ideal gas)	(4)
Internal energy	$dU = dq \ (+dw)$	(First law)	(2.25a)
J	$dU = C_V \ dT$	Constant V^*	(5)
	$dU = 0$	Contant T (ideal gas)	(6)
Enthalpy	$dH = dq \ (+dw_u)$		(2.33a)
J	$dH = C_p \ dT$	Constant p^*	(7)
	$dH = 0$	Constant T (ideal gas)	(8)
Entropy	$dS = dq_{rev}/T$	(Second law)	(2.26a)
J/K	$dS = C_p \ dT/T$	Constant p	(9)
	$dS = C_V \ dT/T$	Constant V	(10)
	$dS = -nR \ dp/p$	Constant T (ideal gas)	(11)
Gibbs free energy	$dG = dH - d(TS)$		(2.41a)
J	$dG = nRT \ dp/p$	Constant T (ideal gas)	(12)

*Restriction relaxed for ideal gas.

Formulae for determining ΔH, ΔS, and other energy parameters are summarized in Table 2.16. When calculating energy changes for availability determinations, the process can be conveniently assumed to occur in two steps:

1. The system temperature changes to T_a at constant pressure (the constant pressure is the initial system pressure)
2. The system pressure changes to atmospheric pressure at constant temperature (T_a)

In the special case of a reaction occurring in a closed system held at constant atmospheric pressure, Eq. (2.44) becomes

$$\phi_0 = -\Delta G° \tag{2.54}$$

Here, ϕ_0 is the *standard chemical availability* and may be used to estimate the maximum work theoretically available from a fuel when computing second law efficiencies. In another special case when the flow process is an ideal heat engine, Eq. (2.52) simplifies to the Carnot expression, Eq. (2.31).

Equation (2.52) can alternatively be used to determine the *minimum theoretical work* done on the system to take it from prevailing atmospheric conditions to some required state. The preparation of pure oxygen by separation of the oxygen/nitrogen mixture in air is a relevant example.

Consider 100 moles of air at 298 K and one atmosphere that is separated into 79 moles of nitrogen, and 21 moles of oxygen each at 298 K and one atmosphere. As each component is compressed from its partial pressure in the air mixture to atmospheric pressure, the entropy decreases according to Eq. (11) in Table 2.16:

For N_2 $\Delta S = -79 \times 8.314 \ln\left(\dfrac{1}{0.79}\right) = -154.8 \text{ J/K}$

For O_2 $\Delta S = -21 \times 8.314 \ln\left(\dfrac{1}{0.21}\right) = -272.5 \text{ J/K}$

Table 2.16b Useful energy formulae: Compression and pumping work*
For a system change from p_1 to p_2 and T_1 to T_2, when work w is done on n kilomoles of fluid of volume V.

Isothermal compression	$w = \Delta H - T\Delta S \ (= \Delta G)$		(2.42a)
(constant T)	$= nRT \ln(p_2/p_1)$	(Ideal gas)	(13)
Isentropic compression	$w = \Delta H$		(2.42b)
(constant S)	$= [\gamma nRT/(\gamma - 1)][(p_2/p_1)^{(\gamma-1)/\gamma} - 1]$	(Ideal gas)	(14)
	$T_2 = T_1(p_2/p_1)^{(\gamma-1)/\gamma}$	(Ideal gas)	(15)

where $\gamma = (C_p/C_v)$ varies with T and p. At 101.3 kPa and 298 K: $\gamma \approx 1.4$ for N_2, O_2, H_2, CO; $\gamma \approx 1.3$ for CO_2, CH_4, NH_3.

For s stages, pressure ratio per stage for minimum work $= (p_2/p_1)^{1/s}$

Pumping liquids	$w = V(p_2 - p_1)$	Constant V	(16)

* Formulae given are for pumps and compressors working reversibly. Actual energy $= w/\eta$, where η is the overall system efficiency.

The total entropy change is therefore $\Delta S = -427.3$ J/K for the 100 moles of air considered. The enthalpy change for this ideal isothermal process is zero. Substituting into Eq. (2.52) yields

$$\phi = -(0 + 298 \times 427.3)\,\mathrm{J} = -127\,\mathrm{kJ} \tag{2.55}$$

The minimum theoretical work for separating 100 moles of air into its components is therefore $+127$ kJ, which converts to 6 kJ per mole of pure oxygen or 189 kJ per kilogram of pure oxygen produced.

Ideal Gasifier Efficiency

It is instructive to obtain some feel for the relative energy rates prevailing in a synthetic fuel conversion system. As an example we have selected the simplified carbon gasification scheme shown in Figure 2.4. The only gasification reaction occurring in the gasifier, which is maintained at 1000 K and 2 atmospheres, is taken to be

$$\mathrm{C} + \mathrm{H_2O} \longrightarrow \tfrac{1}{2}\mathrm{CH_4} + \tfrac{1}{2}\mathrm{CO_2} \qquad \Delta H^\circ_{1000} = +5.8\ \mathrm{kJ/mol} \tag{2.56}$$

In reality other gasification reactions will occur as well, although operation with catalysts at low temperatures as described in Section 4.4 significantly enhances the relative contribution of the idealized reaction, Eq. (2.56). The heat necessary for the gasification reaction as well as that necessary to raise steam is produced by burning part of the carbon in air supplied to the gasifier following the reaction

$$\mathrm{C} + \mathrm{O_2} \longrightarrow \mathrm{CO_2} \qquad \Delta H^\circ_{1000} = -394.5\ \mathrm{kJ/mol} \tag{2.57}$$

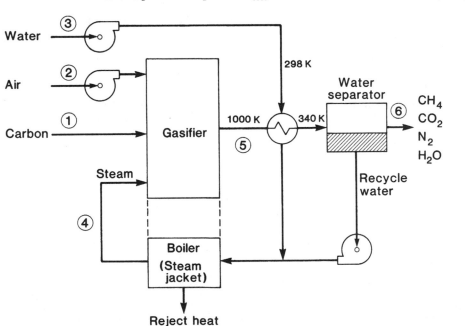

Figure 2.4 Schematic of simplified gasification scheme.

Table 2.17 Material flows in moles for the simplified gasification scheme

Constituent	Stream No. in Figure 2.4					
	1	2	3	4	5	6
Carbon	1.0	—	—	—	—	—
Carbon dioxide	—	—	—	—	0.6	0.6
Nitrogen	—	0.8	—	—	0.8	0.8
Oxygen	—	0.2	—	—	—	—
Water/steam	—	—	1.08	1.6	0.8	0.28
Methane	—	—	—	—	0.4	0.4
Total moles	1.0	1.0	1.08	1.6	2.6	2.08

The number of moles of each constituent, based on a carbon feed of 1.0 mole, is shown in Table 2.17 for the numbered streams in the flow scheme. Note that the values given for the stream leaving the gasifier, stream 5, satisfy the equilibrium composition for Eq. (2.56)

$$K = \frac{(p_{CH_4})^{1/2}\,(p_{CO_2})^{1/2}}{p_{H_2O}} = 0.6 \text{ at } 1000 \text{ K} \tag{2.58}$$

The partial pressures are given by the mole fractions multiplied by the total pressure of 2 atmospheres (Eq. (6), Table 2.13). Substituting the molar quantities of stream 5 from Table 2.17 gives

$$K = \frac{(2 \times 0.4/2.6)^{1/2}(2 \times 0.6/2.6)^{1/2}}{2 \times 0.8/2.6} = 0.6 \tag{2.59}$$

The amount of water vapor remaining in the gas after cooling is determined from the vapor pressure of water at the temperature of the stream. The vapor pressure at 340 K is about 0.27 atmospheres, so the water mole fraction is $0.27/2 = 0.135$; this is the same as $0.28/2.08$ given by the stream 6 compositions in Table 2.17. Completing the water balance in the table shows that half of the 1.6 moles of steam fed to the gasifier are consumed in the reaction, and 0.28 moles leave the system as water vapor, so that $0.8 + 0.28 = 1.08$ moles of water are required as makeup.

A heat balance around the gasifier and heat exchanger system is given in Table 2.18. The heat available to the system is that released on burning 0.2 moles of carbon in the air fed to the gasifier, as well as the heat recovered in the heat exchanger on cooling the gasifier off gases from 1000 K to 340 K. A mean specific heat for the wet gas mixture was used in calculating sensible heat effects; a more accurate value could be obtained by integration of the heat capacity equations for each constituent in the gas mixture. Latent heats for water evaporation or condensation were calculated assuming the change of phase to occur at ambient conditions. The same result would be obtained using latent heat values at any other temperature and including the sensible heat required to heat liquid water to this temperature.

Table 2.18 Heat balance for simplified gasification scheme*

	No. Moles	Reaction or Latent Heat, kJ/mol	Sp. Heat, J/(mol·K)	ΔT, K	Heat Flow, kJ
IN					
Heat of combustion, carbon	0.2	394.5	—	—	78.9
Cooling gasifier off gases:					
sensible heat	2.6	—	41.0	660	70.4
H_2O latent heat	0.52	43.0	—	—	22.4
Total heat available					171.7
CONSUMED					
Gasification reaction, carbon	0.8	5.8	—	—	4.6
Raise steam	1.6	43.0	—	—	68.8
Heat steam	1.6	—	38.0	688	41.8
Heat compressed air	1.0	—	31.0	637	19.8
Heat carbon	1.0	—	15.5	702	10.9
Product gas:					
sensible heat	2.08	—	33.0	42	2.9
H_2O latent heat	0.28	43.0	—	—	12.0
Losses (by difference)					10.9
Total heat required					171.7

* Datum: 298 K, liquid water.

The heat released in the gasifier and heat exchanger is used to raise steam and to heat the steam, air, and carbon feeds to the gasifier temperature of 1000 K. The carbon is initially at 298 K and must be heated through 702 K; the steam is heated from a temperature of 312 K obtained on mixing the condensate at 340 K with makeup water at 298 K, and the air is heated from the compressor outlet temperature of 363 K. The 10.9 kJ of heat lost is about 10 percent of that to raise and heat the steam, suggesting a boiler efficiency of 90 percent.

The thermal efficiency of the overall process is calculated from the total energy input to the plant. This includes the heating value of the carbon plus the electrical energy to compress and pump the gases and water. To more truly represent energy consumption, electrical energy values should be multiplied by a factor of 3 to reflect the approximately 33 percent generation efficiency; this has not been done in this illustrative example. The relevant energy quantities are shown in Table 2.19, including the energy necessary to compress the pure product gas to a pipeline entry pressure of 5.5 MPa. The energy required to separate the nitrogen and carbon dioxide and to dry the product gas was estimated from typical values for purifying conventional gasifier off gases. As shown in the table, the overall efficiency defined by Eq. (2.24) is 83.9 percent.

Another efficiency shown in the table, called the cold gas efficiency, is often quoted in the literature as it is very easy to calculate. One definition of this efficiency is

$$\eta_{\text{cold gas}} = \frac{\text{GCV of desired product}}{\text{GCV of coal fed to gasifier}} \qquad (2.60a)$$

In other definitions the net GCV of the coal fed to the gasifier is used, where the net coal GCV is the GCV of the coal less the GCV of any tars, oils, phenols, and unreacted char leaving the gasifier. In some cases the cold gas efficiency is defined in terms of NCV (Eq. 1.2). The cold gas efficiency for our simplified gasifier as calculated in Table 2.19 is 90.5 percent. In this example there is a significant difference between the overall process thermal efficiency and the gasifier cold gas efficiency. An even bigger discrepancy between these two efficiencies arises if coal were fed to a boiler system for steam raising. While cold gas efficiencies do provide some measure of the performance of the gasifier itself, they can be very misleading if used to evaluate gasifier performance in an integrated plant where other heat effects are usually significant. An efficiency more closely representing the thermal efficiency, while still retaining the advantages of the cold gas concept, might be defined as

$$\eta_{\text{modified cold gas}} = \frac{\text{GCV of desired product}}{\text{Net energy to gasifier}} \qquad (2.60b)$$

We shall use this efficiency later in the book when presenting thermal balances for coal conversion systems.

Table 2.19 Efficiency calculation for the simplified gasification scheme

	No. Moles	GCV, kJ/mol	Energy, kJ	Percent of Carbon GCV
Carbon in feed	1.0	393.5	393.5	100.0
To compress air*	1.0	—	2.7	0.7
To compress methane†	0.4	—	4.8	1.0
To pump water	1.08	—	~0	0
To purify and dry gas‡	—	—	23.6	6.0
Methane product	0.4	890.3	356.1	90.5

$$\text{Overall thermal efficiency (\%)} = \frac{\text{GCV of product gas}}{\text{Energy to plant}} \times 100$$
$$= \frac{356.1}{393.5 + 2.7 + 4.8 + 23.6} \times 100$$
$$= 83.9\%$$

$$\text{Cold gas efficiency (\%)} = \frac{\text{GCV of gasifier off gas}}{\text{Carbon to gasifier}} \times 100$$
$$= \frac{356.1}{393.5} \times 100$$
$$= 90.5\%$$

* Using Eq. (14) in Table 2.16, $\eta = 0.7$.
† From 2 to 55 atm in 6 stages, $\eta = 0.7$.
‡ Estimated.

2.3 REACTION KINETICS AND CATALYSIS

The production of methane and methanol by hydrogenation of carbon monoxide may, for comparison purposes, be represented by the following chemical reactions:

$$CO + 3H_2 \longrightarrow CH_4 + H_2O(g) \tag{2.22}$$

$$CO + 3H_2 \longrightarrow CH_3OH(g) + H_2 \tag{2.50a}$$

The identical synthesis gas having a three-to-one hydrogen-to-carbon monoxide ratio is used as feed in each case, and it is tentatively assumed that conditions in each reactor can be adjusted such that the desired product is formed exclusively. The extent to which this assumption can be realized in practice will now be examined to illustrate the concepts of reaction kinetics and catalysis. First we will determine the conditions that maximize the equilibrium yield of each product using the thermodynamic principles discussed in the previous section.

As both reactions proceed with a decrease in the number of molecules, an increase in pressure will increase yields in both cases. Consequently it is unlikely that pressure alone can be used to significantly promote one reaction relative to another and this is confirmed by the calculated equilibrium concentrations shown in Table 2.20. Other data in Table 2.20 show that both reactions are exothermic, so raising the temperature decreases the equilibrium yield in both cases. However, the methane equilibrium does not shift significantly to the left until above about 700 K, making temperature control a possible means of selecting the desired product. Operation at 700 K, for example, would result in an equilibrium mixture containing virtually no methanol, a result previously demonstrated in Figure 2.3.

Suppose, however, methanol were the desired product. Lowering the temperature would increase methanol equilibrium yields, but at about ambient temperature the equilibrium is completely over to the right for both the methane and the methanol reactions. A further drop in temperature would, therefore, not be of use and in fact would necessitate expensive refrigeration to dissipate the heat evolved. In spite of the adverse effect on equilibrium, exothermic reactions may be most economically carried out at elevated temperatures to permit efficient recovery of the heat released–see for example the discussion on the Carnot cycle in the previous section. Operation of the methane reactor at 700 K, say, would permit some of the waste heat to be recovered by raising superheated steam, whereas operation at ambient temperature would require that all the heat evolved be dissipated to the atmosphere. As we will see later, there are other important reasons for operation at elevated temperature, and in practice methanol production is consequently carried out at about 550 to 650 K. It can be seen from Table 2.20 that even at pressures of up to 100 atmospheres, methanol yields are low at these temperatures and so unreacted gases have to be separated and recycled. In fact, in accordance with Eq. (2.45) for chemical equilibrium, recycling methane to the reactor should inhibit its further production, thereby increasing methanol yields. This argument is, in principle, correct. It turns out, however, that neither methane, methanol, nor anything else is formed in significant quantities under any of the reaction conditions discussed above. This should not be surprising; after all hydrogen and

Table 2.20 Thermodynamic parameters for methane and methanol production

Reaction		$CO + 3H_2 \longrightarrow CH_4 + H_2O(g)$				$CO + 3H_2 \longrightarrow CH_3OH(g) + H_2$		
Temperature, T, °C		25	427	727		25	427	727
kelvin		298	700	1000		298	700	1000
Heat of reaction, ΔH_T°, kJ/mol		− 206.1	− 220.5	− 225.4		− 90.7	− 102.4	− 105.3
Free energy change, ΔG_T°, kJ/mol		− 142.1	− 47.7	+ 27.3		− 24.3	+ 61.6	+ 133
Equilibrium constant, K, at temp. T		$\sim 10^{25}$	3.6×10^3	0.038		1.8×10^4	2.5×10^{-5}	1.1×10^{-7}
Volume percent product in	$p = 0.1$	50	30	0.6		48	9.5×10^{-5}	1.6×10^{-8}
equilibrium gas mixture at	$p = 1.0$	50	43	5.1		50	3.6×10^{-4}	1.6×10^{-6}
p atmospheres	$p = 100$	50	49	39		50	3.2	0.02

carbon monoxide are present in many commercial gas streams and coexist without reacting.

Before proceeding, it is emphasized that the foregoing arguments relating to chemical equilibrium are valid for all reactor conditions. While correct, they are not complete. The rates at which the reactions proceed are not considered in the thermodynamic development. Even though calculated equilibrium yields of products may be acceptable, commercial production is not feasible unless equilibrium can be approached at acceptable rates. As a corollary, the rates of competing side reactions must be relatively insignificant. Many reactions important in synthetic fuel production do not proceed at viable rates under normal conditions, and for most there exist a host of complex side reactions. It is therefore essential to have some understanding of the factors influencing reaction rates.

In presenting chemical reactions thus far, the direction of the desired reaction has been indicated by an arrow. It has further been stated that the reactants do not necessarily all convert to products, and that the fraction of unconverted reactants is determined by the chemical equilibrium. At equilibrium the reaction does not stop, but no net change in concentration of the equilibrium mixture occurs because the rate of the reverse reaction equals the rate of the forward or desired reaction. A chemical reaction which does not proceed rapidly and irreversibly to completion is therefore more correctly presented as follows:

$$CO + 3H_2 \rightleftharpoons CH_4 + H_2O(g) \qquad (2.22a)$$

Further, the species shown in the equation are not necessarily the only ones present. Intermediate, possibly unstable substances, are formed that then break down or combine to form the end products of the written equation. The free radicals formed in the pyrolysis or cracking of petroleum, discussed in Section 2.1, are an example of reaction intermediates. In addition it is postulated that highly unstable transition substances called "activated complexes" are involved.

For two molecules to react it is of course necessary that they collide. However, even though their reaction may be accompanied by a decrease in free energy, their collision is not a sufficient condition for a reaction to occur. This is because the creation of the activated complexes may require an energy input not inherently present in the colliding molecules. Formation of the activated complexes can be thought of as a potential barrier to be overcome in the formation of the end products as illustrated in Figure 2.5.

Two methods of overcoming the potential barrier are available. Firstly, the energy of the molecules in the system may be increased by raising their temperature. An increase in temperature increases the probability that the kinetic energy of two colliding molecules is sufficient to provide the potential energy of the activated complex. Secondly, the potential energy required may be reduced by changing the reaction path via some other activated complex species. This is done by means of catalysts.

The rate of reaction is defined in terms of the production rate of a species as follows:

$$r = \frac{dn}{dt} \qquad \text{mol/s} \qquad (2.61)$$

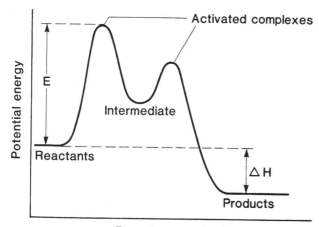

Figure 2.5 Energy profile for an exothermic reaction in which an intermediate is formed.

where r is the reaction rate, n is the number of moles of the species, and t is time. The number of moles of the species is related to its molar concentration c in a reacting mixture of volume V by $n = cV$.

The reaction rate is dependent on the rate of collision of the reacting molecules or, according to the *law of mass action*, is proportional to the active masses of the reacting substances:

$$r = k\Pi\,(c_i^{v_i})_{\text{reactants}} \tag{2.62}$$

where k is the rate constant, c_i is the concentration of species i, and v_i is the stoichiometric coefficient of species i. v_i is also called the order of the reaction with respect to species i, while Σv_i is the overall order of the reaction. The majority of known reactions are of first or second order.

In a reversible reaction such as Eq. (2.22a) the net rate of the reaction r is the difference between the forward and reverse reaction rates

$$r = r_f - r_r \tag{2.63}$$

At equilibrium there is no net rate of reaction so the forward and reverse reaction rates are equal,

$$r_f = r_r \tag{2.64}$$

and

$$k_f\,\Pi\,(c_i^{v_i})_{\text{reactants}} = k_r\Pi\,(c_i^{v_i})_{\text{products}} \tag{2.65}$$

so that

$$K' = k_f/k_r \approx K \tag{2.66}$$

where K' is a dynamic equilibrium constant, and K is the thermodynamic equilibrium constant of Eq. (2.45).

It is in general assumed that corrections made to the law of mass action to account for the actual number of collisions resulting in reaction are such that the dynamic and thermodynamic equilibrium constants become equal. The important point is that the equilibrium constants are calculated from activities, while the rate equations are presented in terms of concentrations. Rate equations presented in the literature embrace a wide spectrum of concentration units and it is essential therefore that the units intended be determined before applying these equations.

The stoichiometric based rate expression of Eq. (2.62) is generally applicable to all reactions. However, the chemical reaction equation as written is often a simplification of a more complex reaction scheme involving series and/or competing side reactions and various intermediates. A combination of the individual stoichiometric rate expressions in terms of the simplified overall reaction may consequently bear little relation to the overall stoichiometry. The order of the reaction with respect to a species i is therefore not necessarily ν_i, its stoichiometric coefficient in the simplified equation, and in fact need not be an integer nor need it be positive. The dimensions of the rate constant k are dependent on the order of the reaction; for an nth-order reaction the dimensions are $(concentration)^{1-n}/time$. In the special case of a first-order reaction, the dimensions become inverse time. In the case of complex reactions, as for example the heterogeneous catalytic reactions discussed later, there is often no well-defined reaction order with respect to the reacting species. However, many of these reactions may be first order with respect to the catalyst concentration.

The rate constant k is not a true constant, but in accord with the description of the potential energy required for reaction, is temperature dependent. It can be shown[9] that

$$k = A \exp\left(\frac{-E}{RT}\right) \tag{2.67}$$

where E is the activation energy, see Figure 2.5, and A is the frequency factor, a constant for a given reaction, with the same dimensions as k. Equation (2.67) is *Arrhenius' law*. As can be seen, $\ln k$ varies as $1/T$. A plot of the logarithm of a rate constant versus $1/T$ is called an Arrhenius plot and a $1/T$ scale an Arrhenius scale.

The rate of a chemical reaction is seen to be a function of the activation energy E. Reactions with relatively low activation energies and which proceed spontaneously at room temperature are only moderately accelerated by increases in temperature. As shown in Figure 2.6, the increase in rate of reactions with relatively high activation energies is dramatic. A doubling of reaction rate for a ten degree increase in temperature at around ambient conditions is typical for many common reactions.

According to the Arrhenius expression, the reaction rate could be increased by providing an alternative reaction path involving intermediates or activated complexes with lower activation energies. One means of providing an alternative reaction path is by means of a solid surface on which the reactants adsorb and then react either with other adsorbed species or with unadsorbed reactants. If, as a consequence, the activation energy is effectively lowered, the reaction rate will be increased according to Eq. (2.67) and the surface is said to catalyze the reaction. The catalyst itself undergoes no net change, although it partakes in the reaction. If the catalyst is in a different

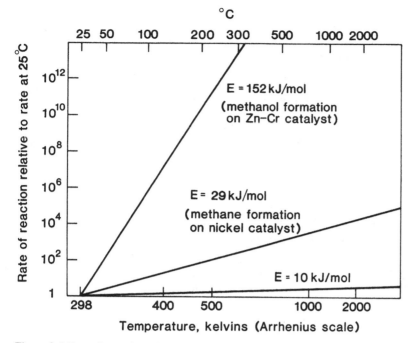

Figure 2.6 Dependence of reaction rate on temperature.

phase from the reactants, as in the case of a solid catalyst and a gaseous reaction, the catalyst is said to be heterogeneous. The rate expression will then in general be more complex than suggested by the simple stoichiometric dependence of Eq. (2.62). This is because the chemical reaction rate need not be the rate determining step of the overall process. Other steps to be considered include:

1. Diffusion of reactants through the gas phase to the catalyst
2. Diffusion of reactants through pores in the catalyst material
3. Adsorption of reactants on the catalyst surface
4. Desorption of products
5. Diffusion of products through the pores
6. Diffusion of the products into the bulk gas phase

Often the overall rate is controlled by one of the mechanisms, which is then called the rate determining or rate limiting step. If this step is the chemical reaction, the rate equation will be related to the types of functions discussed above. If, however, diffusion is rate limiting, then the rate constant will have characteristics of a diffusion coefficient. For Knudsen diffusion, in which the gas molecules collide with the pore walls more frequently than with themselves, the diffusion coefficient increases with the square root of the absolute temperature. The bulk diffusion coefficient, on the other hand, increases with $T^{1.5}$.

In general, rates of catalyzed reactions cannot be based on the stoichiometric equations and are determined by postulating a reaction mechanism and by using experimental data to evaluate the coefficients in the associated rate expression. For instance, the methanation reaction is catalyzed by nickel-containing catalysts and reportedly[16] proceeds at the following rate

$$r = \frac{k(p_{CO})(p_{H_2})^{1/2}}{1 + K_1(p_{H_2}) + K_2(p_{CH_4})} \qquad \frac{\text{moles CH}_4}{\text{(second) (g catalyst)}} \qquad (2.68)$$

where $k = A \exp(-E/RT)$, $A = 7.1 \times 10^{-5}$, $E = 29.1$ kJ/mol, and $K_1 = 0.014$, $K_2 = 0.007$, and the partial pressures are in kilopascals. The temperature range is 550 K to 750 K. The K_1, K_2 are equilibrium adsorption coefficients and in the referenced work were taken to be independent of temperature. Inclusion of their temperature dependence would conceivably extend the range of applicability of the equation. It is noted that the rate is given per mass of catalyst present and reflects the fact that the reaction is constrained by the amount of catalyst available to provide a reaction path. In fact, it is the surface area of catalyst present that is important, so that Eq. (2.68) is strictly valid for a specific catalyst only. Commercial catalysts typically have surface areas of several hundred square meters per gram.

To determine the time required for a reaction to proceed it is necessary to integrate the rate equation. In general, the rate equation may be written

$$r = \frac{dn}{dt} = f(k,n) \qquad (2.69)$$

Integrating this equation gives

$$\int_{n_0}^{n_t} \frac{dn}{f(k,n)} = \int_0^t dt = t \qquad (2.70)$$

For a first-order reaction in which the rate of product formation is expressed in terms of the rate of disappearance of reactant, $r = -kcV$, the integrated expression becomes

$$t = \frac{1}{k} \ln\left(\frac{c_0}{c}\right) \qquad (2.71)$$

where c_0 is the initial reactant concentration, and c is its concentration at time t.

Although integrated rate equations have been tabulated for the more common reaction mechanisms,[12] results for expressions such as Eq. (2.68) are generally not available. The procedure for integrating Eq. (2.68) is therefore given here in some detail and should serve to suggest methods for handling most rate expressions encountered.

The overall reaction is rewritten together with the number of moles of each species in the system after n moles of product have formed. For convenience it is assumed that stoichiometric quantities of reactants are initially present, but this is not a necessary constraint for the method. We may write

$$CO + 3H_2 \rightleftharpoons CH_4 + H_2O(g) \qquad (2.22a)$$

Initial No. moles	1	3	0	0
At equilibrium:				
No. moles	(1-n)	(3-3n)	(n)	(n)

Total moles $\sum n_i = 4 - 2n$

Mole fractions $y_{CH_4} = y_{H_2O} = \dfrac{n}{4-2n}$ $y_{CO} = \dfrac{1-n}{4-2n}$ $y_{H_2} = \dfrac{3-3n}{4-2n}$

The rate expression, Eq. (2.68), is written in terms of the partial pressures, so we apply Eq. (6) of Table 2.13

$$p_i = y_i p$$

Substituting the appropriate partial pressures in Eq. (2.68) yields, on slight rearrangement

$$r = \frac{1.732 k p^{3/2} (1-n)^{3/2} (4-2n)^{-1/2} m_c}{4 - 2n + K_1 p(3-3n) + K_2 pn} \qquad \text{mol CH}_4/\text{s} \qquad (2.72)$$

where m_c is the mass of catalyst present.

Equation (2.72) is now in the form of Eq. (2.69) and can be integrated according to Eq. (2.70) with the result

$$t = \frac{0.577}{k p^{3/2} m_c} \int_0^n \frac{[4 - 2n + K_1 p (3-3n) + K_2 pn] \, dn}{(1-n)^{3/2} (4-2n)^{-1/2}} \qquad (2.73)$$

The quadrature can be readily evaluated by applying Simpson's rule using programs available on many pocket calculators. In this case the integral can be evaluated explicitly by substituting a dummy variable for $(1-n)$. Because the resulting expression is somewhat unwieldy we have not reproduced it here. The numerical results are summarized in Table 2.21.

An examination of the results in Table 2.21 is instructive. The time required to

Table 2.21 Rates of methane formation:* CO + 3H₂ ⇌ CH₄ + H₂O

Temperature, K	Pressure		Methane Concentration in Equilibrium Mixture,† mole fraction	Time to Reach 90% of Equilibrium,‡ s
	MPa	atm		
298	0.01	0.1	0.5	6.5×10^8
	0.10	1	0.5	3.2×10^7
	10.13	100	0.5	1.4×10^6
700	0.01	0.1	0.30	3.5×10^5
	0.10	1	0.43	2.9×10^4
	10.13	100	0.49	1.5×10^3
1000	0.01	0.1	0.006	1.6×10^3
	0.10	1	0.051	7.8×10^2
	10.13	100	0.38	2.5×10^2

* For illustration purposes Eq. (2.68) has been applied outside its temperature limits.

† See Table 2.20.

‡ In a batch system containing initially 1 mole CO, 3 moles H₂, and 1 g catalyst.

approach equilibrium at ambient conditions, even with a catalyst present, is of the order of hundreds of days. On increasing the temperature to 700 K the time decreases to about 8 hours, and on increasing the pressure to 100 atmospheres (10.1 MPa), the time further decreases to about 25 minutes. This is equivalent to reducing the volume of the reactor required for a given methane production by a factor of 22 000. Operation at an elevated temperature does not much decrease the equilibrium yield at temperatures up to about 700 K, and as mentioned, the heat of reaction can then be used for steam raising. There will, of course, be some heat loss to the surroundings due to the high temperature difference, but this loss may be partially reduced by insulation. Operation at elevated pressure not only increases the reaction rate, but also improves the equilibrium yield. However, the cost of the reactor vessel increases with increasing pressure and compression costs can be significant. The pressure selected for the methanation reactor may in part be determined by the operating pressure of upstream process units such as the gasifier. Typical commercial methanation reactors operate at temperatures between 500 K and 750 K, and pressures of up to 70 atmospheres (7 MPa) are not uncommon.

Catalyst Properties

The degree to which a catalyst increases the reaction rate is related to its *activity*, which is defined simply as the rate at which it causes a reaction to proceed to equilibrium. In systems in which the desired product is one of several that may be formed, it is required that the catalyst have high activity for the appropriate reactions only. This characteristic is known as *selectivity*. For example, the adsorbed group [HCOH] is an intermediate in both methane and methanol formation. In methane formation on nickel-containing catalysts, it is postulated[17] that this intermediate is hydrogenated and cleaved to form water and methylene groups as shown in Figure 2.7. The methylene groups then combine with adsorbed hydrogen to form methane. In Fischer-Tropsch synthesis, the iron catalysts used offer more methylene groups for chain polymerization. On the other hand, if the entire [HCOH] group is desorbed and hydrogenated without cleavage, then methanol is formed. Zinc oxide and copper-containing catalysts are used for methanol production. In some cases selectivity is

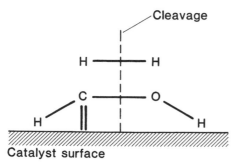

Catalyst surface

Figure 2.7 Hydrogenation and cracking of adsorbed [HCOH] to form methylene and water.[17]

achieved by providing activity under reaction conditions at which the undesired products are not thermodynamically favored. For instance, in the methanation example given above, operation at 700 K and modest pressures excludes the possibility of significant methanol formation on equilibrium grounds; see Table 2.20. It is again stressed that although a catalyst increases the rate at which equilibrium is reached, it does not alter the equilibrium composition.

An example of the effect of temperature and catalysts on product selectivity relevant to fuel production is in pyrolysis reactions, such as the cracking of petroleum. In thermal cracking, the process is initiated by bond rupture at elevated temperature to form free radicals. For instance, in the case of butane, two ethyl radicals are formed

Initiation $$C_4H_{10} \rightleftharpoons 2C_2H_5-$$

Formation of methane, ethane, ethylene and propylene—as in Eqs. (2.3) and (2.4)—occurs in various propagation reactions between the radicals and the feedstock:

Propagation
$$C_2H_5- + C_4H_{10} \rightleftharpoons C_2H_6 + C_4H_9-$$
$$C_4H_9- \rightleftharpoons C_3H_6 + CH_3-$$
$$CH_3- + C_4H_{10} \rightleftharpoons CH_4 + C_4H_9-$$

These chain type reactions are terminated when two radicals react to form stable molecules, as for example

Termination $$2C_2H_5- \rightleftharpoons C_2H_4 + C_2H_6$$

As is to be expected from the discussion of Figure 2.3, the end products of thermal cracking are low-molar-mass gases. Catalytic cracking, on the other hand, is carried out at lower temperatures and the products are in the gasoline boiling range. Catalysts used have acid characteristics, that is, the active sites are associated with hydrogen ions or protons H^+, and they can donate a hydrogen ion to, say, a long chain olefin which then has a positive charge and is called a "carbonium ion:"

Initiation $$CH_2=CH-CH_2-CH_2-R \xrightarrow[\text{acid catalyst}]{+H^+} CH_3-\overset{+}{C}H-CH_2-CH_2-R$$
Carbonium ion

where the + sign indicates the C atom with the positive charge. At the reaction temperature the large carbonium ion is unstable with respect to smaller molecules and undergoes fission at the bond once removed from the C^+

Propagation $$CH_3-\overset{+}{C}H-CH_2-CH_2-R \rightleftharpoons CH_3-CH=CH_2 + \overset{+}{C}H_2-R$$

The carbonium ion can also undergo isomerization to form branched chain species:

$$CH_3-\overset{+}{C}H-CH_2-CH_2-R \xrightarrow{\text{3 steps}} CH_3-\overset{+}{C}-CH_2-R$$
$$\underset{CH_3}{|}$$

Termination is by loss of a proton to other species or to the catalyst.

Selectivity is related to the chemical and physical properties of the catalyst itself. Substances that catalyze synthesis gas reactions should be able to adsorb both hydrogen and carbon dioxide, and must have hydrogenation and carbonylation (the addition of carbon monoxide) activity. Metals with these properties include iron, cobalt, nickel, copper, zinc, rhodium, palladium, and platinum. Those which selectively catalyze specific synthesis gas reactions are shown in Table 2.22. One characteristic imparting selectivity is the spacing of active sites on the catalyst. Active sites are those specific locations on the catalyst surface on which the reaction occurs. These sites should ideally be spaced such that adsorbed reactants can go easily to the desired products.[18] This can be likened to a mechanical or steric effect. Another steric effect determining selectivity is found in molecular sieves such as zeolites (crystalline aluminosilicates), which have well-defined geometries and pore structures. The size and shape of molecules that can be adsorbed, as well as the size and shape of products formed, is constrained by the pore size and so can be predetermined by selection of the appropriate zeolite. An example of the shape selectivity property of zeolites is cracking of only the n-paraffins in a mixture with branched paraffins and aromatics.[8] In the Mobil M process for converting methanol to gasoline, discussed in Section 6.2, the shape of the zeolite effectively constrains the polymerization products to carbon numbers less than 10.

The final catalyst property to be discussed here is that of *stability*. For a catalyst to be commercially viable it must be chemically and physically stable and remain active without need for frequent regeneration or replacement. Physical deterioration of catalysts can occur by mechanical abrasion and by sintering. Attrition of catalysts to fines and dust can lead to plugging of reactors, as well as loss of catalyst and concomitant contamination of the products. Sintering can be thought of as the aggregation or binding of catalyst particles under the action of heat, with consequent loss of surface area. This amounts to loss of catalyst, inasmuch as the rate of the catalyzed reaction is related to the surface area available. Commercial catalysts are supplied on a carrier or support to provide mechanical strength and to reduce sintering effects. Supports are used also to disperse costly catalysts such as platinum, and so effectively increase the available surface area. Typical supports include alumina, silica, and

Table 2.22 Catalysts used in synthesis gas and related reactions

Process	Reaction	Desired Product	Active Catalyst Ingredient
Methanation	$CO + 3H_2 \rightleftharpoons CH_4 + H_2O$	Methane	Nickel
Alcohol synthesis	$CO + 2H_2 \rightleftharpoons CH_3OH$	Methanol	Zinc oxide, copper
Fischer-Tropsch	$nCO + 2nH_2 \rightleftharpoons (CH_2)_n + nH_2O$	Olefinic gasoline	Iron
Gas shift	$CO + H_2O \rightleftharpoons CO_2 + H_2$	Hydrogen	Iron oxide; chrome oxides
Mobil M	$nCH_3OH \rightleftharpoons (CH_2)_n + nH_2O$	Gasoline	Zeolite ZSM-5
Methane reforming*	$CH_4 + H_2O \rightleftharpoons CO + 3H_2$	Hydrogen, carbon monoxide	Nickel

* Reverse of methanation.

activated carbon. Supports are normally inert, but in some cases may be catalytically active. The use of zeolites as supports of metal catalysts offers extensive scope in tailor-making catalysts for specific requirements and has important application in hydrogenation and synthesis gas reactions.[19]

Chemical deactivation of catalysts results from blockage of active sites. So-called *poisoning* occurs when impurities in the feed stream adsorb irreversibly on active sites, rendering them unavailable for further reaction. Sulfur and nitrogen are typical poisons that deactivate cracking catalysts. Sulfur is a poison for many synthesis gas catalysts as well, and necessitates highly efficient desulfurization of these gases prior to the synthesis step. In some instances a guard catalyst, for instance a nearly spent catalyst, is placed upstream of the main catalyst to remove poisons.

Fouling refers to a physical blockage of the catalyst pores or surface by dust or deposited carbon. However, fouling may involve catalyzed reactions, as in the formation of polyaromatic "coke" by polymerization and dehydrogenation on active sites. The coke does not desorb readily, and therefore deactivates the sites and may also cause pore blockage. Regeneration by burning is normal practice, but must be closely controlled to prevent dehydroxylation and sintering. It may be possible to space the active sites sufficiently to reduce coke formation,[20] but this may in some cases modify desired activity and selectivity characteristics.

Further discussion of the properties determining activity, selectivity, and stability is beyond the scope of this text. The classification of catalysts according to the function of the reaction, as shown in Table 2.23, gives some indication of the scope and application of heterogeneous catalysts in synthetic fuel processing. Specific catalysts used in the various process steps are further described in the relevant sections of the book.

2.4 REACTOR CONSIDERATIONS

The reactions which convert a raw carbonaceous material to a synthetic fuel take place in containers or vessels, termed chemical reactors or simply reactors. In Figures 1.1 to 1.4 the reactor vessels are represented by the "black boxes" into which are fed the reactants.

Reactor configurations for synthetic fuel production, as in other chemical or physical processing, generally will have the goal of maximizing the product yield at the lowest cost. However, other factors may also have to be considered, as for example stability, control, reliability, downstream processing requirements, environmental effects, and safety. The study of chemical reactor design is called chemical reaction engineering and a lucid introduction to the topic may be found in the text by Levenspiel.[9] Our purpose here is to set down some of the more important features of reactors and considerations regarding particular reactor choices, insofar as they relate to synthetic fuels manufacture.

Chemical reactions together with physical processes, including heat and mass transfer and fluid flow, will transform the materials brought into the reactor into the desired product. The extent and rate of the transformation characterizes the reactor kinetics and is dependent on the interactions between the chemistry and physics.

Table 2.23 Some examples of heterogeneous catalysts and their reactions

	Metals	Metal oxides			
		Semiconductors		Insulators	
		Anionic (N-type)	Cationic (P-type)	Neutral	Acid
Source and examples	Predominantly transition metals: Pt, Pd, Ag, Au, Cr	Oxides and sulfides of transition metals ZnO, Fe$_2$O$_3$, TiO$_2$, V$_2$O$_5$, CrO$_3$, CuO, NiS, WS	NiO, CoO, Cu$_2$O, SnO, PbO, Cr$_2$O$_3$	MgO, CaO, SiO$_2$, Al$_2$O$_3$	Non-metal and metalloid oxides SiO$_2$-Al$_2$O$_3$
Thermal stability	Sinter above 800°C	Stable in inert atmosphere Reduced by losing oxygen ions	Oxidized by reacting with O$_2$	Relatively stable	Undergo dehydroxylation and loss of activity above 700°C
Catalytic sites	Metal atoms	Negatively charged anionic sites	Positively charged cationic sites	Hydroxyl groups	Al^{3+} ions
Catalytic mechanism	Transfer of electrons between metal and adsorbate	Oxidation; Adsorbate reacts with oxygen ions. Other reactions: Adsorption and activation on anionic sites	Adsorbates react with adsorbed O$_2$	Activation of adsorbate (relatively inactive)	Acid site creates carbonium ion in reactants which react further by cracking, etc.

Functions and typical reactions	*Hydrogenation/dehydrogenation* Gain/loss of H_2 from molecule and so removing/adding olefins. $3H_2$ + benzene $\rightleftharpoons$ cyclohexane over platinum	*Selective oxidation* propylene $\rightarrow$ acrylonitrile in presence of NH_3 over cobalt-molybdenum catalysts	*Deep oxidation* of hydrocarbons to CO_2 and H_2O (combustion)	*Dehydration* ethanol $\rightarrow$ ethylene over alumina (widely used as metal supports)	*Cracking* of heavy oil residues and coal liquids over amorphous SiO_2-Al_2O_3
	Hydrogenolysis or hydrocracking $C_2H_6 + H_2 \rightarrow 2CH_4$	*Fischer-Tropsch synthesis* $nCO + 2nH_2 \rightarrow (CH_2)_n + nH_2O$ over promoted iron oxide catalysts	*Gas shift reaction* $CO + H_2O \rightleftharpoons CO_2 + H_2$ at 500°C over Fe_3O_4/Cr_2O_3		*Cracking* of oils to specific products over zeolites
	Ammonia synthesis $N_2 + 3H_2 \rightleftharpoons 2NH_3$ (promoted iron catalyst)	*Methanol synthesis* $CO + 2H_2 \rightleftharpoons CH_3OH$ over zinc oxide/chromium catalysts			*Polymerization*
	Selective oxidation Ethylene $\rightarrow$ Ethylene oxide over silver	*Hydrodesulfurization* Over tungsten sulfide and cobalt-molybdate			*Isomerization*—straight chain to branch chain
	Ammonia oxidation $NH_3 \rightarrow$ nitrogen oxides over platinum	*Oxidative dehydrogenation* of butane to butene $C_4H_8 + \frac{1}{2}O_2 \rightarrow C_4H_6 + H_2O$ over bismuth-molybdate			*Molecular sieving*
Poisons	O_2, CO, sulfur compounds	Organic bases: nitrogen and sulfur compounds	Relatively immune	Relatively immune	Organic bases

In reactions for producing synthetic fuels, either one, two, or three distinct physical phases will be brought together in the reactor. Not all of the materials will undergo a net change by chemical reaction. For example, in one method of oil shale pyrolysis the heat is supplied by contacting a hot inert solid with the shale, with the purpose of extracting heat from the solid to pyrolyze the shale. In the production of liquid fuels by synthesizing carbon monoxide and hydrogen, a solid catalyst must be present to increase the rate of chemical reaction, but there is no appreciable consumption of the catalyst. To choose a reactor configuration, the types and number of phases involved must be known. When considering the phases involved, it is not necessary that a phase undergoes a net change by chemical reaction, but only that it be present. Those systems in which only one phase is involved are termed "homogeneous," and those with two or three phases "heterogeneous." In this classification the phase type, that is, solid, liquid, or gas, is of importance and not the number of different materials of the same phase.

Table 2.24 shows the principal reacting and contacting materials for producing synthetic fuels, classified by technology and by whether the system is homogeneous or heterogeneous. The technologies shown are those illustrated in Figures 1.1 to 1.4. Not all methods are included, as for example hydropyrolysis, which was mentioned in Section 1.1, or gasification by contacting with molten iron or molten salt. However, the materials and procedures shown in Table 2.24 characterize most of the major methods for producing synthetic fuels.

The only homogeneous system in Table 2.24 is for indirectly heated pyrolysis, where a carbonaceous solid material is brought into contact with a hot inert solid for the purpose of transferring heat from the inert to the carbonaceous material. Inert solids that are used include ceramic balls, sand, ash, or spent shale. Normally the reactor in which the contacting is carried out is a solids mixer such as a tumbler or screw mixer.

As seen in Table 2.24, the bulk of the reactor systems are heterogeneous. The contacting of the phases can be effected in a variety of ways. For heterogeneous systems in which a granular solid phase is present, we may distinguish two broad categories of reactor designs—the packed bed and the fluid bed. In a packed bed the granular solid is packed in a vessel to form a porous bed. The bed may be either fixed or moving with respect to the walls of the container. In a fluid bed the velocity of either the gas or liquid flowing up through the bed of solids is such that the fluid drag force is equal to or greater than the mass of the solids plus the friction of the bed against the walls. If the drag force is equal to the solids mass plus the wall friction force, the fluid bed will be stationary with respect to the container and if it is greater, the bed will be in upward motion.

The drag force on the particle is determined primarily by the velocity of the fluid relative to the particle. At very low velocities the dominant drag force is due to shear, sometimes called skin friction, on the particle surface. This low velocity flow is referred to as Stokesian flow and the force on the particle, if an isolated sphere, is given by Stoke's Law

$$F = 3\pi d_s \mu u \qquad (2.74)$$

Table 2.24 Principal technologies and associated reacting and contacting materials for producing synthetic fuels

Technology	Reactants* and Contactants†				Reacting Phases	Contacting Phases	Common Generic Rector Name
	Solid		Gas	Liquid			
Homogeneous							
Pyrolysis, indirectly heated	Coal, oil shale, or tar sands	Hot inert solid†	—	—	Solid	Solid	Retort
Heterogeneous							
Pyrolysis, indirectly heated	Coal, oil shale, or tar sands		Hot product gas†	—	Solid	Solid/gas	Retort
Pyrolysis, directly heated	Coal, oil shale, or tar sands		Air + steam or product gas	—	Solid, gas	Solid/gas	Retort
Coal gasification	Coal		Steam + air, O$_2$, H$_2$, or CO + H$_2$	—	Solid, gas	Solid/gas	Gasifier
Coal liquefaction, indirect hydrogenation	Catalyst†		CO + H$_2$	—	Gas	Gas/solid	Catalytic reactor
Coal liquefaction, direct hydrogenation	Coal, Catalyst†		H$_2$	Product oil	Solid, liquid, gas	Solid/liquid, gas	Liquefaction reactor; dissolver

* Substances which undergo a net change by chemical reaction.
† Catalysts and heat transfer media which promote chemical reaction.

where F is the total drag force on the sphere, d_s is the sphere diameter, μ is the absolute viscosity, and u is the velocity of the fluid in the undisturbed flow.

The force F is the streamwise force tending to drag the particle along with the flowing fluid. In a vertical reactor with upward flow this force will be acting against the downward force on the particle. The net downward force is the difference between the gravitational and the buoyancy forces, and in a dense-phase or packed system must also include the frictional force between the bed and the tube walls. At low velocities the net downward forces are greater than the drag force and the particle will fall unless held in place as in a packed bed. At higher velocities, where the drag force exceeds the downward forces, the particle is entrained and moves upward with the flow.

The fluid velocity u_f at which an isolated sphere of volume V_s is held stationary in the flow is determined by equating the downward gravitational force with the drag on the particle

$$3\pi d_s \mu u_f = (\rho_s - \rho)g V_s \tag{2.75a}$$

to get

$$u_f = \frac{(\rho_s - \rho)g d_s^2}{18\mu} \tag{2.75b}$$

where g is the acceleration due to gravity, ρ_s is the density of the sphere, and ρ is the fluid density. At fluid velocities greater than u_f the particle will be dragged along or entrained with the fluid.

The fluid velocity appearing in the above equations for single particles is called the ''superficial velocity.'' It is the velocity of the fluid in the undisturbed flow upstream of the particle or particles. In multiparticle systems it is the actual interstitial velocity that governs the drag on the particles and determines whether or not a particle will be entrained in the flow. The actual velocity is greater than the superficial velocity, as a portion of the reactor volume is unavailable for flow due to the presence of the particles and the associated recirculating wakes and stagnant regions. The greater the particle packing density, the greater is the actual or effective velocity relative to the superficial velocity. In a random bed of spherical particles the effective velocity is about five times the superficial velocity. In other words, the particles in a packed bed will tend to become entrained at a superficial velocity about one-fifth of that calculated according to Eq. (2.75b). However, as the particles become entrained, the packed bed expands and the packing density decreases. This results in a decrease in the effective velocity and the particles tend to collapse once more to the packed condition. At velocities of about 0.5 u_f, the particles in the bed are in a constant motion or fluid state, although there can be no net motion of the particles along with the fluid until the superficial velocity approaches u_f.

There are numerous published equations in the literature for estimating the pressure drop across a multiparticle system. Stokes' law is valid for a single particle only; for flow in a porous medium we may apply Darcy's law

$$\frac{\Delta p}{L} = \frac{\mu u}{k} \tag{2.76}$$

Here, k is the permeability of the porous medium with dimensions of length-squared, and $\Delta p/L$ is the pressure gradient in the fluid resulting from the shear drag F on the particle surface or capillary walls. As with Stokes' law, Darcy's law is applicable only at very low flow rates. At higher flows it is necessary to take into account the pressure drops associated with the inertial losses, which occur as the fluid accelerates and changes direction in its tortuous path past the particles. The inertial pressure drop varies more nearly with the square of the velocity. Forchheimer originally suggested that the total pressure drop be written as the sum of the shear and inertial losses, which we here write in the form

$$\frac{\Delta p}{L} = \underset{\substack{\text{Shear} \\ \text{loss}}}{\frac{\mu u}{k}} + \underset{\substack{\text{Inertial} \\ \text{loss}}}{c\frac{\rho u^2}{d_e}} \qquad (2.77)$$

where c is a dimensionless constant on the order of unity, and d_e is a characteristic length associated with the particle size. The characteristic length for a sphere is its diameter.

In order to calculate the drag forces and pressure drops in a packed bed reactor, it is useful to rewrite the terms in Eq. (2.77) as dimensionless numbers. The dimensionless number for the pressure gradient is the friction factor, one definition of which is

$$\lambda = \frac{d_e}{\rho u^2} \frac{\Delta p}{L} \qquad (2.78a)$$

The dimensionless number containing the velocity is the well-known Reynolds number

$$Re = \frac{d_e \rho u}{\mu} \qquad (2.78b)$$

For beds of randomly packed particles with an isotropic permeability, the characteristic length in the friction factor and Reynolds number may be taken to be $\sqrt{k}$, so that in dimensionless form Eq. (2.77) becomes

$$\lambda_{\sqrt{k}} = \frac{1}{Re_{\sqrt{k}}} + c \qquad (2.79)$$

where the subscript $\sqrt{k}$ denotes a characteristic length $\sqrt{k}$. The value of $\sqrt{k}$ is about one-tenth the particle diameter in an isotropic packing. The constant c is not universal, but depends on the packed bed structure and flow; however, its value is typically in the range 0.2 to 0.6.[21,22]

An equation popularly presented in the literature and used for estimation of pressure drops in packed beds is that due to Ergun

$$\frac{\lambda \epsilon^3}{1 - \epsilon} = \frac{150(1 - \epsilon)}{Re} + 1.75 \qquad (2.80)$$

where ϵ is the voidage, or porosity, defined by

$$\epsilon = \frac{\text{volume of voids}}{\text{volume of bed}} = 1 - \frac{\text{volume of particles}}{\text{volume of bed}} \qquad (2.81)$$

The Ergun equation is applicable up to Reynolds numbers of about 1000. The characteristic length in the friction factor and Reynolds number is based on the volume to surface area ratio for the particle

$$d_e = \frac{6 \,(\text{vol. of particle})}{\text{surface area of particle}} \tag{2.82}$$

Note that for spherical particles d_e is again just the sphere diameter. The constants and terms containing the voidage in the Ergun equation can be thought of as correction terms to the characteristic length that are related to the packing structure. The advantage of the Ergun equation is that it can be evaluated from a knowledge of the physical dimensions of the particles and the bed voidage. Equation (2.79) is, however, a more fundamental relationship and its use is preferred in cases where the bed permeability is known.

The above equations for pressure drop are applicable to packed beds. The pressure drop for fluid beds is independent of the velocity but may be estimated using the calculated u_f value in the packed bed pressure drop equations. At much higher velocities where the bed becomes entrained, the resistance to the flow due to the presence of the particles is reduced. With further increases in velocity, the voidage tends to unity and the pressure drop approaches that for flow in an empty tube.

The pressure drop equations discussed here are appropriate to isothermal flow through beds of non-reacting solids. In packed bed coal gasifiers, however, there is invariably a strong temperature gradient in the bed, the quantity of fluid along the bed changes due to the reactions occurring, and in addition there is significant mass transfer between the particles and the fluid. Application of the above equations can therefore be expected to result in at best a rough estimation of the pressure drop.

The different categories of the two broad classes of reactors are illustrated diagramatically in Figure 2.8. The dense-phase fluidized bed refers to the case in which

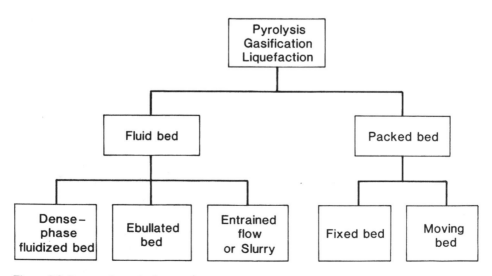

Figure 2.8 Reactor classes for heterogeneous systems.

the packed bed has expanded uniformly in the vessel so as to allow the gas or liquid to flow past the granules with the pressure drop through the bed just balanced against the solid mass. Typically the bed expands to twice its settled height. The particles will move about locally in a semi-stable arrangement resembling a boiling liquid and, as mentioned, there is no net solids flow. This quiescent, dense-phase fluidized state is what is usually called "fluidization," and the fluid-particle arrangement a "fluidized bed." We shall follow this terminology in the book. In the ebullated bed, all else being the same, the velocity of the fluid through the bed is larger than for the fluidized bed. The space between the granules enlarges, there is no semi-stable arrangement, and there is an overall turbulent circulation of the bed with the particles entrained in high velocity regions and then falling back in regions of lower fluid velocities. There is also some entrainment and carry-over of the solids in the gas or liquid stream leaving the reactor. The active, bubbling, ebullient character of the bed has given rise to the name "ebullated bed." In the entrained flow or slurry configuration there is no longer a distinct bed; the fluid velocity being large enough that the particles are entrained by the gas or liquid and carried upward in a relatively dilute phase. The entrained flow or slurry reactor need not necessarily be oriented vertically, so long as the gas or liquid flow maintains the solids in an entrained or slurried state. This reactor type is also called a "dilute-phase transport reactor" or just "transport reactor."

In the packed bed configuration the bed can also be either fixed or moving with respect to the container walls. In this regard there is an analogy between the packed and fluid bed reactors. However, in the fluid bed the gas or liquid flow is always cocurrent with any net solids flow. On the other hand, in the packed bed the gas or liquid may flow cocurrent, countercurrent, or even crosscurrent to the flow of solids. Figure 2.9 shows examples of fluid bed and packed bed reactor configurations.

To assess the performance of different reactors, they are often compared with the two basic reactor types, the batch reactor and the continuous flow reactor. In the ideal batch reactor all chemical and thermal changes are only with respect to time, it being assumed that the system is spatially uniform. The continuous flow reactor is one in which there is a steady flow of the reactants into the reactor, and products out of the reactor. One ideal limiting configuration is the perfectly mixed reactor in which complete uniformity of concentration and temperature is maintained throughout, with the product stream having the same composition and temperature as the mixture in the reactor. Another limiting configuration is the plug flow reactor in which the reactants move through a tubular vessel with a uniform velocity along parallel streamlines; that is, there is a flat velocity profile. In this case there is no intermixing of the fluid elements entering the vessel. For simple first-order decompositions, where the reaction rate declines as the reaction proceeds and the reactant disappears, plug flow gives the least reactor volume for a specified processing rate and extent of conversion. The perfectly mixed reactor gives the largest reactor volume. The real reactors we have described will generally fall somewhere in between.

The choice of a particular reactor type for different synthetic fuel processes is a complex one, and one that is not unique. It is generally difficult to make a choice from "first principles," since the behavior of the reactants in different reactors must usually be evaluated empirically. We may, however, set down some general guidelines

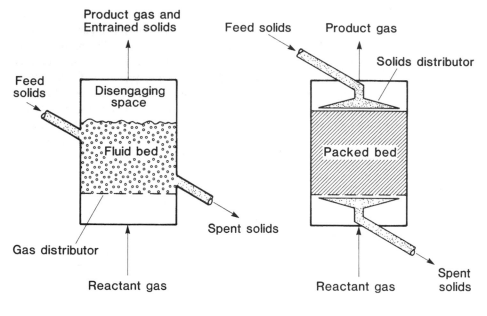

Figure 2.9 Examples of fluid bed and packed bed reactors.

regarding packed and fluid bed characteristics, with the caveat that they are subject to the constraints of the particular system and design.

In packed beds, gases flow through the bed in nearly plug flow as do liquids when the interstices are filled. The reactor volume, when compared with other types, is therefore smaller for the same conversion rate. A packed bed reactor where the heat evolved (or absorbed) by the reaction is sufficiently small that temperature control by the transfer of heat from (or to) the reactor is not required is termed an "adiabatic reactor." In this type of reactor the temperature change through the reactor is proportional to the fractional conversion multiplied by the heat of reaction (see Section 5.2). For relatively fast reactions involving large heat effects, adiabatic reactors are generally not appropriate. Temperature control in packed beds can be difficult and hot spots can occur. For packed bed coal gasifiers, cooling of the outside reactor walls is generally sufficient. However, for highly exothermic reactions, as in indirect liquefaction, internal tubes are generally required to increase the cooling surface for an externally supplied coolant. Moving packed beds usually employ countercurrent flow and heat exchange, which lead to a high thermal efficiency. If there are small particles present, as for example coal and shale fines or small size catalysts, plugging can result and the pressure drop through packed beds can become large. Plugging of packed beds can also result from agglomerating coals, which leads to a decrease in permeability, and to high pressure drops in packed bed gasifiers. There is generally a low contamination of the fluid product with solids; bringing the solids in and out of packed beds can sometimes be a problem.

The flow through fluid beds has a behavior lying somewhere between that of a plug flow and a perfectly mixed flow. A fluidized bed may be closer to a perfectly mixed flow, whereas a turbulent flow in a tubular slurry reactor would be much closer to a plug flow. For a fixed conversion rate, fluid beds will therefore generally be larger than packed beds. Because of the good heat exchange between phases, fluid beds have a much more uniform temperature than packed beds, a feature which makes them particularly useful for highly exothermic processes. However, the cocurrent flow pattern between phases in entrained flow and slurry reactors makes them less thermally efficient than countercurrent flow packed bed reactors. Small-size particles can be handled in fluid beds without incurring excessive pressure drops. Solids addition and removal are normally easier in fluid than in packed beds, but contamination of the fluid product with solids and the need for subsequent separation equipment is an added problem. Moreover, when the contacting solid phase is a catalyst which is to be reused, it must be resistant to attrition. A design problem related particularly to fluidized and ebullated bed reactors is that scale-up procedures are not fully established because of the complexity of the hydrodynamics.

In comparing the performance of different reactors for a given process and degree of conversion it is necessary to compare parameters which provide a measure of the same quantity in each reactor. One measure of reactor capacity is the mean residence time of the reacting phase or mixture. Residence time is a useful measure for a batch reactor when the density of the mixture is constant throughout, as with a liquid. However, when there are density changes it is no longer satisfactory since it is then not easily determined. The density changes may be brought about by gas phase reactions in which the number of moles changes, or by gradients in temperature and pressure as may occur in a continuous flow reactor. Moreover, the residence time is defined in terms of the reactor volume occupied by the reaction mixture, and this may be unknown for packed or fluidized beds.

An alternative parameter used to characterize reactor capacity, one which is directly quantifiable, is "space time," defined as the time elapsed in processing one reactor volume of feed at known conditions. One accepted definition for space time, but not the only one, is

$$\tau' = \frac{V}{Q_0} \tag{2.83}$$

where V is the reactor volume and Q_0 is the influent volumetric flow rate of reactants. When the temperature, pressure, and total number of moles is constant in the reactor, the space time and mean residence time correspond. It is common to use as a parameter the reciprocal of the space time, termed the "space velocity." Sometimes in place of the reciprocal of the definition (2.83), the space velocity is defined by the group F_{A_0}/V, where F_{A_0} is the flow rate of reactant A in the feed.

Residence times given later in the book for coal gasification and liquefaction reactors are calculated using

$$\tau = \frac{V(1 - \epsilon)}{F_{\text{coal}}} \tag{2.84}$$

where F_{coal} is the volumetric feed rate of coal to the reactor. The reactor volume V is here multiplied by the factor $(1 - \epsilon)$ to give the actual volume occupied by the coal particles. Coal residence times in entrained flow reactors gasifying pulverized coal are on the order of seconds, whereas in packed and fluidized bed reactors gasifying larger size coal, residence times may be up to an hour or more. Coal residence times in direct liquefaction reactors range from about 0.5 to 1.5 hours.

REFERENCES

1. Morrison, R. T., and Boyd, R. N., *Organic Chemistry*. 3rd Edition, Allyn and Bacon, Boston, 1973.
2. Cotton, F. A., and Wilkinson, G., *Advanced Inorganic Chemistry: A Comprehensive Text*. 3rd Edition, Interscience Publishers, New York, 1972.
3. Barrow, G. M., *Physical Chemistry*. 3rd Edition, McGraw-Hill, New York, 1973.
4. Denbigh, K., *The Principles of Chemical Equilibrium*. 3rd Edition, Cambridge University Press, London, 1971.
5. Hatsopoulos, G. N., and Keenan, J. H., *Principles of General Thermodynamics*. Wiley, New York, 1965.
6. Carberry, J. J., *Chemistry and Catalytic Reaction Engineering*. McGraw-Hill, New York, 1976.
7. Bond, G. C., *Heterogeneous Catalysis: Principles and Applications*. Clarendon Press, Oxford, 1974.
8. Satterfield, C.N., *Heterogeneous Catalysis in Practice*. McGraw-Hill, New York, 1980.
9. Levenspiel, O., *Chemical Reaction Engineering*. 2nd Edition, Wiley, New York, 1972.
10. Froment, G. F., and Bischoff, K. G., Chemical Reactor Analysis and Design. Wiley, New York, 1979.
11. Nelson, W. L., *Petroleum Refinery Engineering*. 4th Edition, McGraw-Hill, New York, 1958.
12. Perry, R. H. and Chilton, C. H. (eds.), *Chemical Engineers' Handbook*. 5th Edition, McGraw-Hill, New York, 1973.
13. Gaggiolli, R. A., and Petit, P. J., "Use the Second Law First," *ChemTech* **7**, 496-506, 1977.
14. Massey, L. G., "Coal Gasification for High and Low Btu Fuels," in *Coal Conversion Technology* (C. Y. Wen and E. S. Lee, eds.), pp. 313-427. Addison-Wesley, Reading, Mass., 1979.
15. Reid, R. C., Prausnitz, J. M., and Sherwood, T. K., *The Properties of Gases and Liquids*. 3rd Edition, McGraw-Hill, New York, 1977.
16. Lee, A. L., Feldkirchner, H. L., and Tajbl, D. G., "Methanation for Coal Gasification," *Am. Chem. Soc., Div. Pet. Chem., Preprints* **15**(4), A93-A105, 1970.
17. Seglin, L., *et al.*, "Survey of Methanation Chemistry and Processes," in *Methanation of Synthesis Gas* (L. Seglin, ed.), pp. 1-30. Advances in Chemistry Series No. 146, American Chemical Society, Washington, D.C., 1975.
18. Trimm, D. L., "Designing Catalysts," *ChemTech* **9**, 571-577, 1979.
19. Leith, I. R., "Hydrogenation and Fischer-Tropsch Synthesis on Zeolite Group VIII Metal Catalysts," *ChemSA* **4**, 70-79, 1978.
20. Manton, M. R. S., and Davidtz, J. C., "Controlled Pore Sizes and Active Site Spacings Determining Selectivity in Amorphous Silica-Alumina Catalysts," *J. Catalysis* **60**, 156-166, 1979.
21. Shwartz, J., and Probstein, R. F., "Experimental Study of Slurry Separators for Use in Desalination," *Desalination* **6**, 239-266, 1969.
22. Hicks, R. E., "Effect of Packing Geometry on Transport Phenomena in Packed Beds of Spheres," Ph.D. Thesis, Faculty of Engineering, University of the Witwatersrand, Johannesburg, South Africa, 1975. (NTIS Catalog No. PB 242 627).

THREE

CONVERSION FUNDAMENTALS

3.1 PYROLYSIS

Pyrolysis refers to the decomposition of organic matter by heat in the absence of air. Thermal decomposition is frequently used to mean the same, although it generally connotes the breakdown of inorganic compounds.

When coal, oil shale, or tar sands are pyrolyzed, hydrogen-rich volatile matter is distilled and a carbon-rich solid residue is left behind. A common synonym for pyrolysis is "devolatilization." The carbon and mineral matter remaining behind is the residual char. In this regard, the term "carbonization" is sometimes used as a synonym for coal pyrolysis. However, carbonization has as its aim the production of a solid char, whereas in synthetic fuel production greatest interest centers on liquid and gaseous hydrocarbons.

Pyrolysis is one method to produce liquid fuels from coal, and it is the principal method used to convert oil shale and tar sands to liquid fuels. Moreover, as gasification and liquefaction are carried out at elevated temperatures pyrolysis may be considered a first stage in any conversion process.

The study of pyrolysis, particularly that of coal, can be dated from the end of the eigthteenth century. Despite the many advances in understanding achieved since that time, there is still not a unified fundamental picture of the complex chemical and physical phenomena that take place, which given the raw material and the physical conditions of heating would enable a unique determination of the products. Our purpose here is to elucidate the fundamentals of pyrolysis to the extent that may be helpful in the design of synthetic fuel plants, even though this may place greater emphasis on empirical and semi-empirical data. Despite considerable effort expended on fundamental laboratory scale studies of pyrolysis, a clear link between many of the results and the data base needed to design and scale-up synthetic fuel plants generally has not been established.

Of most interest in the production of synthetic fuels is the prediction of the rate and amount of volatile yield and product distribution for a given raw material and set of pyrolysis conditions. Among the important chemical variables are the elemental and functional composition of the organic and inorganic matter, as well as the composition of the ambient gas in which the pyrolysis takes place. Among the more important basic physical variables are the final temperature, the time and rate of heating, the particle size distribution, the type and duration of any quenching, and the pressure. An indication of the uncertainty existing in this field is that at present there is no agreement on whether yield, that is the loss in mass of the raw material from pyrolysis, is changed with heating rate.[1,2]

Perhaps the best understood pyrolysis processes are the cracking and coking of petroleum, discussed in Section 2.1. We recall that cracking means the breaking down of the hydrocarbons in the vapor phase to smaller molecular units, plus possibly some carbonaceous residue from the oil vapors. The most certainty in predicting product yield and distribution might be expected for tar sands, at least to the extent that the bitumen in the tar sands is related to conventional petroleum. The least certainty regarding product yield can be expected with coal pyrolysis. This is in part attributable to the fact that coal is not a unique chemical substance, but rather a complicated inhomogeneous mixture of different organic and inorganic materials of different petrographic compositions. The predictive capability for oil shale may fall somewhere between that of coal and tar sands. However, much less attention has been devoted to fundamental studies of oil shale pyrolysis than has been given to coal pyrolysis, a fact related to the relative importance of these materials in the past.

There are a number of features of pyrolysis common to coal, oil shale, and tar sands which we discuss first before separately presenting data specific to each of these materials. The *composition of the raw material* is important in determining the yield of volatile matter. The principal material property defining the yield is the hydrogen-to-carbon ratio. For lignite, subbituminous, and bituminous coals, the proximate volatile matter, denoted by the letters VM, will generally range between 30 and 50 mass percent, as shown in Figure 1.6. It is to be emphasized that this yield is one obtained under controlled, slow heating assay conditions and is not necessarily optimum. The VM value is, however, a convenient and standardized measure. When the kerogen from Green River shale is pyrolyzed by the Fischer assay at 500°C, the volatile matter yield is about 80 to 85 percent.[3] The volatile matter yield from the slow pyrolysis of Canadian and U.S. tar sand bitumens at 600°C is in the range of 83 to 88 percent.[4]

The composition of the volatile products evolved in pyrolysis is largely determined by the raw organic material. For coals that are pyrolyzed at 500°C in an inert atmosphere under assay conditions of slow heating, the volatiles are composed of about 65 mass percent gas and water, and about 35 percent tar and light oils. Oil shale kerogen volatiles are about 80 percent oil and 20 percent gas and water, while tar sand bitumen volatiles at 600°C are 90 to 95 percent liquids. In all cases, at 500°C the major constituents of the gas include carbon oxides, methane, hydrogen, and light hydrocarbons along with hydrogen sulfide and ammonia.

The *pyrolysis temperature* affects both the amount and composition of the volatile yields. When coal, oil shale, and tar sand bitumen are heated slowly, rapid evolution of volatile products begins about 350 to 400°C, peaks sharply at about 450°C, and drops off very rapidly above 500°C. This is termed the stage of "active" thermal decomposition. There are three principal stages of pyrolysis. In the first stage, above

100°C and below, say, 300°C the evolution of volatile matter is not large and what is released is principally gas composed mainly of oxides of carbon and water. In the active or second stage of decomposition, about three-quarters of all the volatile matter ultimately released is evolved. The third stage is one most appropriately defined for coal, in which there is a secondary degasification associated with the transformation of the char, accompanied by the release of noncondensable gases, mainly hydrogen.

In all stages of pyrolysis, but particularly the higher temperature ones, it is important to distinguish between the "primary" volatile products of decomposition and those products which result from secondary cracking and coking reactions and from gasification reactions such as between the char and evolved water. Figure 3.1 is a sketch for coal of a typical cumulative mass loss and differential mass loss with temperature.

The *effect of heating rate* on the amount and composition of pyrolysis volatile yields has been an important question, particularly for coal, since it was demonstrated that when finely ground coal particles are heated rapidly, larger yields than the

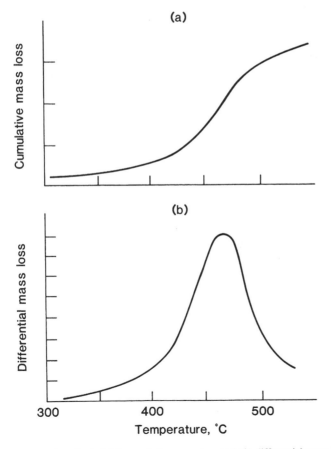

Figure 3.1 Typical (*a*) cumulative mass loss and (*b*) differential mass loss with temperature for coal.[5]

proximate volatile matter yields could be obtained. Moreover, rapid heating to the peak decomposition temperatures around 500°C produced significantly larger tar fractions, suggesting the possibility of a means of converting coal to liquid hydrocarbons simply and with relatively large liquid yields. Traditional laboratory and assay heating conditions have been characterized by rates of temperature increase around 20°C/s during the period of rapid evolution of the volatile matter, so slow heating is defined to correspond to gradients about equal to or smaller than this value. Rapid heating is loosely defined in the literature as being larger than 10^3 to 10^5 °C/s.[6] The reason for this broad range is clarified in what follows.

Let us define the pyrolysis of any organic material as "slow" or "fast," depending on whether the characteristic time required for heating the material to a specified final pyrolysis temperature is very much greater than, or very much less than, the characteristic pyrolysis reaction time. Such a definition presumes that a single characteristic pyrolysis time can be defined. It is known that the thermal decomposition of simple organic species is describable as a first-order reaction in the amount of unreacted material remaining. This has been carried over with some success to the pyrolysis of complex organic materials, where the overall pyrolysis process has been represented as a first-order decomposition occurring uniformly throughout the material, with the rate of volatile yield assumed to be proportional to the remaining volatile matter

$$\frac{dV}{dt} = k(V_\infty - V) \tag{3.1}$$

Here, V is the volatile matter released and V_∞ is the total amount of volatile matter evolved after a long reaction time. In the simplest model, the rate constant k is correlated with temperature by an Arrhenius expression

$$k = A\,e^{-E/RT} \tag{3.2}$$

where A is the frequency factor and E the activation energy. This representation is consistent with the concept of a characteristic reaction time. Discrepancies in observed rate-constants have led to modifications of this model and one successful approach has been to include a large number of parallel decompositions by considering only one frequency factor but a distribution of activation energies.[1] A characteristic reaction time can also be defined with this model.

We may define the characteristic reaction time τ_R for a single reaction as the reciprocal of the rate constant k evaluated at an appropriate characteristic temperature, for example, the final temperature. With a distribution of activation energies, an appropriate mean can be defined. In any case,

$$\tau_H \gg \tau_R \quad \text{(slow)} \tag{3.3}$$

$$\tau_H \ll \tau_R \quad \text{(fast)} \tag{3.4}$$

where τ_H is the characteristic heating time.

In slow pyrolysis the reactions taking place are always in "local equilibrium," the heating time being sufficiently slow to allow equilibration along the temperature path. For this case the ultimate yield and product distribution depends on the temperature history. Whether the products remain in the reactor is also important since their presence can influence both the primary and secondary reactions. In fast pyrolysis there is a negligible amount of reaction during the heat-up period and the material is "frozen" during this time. Whatever pyrolysis reactions occur take place isothermally at the final temperature and there can be no effect of past history.[7]

To clarify these ideas let us consider coal pyrolysis, which is a rapid process and one for which equilibrium is usually reached in times on the order of a second or less. However, the equilibration rate is very dependent on temperature, as follows from Eq. (3.2). If we adopt 1 second as the characteristic reaction time then the criteria in seconds for slow and fast pyrolysis become

$$\tau_H > 10 \text{ s} \quad \text{(slow)} \tag{3.5}$$

$$\tau_H < 0.1 \text{ s} \quad \text{(fast)} \tag{3.6}$$

where we have taken 10 to mean large and 0.1 to mean small in Eqs. (3.3) and (3.4). Using these criteria, it can be seen that were we to pyrolyze at 1000°C, the characteristic rate of temperature increase below which the pyrolysis would be defined as slow is 10^2 °C/s, while fast would be defined above 10^4 °C/s. These values are less dependent on the final temperature *per se* than they are on the effect of temperature in decreasing or increasing the reaction rate exponentially. It is probably for this reason that such wide ranges have been reported in °C/s for defining slow and fast pyrolysis. The appropriate definitions are given by the dimensionless ratios that follow from Eqs. (3.3) and (3.4), though the key element to define is the characteristic reaction time.

Particle size is known to influence pyrolysis yield. This effect may be related to heating rate, in that larger particles will heat up more slowly, so the average particle temperatures will be lower, and hence volatile yields may be expected to be less. If the particle size is sufficiently small, it will be heated uniformly. Data[2] on coal particles indicate that size no longer effects yield below about 50 μm. Similar results should apply to oil shale and the direct pyrolysis of tar sands. It should be noted, however, that for coal and shale in particular, the grinding of particles to this size requires a considerable expenditure of energy. In the case of shale, a reduction to a size in this range would probably be warranted only if the kerogen were to be separated from the shale, not by pyrolysis, but by physical means such as gravity separation in water.[8]

Pressure may be expected to affect volatile yields, with higher pressures reducing the yields and lower ones increasing them. However, higher pressures favor cracking reactions, so high pressure will produce a larger volume of light hydrocarbon gases, whereas lower pressure will lead to larger tar and oil fractions.

Pyrolysis in a hydrogen atmosphere, termed hydropyrolysis, can increase the volatile yield and proportion of lower-molar-mass hydrocarbons. Hydropyrolysis is defined to include thermal decomposition and the attendant hydrogenation but not the hydrogasification of the char. Hydrogen reacts much faster with decomposing organic material and with the primary volatiles than with residual char. For coal, the period of highest reactivity to hydrogen is postulated to correspond to the occurrence of pyrolysis and to last only a few seconds or less at temperatures approaching 800°C or higher.[2] The increase in volatile yield is attributable to the hydrogenation of free radical fragments sufficient to stabilize them before they repolymerize and form char.

Coal

The standard laboratory assays of liquid and gaseous yields from coal by pyrolysis are all carried out under conditions of slow heating, and as such, the results obtained are specific to the assay procedure. In detailing coal pyrolysis products evolved under conditions of slow heating, it is to be emphasized again that the yield and product distribution depend on the temperature-time history. The products are not necessarily the result of single decomposition reactions but instead may be formed from products evolved at low temperatures that are exposed to high temperatures and a different chemical environment. Examples include decomposition products from primary volatiles and products from char-gas reactions, some of which might take place in the char matrix with reactives deposited inside the particles. Moreover, in any history-dependent process the physical and flow characteristics of the reactor will enter.

One of the more common assay procedures is the Fischer assay, described in Section 1.3 in connection with the properties of oil shale. Another widely used procedure is the Gray-King assay. These assays are carried out in the absence of air, usually at 500 or 600°C, with the temperature held at the final level until the evolution of volatiles stops. The volatiles are drawn off and condensed. At these temperatures, under conditions of slow heating, the yield of tar plus light oils tends to a maximum, although the gas volumes are not a maximum. In Table 3.1 are shown typical Fischer assay data on different rank as-received coals, which we recall means coals for which the moisture has not been removed by drying. Typical distributions on a dry and ash-free (daf) basis of the pyrolysis products from various carbonaceous materials assayed at 500°C are shown in Figure 3.2. The product yields from wood and peat are included for comparison purposes (cf. Table 1.6).

Several features may be observed from the data in Table 3.1 and Figure 3.2 concerning the product distribution in relation to the chemical composition of the coal. An important one is that the total volatile matter yield increases with the hydrogen-to-carbon ratio in the coal. From a correlation of data on British and American coals, Dryden[10] derived an empirical relation for the standard proximate volatile matter yield at 900°C, which if we neglect the contribution of the coal oxygen may be written approximately as

$$\text{VM (mass \%)} \approx 97.3 \frac{H}{C} - 40.4 \tag{3.7}$$

Table 3.1 Typical Fischer assay yields at 500°C from different rank as-received coals[3]

	Char, mass %	Tar, L/t	Light Oil, L/t	Gas, m³/t	Water, mass %
Medium volatile bituminous	83	79	7	61	4
High volatile A bituminous	76	129	10	61	6
High volatile B bituminous	70	126	9	63	11
Subbituminous A	59	86	7	83	23
Subbituminous B	58	64	5	71	28
Lignite A	37	63	5	66	44

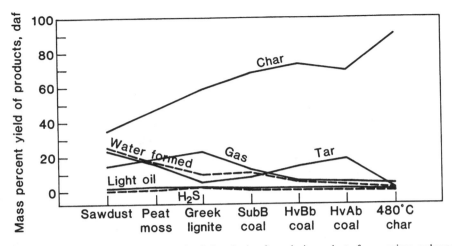

Figure 3.2 Distributions on a dry and ash-free basis of pyrolysis products from various carbonaceous materials assayed at 500°C.[9]

where VM is the standard proximate volatile matter expressed in mass percent of the dry and ash-free coal, and H/C is the hydrogen-to-carbon atom ratio of the coal. This relation is satisfactory for coals with hydrogen-to-carbon atom ratios between 0.5 and 0.9.

That the volatile matter should depend principally on the hydrogen-to-carbon ratio is dimensionally consistent with the coal composition. In this regard it may be noted that coal is a polymer-like molecule consisting of fused aromatic and hydroaromatic ring clusters linked by relatively weak aliphatic bridges. From 60 to 75 percent of the carbon is contained in aromatic configurations, about 15 to 25 percent in hydroaromatic structures, and 5 to 10 percent in aliphatic carbon. The ring clusters contain the heteroatoms—oxygen, sulfur, and nitrogen. During active decomposition at temperatures around 500°C, it is the weak aliphatic links that are ruptured with subsequent release of the large molecular ring free radicals. The tar and light oils are formed from the stabilization of these free radical fragments by the hydrogen evolved. Solomon[11] has pointed out that the abstracted hydrogen is most likely to come from the aliphatic portion of the coal, since the unsaturated aliphatics are less stable and more reactive than the aromatic compounds. He has shown a strong correlation between the tar yield and the aliphatic hydrogen concentration. The light molecules of the gas arise simultaneously with the evolution of the tar molecules from competitive thermal cracking of the bridges, ring clusters, and attached functional groups.

Writers on pyrolysis have noted the strikingly similar structure and composition of the tar and the parent coal except, of course, for the extra aliphatic derived hydrogen in the tar. It is therefore reasonable to expect that the mass fraction of the primary low temperature tar would be directly proportional to the hydrogen-to-carbon ratio in the original coal. A similar observation has been made by Tyler.[12] We have carried out a linear least-squares fit of the yields of tar plus light oils from 130 coals,[13] mostly American but including a variety of other coals, and find that with respect to the daf coal the percent of tar plus light oils evolved at 500°C assay conditions may be closely estimated by the relation

$$\text{Tar + Light Oils (mass \%)} = 29.1\frac{H}{C} - 12.1 \tag{3.8}$$

Comparison of Eqs. (3.8) and (3.7) shows that there is a large margin for increase in the tar yields, since essentially all of the hydrogen originally in the coal is to be found in the volatile matter. It is interesting that the ratio of Eq. (3.8) to (3.7) almost equals 30 percent. The tar and light oil fraction remains approximately constant up to 1000°C. However, at the temperature of 900°C corresponding to the temperature of the proximate volatile matter analysis, the gas fraction increases over that at 500°C, so the tar plus light oils represent closer to 25 percent of the volatile matter at that temperature. The increase in the gas fraction is associated principally with char decomposition.

Berkowitz[5] describes the low temperature tar and light oils formed from subbituminous and bituminous coals at temperatures below about 700°C as relatively fluid, dark brown oils that contain phenols, pyridines, paraffins, and olefins. The oils are heterogeneous, with any one component constituting only a fraction of a percent of the total mass. The lignite tars may also contain up to 10 percent of paraffin waxes, so the product has a "butterlike consistency" and solidifies at temperatures as high as 6 to 8°C. The primary high temperature tar vapors formed above 700°C are more homogeneous. The light oils are predominantly benzene, toluene, and xylenes (BTX) and the tars are bitumen-like viscous mixtures that contain high proportions of polycondensed aromatics. For the most part, the pyrolysis tars and oils are not suitable final fuel products. Often they are unstable and, when warmed, they polymerize and become more viscous. Ash and mineral matter are removed in pyrolysis, which increases the heating value, but sulfur and nitrogen are not completely removed. A more stable and useful product is obtained by hydrogenating and by removing the sulfur and nitrogen in the fuel as hydrogen sulfide and ammonia. These procedures are, as noted previously, similar to the various refinery procedures used to upgrade natural crude oils.

The chemically formed water vapor that distills off during pyrolysis in an inert atmosphere can be expected to correlate with the oxygen content of the coal. Dimensional considerations suggest that the mass fraction of water should be a function of the oxygen-to-carbon ratio. From a linear least-squares fit of the 500°C assay data on 130 coals reported in Ref. 13, we find with a correlation coefficient of about 0.9 that

$$H_2O \text{ (mass \%)} = 19.7\frac{O}{C} + 1.64 \tag{3.9}$$

where H_2O is the water expressed in mass percent of the daf coal, and O/C is the oxygen-to-carbon atom ratio of the coal. The water vapor fraction, like the tar fraction, remains approximately constant up to 1000°C. Assuming the carbon content of the daf coal to be between 70 and 80 percent, and with 18 g of water produced for every 16 g of oxygen, it may be seen from Eq. (3.9) that about one-third of the oxygen originally in the coal is consumed in forming water.

Compositions of the noncondensable pyrolysis gases from various carbonaceous materials assayed at 500°C are shown in Figure 3.3. The raw materials are the same as those of Figure 3.2. At this temperature the principal non-inert gaseous component for all coals is methane, with the fraction larger the higher the coal rank. On the other hand, total carbon oxides decrease with increasing rank, consistent with the decreasing oxygen-to-carbon ratio (see Table 1.6). Illuminants are the unsaturated hydrocarbons including acetylenes, olefins, and aromatics, and their quantities are never very large.

The gaseous yields, and particularly the relative amounts of the constituents, are strongly affected by the assay pyrolysis temperature and to a lesser extent by the coal rank. The variation in gas composition for a Wyoming subbituminous B coal at temperatures up to 1000°C is given in Figure 3.4. The trend of gaseous product yield and distribution is generally similar for most coals. The increase in gas yield at higher temperatures is mainly a consequence of char decomposition in the form of a dehydrogenation reaction, resulting in a concomitant increase in hydrogen yield. The decrease in methane and carbon oxides is a consequence of the higher temperature equilibrium distribution of the gaseous products.

Above about 1500°C acetylene, C_2H_2, is the only thermodynamically stable hydrocarbon, its stability decreasing rapidly below about 1200°C (see Figure 2.3a). Acetylene is the principal pyrolysis product in high temperature arc or plasma pyrolysis.

The discussion so far on the yields and composition of coal pyrolysis products has not concerned itself with the rate of volatiles production, beyond the restriction to "slow pyrolysis" in the sense defined earlier. The kinetics of the thermal decomposition of coal have been the subject of much attention and controversy.[2,6] Differences arising from the chemical kinetics, reactor characteristics, and flow and heating factors have sometimes been difficult to distinguish.

We have already observed that slow pyrolysis yields depend on the temperature-time history, as distinct from fast pyrolysis where the products are the result of an essentially isothermal decomposition and only the chemical times are important.

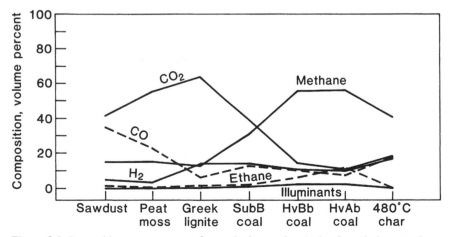

Figure 3.3 Compositions on an oxygen-free and nitrogen-free basis of pyrolysis gases from various carbonaceous materials assayed at 500°C.[9]

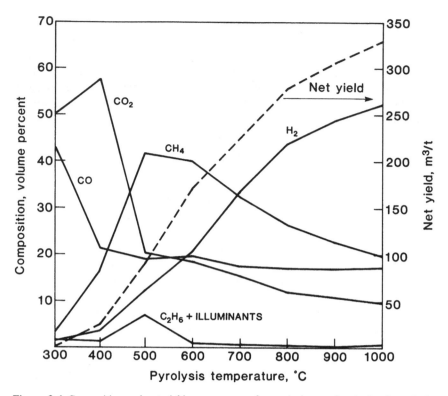

Figure 3.4 Composition and net yield on an oxygen-free and nitrogen-free basis of pyrolysis gas from Wyoming subbituminous B coal assayed at different temperatures.[14]

Indeed, if a slow pyrolysis process is arrested at any point and the temperature subsequently held constant, the coal would asymptotically approach an equilibrium mass dependent on the temperature and the coal. Further devolatilization occurs only if the temperature is raised. The more general chemical kinetic nature of the pyrolysis process may be seen in the fact that the total volatile matter and tar produced depend, among other things, on the hydrogen-to-carbon ratio in the coal, which changes as devolatilization proceeds. The low temperature tar initially distilled will bear the closest resemblance to the original coal. Subsequently produced tars and oils should bear less resemblance. Moreover, if the primary pyrolysis products are not removed on a time scale short compared to the time for secondary gas phase and gas-solid phase reactions to take place, this will introduce yet another chemical history effect.

Pyrolysis kinetics is clearly many sided and it is perhaps surprising that the thermal decomposition of coal is representable by the simple pseudo first-order reaction of Eq. (3.1), wherein the rate of volatiles production is proportional to the remaining volatile fraction of the undecomposed coal. However, the difficulties with this simple reaction picture may be seen in the fact that at the same temperature, rate differences as large as four orders of magnitude are observed for coals of widely different rank, while differences between slow and fast pyrolysis experiments lead to as much as two orders of magnitude difference in rates.[7] The apparent activation energies range from about 8 to 210 kJ/mol.[1] More general models have considered coupled first-order reactions with different rate parameters for char, tar, and gas,[6,11] while still another model assumes a large series of parallel reactions all with the same rate constants, but different activation energies.[1]

Evidently much work remains to achieve a fundamental understanding of coal pyrolysis. Our view for the purposes of the book is, however, more pragmatic and concerned with the information needed for understanding the design and performance of synthetic fuel plants. With this in mind, we adopt the view that adequate empirical information exists on the yields and product composition from slow pyrolysis, and that this data base may be used to compare the higher volatile yields and higher gas and liquid yields observed in practice in both laboratory and pilot plant fast pyrolysis reactors.

That fast pyrolysis rates can result in greater yields and more favorable product distributions is of great importance for synthetic fuel plant design, although a note of caution is in order. Fluidized beds, entrained flow reactors, free-fall reactors, and flash heating, among other methods, are necessary to achieve fast pyrolysis. The products must also be rapidly quenched. All of these procedures use finely ground coal, generally less than 200 μm in size, with data indicating that sizes of 50 to 100 μm are required to achieve uniform temperature distributions in the coal particles at high heating rates. This involves a considerable processing expense. The extra expense may be made up in part by smaller reactor volumes, since the residence times of the particles will be short, typically less than 50 to 500 milliseconds for very rapid heating. Generally, such fast pyrolysis equipment will be more complex.

Relative yields and product distributions for coal pyrolysis in an inert atmosphere are shown in Figure 3.5 as a function of temperature and time. The regions of slow and fast pyrolysis are defined by the limits of Eqs. (3.5) and (3.6). This picture is taken in part from Ref. 6, but differs importantly from the one there in that no dependence on heating rate *per se* is indicated. Moreover in Ref. 6, four classifications of pyrolysis heating rate are defined, whereas we have stressed that there are two limiting pyrolysis rates, slow and fast. We have arbitrarily classified the range in between as "intermediate." In the figure, the relative yield and total yield of volatiles are shown qualitatively along with the principal gaseous component in the given temperature range. We assume that the solid is held at the final temperature until decomposition is complete, while the vapor residence time is kept to a minimum, consistent with the overall pyrolysis time.

What is seen from Figure 3.5 is that decreasing the heating time to reach the assay temperature of 500°C will increase both the total yield as well as the ratio of tar-to-gaseous fractions. At 1000°C, decreasing the heating time also increases the total yield, but the ratio of tar-to-gaseous fractions decreases. In the fast pyrolysis limit of short residence time and high heating rate the pyrolysis reactions essentially take place isothermally. At the lower temperature, the formation of liquids is thermodynamically favored, while at the higher temperatures the formation of gas is thermodynamically favored. Although this explains the difference in relative compositions in the two

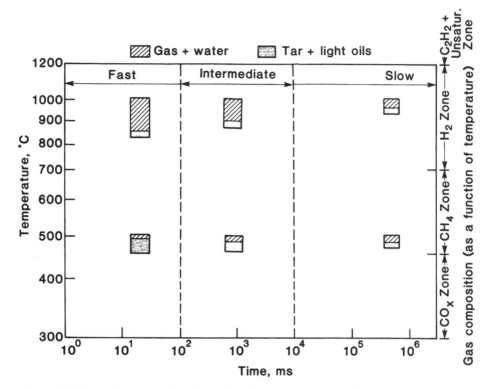

Figure 3.5 Relative yields and product distributions for coal pyrolysis in an inert atmosphere as a function of temperature and time.

temperature ranges, it does not explain why the liquid and total volatile yields are higher for fast pyrolysis than for slow pyrolysis.

One qualitative explanation for the difference between slow and fast pyrolysis yields relates to secondary decomposition and repolymerization reactions. In a time-dependent non-isothermal pyrolysis, the decomposition occurs over the entire temperature range. When the pyrolysis is slow enough, the evolved tars have time to undergo secondary reactions producing coke and gas. This explains why at 500°C there is more gas in the yield from slow pyrolysis than in fast pyrolysis. That the tar yield is higher at 500°C for the fast pyrolysis follows from the reverse of the argument, that there is insufficient time to repolymerize some of the tar to form coke, resulting in a higher liquid yield. This picture is also consistent with the fact that the fast pyrolysis tars are generally found to have a higher density than the tars from slow pyrolysis. We have observed that the tar composition is proportional to the coal hydrogen-to-carbon ratio, however, this ratio is constant for the coal so that the greater the tar yield the lower its hydrogen-to-carbon ratio and hence the higher its density.

The same arguments may be extended to the higher temperature of 1000°C illustrated in Figure 3.5. The higher temperature results in additional gas evolution due to char decomposition. However, in slow pyrolysis, coking reactions most probably lead to deposition in the char matrix, which tends to reduce this gas evolution. On the other hand, with fast pyrolysis these secondary reactions are less important so that the gas evolution is relatively larger.

Although the argument we have presented is self-consistent, other models could be devised and there remain, moreover, questions concerning whether or not the observed yield increases are in fact attributable to heating rate. It is interesting to conjecture that since devolatilization takes place isothermally for fast pyrolysis, as distinct from a series of decomposition and repolymerization reactions in slow pyrolysis, that on thermodynamic grounds the heating value of the product per unit of heat input should be larger for the fast pyrolysis volatiles than the slow pyrolysis volatiles.

A preponderance of data supports the idea that short residence times (fast heating rates) can increase volatile yields, as illustrated in Figure 3.5. The yields do appear to be directly proportional to the hydrogen-to-carbon ratio of the coal. In the range 500 to 600°C, tar yields over 50 percent higher than slow pyrolysis assay yields are reported for fast pyrolysis, while at 1000°C, total yields up to 40 percent in excess of proximate volatile matter are indicated.

Oil Shale

Oil shale is composed mainly of inorganic mineral matter, so there is little economic incentive to grind it down for the purpose of rapidly pyrolyzing it. Interest in oil shale pyrolysis is therefore in slow pyrolysis and we may expect that different retorting procedures, as manifested by different temperature-time programs, will result in different yields and product compositions.

In Section 1.3, the oil yield from the pyrolysis of shale was shown to be directly related to the organic matter content of the shale. When the organic matter in Green River shale, a typical composition of which is given in Table 1.9, is pyrolyzed at 500°C it yields approximately 66 percent oil, 9 percent gas, 5 percent water, and 20 percent carbon residue.[3]

Active devolatilization of oil shale begins at about 350 to 400°C, with the peak rate of oil evolution at about 425°C, and with devolatilization essentially complete in the range of 470 to 500°C. As with slow coal pyrolysis, if the temperature is arrested at a value below the final pyrolysis temperature, the volatile matter mass loss will approach a value characteristic of that temperature. A van't Hoff-type plot of the mass fraction of organic matter volatilized as a function of reciprocal temperature is shown in Figure 3.6. The points at which the slope of the curve changes may be interpreted in terms of the products evolved, the principal ones of which are indicated for the different temperature ranges. For temperatures near 500°C, the mineral matter, consisting mainly of calcium/magnesium and calcium carbonates, begins to decompose yielding carbon dioxide as the principal product.

The properties of crude shale oil evolved from the slow pyrolysis of oil shale are dependent on the retorting temperature, but more importantly on the temperature-time history because of the secondary reactions accompanying the evolution of the liquid and gaseous products. Shale oils are dark brown, odoriferous, and waxy oils that are

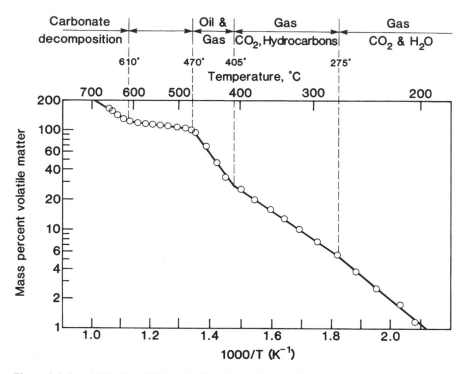

Figure 3.6 Devolatilization of Colorado oil shale showing the effect of temperature on mass loss.[15]

more viscous and have a higher pour point (congealing temperature) than typical petroleum crudes. They also contain a lot of nitrogen, about 2 mass percent, and a lot of oxygen, typically about 1.5 percent. The H/C atom ratio in the crude shale oil evolved in the Fischer assay is about 1.63 compared with 1.56 in the original organic matter. This ratio is about the same for most crude shale oils retorted near 500°C. With decreasing temperature the H/C ratio in the product oil increases but, of course, the yield decreases. This is consistent with the picture given for coal pyrolysis, in which at temperatures below 500°C the cumulative tar yield is lower and the hydrogen levels in the tar higher. In line with the hydrogen content, the shale oil is found to become denser and the pour point higher with increasing pyrolysis temperature. Because of the deleterious characteristics noted, crude shale oil is not normally considered a synthetic crude suitable as a feedstock for a conventional refinery, although it can be upgraded by refining procedures for this purpose.

The composition of the noncondensable gases evolved from the pyrolysis of oil shale, like that of the crude shale oil, depends on the retorting temperature and procedure. Shown in Table 3.2 is the gas composition from a Fischer assay of 111 L/t oil shale. Data on the effect of temperature on the gas composition show the rate of evolution of carbon dioxide increases sharply at about 500°C as a result of carbonate decomposition. The evolution of both hydrogen and methane rises sharply at about 450°C and falls equally sharply at about 480°C, with the cumulative hydrogen levels in this range about twice those of the methane.

Table 3.2 Gas composition from Fischer assay of 111 L/t oil shale[3]

Gas	Volume %
Carbon monoxide	5.5
Carbon dioxide	36.5
Hydrogen sulfide	3.7
Hydrogen	18.7
Hydrocarbons	35.6
Total	100.0
GCV, 28.2 MJ/m³	

Of more practical consequence for the composition of the noncondensable gases is whether the pyrolysis vessel or retort is directly or indirectly heated, as discussed in Section 2.4. In all assay procedures, the shale is pyrolyzed in an inert atmosphere with the heat supplied indirectly from an external source. In many retorting procedures, however, the heat is supplied directly by burning the residual carbon in the presence of air and often in the presence of steam. In these procedures the gases are not true pyrolysis products; they are diluted with nitrogen and have a low calorific value, and may be similar to gases produced by gasification.

Studies on the kinetics of the primary pyrolysis of oil shale indicate that below 500°C the kerogen (organic matter) decomposes into bitumen with subsequent decomposition into oil, gas, and carbon residue. There is some question as to whether or not the carbon residue is formed during the bitumen conversion. Unlike coal pyrolysis, where equilibrium is reached in the order of a second, with oil shale the overall decomposition rates are measured in minutes. This can be ascribed to the longer time to heat the organic material which is dispersed throughout the mineral matrix, and to the increased resistance to the outward diffusion of the products by the matrix which does not decompose. Figure 3.7 shows for a 111 L/t Colorado oil shale the rate of disappearance of kerogen and the rate of formation of oil, gas, and carbon residue on the spent shale at 475°C. The heating rates, although varied in many of the kinetic studies, are characteristic of slow pyrolysis assay conditions with values on the order of 0.1 °C/s or less.

From the practical standpoint of oil shale retorting, the rate of oil production is the important aspect of kerogen decomposition. As in coal pyrolysis, the volume of oil generated has been shown by a number of authors to be closely represented by the pseudo overall first-order reaction of Eq. (3.1), where the volatile production V is here interpreted as the volume of oil produced per unit mass of oil shale pyrolyzed. The total volume of oil produced per unit mass of oil shale is V_∞. For the slow pyrolysis we have been discussing, Shih and Sohn[17] find the first-order rate constant in inverse seconds to be expressable by

$$k = 8.25 \times 10^{11} \exp\left(-\frac{24\,000}{T}\right) \quad \text{s}^{-1} \quad (3.10)$$

where T is in kelvins.

Care must be exercised when applying laboratory determined rate expressions,

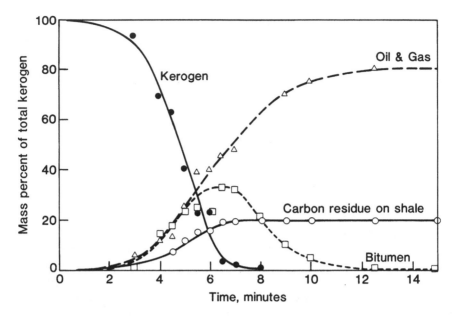

Figure 3.7 Data for a 111 L/t Colorado oil shale on the rate of disappearance of kerogen and the rate of formation of oil, gas, and carbon residue on spent shale at 475°C.[16]

such as the one given above, because of differences which can result from differing temperature-time histories. This is particularly so if the temperatures exceed 500°C and competing secondary cracking and coking reactions become important. For example, as discussed for coal, high liquid yields will be obtained by holding down the residence time of the oil in the liquid phase to minimize coking. On the other hand, higher average temperatures will result in higher gas yields. Further, oil loss by coking will yield a product with a higher H/C ratio, while cracking to the gas phase will decrease the H/C ratio. In addition, there are effects of heat transfer limitations associated with shale size, wherein the larger sizes heat up more slowly than the smaller ones. The resulting non-uniform temperature distributions can lead to different products formed during the pyrolysis reactions. This is an important consideration in practical oil shale retorting, in which the shale may be relatively small-size granules or widely distributed large blocks of material, as in *in situ* retorting.

Tar Sands

The bitumen that makes up the organic component of tar sands is of the petroleum family and when separated from the sand its decomposition by pyrolysis is similar to the cracking of petroleum crude. Slow thermal pyrolysis of Athabasca bitumen at 600°C produces an ultimate yield of about 77 mass percent liquids, 16 percent coke, and 7 percent gaseous material. The corresponding values for a typical Asphalt Ridge tar sand bitumen from the Uinta Basin of Utah are 83, 12, and 5 percent, respectively, and for a typical bitumen from the Tar Sand Triangle of southeastern Utah 73, 22, and 5 percent. The higher liquid yields are associated with bitumens of higher hydrogen content. These yields depend, of course, on the pyrolysis temperature and fall off sharply below about 425°C, although the product quality is not much dependent on the temperature.[4] The distillate is mainly paraffinic and the hydrocarbon gases are predominantly C_3 and C_4 compounds. The sulfur levels in the crude may be too high to be acceptable for conventional refining, and hydrocracking may be required.

The fraction of coke and distribution of products also depend on the pyrolysis procedure. Under conditions of flash pyrolysis, higher synthetic crude yields are reported than for equivalent slow pyrolysis procedures at the same temperature. Tar sands may also be pyrolyzed without first separating the bitumen and this will influence the product composition depending on the procedure, although the yields are generally comparable with those cited above.

3.2 GASIFICATION

Gasification is the conversion of a solid or liquid into a gas. In a broad sense it includes evaporation by heating, although the term is generally reserved for processes involving chemical change. Also in a broad sense, coal gasification refers to the overall process of converting coal to a product gas, including the initial pyrolysis and subsequent gas upgrading steps. In this section the term is applied more narrowly to the gas phase and gas-solid chemical reactions that occur at elevated temperatures within the gasifier itself.

As shown in the simplified representation, Figure 3.8, coal entering a gasifier is first dried by rising hot gases. Further heating results in the devolatilization and pyrolysis reactions discussed in the previous section. In the next stage, temperatures are sufficiently high for the gasification reactions to proceed. In discussing these reactions we shall for convenience and clarity of presentation represent coal by pure carbon (graphite). This is a reasonable assumption, since the coal that is fed to the

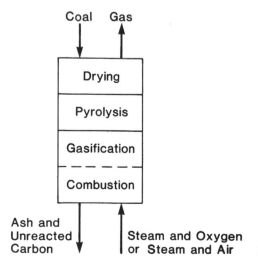

Figure 3.8 Schematic of moving bed gasifier.[18]

gasifier is not what is reacted directly; rather it is the carbonaceous char product of the pyrolysis. The assumption that coal behaves as graphite is examined briefly at the end of the section. The principal gasification reactions are:

1. *Oxygenolysis*, or reaction with oxygen to form carbon monoxide and carbon dioxide. This is actually controlled combustion in an oxygen-depleted atmosphere. Much of the oxygen fed to the gasifier, either as pure oxygen or in air, is used up in the initial combustion zone that provides the heat necessary for conversion:

$$C + O_2 \rightarrow CO_2 \qquad \Delta H^\circ_{298} = -393.5 \text{ kJ/mol} \qquad (3.11)$$

In the oxygen-depleted gasification zone, the coal may "burn" in the CO_2 atmosphere according to the Boudouard reaction

$$C + CO_2 \rightleftharpoons 2CO \qquad \Delta H^\circ_{298} = +172.5 \text{ kJ/mol} \qquad (3.12)$$

Although the heat of reaction is dependent on temperature, this dependence is not large and we have chosen to present heats of reaction at 298 K for comparison purposes. Later in analyzing specific reactor conditions, the values appropriate to the prevailing temperatures are used (see also Tables 2.14 and 2.15).

2. *Hydrogenolysis*, or reaction with hydrogen to form methane

$$C + 2H_2 \rightleftharpoons CH_4 \qquad \Delta H^\circ_{298} = -74.8 \text{ kJ/mol} \qquad (3.13)$$

Hydrogen also reacts with the nitrogen and sulfur in the coal to form ammonia and hydrogen sulfide gases. Since the hydrogen-to-carbon ratios in the product gases are higher than in coal, hydrogen must be added to the system. Hydrogen is generally added in the form of water.

3. *Hydrolysis*, or reaction with water. A typical hydrolysis reaction under gasifier conditions is the endothermic formation of carbon monoxide and hydrogen

$$C + H_2O \rightleftharpoons CO + H_2 \qquad \Delta H^\circ_{298} = +131.3 \text{ kJ/mol} \qquad (3.14)$$

The water, or more specifically, the steam fed to the gasifier provides oxygen as well as hydrogen and under some conditions may react to form carbon dioxide directly

$$C + 2H_2O \rightleftharpoons CO_2 + 2H_2 \qquad \Delta H^\circ_{298} = +90.1 \text{ kJ/mol} \qquad (3.15)$$

or

$$C + H_2O \rightleftharpoons \tfrac{1}{2}CO_2 + \tfrac{1}{2}CH_4 \qquad \Delta H^\circ_{298} = +7.7 \text{ kJ/mol} \qquad (3.16)$$

While both these reactions oxidize carbon to carbon dioxide, they are endothermic and require a heat input. Consequently, oxygen added to the gasifier to provide the energy for the conversion should not be combined as water.

Relation Between Product Gas and Gasifier Conditions

It is not difficult to write down a plethora of reactions of the above type between carbon, hydrogen, oxygen, and water and so be faced with a confusing picture of gasification. Unlike pyrolysis, however, the gasification process can be roughly represented by equilibrium conditions, and information such as presented in Figures 2.2 and 2.3 may be used to determine those species likely to be present in significant quantities. A limited number of reactions can then be selected and used to determine gasifier characteristics. A characteristic we are specifically interested in here is the degree to which projduct gas composition can be controlled by varying the gasifier temperature, pressure, and feed composition. A method for determining the product gas composition will now be developed.

Gasifiers are commonly operated in the temperature range 800 to 1500 K, so that in accordance with Figure 2.3 the species most likely to be present are carbon monoxide, carbon dioxide, hydrogen, methane, and water. In selecting the representative chemical equations, it is convenient to start with the relevant formation reactions

$$C + \tfrac{1}{2}O_2 \rightleftharpoons CO \tag{3.17}$$

$$C + O_2 \rightarrow CO_2 \tag{3.11}$$

$$C + 2H_2 \rightleftharpoons CH_4 \tag{3.13}$$

$$H_2 + \tfrac{1}{2}O_2 \rightleftharpoons H_2O \tag{3.18}$$

The standard state for both hydrogen and oxygen at these relatively low temperatures is the molecular state. At higher temperatures it would have been necessary to write the formation reactions in terms of the atomic states, and to include the formation reactions of molecular oxygen and hydrogen from their respective atoms.

As oxygen is not present in the equilibrium mixture, it is eliminated from the above equations leaving

$$C + CO_2 \rightleftharpoons 2CO \tag{3.12}$$

$$C + 2H_2 \rightleftharpoons CH_4 \tag{3.13}$$

$$C + H_2O \rightleftharpoons CO + H_2 \tag{3.14}$$

The procedure of writing down the formation reactions and eliminating those species not present results in the least number of independent equations sufficient for describing the stoichiometry of the system under consideration. These equations will generally not be sufficient to determine the system kinetics. However, their corresponding equilibrium equations and the elemental mass balance equations may be solved to determine the equilibrium gas compositions for any conditions of temperature and pressure.

To illustrate the procedure for the mass balance and equilibrium calculations without undue complexity, we will specify a simplifying yet appropriate condition, namely that the carbon in the coal is completely gasified. This requires that the amount of steam and oxygen fed to the gasifier is sufficient for complete conversion of the carbon. In this special case, by eliminating carbon we can reduce the number of independent equations to two

I
$$CH_4 + H_2O \rightleftharpoons CO + 3H_2 \tag{3.19}$$

II
$$CO_2 + CH_4 \rightleftharpoons 2CO + 2H_2 \tag{3.20}$$

The corresponding equilibrium equations are

$$K_I = \frac{(y_{CO})(y_{H_2})^3 p^2}{(y_{CH_4})(y_{H_2O})} \tag{3.21}$$

and

$$K_{II} = \frac{(y_{CO})^2(y_{H_2})^2 p^2}{(y_{CO_2})(y_{CH_4})} \tag{3.22}$$

where the y_i are the mole fractions and p is the total pressure of the system. Since the chemistry can be expressed in terms of the formation reactions, it is clear that the equilibrium constant K can also be written in terms of those reactions (see, e.g., footnote Table 2.15b). In this example the relevant equilibrium constants are

$$K_I = \frac{K_{CO}}{K_{CH_4} \cdot K_{H_2O}} \tag{3.23}$$

and

$$K_{II} = \frac{K_{CO}^2}{K_{CO_2} \cdot K_{CH_4}} \tag{3.24}$$

where the K_i are the equilibrium constants for the subscripted formation reactions. The data in Table 2.14 can consequently be used to evaluate K_I and K_{II} for any temperature in the range of interest.

To completely specify the system, we introduce the elemental mass balance equations in which the number of moles of each element in the equilibrium mixture is equated with the amount originally present in the gasifier feed.

For H_2: $\qquad (2y_{CH_4} + y_{H_2} + y_{H_2O}) n = n_{H_2,F} \tag{3.25}$

For O_2: $\qquad (\tfrac{1}{2}y_{CO} + y_{CO_2} + \tfrac{1}{2}y_{H_2O}) n = n_{O_2,F} \tag{3.26}$

For C: $\qquad (y_{CH_4} + y_{CO} + y_{CO_2}) n = n_{C,F} \tag{3.27}$

where $n_{H_2,F}$, $n_{O_2,F}$, and $n_{C,F}$ are the number of moles of hydrogen, oxygen, and carbon fed to the gasifier and n is the total number of moles in the equilibrium gas mixture. One more equation is required:

$$y_{CO} + y_{CO_2} + y_{CH_4} + y_{H_2} + y_{H_2O} = 1 \tag{3.28}$$

which states that the sum of the mole fractions is unity. Specifying the gasifier temperature fixes the equilibrium constants K_I and K_{II}. We may further specify the gasifier pressure as well as the feed rates $n_{C,F}$, $n_{O_2,F}$, $n_{H_2,F}$. This leaves us with six unknowns, namely the five equilibrium gas compositions, y_{CO}, y_{CO_2}, y_{H_2}, y_{CH_4}, and y_{H_2O}, as well as n, the total number of moles in the equilibrium gas mixture. Values for these six unknowns may be obtained for the specified feed rates and reactor conditions using the six equations: Eqs. (3.21) and (3.22), and Eqs. (3.25) to (3.28).

The elemental mass balance equations (3.25) to (3.27) and Eq. (3.28) can be used to obtain expressions for the mole fractions in terms of the carbon dioxide mole fraction and n:

$$y_{CO} = \frac{n_{C,F} - n_{H_2,F} + n}{2n} - y_{CO_2} \tag{3.29}$$

$$y_{CH_4} = \frac{n_{H_2,F} + n_{C,F} - n}{2n} \tag{3.30}$$

$$y_{H_2} = \frac{3n - 4n_{O_2,F} - n_{C,F} - n_{H_2,F}}{2n} + y_{CO_2} \tag{3.31}$$

$$y_{H_2O} = \frac{4n_{O_2,F} + n_{H_2,F} - n_{C,F} - n}{2n} - y_{CO_2} \tag{3.32}$$

Substituting Eqs. (3.29) to (3.32) in the quotient of Eqs. (3.21) and (3.22) yields a quadratic equation in y_{CO_2} and n. Finally, substitution of the mole fractions into Eq. (3.21) results in an implicit equation for n that may be solved using the "zeroes of functions" routines available on many pocket calculators:

$$K_1(n_{H_2,F} + n_{C,F} - n)[4n_{O_2,F} + n_{H_2,F} - n_{C,F} - n(1 + 2y_{CO_2})]$$

$$- \frac{p^2[n_{C,F} - n_{H_2,F} + n(1 - 2y_{CO_2})][n(3 + 2y_{CO_2}) - n_{C,F} - 4n_{O_2,F} - n_{H_2,F}]^3}{4n^2} = 0 \tag{3.33}$$

where

$$y_{CO_2} = \frac{b - \{b^2 - 16n^2(K_1 - K_{II})K_1[4n_{O_2,F} + n_{H_2,F} - n_{C,F} - n][n_{C,F} - n_{H_2,F} + n]\}^{1/2}}{8n^2(K_1 - K_{II})} \tag{3.34}$$

and

$$b = 8nK_1 n_{O_2,F} + 2nK_{II}(3n - n_{C,F} - 4n_{O_2,F} - n_{H_2,F}) \tag{3.35}$$

In order that the reader may check the use of Eq. (3.33), we consider the case in which the feed to the gasifier at 1000 K and 1 atmosphere is 0.5 moles steam and 0.25 moles oxygen per mole of carbon. In this system $n_{C,F} = 1.0$, $n_{H_2,F} = 0.5$, and $n_{O_2,F} = 0.50$, and using $K_1 = 25.82$ and $K_{II} = 19.41$, results in a value of $n = 1.378$ with $y_{CO} = 0.648$, $y_{CO_2} = 0.034$, $y_{H_2} = 0.264$, $y_{CH_4} = 0.044$, and $y_{H_2O} = 0.010$.

Constraints on Feed

In accordance with the development leading to Eq. (3.33), there must be neither residual oxygen nor residual carbon in the system at equilibrium. The feed rates selected for use with Eq. (3.33) must therefore conform to this imposed stoichiometry. The relative feed rates of steam, oxygen, and carbon to the gasifier consistent with this constraint can most instructively be examined by means of a triangular diagram.[18] In the diagram shown in Figure 3.9, the apices of the triangle are taken to represent pure carbon, oxygen, and steam. Points within the triangle represent compositions of

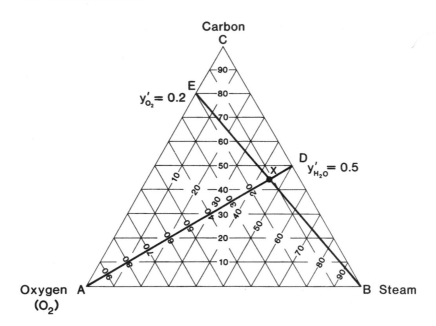

Figure 3.9 Triangular diagram for representing ternary compositions.

ternary mixtures of the three substances, while the sides represent binary mixtures. The side BC, for example, represents a binary carbon-steam system with no oxygen present, and AB and AC similarly represent oxygen-steam and oxygen-carbon mixtures. The distance along a side gives the relative amounts of the two substances present in the usual way. The midpoint D on BC represents a 50 percent mixture of carbon and steam, so the mole fraction of steam in this binary mixture is $y_{H_2O} = 0.5$. Writing the mole fraction in terms of the number of moles n present, according to Eq. (3) in Table 2.13, $y_{H_2O} = n_{H_2O}/(n_{H_2O} + n_C)$, leads to an expression for the relative number of moles present:

$$\frac{n_{H_2O}}{n_C} = \frac{y_{H_2O}}{1 - y_{H_2O}}$$
$$= \frac{0.5}{1 - 0.5} = 1$$

Similarly, the point E on line AC represents a 20 percent mixture of oxygen in carbon, or $y_{O_2} = 0.2$, and using the above procedure, $n_O/n_C = 0.25$.

A most useful property of the triangular diagram is that a line drawn to an apex represents a constant ratio of two substances in the ternary system. Line AD, for example, is the locus of all ternary mixtures containing equal quantities of steam and carbon. Point X consequently represents a mixture with $n_{H_2O} = n_C$ and $n_{O_2} = 0.25 \, n_C$, or in terms of mole fractions, $y_{H_2O} = y_C = 0.444$ and $y_{O_2} = 0.111$ in the ternary mixture. This property allows the composition of any point in the diagram to be established readily by extending lines to the sides and apices and reading off the binary mole fractions from the graduated scales.

We are now in a position to delineate those feed compositions falling within the constraints imposed in deriving Eq. (3.33). One constraint is that there be no free oxygen in the equilibrium mixture. The maximum oxygen utilization occurs when the carbon burns to CO_2, consuming one mole of oxygen per mole of carbon. The corresponding composition is shown as point E in Figure 3.10. The oxygen in any steam present will not appear in the gas mixture provided it stays combined, so the line EB represents a limit to the amount of oxygen allowed in the feed. Points below this line have too much oxygen and this area of the diagram has therefore been crossed out.

We further require that all the carbon be utilized. Maximum carbon consumption by oxygen occurs when the product formed is CO. Here, one mole of carbon is gasified for each half mole of oxygen, corresponding to a carbon-oxygen feed having $y_{O_2} = 0.5/1.5 = 0.33$, as shown by point F on Figure 3.10. In a steam-carbon system maximum carbon consumption occurs when[18]

$$3C + 2H_2O \rightleftharpoons 2CO + CH_4 \qquad (3.36)$$

Point D on the Figure represents the stoichiometric quantities of steam and carbon in Eq. (3.36), and points above line DF have compositions in which there is insufficient steam and oxygen to convert the carbon present.

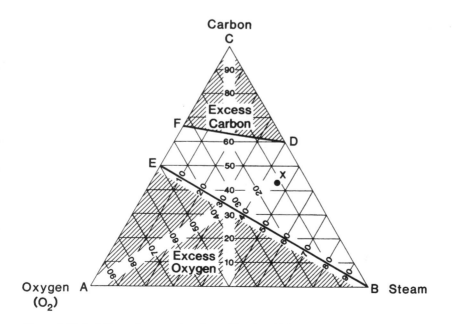

Figure 3.10 Stoichiometric constraints for gasifiers in which all the carbon and oxygen is consumed.

Product Gas Characteristics

Feed compositions satisfying the constraint of complete carbon and oxygen consumption are selected, as discussed above, to determine equilibrium gas compositions using Eq. (3.33). The results for a pressure of 1 atmosphere and a temperature of 1000 K are shown in Figure 3.11, while the effects of temperature and pressure are shown in Figure 3.12. Although it is customary to show these data in terms of concentrations, we have elected to use the total number of moles of each substance in the gas mixture per mole of carbon converted. The absolute quantity of each substance produced under the various feed and operating conditions can then be directly compared on the basis of unit quantity of carbon consumed. This cannot be done with the usual equilibrium plots of concentration data, as the total number of moles in the system is *not constant*.

Figure 3.11 covers steam-to-carbon ratios of 0.5 to 5.0. For each steam-to-carbon ratio selected there is a limited range in the ratio of total oxygen to carbon (abscissa in Figure 3.11) that needs to be considered. It follows from Eq. (3.36) that for steam-to-carbon ratios less than 0.67, oxygen must be added to the system if all the carbon is to be consumed. The minimum oxygen required can be obtained from line DF in Figure 3.10. The maximum oxygen that can be consumed sets the upper bound for each steam-to-carbon ratio in Figure 3.11, and can be obtained from line BE in Figure 3.10.

The essential characteristics of Figures 3.11 and 3.12 may be summarized as follows:

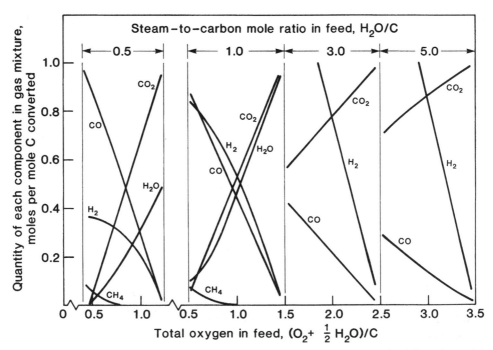

Figure 3.11 Equilibrium gas compositions at 1 atmosphere and 1000 K. Data calculated for a range of gasifier feed conditions for which all oxygen and carbon is reacted.

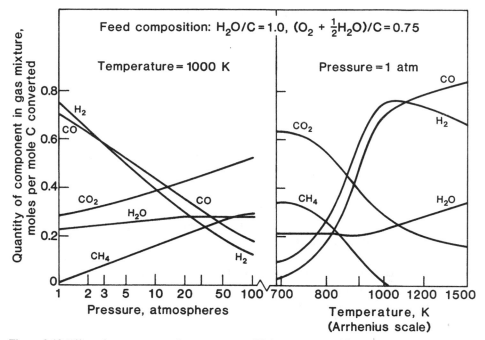

Figure 3.12 Effect of temperature and pressure on equilibrium gas composition.

1. *Hydrogen and carbon monoxide* production increases with decreasing oxygen in the feed, with decreasing pressure, and with increasing temperature. As might be expected, hydrogen increases and carbon monoxide decreases with increasing steam rates. This effect diminishes at high steam rates and there appears to be little incentive for selecting steam rates above two to three moles per mole of carbon. The extra water passes through the gasifier unchanged and much of it is condensed in scrubbing the raw gas leaving the gasifier. Additional steam must then be added in a "shift" converter in which the hydrogen-to-carbon monoxide ratio is increased by oxidizing some of the carbon monoxide following the reaction

$$CO + H_2O \rightleftharpoons CO_2 + H_2 \qquad (3.37)$$

The hydrogen-to-carbon monoxide ratio is seen to be strongly dependent on feed composition, although it is not much affected by temperature and pressure. A gasifier producing a synthesis gas for methanol production would ideally operate with a H_2/CO ratio of 2, and even higher for methane production. However, if methane is the desired end product, it may be preferable to produce it directly in the gasifier (see Section 5.5).

2. *Methane production* is favored by low temperatures and high pressures. Variations in feed conditions do not significantly affect its formation.

3. *Carbon dioxide formation*, as expected, increases with increasing oxygen in the feed. It also increases with increasing pressure. However, at high temperatures carbon dioxide formation is suppressed in favor of carbon monoxide.

In summary we can say that high temperature gasification at low pressures produces mainly a synthesis gas and is governed by the equation

$$C + H_2O \rightleftharpoons CO + H_2 \tag{3.14}$$

while low temperature-high pressure operation will increase methane production according to

$$C + H_2O \rightleftharpoons \tfrac{1}{2}CO_2 + \tfrac{1}{2}CH_4 \tag{3.16}$$

The underlying assumptions to the foregoing discussion are that the prevailing kinetics do in fact allow equilibrium to be approached, and that the required energy input and temperature control are attainable for all feed conditions considered. Rate equations for individual gasification reactions have been reported,[20,21] while a method of correlating the overall gasification rate for a wide variety of coals has been developed.[22] Some relative gasification rates are summarized in Table 3.3 and show that the rates of the reactions producing carbon monoxide and hydrogen are of the same order of magnitude. Gasifiers will typically be designed to produce these substances at concentrations approaching their equilibrium values. Oxygen is consumed extremely rapidly and justifies our assumption, within the applied stoichiometric constraints, that free oxygen will not be present. Methane formation is slow, however, and concentrations may therefore be even less than indicated in Figures 3.11 and 3.12. Increasing the gasifier temperature to accelerate its reaction rate is not useful in view of the accompanying decrease in equilibrium yield. The use of catalysts to enhance methane production is being investigated, since its direct formation at low temperatures within the gasifier is more economical than having to first produce carbon monoxide and hydrogen and then synthesize methane in a separate process step. Catalytic gasification is discussed in Section 4.3.

The heat of the overall gasification reaction, ΔH_T°, is given by the sum of the heats of formation of the components in the equilibrium gas mixture

$$\Delta H_T^\circ = n_{CO}\Delta H_{CO}^\circ + n_{CO_2}\Delta H_{CO_2}^\circ + n_{CH_4}\Delta H_{CH_4}^\circ + (n_{H_2O} - n_{H_2O,F})\Delta H_{H_2O}^\circ$$

$$\text{kJ/mol carbon converted} \tag{3.38}$$

where ΔH_i is written as a shorthand notation for $(\Delta_f H_T^\circ)_i$. The heats of formation at thereaction temperature may be obtained using the data given in Table 2.14. Heats of reaction calculated using Eq. (3.38) for the feed conditions examined in Figure 3.11 are in general exothermic, as shown in Figure 3.13. Part of the heat evolved goes into

Table 3.3 Relative gasification rates at 10 kPa and 800°C[19]

Reaction	Relative Rate
$C + O_2$	1×10^5
$C + H_2O$	3
$C + CO_2$	1
$C + H_2$	1×10^{-3}

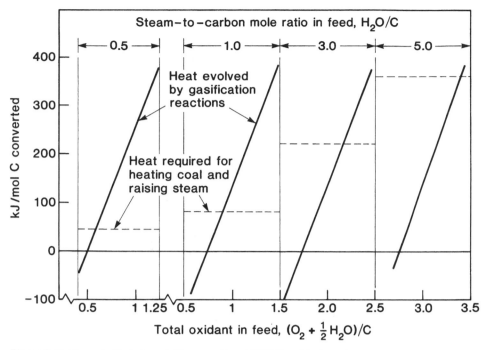

Figure 3.13 Thermal effects in a gasifier operating at 1000 K and 1 atmosphere.

heating and drying the coal feed, while any excess heat may be used for water preheating or steam raising in the gasifier jacket. Assuming the coal to be externally dried and to have a mean specific heat based on that of carbon of 16.6 J/(mol·K), then the energy required for heating the coal is approximately

$$16.6 \times (1000 - 298) = 11.7 \quad \text{kJ/mol carbon}$$

Raising steam and heating it to the reaction temperature of 1000 K requires a further 69.7 kJ/mol steam. These values were used to estimate the energy required for coal heating and steam raising for the feed conditions in Figure 3.13, and have been included in the figure as dashed horizontal lines.

While there are other energy requirements, the dashed lines in Figure 3.13 may be taken to represent a *thermal constraint* on gasifier feed and operating conditions. The heat evolved in excess of these values has to be removed by cooling and will be wasted unless it can somehow be recovered and utilized elsewhere in the plant. The actual thermal constraint must, of course, be determined by a more detailed evaluation of process heat effects and the limitations imposed by thermal losses and practical heat exchanger surface areas. It is not unusual to operate the gasifier below the thermal constraint indicated in Figure 3.13, and to raise steam separately. This reduces the oxygen requirements in the gasifier and increases the production of useful gases relative to carbon dioxide. Another means of reducing oxygen requirements is to provide the energy for gasification by electrical heating or by use of an inert heat transfer medium.

Apart from the use of alternative heating procedures, other methods are available for increasing the formation of desired products. For example, the gasifier may be divided into several stages, each with distinct temperature and composition characteristics. A high temperature combustion stage for production of carbon monoxide and hydrogen may be followed by a lower temperature stage with introduction of steam, or even hydrogen, to promote methane production. The Hygas gasifier (discussed in Chapter 4) has two hydrogasification stages in which the coal reacts with a hydrogen-rich gas to produce methane. The hydrogen-rich gas is produced in a final steam/oxygen-gasification stage. The type of reactor, whether cocurrent or countercurrent, fixed or fluidized bed, has a strong bearing on temperature distributions and is selected to promote the product desired. For example, in a cocurrent operation the maximum temperature will occur near the gas outlet, and the product can be expected to contain little methane. In a well-mixed or fluidized bed reactor, temperatures are more uniform and can be better controlled to conditions suited for methane production.

Another method for promoting methane production according to Eq. (3.22) is to remove carbon dioxide. Calcium oxide has been used to absorb carbon dioxide within the gasifier:

$$CaO + CO_2 \rightleftharpoons CaCO_3 \qquad \Delta H^\circ_{298} = -183 \text{ kJ/mol} \qquad (3.39)$$

Apart from absorbing carbon dioxide, the lime can be used as a heat transfer medium to introduce sensible heat into the gasifier, and also to generate significant heat on reaction with the carbon dioxide. In addition, the lime removes some of the sulfur gases by formation of calcium sulfide. The product gas can therefore be expected to

(a) have relatively low concentrations of the undesirable acid gases, H_2S and CO_2;

(b) have a high H_2/CO ratio, as much of the carbon is in the species on the left hand side of Eqs. (3.19) and (3.20).

Unfortunately, further development of this attractive looking process is not being pursued.

It is not necessary nor always desirable to convert all of the carbon in the gasifier. As discussed in Chapter 1, two types of carbon with different reactivities coexist in most coals. The highly reactive carbon is rapidly reacted, while longer residence times are required for complete carbon conversion. Steam and oxygen rates may be reduced to levels sufficient for conversion of the more reactive carbon, with the unconverted portion removed from the gasifier as a char. This char may be burned in separate units to provide steam or other plant energy requirements. Removal of a carbonaceous char effectively increases the hydrogen-to-carbon ratio of the remaining coal.

When carbon is present in the gasifier, the overall equilibrium is additionally determined by the Boudouard reaction

$$C + CO_2 \rightleftharpoons 2CO \qquad (3.12)$$

which was eliminated in the foregoing equilibrium evaluation. Gruber[23] has discussed in some detail the operating regimes in which a solid carbon phase is present, as well as the implications of occurrence of a non-ideal or "Dent" carbon that behaves differently from graphite.

The assumption of coal behaving as graphite appears to be conservative, in that coal contains hydrogen and can produce more hydrogen product than graphite for the equivalent energy input. In fact, the overall gasification reaction in terms of graphite

$$C + H_2O \rightleftharpoons \tfrac{1}{2}CH_4 + \tfrac{1}{2}CO_2 \qquad \Delta H^\circ_{298} = +7.7 \text{ kJ/mol} \qquad (3.16)$$

is endothermic, and more endothermic if written in terms of coal

$$CH_{0.72} + 0.82H_2O \rightleftharpoons 0.41CO_2 + 0.59CH_4 \qquad \Delta H^\circ_{298} = +17.6 \text{ kJ/mol} \qquad (3.40)$$

This and other differences between coal and graphite are discussed by Massey.[24] The behavior of different coals in commercial-type gasifiers is considered further in Chapter 4.

3.3 GAS SHIFT AND SYNTHESIS

The raw gas produced on gasification of coal has a low-to-medium-calorific value, depending on whether air or oxygen is used as the oxidant in directly heated gasifiers. The product from indirectly heated gasifiers, in which an inert material is typically used to transfer the heat from an external source, is generally a medium-CV gas. The energy in these gases could theoretically be utilized directly, for example by generation of electricity, either by raising steam in a boiler or by combustion in a gas turbine. However, the raw coal itself could be used to raise steam for electricity production, and little, if any, benefit would have been gained by the gasification step.

We recall that there are two major reasons for producing synthetic fuels. Firstly, it is economically desirable to replenish dwindling traditional fuels such as natural gas and petroleum products, on which virtually the entire energy infrastructure of the United States is currently based. A second reason for fuel conversion is to eliminate pollutants and inert materials, and so provide a clean-burning fuel as well as a fuel that is cheaper to handle and transport than the original raw material. The gases produced on gasification, as discussed in the previous section, clearly do not satisfy the above criteria, and so further processing is necessary. Removal of pollutant gases such as ammonia and hydrogen sulfide, and inert gases such as carbon dioxide, is an obvious requirement. Gas purification processes do not involve reactions of the active gas components, and are discussed in Section 5.1. In this section we examine the reactions of carbon monoxide, hydrogen, and steam to produce substitute natural gas, gasoline, and methanol. Production of liquid fuels from coal after first completely breaking down the coal structure in a gasification step is known as *indirect liquefaction*. In direct liquefaction, the coal structure is only partially broken down by hydrogenation to the level of the desired liquid product. Direct liquefaction reactions are not related to the gaseous reactions considered here, and are presented separately in the next section.

Gas Shift

A gas in which the major active components are carbon monoxide and hydrogen is called a synthesis gas, as these two compounds can be made to combine, or synthesize, to form a myriad of products. The products formed from the synthesis gas depend both on the hydrogen-to-carbon monoxide ratio in the gas as well as on the catalyst and reactor conditions selected. Hydrogen-to-carbon monoxide mole ratios range from 0.5 to 1 for gasoline production with rejection of carbon dioxide, to 3 for methane production with rejection of water. The required hydrogen-to-carbon monoxide ratio can sometimes be achieved directly in the gasifier, although a H_2/CO ratio as high as 3 is normally not produced in commercial systems. In fact, many gasifiers produce a gas having a H_2/CO ratio less than 1. In these cases an adjustment to the H_2/CO ratio is normally required, and is done by adding steam to the synthesis gas and reacting it with the carbon monoxide to form hydrogen and carbon dioxide:

$$CO + H_2O(g) \rightleftharpoons CO_2 + H_2 \qquad \Delta H^\circ_{298} = -41.2 \, \text{kJ/mol} \qquad (3.41)$$

This is the *gas shift reaction*, also called the water-gas shift, or just shift reaction.

The need to "shift" the gas introduces an additional process step, so increasing overall process complexity. In cases where the required synthesis gas composition can be achieved directly in the gasifier, this may be preferred in the interest of process simplicity. On the other hand, it has been argued that hydrogen can be more efficiently produced in a separate reactor under controlled conditions than in the gasifier where other conflicting operating conditions prevail. In practice, these minor energy considerations are probably less relevant than factors such as process reliability and the availability of proven gasifiers for the coal type to be converted.

The gas shift reaction is moderately exothermic and equilibrium yields decrease slowly with increasing temperature. Although optimum yields are obtained at low temperatures, yields are still favorable at temperatures up to about 225°C as shown in Figure 2.2b. Pressure has no effect on yield. The reaction does not proceed appreciably under typical reactor conditions unless catalyzed, traditionally with a chromium promoted iron formulation. More recently, attention has been directed towards copper-zinc and cobalt-molybdenum catalysts which remain active at lower temperatures than do the iron catalysts.

Carbon Deposition

Before proceeding to a discussion of the specific synthesis reactions, it is instructive to first discuss a problem common to all synthesis reactors, namely carbon formation. This problem is of considerable commercial importance as it not only represents a loss of carbon, but more importantly also results in deactivation of the catalyst due to carbon deposition on active sites.

Carbon deposition is governed by the following equations:

$$C + CO_2 \rightleftharpoons 2CO \qquad \text{(Boudouard)} \qquad (3.12)$$

$$C + 2H_2 \rightleftharpoons CH_4 \qquad \text{(carbon-hydrogen)} \qquad (3.13)$$

$$C + H_2O \rightleftharpoons CO + H_2 \qquad \text{(carbon-steam or gasification)} \qquad (3.14)$$

Equilibrium constants for these reactions are shown in Figure 2.2, and a further indication for equilibrium carbon deposition is given in Figure 2.3. In these equilibrium representations the solid carbon has been taken to be graphite, whereas in practice a type of carbon known as "Dent" carbon may be deposited.[23] In the low temperature range, carbon is slightly less likely to be deposited than calculations based on graphite suggest, although by 825 to 925°C the difference between the equilibrium constants for Dent carbon and graphite formation is small.

According to the above three equations, carbon deposition can be reduced by increasing steam, hydrogen, and carbon dioxide concentrations. The relative concentration of carbon, hydrogen, and oxygen in mixtures in which solid carbon is present at equilibrium can be determined by simultaneous solution of the relevant chemical equilibrium equations. This procedure was used in determining the equilibrium compositions presented in Figure 2.3. It is convenient here to present the compositions at which solid carbon may exist in the coordinate system shown in Figure 3.14. As can be seen in the figure, no solid carbon exists at low temperatures in binary mixtures containing at least 2 atoms of oxygen or 4 atoms of hydrogen per atom of carbon. In these systems all the carbon is present either as CO_2 or CH_4. In ternary mixtures some

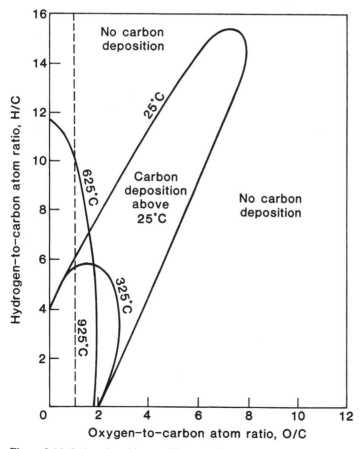

Figure 3.14 Carbon deposition equilibrium isotherms at atmospheric pressure.

of the hydrogen and oxygen combine to form water so that additional hydrogen and/ or oxygen is required to react with the carbon.

At higher temperatures less oxygen and more hydrogen are required to prevent carbon deposition. This is because at these conditions CO becomes favored thermodynamically, requiring only one atom of oxygen per atom of carbon, while CH_4 is less favored and much of the hydrogen is present as H_2. In shift reactors, temperatures are usually in the range 325 to 625°C and there is a large excess of steam, oxygen-to-carbon ratios are consequently greater than about 3, and carbon deposition is therefore not a problem.

Detailed equilibrium studies of carbon deposition[23,25] indicate that, in general, increasing the H_2/CO ratio and decreasing the temperature reduces the likelihood of carbon deposition in synthesis reactors. In practice, kinetic factors appear to dominate. For instance, the hydrogenation of carbon monoxide occurs significantly faster than its bipartition by the Boudouard reaction.[26] Carbon deposition is consequently significantly less than might be expected from equilibrium calculations. Nevertheless, it does occur in many synthesis systems and has to be controlled to avoid catalyst losses and the need for too frequent catalyst regeneration.

Synthesis

The simplest synthesis reaction is the combination of one mole of carbon monoxide with two moles of hydrogen to form methanol

$$CO + 2H_2 \rightleftharpoons CH_3OH(l) \qquad \Delta H^\circ_{298} = -128.2 \text{ kJ/mol} \qquad (3.42)$$

The stoichiometric H_2/CO ratio required for methanol synthesis according to Eq. (3.42) is 2. A wide spectrum of organic species in addition to methanol can be formed from this synthesis gas, so selectivity control is of importance. However, considerable industrial experience since the 1920s has demonstrated that commercial catalysts are highly selective and are able to yield methanol almost exclusively.

As shown in Section 2.3, equilibrium formation of methanol is favored by low temperatures and high pressures. Earlier reactors used zinc oxide-chromic oxide catalysts that required temperatures of about 400°C and pressures from 30 to 38 MPa for economical yields. More recently, copper-containing catalyst formulations have been developed that are active at lower temperatures of about 260°C, and have been used commercially at pressures down to about 5 MPa. Copper-containing catalysts are, however, extremely sensitive to sulfur poisoning, so more thorough sulfur control is required with the low pressure processes.

Methanol catalysts have the ability to convert some carbon dioxide in addition to the carbon monoxide:

$$CO_2 + 3H_2 \rightleftharpoons CH_3OH(g) + H_2O(g) \qquad \Delta H^\circ_{298} = -49.5 \text{ kJ/mol} \qquad (3.43)$$

It is assumed that this reaction proceeds via the reverse of the gas shift reaction, Eq. (3.41), and the usual methanol reaction, Eq. (3.42). The heat of reaction is smaller than for carbon monoxide conversion, so the presence of carbon dioxide can reduce thermal problems in the reactor. Complete removal of carbon dioxide from the synthesis gas is also not required. Disadvantages of carbon dioxide conversion are that it consumes more hydrogen and rejects steam which must be separated from the product.

Side reactions are not the problem they are in Fischer-Tropsch synthesis discussed below. Low temperatures and excess hydrogen tend to suppress undesirable side reactions. Although conversion yields of only a few percent are possible per pass (see Table 2.20), the absence of significant byproducts makes it easy to separate the methanol and recycle the unused reactants and still produce methanol economically. Kinetics of the methanol reaction, as well as its commercial production and use as a fuel, have been reviewed by Hagen.[27]

In forming products that do not contain oxygen, the oxygen in the carbon monoxide may be rejected either as water or as carbon dioxide. For instance in the formation of methane:

$$CO + 3H_2 \rightleftharpoons CH_4 + H_2O(g) \qquad \Delta H^\circ_{298} = -206.1 \text{ kJ/mol} \qquad (3.44)$$

$$2CO + 2H_2 \rightleftharpoons CH_4 + CO_2 \qquad \Delta H^\circ_{298} = -247.3 \text{ kJ/mol} \qquad (3.45)$$

In the first reaction a high H_2/CO ratio of 3 is required in the synthesis gas, and one-third of the hydrogen content is wasted in the rejected steam. Rejection of the oxygen with water is further associated with a heat loss, as much of the latent heat content of the steam produced is dissipated in drying the product methane. If the oxygen is rejected with carbon dioxide, as in the second reaction, there is a carbon loss from the system. The greater exothermicity of this second reaction is attributable to the greater heat loss associated with the combustion of carbon to form carbon dioxide in relation to that for hydrogen combustion to form steam. Although the quantity of heat rejected is larger, in this case it is rejected at the high temperature of the methane reactor and much of it can be recovered by raising process steam. There are additional advantages of a carbon dioxide rejection process. A second generation, high efficiency gasifier could be used, a specific gas shift step would not be required, and problems of heat recovery on drying the product would be eliminated.

In selecting between the two methanation routes, factors, including the cost of producing the synthesis gas with the desired H_2/CO ratio, the ease of separating the rejected carbon dioxide or water from the product, and the relative kinetics and catalyst costs must be considered. As mentioned previously, the use of a synthesis gas with low H_2/CO ratios invariably results in carbon deposition and the associated catalyst deactivation. Several attempts have nevertheless been made to avoid carbon deposition problems with low hydrogen synthesis gases by using slurry reactors in which greater temperature control can be exercised. This work is based on the original work of Kölbel but to date has met with only limited success.[28] The theoretically less efficient methanation route with steam rejection is consequently the basis for commercial methanation.

The methanation reaction, Eq. (3.44), is highly exothermic and its equilibrium is strongly temperature dependent, as shown in Figure 2.2b. Low temperatures and high pressures favor methane formation, although satisfactory yields can still be obtained at temperatures up to about 525°C. Nickel catalysts are almost exclusively used for methanation, and economical reaction rates are obtained at 300 to 400°C.

Other metals including ruthenium, iridium, and rhodium have greater activity but are less selective to methane and more expensive.

Carbon deposition is thermodynamically possible in the entrance region of methanators based on Eq. (3.44), but not near the reactor outlet where steam concentrations are high.[23] Catalyst formulations are available that avoid carbon deposition. Excess steam, however, cannot be used as it is formed within the reactor and at high concentrations poisons the nickel catalysts.[29] The steam also serves to control the large heat of reaction released in the reactor. Other means of temperature control are presented in Section 5.2, while the kinetics of the methanation reaction are discussed in Section 2.3.

A further synthesis gas reaction of commercial importance is the Fischer-Tropsch synthesis to produce gasoline, Eqs. (2.5) and (2.6). As most of the gasoline produced is olefinic, we will here consider only a simplified version of the olefin producing reaction

$$CO + 2H_2 \rightleftharpoons (-CH_2-)(l) + H_2O(l) \qquad \Delta H^\circ_{298} = -231.1 \text{ kJ/mol} \qquad (3.46)$$

The heat of reaction is calculated assuming the hypothetical product $(-CH_2-)$ is one-eighth of a 1-octene, C_8H_{16}, molecule. Oxygen is normally rejected as steam in Fischer-Tropsch synthesis. The reaction with CO_2 rejection, sometimes called the Kölbel reaction, is

$$2CO + H_2 \rightleftharpoons (-CH_2-)(l) + CO_2 \qquad \Delta H^\circ_{298} = -228.3 \text{ kJ/mol} \qquad (3.47)$$

Both the methane and Fischer-Tropsch reactions in which carbon dioxide is rejected are stated to occur as a combination of the usual reaction with steam rejection, followed by steam consumption by gas shift.[28] In the case of the Fischer-Tropsch process the mechanism is as follows:

$$CO + 2H_2 \rightleftharpoons (-CH_2-) + H_2O \qquad (3.46)$$

$$\underline{+ \ CO + H_2O \rightleftharpoons CO_2 + H_2} \qquad (3.41)$$

to give $\qquad\qquad 2CO + H_2 \rightleftharpoons (-CH_2-) + CO_2 \qquad (3.47)$

According to the Kölbel reaction, a synthesis gas with a H_2/CO ratio as low as 0.5 could be used directly to produce a liquid fuel. Such a process is of considerable interest in view of its greater hydrogen economy, but problems associated with carbon deposition and catalyst deactivation have yet to be resolved.[28]

The Fischer-Tropsch reactions produce a wide spectrum of oxygenated compounds, in particular, alcohols and aliphatic hydrocarbons ranging in carbon numbers from C_1–C_3 (gases) to C_{35+} (solid waxes). Much work has been directed towards determining reaction conditions and developing catalysts to improve the specificity of the synthesis within the desired product spectrum. In synthetic fuel production the desired products are olefinic hydrocarbons in the C_5 to C_{10} range. On the other hand, if the objective is to replace petroleum-derived chemicals with petrochemicals derived from coal, then the lower- and higher-boiling hydrocarbons, and the alcohols, ketones, and acids that are coproduced with the gasoline may, after some refining, be considered essential products.

Fischer-Tropsch reactors operate within a temperature range of 225 to 365°C and at pressures from 0.5 to 4 MPa. Low temperatures favor high-molar-mass compounds, while temperatures towards the upper end of the range are used for gasoline production. Further control over selectivity is obtained by choice of synthesis gas composition, reactor resistance time, and catalyst formulation.[30] Iron-, cobalt-, and nickel-based catalysts have been studied extensively and some attention has been given to other metals such as ruthenium. However, iron catalysts are least expensive and have consequently been used commercially. Key factors which control the selectivity of iron catalysts are the type of iron—sintered, fused, or precipitated—the basicity or amount of alkali present, and the amount of other promoters/impurities present. Potassium, aluminum, and silicon oxides are common promoters.[30]

When the process and catalyst are aimed at gasoline production, neither the H_2/CO ratio of the feed gas nor temperature within the upper range appear to have a significant effect on product selectivity. A strong correlation was found, however, between the carbon dioxide concentration and product selectivity, with distillate fuel yields increasing with increasing carbon dioxide partial pressure. Carbon dioxide may be introduced with the feed gas, but is also formed within the reactors by the gas shift reaction which is in equilibrium in the high temperature reactors using iron catalysts. Carbon dioxide production—and associated carbon deposition—by the Boudouard reaction (3.12) is not thought to be significant. Also the reaction of carbon dioxide to form a useful product,

$$CO_2 + 3H_2 \rightleftharpoons (\text{—CH}_2\text{—}) + 2H_2O \tag{3.48}$$

is not presented as contributing to Fischer-Tropsch synthesis.

Carbon deposition is a problem, particularly in higher temperature synthesis, and leads to swelling and attrition of the catalyst. The rate of carbon deposition has been found to increase linearly with the ratio $p_{CO}/p_{H_2}^3$. Increasing the hydrogen content of the feed gas consequently reduces carbon deposition. Increasing the reactor pressure decreases the $p_{CO}/p_{H_2}^3$ ratio for a fixed feed gas composition, and so reduces carbon deposition. Increasing the pressure also increases the carbon dioxide partial pressure, thus improving product selectivity and increasing reactor throughout.[30] At the Sasol II commercial complex where the main objective is to produce distillate fuels, fluidized bed reactors in conjunction with a fused-iron catalyst operating at the higher end of the temperature range are used. The feed gas H_2/CO ratio is close to 3, which is achieved by blending a high-hydrogen-content reformed tail gas with the purified gasifier synthesis gas. A specific shift step is not required when using gasifiers producing a synthesis gas with a H_2/CO ratio of about 2.

Product selectivity in Fischer-Tropsch processes is controlled by kinetics rather than equilibrium. At conditions prevailing in the reactors, olefins and oxygenated compounds are in fact thermodynamically unstable relative to paraffins. Primary products appear to be olefins and alcohols; increasing the residence time with further conversion to paraffins reduces alcohols but also reduces desirable olefins. Dry[30] has reviewed the kinetics and product selectivity on iron catalysts. Maximum selectivities were estimated to be about 40 percent for gasoline and 20 percent for diesel fuel. On the other hand, over 70 percent methane selectivity is possible. In South Africa, methane was an undesirable product and had to be reformed back to carbon monoxide and hydrogen, and recycled. This is an expensive process and consequently efforts were made to establish a market for a high calorific gas. In the United States, both methane and gasoline are desired products. Operation of a Fischer-Tropsch system to produce methane with up to 70 percent conversion would produce mainly marketable gasoline as a byproduct. Higher hydrocarbons are not produced significantly in systems designed for methane production.

While methanol can be readily produced from coal, its use as a motor fuel is not without problems. It is toxic, corrosive, has an affinity for water, and on a volume basis has only half the energy content of gasoline. Further processing of methanol to gasoline may therefore be desirable, and could prove to be more economical than the more direct Fischer-Tropsch route if the large production of byproducts associated with Fischer-Tropsch could be eliminated. Mobil has developed the shape-selective zeolite catalyst ZSM-5 which forms the basis of their Mobil M process for converting methanol to gasoline.[31] The catalyst is claimed to have sufficient activity and selectivity for gasoline to make the methanol route commercially attractive. It is further stated that the gasoline product has a higher octane rating than Fischer-Tropsch gasoline due to the presence of aromatics and branched olefins, so extensive refining is not required.[28]

Although the process of converting methanol to gasoline is not a synthesis gas reaction, it can be considered an extension of the methanol reaction, Eq. (3.42), and is conveniently discussed here for purposes of comparison with the Fischer-Tropsch indirect liquefaction route. The methanol need not, of course, be derived from coal, but could be from reforming of natural gas, or produced from biomass. The overall reaction from a synthesis gas is identical to that for Fischer-Tropsch synthesis

$$CO + 2H_2 \rightleftharpoons CH_3OH \rightleftharpoons (\text{---}CH_2\text{---})(l) + H_2O(g) \quad \Delta H^\circ_{298} = -187.1 \text{ kJ/mol} \quad (3.49)$$

The conversion takes place in two stages. In the first stage, methanol vapor is converted to dimethylether at about 300 to 325°C

$$2CH_3OH \rightleftharpoons CH_3OCH_3 + H_2O \qquad \Delta H^\circ_{600} = -20.6 \text{ kJ/mol} \qquad (3.50)$$

The dimethylether is then converted to gasoline at above about 350°C

$$CH_3OCH_3 \rightleftharpoons 2(\text{---}CH_2\text{---}) + H_2O \qquad \Delta H^\circ_{700} = -75.3 \text{ kJ/mol} \qquad (3.51)$$

The selectivity of the catalyst towards dimethylether or gasoline hydrocarbons is determined essentially by temperature. The ability to stage the process is convenient for controlling the heat released. The hydrocarbons produced in the high temperature reactor exhibit a relatively narrow range of molar masses terminating abruptly at about C_{10}, as constrained by the size of the catalyst pores. The initial products are light olefins which subsequently polymerize to heavier olefins, paraffins, and aromatics.

In addition to methanol, several other oxygenated feeds including higher alcohols, aldehydes, and ketones have been successfully converted to gasoline in laboratory scale ZSM-5 reactors. As these substances are typical Fischer-Tropsch byproducts, there is some interest in combining the Fischer-Tropsch and Mobil M processes.[28] Further information on the aging characteristics and product distribution with these feeds are needed to assess the economics of such a two-step process. The difference in optimum operating temperatures, 500 to 640 K for Fischer-Tropsch and 650 to 700 K for the Mobil M, probably excludes the possibility of a single stage mixed catalyst system.

Some attention has been given to the relative efficiencies and economics of the two processes considered separately. Although Fischer-Tropsch is a demonstrated process, at the time of writing there is no commercial experience with the Mobil M and so a clear-cut comparison is not possible. The success of the Mobil M will depend largely on the selectivity and long-term stability of the catalyst. If in practice, side reactions turn out to be significant and/or the product spectrum varies with catalyst age, then there will be little incentive to follow the methanol route. Otherwise it may well become the preferred route for indirect gasoline production, with Fischer-Tropsch reserved mainly for petrochemicals. In reality, the selection of either process will depend largely on local energy resources and market conditions as well as on factors such as process reliability and available technological expertise.

3.4 DIRECT LIQUEFACTION

The two principal routes for the direct hydrogenation of coal to form a liquid involve the addition of hydrogen to the coal either directly from the gas phase or from a donor solvent. These direct liquefaction procedures are hydrocracking processes applied to coal. When the hydrogen is added directly from the gas phase, it is mixed together with a slurry of pulverized coal and recycled coal-derived liquid in the presence of suitable catalysts. This is called hydroliquefaction or catalytic liquefaction. It is descended from the hydrogenation technology developed by Bergius in Germany starting around the beginning of World War I.[5,32] In the solvent extraction procedure a coal-derived liquid, which may or may not be separately hydrogenated, transfers the hydrogen to the coal without external catalyst addition. The forerunner of this technique is attributable to Pott and Broche, who began their work, also in Germany, in the late 1920s.[5,32]

The direct liquefaction of coal may be modeled, albeit simplistically, by the relation

$$C + 0.8\,H_2 \longrightarrow CH_{1.6} \qquad (3.52)$$

At liquefaction temperatures this reaction is only slightly exothermic, so that from stoichiometry alone the thermal efficiency would be high, although the reaction proceeds slowly. However, hydrogen must be supplied and its manufacture accounts for an important fraction of the process energy consumption and cost of producing the liquid fuel. The source of the hydrogen for all of the hydrogenation processes is water from which it may be produced, for example, by gasification with carbon (Eq. 3.14). This reaction is highly endothermic and requires heat which may be supplied by combustion of byproduct gas, coal, char, or residual oil. In Section 5.4 we shall discuss the various means by which hydrogen can be produced, but for our purposes here we should recognize it as a fairly energy intensive process, with the heat to produce hydrogen from steam equal to about 130 MJ/kg for an ideal process.

Hydrogen goes not only into liquefying the coal, but is also consumed in reducing the oxygen, sulfur, and nitrogen in the coal. From the typical ultimate analyses in Table 1.8, we may represent bituminous coal by the approximate molar formula

$$CH_{0.8}O_{0.1}N_{0.02}S_{0.02}$$

Assuming complete reduction of all the constituents to liquid fuel, water, ammonia, and hydrogen sulfide, we may write

$$CH_{0.8} + 0.4H_2 \longrightarrow CH_{1.6} \tag{3.53}$$

$$0.05O_2 + 0.1H_2 \longrightarrow 0.1H_2O \tag{3.54}$$

$$0.01N_2 + 0.03H_2 \longrightarrow 0.02NH_3 \tag{3.55}$$

$$0.02S + 0.02H_2 \longrightarrow 0.02H_2S \tag{3.56}$$

The stoichiometry shows that 1.1 kg of hydrogen and 15.32 kg of coal go into producing 13.6 kg of liquid fuel.

From the Dulong formula, Eq. (1.1), the GCV of the typical bituminous coal considered is 32.5 MJ/kg. It follows that about 30 percent of the calorific value of the coal is required to produce the hydrogen. This in part explains why, although the direct hydrogenation of coal to a liquid fuel is close to thermoneutral, the overall thermal efficiency is sharply reduced to the range shown in Table 1.16.

Direct liquefaction processes under development are carried out at temperatures from about 450 to 475°C and at high pressures from 10 to 20 MPa and up to 30 MPa. Char and gas are both undesired products, and to limit their formation it is necessary to operate at these high pressures. The reactor residence times for conversion are quite long, on the order of an hour. Carbon conversions of about 90 to 95 mass percent of the daf coal are typical. Distillable liquids constitute only about 50 to 60 percent of the product for the lower pressure processes now being developed, whereas in the older German processes that operated at pressures of 70 MPa and higher, the product was mostly distillable liquids.

Bituminous coals are the favored feedstock for direct liquefaction because they produce the highest liquid yields. The primary coal liquids, like those from pyrolysis, are naphthenic and aromatic, with their composition dependent on the degree of hydrogenation. In the original solvent refining of coal, where only 1 to 2 mass percent hydrogen is added, the product is a solid at room temperature, with a melting point in the range of 175 to 200°C. Indeed, the hydrogen-to-carbon ratio in the "solvent refined coal" is lower than that in the original feed coal. This is because the hydrogen is consumed in the removal of the oxygen, nitrogen, and sulfur, and in the production of hydrocarbon gases. If, on the other hand, 3.5 to 4 mass percent hydrogen is added, the product, after separation from the solvent, will be a heavy liquid at room temperature. In hydroliquefaction it is usual to add 4 to 5 percent or more of hydrogen. Solids separation of ash, unreacted coal, and any catalysts that may be present is an important step in all of the processes.

Either of the direct liquefaction procedures may be considered as hydrogen donor processes in which the coal is thermally pyrolyzed, producing low boiling materials and free radicals that are stabilized with hydrogen to produce the liquid product. In solvent extraction, the hydrogen is abstracted from the donor solvent and from the hydrogen originally in the coal. In hydroliquefaction, hydrogen also comes from catalytically activated molecular hydrogen added directly to the coal, or added indirectly to the coal through the medium of the coal liquids which "accept" and then "donate" the hydrogen. It is generally thought that the latter mechanism is the predominant reaction.

Although the course of reactions characterizing coal dissolution is still not clear, there is a consensus that the coal first forms intermediate products consisting of "preasphaltenes," also called "asphaltols," and "asphaltenes," which subsequently convert to oil. Unreacted coal, oil precursors, and oil are defined by whether or not they are soluble in particular organic solvents, of which benzene is the most commonly used. The operational definitions are diagrammed in Figure 3.15, where it can be seen that the benzene soluble portion is further delineated by whether or not it is soluble in pentane, with the pentane soluble fraction called "oil" and the insoluble fraction "asphaltenes." Part of the benzene insolubles are pyridine soluble and therefore considered "reacted coal." This fraction is termed the "preasphaltenes," which is distinguishable by its high viscosity. The pyridine insolubles are regarded as the unreacted coal.

One important parameter in characterizing process performance and the effect of various process variables is the extent of the coal conversion. Among the important factors affecting the conversion, as well as the liquid yield and product distribution, are coal rank and type, solvent or coal liquid properties, temperature and pressure, residence time and reactor conditions, and catalysts. In presenting the effect of the process variables, our remarks will be general to direct liquefaction and not specific to hydroliquefaction or solvent extraction, except as necessary.

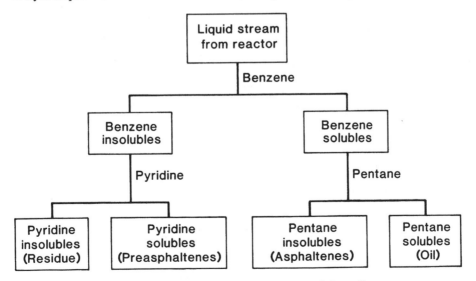

Figure 3.15 Scheme for identification of oil, asphaltenes, and preasphaltenes.[33]

Coal rank and type importantly influence the yield and products of direct lique-faction. Anthracites produce low liquid yields and are not suitable for hydrogenation. Lower-rank coals and lignites liquefy most readily, but they give lower liquid yields than bituminous coals. They also contain a lot of oxygen which consumes hydrogen and the molecules of which are believed to be responsible for high viscosities of the product. The highest liquid yields are obtained with high-volatile bituminous coals, and as such they lend themselves most readily to liquefaction.

The petrographic composition of coal (see Table 1.5) has for a long time been known to influence its liquefaction characteristics.[32] A higher fraction of the reactive macerals, vitrinite and exinite, gives a higher liquid yield. On the other hand, inertinite is very resistant to liquefaction. However, it has been concluded by some authors[34,35] that attempts to correlate coal liquefaction behavior with coal rank or coal type alone or even together may be inadequate. An example of what has led to this reasoning can be seen in Figure 3.16, which gives the yield of distillable liquids as a function of coal and reactor residence time for the Exxon Donor Solvent process. This is a solvent extraction process where the solvent, a distillate product, has been externally hydro-genated. The reactor temperature is 450°C, the reactor pressure is 10 MPa, and the reactants are in cocurrent upward flow. The liquid yield generally follows the trend with rank noted above, except for the anomalously low yield from the one Illinois No. 6 coal. Apart from the fact that the two Illinois coals are from different mines, there is on the surface little else to distinguish them, their ultimate analyses, proximate

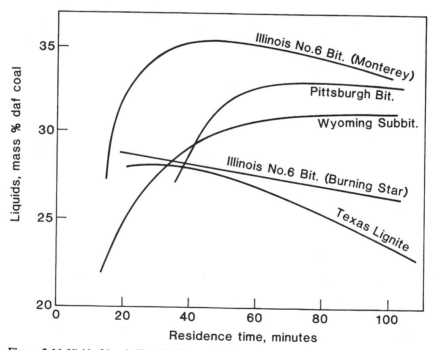

Figure 3.16 Yield of low-boiling liquids as a function of reactor residence time and coal in Exxon Donor Solvent process at $T = 450°C$ and $p = 10$ MPa.[36]

analyses, and reactive maceral contents being reasonably similar. This example illustrates some of the difficulties encountered when trying to predict coal conversion performance in practice from "first principles."

A *solvent or coal liquid* is used to transport the coal into the high pressure reactor, and its properties have an important effect on the dissolution or hydroliquefaction of the coal. In solvent extraction, where catalysts are not added to the system, the hydrogen required to cause the coal to dissolve comes from the solvent. Hydrogenation of the solvent may be carried out by mixing the hydrogen with the slurry prior to dissolution of the coal, or the solvent may be hydrogenated separately prior to its use as a donor. Good hydrogen-donor properties are associated with hydroaromatic compounds such as tetralin, which exists in coal-derived liquids in appreciable quantities. The least effective solvents are high boiling aromatic compounds such as naphthalene and cresol. The hydrogen-donor ability of tetralin and the reverse solvent hydrogenation reaction may be illustrated by the equation (see Eq. 2.2)

$$\begin{array}{c} \underset{\text{solvent}}{\overset{\text{liquefaction}}{\rightleftharpoons}} \end{array} + 4H \qquad (3.57)$$

Tetralin — hydrogenation — Naphthalene

Hydrogen-donor solvents can increase the hydrogen-donor capabilities of non-donor mixtures significantly. Thus a mixture of equal parts of tetralin and cresol has a higher hydrogen-donor capability than the pure solvents. Empirically, this serves to explain why creosote oil, which is a mixture of poor and good solvents, is a very effective donor solvent.[32]

The *effect of temperature* manifests itself first in the physical disintegration of the coal at about 350°C. This is mainly a thermal solubilization and involves little chemical action. The hydrogen-donor properties of the organic solvent or slurry liquid do not appear to be important. Consistent with the physical disintegration which takes place, there is little influence of the initial particle size below about 2 to 3 mm.[37]

In the discussion on pyrolysis, it is pointed out that active thermal decomposition of coal only begins above a temperature of about 400°C. Since hydrogenation depends on the rate of radical formation, this places a lower limit of 400°C for the temperature at which direct liquefaction procedures would be operated. In the reaction of coal with hydrogen the formation of aliphatic and alicyclic liquid products is thermodynamically favored at low temperatures, while at higher temperatures the production of gases and aromatics is favored, as with pyrolysis. Thermodynamic considerations suggest that the optimum liquid yield should be obtained in the neighborhood of the temperature for the maximum rate of reactive fragment formation, which is about 450°C.

Figure 3.17 illustrates the effect of temperature on the hydroliquefaction of a high-volatile bituminous coal in a fixed bed reactor, with a high temperature tar as a solvent. Above 450°C the conversion to oil is seen to drop off sharply along with the hydrogen-to-carbon ratio in the liquid products. This is consistent with the equilibrium considerations discussed, and with the characteristics of coal pyrolysis outlined in Section 3.1.

The *effect of pressure* is beneficial to direct liquefaction both from equilibrium and non-equilibrium considerations.[32] As noted before, high pressure inhibits the coke- and gas-forming dehydrogenation reactions. Also, if the hydrogenation is to be from the gas phase, the hydrogen must first dissolve in the liquid and the extent of the dissolution according to Henry's law is proportional to the partial pressure of the hydrogen, which therefore should be as large as possible. Increased dissolved hydrogen will increase the rate of hydrogen consumption. Additionally, high pressures increase the chemical reaction rates. The effect of pressure on hydroliquefaction at a temperature of 425°C is shown in Figure 3.18 for the same coal and reactor as in Figure 3.17. The increase in oil production and hydrogen content in the liquid products with increasing pressure is almost linear for the 7 to 28 MPa range shown. However, the overall degree of conversion based on benzene insolubles increases much more slowly, with the increase going more like the square root of the pressure.

Much effort has been directed toward lower pressure operation because of the economic penalties of high pressure. This has been accomplished mostly by increasing the hydrogen transfer from the slurry liquid directly. However, no alternative to high pressure has been found to inhibit coke formation and to keep any catalysts which may be present stable and active by keeping hydrogen adsorbed on the catalyst surface to minimize carbon deposition.

Residence time of the coal particles is an important process variable in direct liquefaction. Generally it is quite long, on the order of an hour, to obtain appreciable conversion to benzene-soluble liquids. The conversion of the coal to the heavy asphaltenes is relatively rapid. Longer residence times reduce the asphaltene content, but can also lead to reduced oil production because of thermal decomposition. The difference in yield of low-boiling liquids as a function of reactor residence time is illustrated in Figure 3.16 for the Exxon Donor Solvent process. It is important to keep in mind that the yields shown can be altered significantly by changes in other process variables.

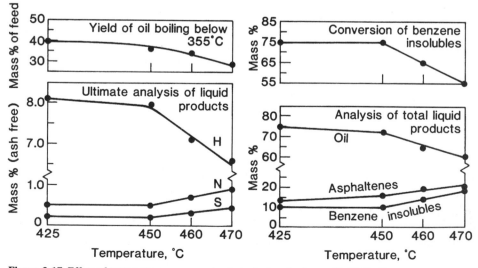

Figure 3.17 Effect of temperature on conversion in hydroliquefaction at 27.6 MPa.[38]

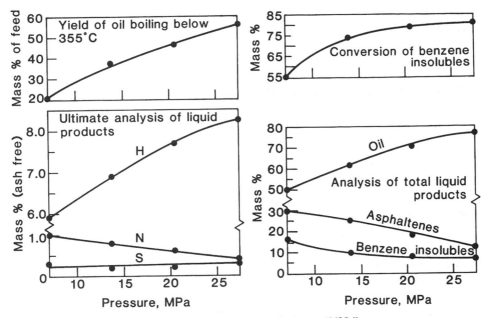

Figure 3.18 Effect of pressure on conversion in hydroliquefaction at 425°C.[38]

A large number of kinetic models have been put forward to characterize the rate of conversion, although there is no general agreement on the reaction paths. It is evident that there are a number of basic rates involved including physical disintegration of the particles, thermally induced bond rupture, repolymerization, and free radical stabilization from the coal hydrogen inventory and the donor hydrogen, which may or may not be catalytically activated. In "dry" pyrolysis or hydropyrolysis the time to reach equilibrium is measured in seconds or less. The long conversion times associated with the coal slurry liquefaction systems are most probably associated with the heat and mass transfer effects of longer times to heat up the particles and increased diffusional resistance to reactants because of the liquid phase surrounding the coal particles. The thermal and flow conditions in the reactor may therefore be expected to be of importance.

Catalyst addition in direct liquefaction leads to larger light oil fractions and lower sulfur content in the product oil.[32] Ideally, catalysts will increase the production of light oils without significantly increasing the consumption of hydrogen. To do this the catalysts must selectively hydrocrack the coal and remove the nitrogen and sulfur without being wasted on producing water (Eq. 3.54) and gases. Further developments are required to produce such an ideal catalyst.[39] The influence of the catalyst on the liquefaction rate is not very marked since the reactions take place in the absence of the catalyst, although it is generally thought that some of the minerals present in the coal provide a catalytic function. Because of the need for catalyst recovery, hydroliquefaction systems are more complex, in addition to having the added catalyst cost. A critical problem is catalyst deactivation and poisoning as a result of coke formation and metal deposition on the catalyst from the minerals in the coal. Many metals act as catalysts in coal liquefaction, including Co, Mo, Fe, Sn, Ni, and Zn. Commercial hydrodesulfurization catalysts, like cobalt molybdate on alumina, are found to be effective for hydroliquefaction.

3.5 COMPARISON OF SYNTHETIC FUEL ROUTES

Several synthetic fuel routes will now be compared on the basis of their steam and carbon consumption per unit of fuel produced. The procedure followed is essentially that reported by Wei.[40] It is stressed that the comparison is a theoretical one based on ideal reactions and hypothetical and somewhat optimistic heat transfer/recovery schemes. Several simplifying assumptions are made here and applied uniformly to all systems considered to facilitate the comparison. These assumptions may be compared with actual operating conditions as described in later chapters.

Carbon, water, and air are the starting materials in all cases. Heat requirements are met by burning carbon in air and supplying the heat via an indirect heat transfer medium. This avoids the complication of considering pure oxygen production. One mole of carbon is taken to provide 395 kJ of heat or 6.7 moles of steam at 700 K, or 4.3 moles of steam at 1500 K. Heat recovery from all off gases or product gases is restricted to 80 percent of the heat of condensation of the equilibrium steam in these gases. Similarly, 80 percent of the heat of reaction of synthesis reactions is assumed recoverable. A pressure of 1 atm is assumed. Other heat effects, and energy for compression, separation, etc., are not considered. The reaction schemes are described below.

(a) *Direct liquefaction using steam.*

$$3C + 2H_2O(g) \rightleftharpoons 2(\text{—CH}_2\text{—})(g) + CO_2$$

$$\Delta H^\circ_{1500} = +67.9 \text{ kJ}; \qquad K_{1500} = 1.19 \times 10^{-6}$$

This is a very unfavorable equilibrium, necessitating the combustion of excessive carbon to supply the required steam. The equilibrium of this endothermic reaction becomes even less favorable at lower temperatures.

(b) *Direct liquefaction using hydrogen.*

The hydrogen is produced directly in a gasifier:

$$\tfrac{1}{2}C + H_2O(g) \rightleftharpoons \tfrac{1}{2}CO_2 + H_2$$

$$\Delta H^\circ_{1500} = +53.1 \text{ kJ}; \qquad K_{1500} = 14.7$$

The hydrogen then reacts with additional carbon in a separate reactor and is subsequently recycled to extinction:

$$C + H_2 \rightarrow (\text{—CH}_2\text{—})(g)$$

$$\Delta H^\circ_{700} = -14.6 \text{ kJ}$$

(c) *Indirect liquefaction by Fischer-Tropsch.*

A synthesis gas having a H_2/CO ratio of 2 is first produced directly in a gasifier:

$$3C + 4H_2O(g) \rightleftharpoons 2CO + 4H_2 + CO_2$$

$$\Delta H^\circ_{1500} = +374.9 \text{ kJ}; \qquad K_{1500} = 7.7 \times 10^7$$

The synthesis gas is then reacted over an iron catalyst with rejection of steam, with the gases recycled to extinction:

$$CO + 2H_2 \rightarrow (-CH_2-)(g) + H_2O(g)$$

$$\Delta H^\circ_{700} = -149.7 \text{ kJ}$$

(d) *Indirect liquefaction via methanol and Mobil M conversion.*
A synthesis gas having a H_2/CO ratio of 2 is first produced as in (c). The synthesis gas is then reacted to form methanol with complete consumption of the gases:

$$CO + 2H_2 \rightarrow CH_3OH$$

$$\Delta H^\circ_{700} = -101.9 \text{ kJ}$$

Finally the methanol is converted to gasoline over the zeolite ZSM-5 catalyst with rejection of steam:

$$CH_3OH(g) \rightarrow (-CH_2-)(g) + H_2O(g)$$

$$\Delta H^\circ_{700} = -47.3 \text{ kJ}$$

(e) *Methanol production.*
A synthesis gas with a H_2/CO ratio of 2 is produced in a gasifier followed by reaction over a zinc oxide catalyst to produce methanol as in (d).
(f) *Methane production with CO_2 rejection.*
A low H_2/CO ratio synthesis gas is produced in a gasifier

$$C + H_2O(g) \rightleftharpoons CO + H_2$$

$$\Delta H^\circ_{1500} = +134.4 \text{ kJ} \qquad K_{1500} = 594$$

The synthesis gas is reacted over an iron-containing catalyst to form methane; unreacted gases are recycled to extinction:

$$CO + H_2 \rightarrow \tfrac{1}{2}CH_4 + \tfrac{1}{2}CO_2$$

$$\Delta H^\circ_{700} = -129.2 \text{ kJ}$$

(g) *Methane production with H_2O rejection.*
The H_2/CO ratio in the synthesis gas produced under (f) is increased to 3 using the gas shift reaction over an iron or copper-zinc catalyst

$$CO + H_2O \rightleftharpoons CO_2 + H_2$$

$$\Delta H^\circ_{700} = -37.9 \text{ kJ} \qquad K_{700} = 8.8$$

The resulting synthesis gas is reacted over nickel to form methane:

$$CO + 3H_2 \rightarrow CH_4 + H_2O$$

$$\Delta H^\circ_{700} = -220.4 \, kJ$$

The results summarized in Table 3.4 indicate direct liquefaction using hydrogen to be the most effective route for synthetic fuel production. The two indirect liquefaction routes to gasoline are virtually identical when following the hypothetical reaction schemes, and consume nearly 10 percent more carbon for heat requirements than the direct liquefaction route. In all these cases additional energy will be required for refining the liquid products produced. Production of methane with its high hydrogen content consumes more steam and more carbon. Methanation with rejection of CO_2 is slightly more efficient than with rejection of steam in the hypothetical schemes, and can be expected to be more economical in practice as the shift step is eliminated. Methanol liquid has a lower heating value than methane, and the production appears to be one of the least attractive of the hypothetical processes considered. It is, however, probably the most developed of all the processes and does not have some of the kinetics related problems, such as byproduct formation, that are associated with the other processes. In addition, there is little further product upgrading required with methanol, whereas in the case of the methanation and indirect liquefaction routes, the products require, at least, separation of either water or carbon dioxide.

It is again emphasized that the results summarized in Table 3.4 are based on idealized hypothetical reaction schemes. In practice, other process steps not considered here will be required in the production of the final product, and further, cost determining factors other than steam and carbon consumption will have an important bearing on process viability. These factors include:

Table 3.4 Comparison of idealized synthetic fuel processes

Reaction Scheme	Product	Net Consumption mol/MJ of product (NCV)	
		Steam*	Carbon†
(a) Direct liquefaction using steam	Gasoline	1.60	8.53
(b) Direct liquefaction using hydrogen produced by gasification	Gasoline	1.60	2.94
(c) Indirect liquefaction using Fischer-Tropsch	Gasoline	1.60	3.23
(d) Indirect liquefaction via methanol and Mobil M	Gasoline	1.60	3.23
(e) Indirect liquefaction via synthesis gas	Methanol	3.13	3.49
(f) Gasification followed by methanation with rejection of CO_2	Methane	2.49	3.27
(g) Gasification followed by gas shift and methanation with rejection of H_2O	Methane	2.49	3.42

* Steam provides the hydrogen in the product.
† Carbon in product plus carbon burned to provide heat for the conversion.

1. Reactor size as determined by reaction kinetics
2. Byproduct formation as determined by reaction selectivity
3. Reactor pressure and temperature as determined by equilibrium and kinetics
4. The cost of product purification and refining
5. The cost of catalysts
6. Process complexity

Other considerations include the stage of development and reliability of a process as well as the demand for a particular fuel and its storage, handling, and transportation costs per unit energy content. These factors are examined in some detail in subsequent chapters in the book.

REFERENCES

1. Anthony, D. B., and Howard, J. B., "Coal Devolatitilization and Hydrogasification," *AIChE J.* **22**, 625-656, 1976.
2. Howard, J. B., "Fundamentals of Coal Pyrolysis and Hydropyrolysis," in *Chemistry of Coal Utilization, Second Supplementary Volume* (M. A. Elliott, ed.), pp. 665-784. Wiley, New York, 1981.
3. Baughman, G. L., *Synthetic Fuels Data Handbook.* 2nd Edition, Cameron Engineers, Inc., Denver, Colorado, 1978.
4. Bunger, J. W., Cogswell, D. E., and Oblad, A. G., "Thermal Processing of Utah Tar Sand Bitumen," in *The Oil Sands of Canada-Venezuela 1977* (D. A. Redford and A. G. Winestock, eds.), pp. 177-182. CIM Special Vol. 17, The Canadian Institute of Mining and Metallurgy, Montreal, 1977.
5. Berkowitz, N., *An Introduction to Coal Technology.* Academic Press, New York, 1979.
6. Wen, C. Y., and Dutta, S., "Rates of Coal Pyrolysis and Gasification Reactions," in *Coal Conversion Technology* (C. Y. Wen and E. S. Lee, eds.), pp. 57-170. Addison-Wesley, Reading, Mass., 1979.
7. Kobayashi, H., "Devolatilization of Pulverized Coal at High Temperatures," Ph.D. Thesis, Department of Mechanical Engineering, M.I.T., Cambridge, Mass., 1976.
8. Fahlstrom, P. H., "A Physical Concentration Route in Oil Shale Winning," in *Proc. 12th Oil Shale Symposium* (J. H. Gary, ed.), pp. 252-277. Colorado School of Mines Press, Golden, Colo., 1979.
9. Goodman, J. B., *et al.*, "Low-Temperature Carbonization Assay of Coal in a Precision Laboratory Apparatus," U.S. Bureau of Mines Bulletin 530, Government Printing Office, Washington, D.C., 1953.
10. Dryden, I. G. C., "Chemistry of Coal and Its Relation to Coal Carbonization," *J. Institute of Fuel* **30**, 193-214, 1957.
11. Solomon, P. R., "Characterization of Coal and Coal Thermal Decomposition," in *Pulverized Coal Combustion: Pollutant Formation Control*, Ch. III. U.S. Environmental Protection Agency, Project Decade Monograph. Government Printing Office, Washington, D.C., 1981 (to appear).
12. Tyler, R. J., "Flash Pyrolysis of Coals. Devolatilization of Bituminous Coals in a Small Fluidized-Bed Reactor," *Fuel* **29**, 218-226, 1980.
13. Landers, W. S., Goodman, J. B., and Donaven, D. J., "Low-Temperature Carbonization Assays of Coals and Relation of Yields to Analyses," U.S. Bureau of Mines Report of Investigations 5904, Government Printing Office, Washington, D.C., 1961.
14. Goodman, J. B., Gomez, M., and Parry, V. F., "Laboratory Carbonization Assay of Low-Rank Coals at Low, Medium and High Temperatures," U.S. Bureau of Mines Report of Investigations 5383, Government Printing Office, Washington, D.C., 1958.
15. Allred, V. D., "Kinetics of Shale Oil Pyrolysis," *Chem. Eng. Prog.* **62**(8), 55-60, 1966.
16. Hubbard, A. B., and Robinson, W. E., "A Thermal Decomposition Study of Colorado Oil Shale," U.S. Bureau of Mines Report of Investigations 4744, Government Printing Office, Washington, D.C., 1950.

17. Shih, S. M., and Sohn, H. Y., "Nonisothermal Determination of the Intrinsic Kinetics of Oil Generation from Oil Shale," *Ind. Eng. Chem. Process Des. Dev.* **19**, 420-426, 1980.

18. Yoon, H., Wei, J., and Denn, M. M., "Feasible Operating Regions for Moving Bed Coal Gasification Reactors," *Ind. Eng. Chem. Process Des. Dev.* **18**, 306-312, 1979. See also Wei, J., "A Stoichiometric Analysis of Coal Gasification," Ibid pp. 554-558.

19. Walker, P. L., Rusinko, F., Jr., and Austin, L. G., "Gas Reactions of Carbon," in *Advances in Catalysis and Related Subjects Vol. XI* (D. D. Eley, P. W. Selwood and P. B. Weisz, eds.), pp. 134-221. Academic Press, New York, 1959.

20. Johnson, J. L., "Kinetics of Bituminous Coal Char Gasification with Gases Containing Steam and Hydrogen," in *Coal Gasification* (L. G. Massey, ed.), pp. 145-178. Advances in Chemistry Series No. 131, American Chemical Society, Washington, D.C., 1974.

21. Katta, S., and Keairns, D. L., "Study of Kinetics of Carbon Gasification Reactions," *Ind. Eng. Chem. Fundam.*, **20**, 6-13, 1981.

22. Mahajan, O. P., Yarzat, R., and Walker, P. L., "Unification of Coal Char Gasification Reaction Mechanisms," *Fuel* **57**, 643-666, 1978.

23. Gruber, G., "Equilibrium Considerations in the Methane Synthesis System," in *Methanation of Synthesis Gas* (L. Seglin, ed.), pp. 31-46. Advances in Chemistry Series No. 146, American Chemical Society, Washington, D.C., 1975.

24. Massey, L. G., "Coal Gasification for High and Low Btu Fuels," in *Coal Conversion Technology* (C. Y. Wen and E. S. Lee, eds.), pp. 313-427. Addison-Wesley, Reading, Mass., 1979.

25. White, G. A., Roszkowski, T. R., and Stanbridge, D. W., "The RMProcess," in *Methanation of Synthesis Gas* (L. Seglin, ed.), pp. 138-148. Advances in Chemistry Series No. 146, American Chemical Society, Washington, D.C., 1975.

26. Moeller, F. W., Roberts, H. and Britz, B., "Methanation of Coal Gas for SNG," *Hydrocarbon Processing* **53**(4), 69-74, April 1974.

27. Hagen, D. L., "Methanol: Its Synthesis, Use as a Fuel, Economics and Hazards," M.Sc. Thesis, Graduate School, University of Minnesota, Minneapolis, Minn., Dec. 1976. (NTIS Catalog No. NP-21727).

28. Poutsma, M. L., "Assessment of Advanced Process Concepts for Liquefaction of Low H_2:CO Ratio Synthesis Gas Based on the Kölbel Slurry Reactor and the Mobil-Gasoline Process," Report No. ORNL-5635, Oak Ridge National Laboratory, Oak Ridge, Tennessee, February 1980.

29. Seglin, L., and Geosits, R., "Survey of Methanation Chemistry and Processes," in *Methanation of Synthesis Gas* (L. Seglin, ed.), pp. 1-30. Advances in Chemistry Series No. 146, American Chemical Society, Washington, D.C., 1975.

30. Dry, M., "Advances in Fischer-Tropsch Chemistry," *Ind. Eng. Chem. Product Res. Dev.* **15**, 282-286, 1976.

31. Meisel, S. L., McCullough, J. P., Lechthaler, C. H., and Weisz, P. B., "Gasoline from Methanol in One Step," *ChemTech* **6**, 86-89, 1976.

32. Lee, E. S., "Coal Liquefaction," in *Coal Conversion Technology* (C. Y. Wen and E. S. Lee, eds.), pp. 428-545. Addison-Wesley, Reading, Mass., 1979.

33. Wen, C. Y., and Tone, S., "Coal Conversion Reaction Engineering," in *Chemical Reaction Engineering Reviews-Houston* (D. Luss and V. W. Weekman, Jr., eds.), pp. 56-109. ACS Symposium Series No. 72, American Chemical Society, Washington, D.C., 1978.

34. Neavel, R. C., "Coal Science and Classification," in *Scientific Problems of Coal Utilization* (B. R. Cooper, ed.), pp. 77-79. U.S. Department of Energy Symposium Series No. 46, Technical Information Center, U.S. Department of Energy, Washington, D.C., 1978.

35. Given, P. H., *et al.*, "Some Proved and Unproved Effects of Coal Geochemistry on Liquefaction Behavior with Emphasis on U.S. Coals," in *Coal Liquefaction Fundamentals* (D. D. Whitehurst, ed.), pp. 3-34. ACS Symposium Series No. 139, American Chemical Society, Washington, D.C., 1980.

36. Epperly, W. R., and Taunton, J. W., "Exxon Donor Solvent, Coal Liquefaction Process Development," in *Coal Conversion Technology* (A. H. Pelofsky, ed.), pp. 71-89. ACS Symposium Series No. 110, American Chemical Society, Washington, D.C., 1979.

37. Neavel, R. C., "Liquefaction of Coal in Hydrogen-Donor and Non-Donor Vehicles," *Fuel* **56**, 237-242, 1976.

38. Akhtar, S., Friedman, S., and Hiteshue, R. W., "Hydrogenation of Coal to Liquids in Fixed Beds of Silica Promoted Cobalt Molybdate Catalyst," *Am. Chem. Soc., Div. Fuel Chem., Preprints* **14**(4), Part I, 27-41, September 1970.
39. Cusumano, J. A., Dalla Betta, R. A., and Levy, R. B., *Catalysis in Coal Conversion.* Academic Press, New York, 1978.
40. Wei, J., "Stoichiometric Analysis of Synthetic Fuels and Chemicals from Coal," *Ind. Eng. Chem. Process Des. Dev.* **20,** 294-298, 1981.

FOUR

GAS FROM COAL

4.1 GASIFICATION TECHNOLOGIES

The characteristics of coal gasification have thus far been assessed principally in terms of the ideal stoichiometry of the main reactions occurring in a gasifier, and in terms of idealized thermal constraints. The ideal characteristics so derived can serve as a relative guide but not as an absolute measure of performance, mainly because of additional practical constraints to which actual gasifiers are subjected. These may include the reactivity and caking characteristics of the coal; product throughput and its relation to reactor pressure; size and channeling characteristics; clinkering and ash removal; material and mechanical limitations; interfacing with downstream cleanup and processing equipment; and, of course, the desired end product.[1]

One of the major end products is high-calorific value (high-CV) pipeline quality gas as a substitute for natural gas. Synthesis gas, consisting primarily of a mixture of carbon monoxide and hydrogen, is of medium-calorific value. It is needed for the chemical process industries, for the indirect liquefaction of coal, and to manufacture hydrogen for the direct liquefaction of coal and the upgrading of pyrolysis liquids. Industrial consumers use a low-CV gas for the generation of steam and process heat, while electric utilities use this gas for the generation of electricity with gas turbines or combined cycle systems. As we showed in Figure 1.1, the plants to manufacture low-, medium-, or high-CV gas are different, but at the heart of each is the gasifier.

Many gasifier systems are commercially available or have the potential to become so. There are a number of ways to characterize the different systems, though here we distinguish between an ''independent'' characteristic and a ''dependent'' one. The main independent characteristics are the method of supplying the heat, the gasifying medium, and the reactor type. Other characteristics often used to distinguish gasifiers, that are really dependent characteristics once the independent ones have been specified, include: whether the solid residue is ash or slag, the raw gas composition and calorific value, and the gasification temperature. Table 4.1 lists both types of characteristics together with the subgroups by which they are generally categorized. In this section,

Table 4.1 Major independent and dependent gasifier characteristics

Independent	Dependent
1. Method of supplying heat Direct Indirect	1. Reaction temperature High Medium Low
2. Gasifying medium and amount Steam with air or oxygen Air Hydrogen with or without steam Steam with catalyst	2. Raw gas properties Composition H_2/CO ratio Tar and oil content Calorific value
3. Reactor type Moving packed bed Fluidized bed Entrained flow Molten media	3. Solid residue Ash Slag Unconverted carbon

we briefly describe the various methods of supplying gasifier heat and the various reactor types. Emphasis is placed on reactor type, with the gasifying medium discussed in context, because it is chiefly here that the main differences exist. In the following sections, where specific gasifiers are considered in detail, we differentiate mainly by the gasifying medium, which relates more closely to the detailed chemistry and thermodynamics of the process.

One gasification technology we shall treat separately (Section 4.5) because the procedure is sufficiently different, although the chemistry is not, is that of underground, or *in situ*, gasification. In this method, the gasification is carried out directly in the unmined coal deposit, which by appropriate preparation is turned into a fixed packed bed. The reactants are brought down to the coal bed and the gases formed are brought up to the surface through holes drilled into the deposit.

Method of Supplying Heat

In most gasifiers the heat needed to drive the endothermic carbon-steam reaction (3.14) and the Boudouard reaction (3.12) is generated directly by burning the coal or char in the gasifier. One problem of supplying heat in this manner is that if air is used the product gases will be diluted with nitrogen and have a low calorific value, less than 10 MJ/m^3 and generally closer to 5 MJ/m^3. If a medium-CV synthesis gas, consisting mainly of hydrogen and carbon monoxide, with a calorific value, say, between 10 and 15 MJ/m^3 is desired, two direct methods can be used. One method is to remove the nitrogen before the process, that is, to use relatively pure oxygen in place of air. The other one is to remove the nitrogen after the process, for example, by cryogenic distillation wherein the gas is liquefied and the components are fractionally distilled (see Section 5.3). At present no gasifier system has been developed based on removing the nitrogen from the product stream. Supplying relatively pure oxygen, on the other hand, is very costly and accounts for a large fraction of both the plant and product cost when it is used.

An indirect method to get around the nitrogen barrier is to remove it during the process by rejecting the nitrogen in a separate stream of fully combusted gas. In fact, this was essentially the procedure used from the middle of the last century through World War II to manufacture "blue water gas" composed of about 50 percent hydrogen and 40 percent carbon monoxide (see Section 1.2). The method was to first convert caking coals to coke in coke ovens and then gasify the coke in "cyclic" gas generators. The gasifiers were cyclic in that the coke was blasted with air to provide heat until the carbon reached incandescence, after which the air was shut off and steam was blown through the bed of hot carbon. The endothermic reaction forming the carbon monoxide and hydrogen dropped the bed temperature, following which the steam was shut off and air was again blown in to heat up the coke. This was the method used in Germany to produce synthesis gas for the manufacture of liquid fuels by Fischer-Tropsch synthesis during World War II. In addition to requiring caking coals, the cyclic gas generator's principal drawback is that it is first necessary to pyrolyze the coal into coke and then to gasify the coke, with all the attendant losses of a two-stage process and with the need to process the byproduct tars and oils formed in the pyrolysis operation. Although coal could be gasified in a cyclic generator, the gas output would only be 70 to 80 percent of that using coke. It was in fact the reduction in cost of oxygen brought about by large-scale cryogenic manufacturing methods (Section 5.3) that led to the abandonment of cyclic gas generators in favor of continuous oxygen-blown gasifiers.

One relatively recent continuous indirect procedure we shall describe, termed the COGAS process, gets around the nitrogen barrier by making use of pyrolysis in a sequence of staged reactors. The product char is subsequently gasified with steam in a separate reactor. The tars and oils that are produced in the pyrolysis stages are considered important coproducts to be hydrotreated into synthetic liquid fuels.

The indirect heating techniques presently being developed to remove the nitrogen barrier during the gasification process are all continuous ones in which the aim is nearly complete conversion of the coal to gas. In these procedures, the heat is supplied from an external source, generally an inert solid or fluid carrier, which has been heated in a separate furnace. In other indirect procedures the heat may be supplied by heat transfer through the tube walls of a furnace, electrically, or from a nuclear reactor. The use of a separate furnace, with heat supplied by transfer through the walls of the reactor, is restricted to relatively low-throughput systems, because of the rate at which heat can be transferred to the gasifier. When the heat supply is from a nuclear source and in the case of electricity generated by nuclear means, nitrogen removal in a combusted gas stream is not a consideration.

In one other indirect heat transfer process, described briefly in Section 3.2, the heat carrier is calcium oxide in the form of heated (calcined) dolomite or limestone. Not only is the calcined material the sensible heat carrier, but it also reacts exothermically with the CO_2

$$CaO + CO_2 \rightleftharpoons CaCO_3 \qquad \Delta H^\circ_{298} = -183 \text{ kJ/mol} \qquad (4.1)$$

The accompanying heat release together with the indirect sensible heat provides the process heat for the gasification reaction. This process is termed the CO_2 Acceptor process, since the calcium oxide "accepts" the CO_2. Of course, the acceptor must be regenerated by heating to drive off the CO_2 and this is done in a separate furnace using air and char that has been produced in the gasifier.

Table 4.2 summarizes the major procedures for the indirect gasification of coal along with their main limitations. The last two continuous procedures listed in the table are presently under active development in the United States.

Reactor Type

Three principal reactor types are employed in gasifier design: the moving packed bed, the fluidized bed, and the entrained flow reactor (see Section 2.4). A fourth type, the molten media reactor, could be grouped together with the entrained flow reactor, but is sufficiently distinctive in its operation that we treat it separately.

The reactor type strongly influences the temperature distribution, and in this way the gas and residue products. The reaction temperature varies from about 815 to 1925°C, with each type of gasifier covering a specific range. The exception to this is the molten media gasifier whose temperature and, to some extent, characteristics of operation are determined by the melt employed.

At high temperatures a synthesis gas is produced and at low temperatures methane formation is favored. As the overhead product is affected, so too is the residue. Gasifiers in which the temperature is low enough that the ash does not melt are sometimes referred to as "dry bottom gasifiers." High temperature gasifiers in which molten ash (slag) is formed are called "slagging gasifiers." The slagging temperature is dependent on the ash composition, but for most coals lies roughly in the range 1200 to 1800°C.

Table 4.2 Major procedures and limitations for the indirect gasification of coal with steam

Procedure	Limitations
Alternate heating and gasification of separately produced coke.	Two-stage process limited to caking coals, tar by-products, incomplete gasification.
Alternate heating and gasification of coal.	Low gas yield.
Continuous coal gasification in externally heated gasifiers.	Low throughput, limited by indirect heat transfer rate.
Continuous coal pyrolysis integrated with separate gasification of char.*	Two or more integrated stages, tar byproducts, incomplete gasification.
Continuous coal gasification using separately heated fluid or solid heat carrier.*	Solids handling, two or more stages.

* Under development.

Moving Bed

In the terminology of gasifier technology the term "fixed bed" gasifier refers to a gasifier in which a packed bed of fuel is supported by a grate or by other means and maintained at a constant depth above the support. The fuel moves downward under the action of gravity from the top of the gasifier, through the gasification and combustion zones, and the residue is discharged at the bottom (see Figure 3.8). Within the reactor classes defined in Section 2.4, this is clearly a "moving packed bed" or more succinctly a "moving bed" reactor. Apparently because the upper and lower extremities of the fuel bed are fixed in space the name "fixed bed" has become strongly, though not universally, attached to this type of gasifier. We adopt the minority view in this book and refer to it as a "moving bed gasifier." The reader is cautioned, however, to be aware that the same reactor type is called by two different names.

Moving bed coal gasifiers operate with countercurrent flow, use either steam and oxygen or steam and air, and the residue may be either slag or dry ash plus any unconverted carbon. Undried coal particles that range in size between 3 and 50 mm are fed into the top of the gasifier, with the size range dependent on the particular process. The coal particle sizes must be sufficiently large (>3 mm) in order that they not be carried out overhead by the exiting product gases (see Eq. 2.75). Smaller particles, termed "fines," must be removed from the coal feed, not only because of the carryover problem, but because they tend to clog the bed and offer a high resistance to the upward flowing gases.

The coal passes downward through the upward flowing gases, first being heated, then devolatilized, and then still lower down gasified. The bottom of the coal bed is the combustion zone, with the residue of the combustion cooled by the steam/oxygen or steam/air mixture entering from the bottom before it is discharged. The downward movement of the bed is relatively slow, with average linear velocities on the order of 0.5 m/h in atmospheric steam/air gasifiers and 5 m/h in high pressure steam/oxygen gasifiers. Coal residence times are correspondingly shorter with the higher velocity beds, and are typically on the order of 0.5 to 1 hour for the high pressure steam/oxygen gasifiers and some ten times longer for atmospheric steam/air gasifiers.

Temperatures in the combustion zone are mainly a function of the relative level of the oxidant. In atmospheric steam/air gasifiers, temperatures do not exceed the ash melting point. Steam/oxygen gasifiers may, however, be either dry ash or slagging depending on the amount of steam blown into the gasifier in relation to the amount of oxygen. In dry ash gasifiers, a large quantity of steam is blown in, which holds down the temperature by virtue of its high heat capacity, and because the steam-carbon gasification reaction is endothermic. The excess steam does not participate in the reaction and is recovered from the product gases as quite a dirty water. The maintenance of lower combustion temperatures by excess steam addition is critical for gasifiers designed for dry ash operation, since if the ash melts it then cannot be handled on a supporting grate.

The excess steam requirement can be overcome by operating at high temperatures in a slagging mode. By deliberately operating at high temperatures, excess steam is no longer required to hold down the temperature. A second, but important, effect is that the rate of the carbon-steam reaction is enhanced at elevated temperature resulting in improved steam utilization and little undecomposed steam in the product gases. The absence of undecomposed steam coupled with lower carbon dioxide contents, because of the higher combustion temperatures, minimizes the amount of diluent gases that pass through the gasifier. The result is a higher product gas output, all else being equal. The penalty for this mode of operation is that handling of molten slag is required. The Lurgi gasifier, which was first run commercially in 1936, has been developed to operate in both the dry ash and slagging modes.

Combustion temperatures are also dependent on the reactivity of the coal (Eq. 1.3) as shown in Figure 4.1. This behavior has been confirmed theoretically.[3] Less reactive coals attain higher maximum combustion temperatures and require a higher steam/oxygen ratio to hold the temperature below a given ash melting point.

An advantage of the moving bed gasifier is that the countercurrent direct contact heat exchange that takes place within the gasifier results in a relatively high overall conversion efficiency with a minimum of heat loss. A limitation of the moving bed gasifier is that it is not easy to use caking and swelling coals without a mild oxidative pretreatment to render them non-agglomerating or without altering the mechanical design, for example, to include internal agitation. Moreover, the fraction of fines in mined coal has been increasing with the increasing use of large-scale mining machinery, and the utilization of these fines must be considered when moving bed gasifiers are employed.

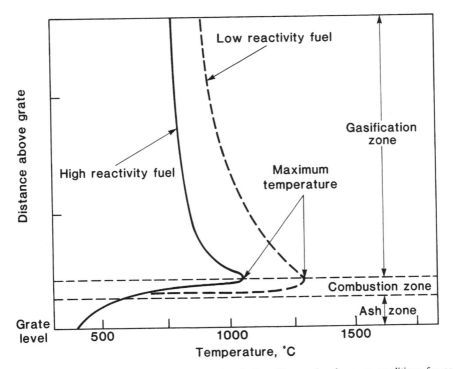

Figure 4.1 Temperature distributions in moving bed gasifiers under the same conditions for coals of different reactivity.[2]

Tables 4.3 and 4.4 give typical operating characteristics and gas compositions, respectively, for moving bed gasifiers. All are commercial except for the Lurgi slagger which has, however, undergone large-scale demonstration. The data shown are drawn largely, but not entirely, from a handbook of gasifier systems compiled by the Dravo Corporation[4] and from the synthetic fuels data handbook of Cameron Engineers.[5] Although the values are representative, it must be emphasized that wide ranges are possible for each gasifier, depending on the coal and the manner of operation. The cold gas efficiency shown in Table 4.3 is taken to be the gross calorific value of the product gas divided by the gross calorific value of the feed coal (Eq. 2.60).

Entrained Flow

Entrained flow gasifiers all use coal (or char) pulverized to a size on the order of 75 μm. Oxygen or air together with steam generally is used to entrain the coal, which is injected through nozzles into the gasifier burner. Hot product gas, or in the case of hydrogasification hot hydrogen, may also be used to entrain the coal and at the same time gasify it. In the Texaco gasifier the solids are carried in a water slurry. However, in a preheater the water is turned into steam, and it is the steam plus coal that is blown into the reactor.

The most important feature of entrained flow gasifiers is that as a group they operate at the highest temperatures under conditions where the coal slags. The aim is to gasify the coal rapidly, so that the residence times in these reactors are measured

Table 4.3 Typical operating characteristics of moving bed gasifiers using bituminous coals and blown with steam and oxygen or air

Characteristic	Steam/Oxygen		Steam/Air	
	Lurgi Slagging	Lurgi Dry Ash	Woodhall Duckham/ Gas Integrale	Wellman-Galusha
Pressure, MPa	2.1	2.5	0.1	0.1
Temperature,* °C				
Combustion zone (max.)	~2000	980–1370	N/A	~1300
Gasifier zone	~1475	650–815	1200	N/A
Gas exit	350–450	370–540	650	590–650
Steam/Oxidant, kg/kg	1	4	0.11	0.12
Oxidant, kg/GJ gas	20	17	110	130
Raw coal throughput, t/(m² · h)	5.9	2.9	0.37	0.44
Bed velocity,† m/h	8.2	4.0	0.51	0.61
Coal residence time,‡ h	0.4	1	23	3
Cold gas efficiency, %	90	80	77	75

* Dependent on coal.
† Assuming coal bed density of 720 kg/m³.
‡ Based on estimated coal bed heights.
  N/A Not available.

Table 4.4 Typical product gas composition* in mole percent† from gasification of bituminous coals in moving bed gasifiers blown with steam and oxygen or air

	Steam/Oxygen		Steam/Air	
Constituent	Lurgi Slagging	Lurgi Dry Ash	Woodhall Duckham/ Gas Integrale	Wellman- Galusha
CO	61	18	28	29
CO_2	3	30	5	3
H_2	28	40	17	15
CH_4	7	9	3	3
N_2	1	1	47	50
Other	—	2	—	—
Total	100	100	100	100
GCV, MJ/m^3	13.8	11.3	6.5	6.3

* Dry basis after scrubbing and cooling.
† Identical with volume percent.

in seconds. Because of the high temperature, all of the volatile matter in the coal is oxidized, which leads to relatively high oxygen consumptions. At high temperatures negligible methane is produced, the CO/CO_2 ratio is high, the conversion of the carbon in the coal is high (on the order of 90 to 100 percent), and there are few tars or heavy hydrocarbons in the product gas and no carbon in the slag. The low methane formation makes the entrained flow gasifier suited to hydrogen production (see Section 5.4). On the other hand, the H_2/CO ratio is lowest in these gasifiers, thereby requiring more shift conversion, especially if the gas is to be used for SNG production.

Entrained flow reactors have the advantage of being able to accept any coal and have high throughput rates, especially with higher pressure operation. But because they are cocurrent flow systems they are not the most efficient thermally. Moreover, the high off gas temperature, that is the temperature of the gas coming off the top of a gasifier, introduces additional losses through the requirement for irreversible quenching. The high temperature environment also imposes severe operating conditions, including the problem of slag handling and removal.

Tables 4.5 and 4.6 list typical operating characteristics and product gas compositions for entrained flow gasifiers that are either commercial or being demonstrated on commercial scale. The Koppers-Totzek gasifier was first introduced in the early 1950s, and many of these units are in operation throughout the world. Both the Combustion Engineering and Foster Wheeler gasifiers are intended mainly for the production of low-CV gas. The Combustion Engineering unit can be blown with or without steam and we have shown the products where air is blown alone. In this case, the principal gasification reaction is the reduction of the carbon dioxide produced by the combustion to carbon monoxide in accordance with the Boudouard reaction (3.12). The general entrained flow characteristics mentioned, including high oxygen con-

Table 4.5 Typical operating characteristics of entrained flow gasifiers using bituminous coals and blown with oxygen or air

Characteristic	Steam/O_2 (Koppers-Totzek)	Water/O_2 (Texaco)[†]	Steam/Air (Foster Wheeler)	Air (Combustion Engineering)
Pressure, MPa	0.13	4.0	2.5	0.1
Temperature,* °C				
Combustion	1925	~1400	1370–1540*	1750
Gas exit	1480	~ 230‡	925–1150§	925
Steam (Water)/Oxidant, kg/kg	0.4	0.5	0.05	0
Oxidant, kg/GJ gas	52	37	111	139
Coal residence time, s	~1	~3	N/A	2.5
Cold gas efficiency, %	75	~75	90	69

  * Lower stage.
  † Low water feed case.
  ‡ After quenching.
  § Upper stage.
  N/A Not available.

sumptions, short residence times, and low H_2/CO ratios are evident in the tables. A number of sources have been used to compile these tables, including Refs. 4 to 6.

Fluidized Bed (Directly Heated)

Fluidized bed gasifiers are fed with pulverized or crushed coal that is lifted in the gasifier by feed and product gases. In single stage, directly heated gasifiers a steam/oxygen or steam/air mixture is injected near the bottom of the reactor, either cocurrently or countercurrently to the flow of coal (or char). The rising gases react with the coal and at the same time maintain it in a fluidized state. As the coal is gasified, the larger size mineral particles, which are about twice as dense as the carbonaceous material, fall down through the fluidized bed together with the larger char particles. The smaller and lighter particles of unconverted carbon, char, and ash are carried upward with the product gases. This description holds generally, although the specific operating characteristics of each gasifier can be expected to vary somewhat from that given here.

Advantages of fluidized beds have been described in Section 2.4. These include good mixing of the solids and a uniform temperature bed, with rapid temperature equilibration between the solids and gases. As a consequence, fluidized bed gasifiers have the advantage that there is efficient heat transfer from the exothermic to the endothermic regions and therefore the gasification reactions reach equilibrium quickly, resulting in relatively high throughputs. Moreover, there are no hot spots with attendant formation of fused ash particles, called "clinkers."

A disadvantage of most fluidized bed gasifiers is that without oxidative pretreatment of the coal or special design configurations, they encounter difficulties in handling caking and swelling coals, which agglomerate and form larger size particles. The

Table 4.6 Typical product gas composition* in mole percent from gasification of bituminous coals in entrained flow gasifiers blown with oxygen or air

Constituent	Steam/O$_2$ (Koppers-Totzek)	Water/O$_2$ (Texaco)	Steam/Air (Foster Wheeler)	Air (Combustion Engineering)
CO	53	52	29	23
CO$_2$	10	12	3	5
H$_2$	36	35	15	12
N$_2$	—	—	49	59
Other	1	1	4†	1
Total	100	100	100	100
GCV, MJ/m^3	11.3	11.1	6.6	4.2

* Dry basis after scrubbing and cooling.
† Mostly methane.

reason for this difficulty is that the larger agglomerated particles fall, and in so doing can lead to a collapse of the dense-phase fluidized state, with resultant coking of the coal in and on the walls of the gasifier. Another limitation of these gasifiers is that there is carryover of solids in the product gas, and a means of removing this carryover has to be incorporated.

Although fluidized bed gasifiers are thought of as a relatively recent development, it is interesting that work on the Winkler fluidized bed gasifier began in 1921 in Germany, and the first commercial unit went into operation in 1926. The Winkler gasifier operated at atmospheric pressure. Examples of more recent fluidized bed gasifiers proposed for demonstration on commercial scale are the U-Gas gasifier developed by the Institute of Gas Technology and the Westinghouse gasifier, both of which operate at elevated pressures. These gasifiers have as their main aim the manufacture of a low-CV gas in an air-blown mode. However, they could in principle be oxygen-blown, and work on the Westinghouse gasifier is also proceeding in this direction. The Winkler gasifier is operable with either oxygen or air, and in the past it has been largely used for producing synthesis gas by blowing with steam and oxygen. A fluidized bed gasifier that produces a high methane content gas by operating in several vertically stacked stages is the Hygas gasifier, which has been developed to the pilot plant stage by the Institute of Gas Technology. The bottom stage is where the gasification with falling char takes place. In the upper stages the rising product gases "hydrogasify" the incoming countercurrently falling coal, the residual char then being gasified in the bottom stage as just noted.

In Tables 4.7 and 4.8 we compare the operating characteristics and product gas compositions of the four gasifiers mentioned. These data have been drawn or derived mostly, but not exclusively, from information in Ref. 4. What can be seen is that these gasifiers operate at the lowest temperatures, are in the upper range of efficiencies for the different gasifier types, and have methane levels about the same or somewhat lower than for moving bed gasifiers, except with hydrogasification.

Table 4.7 Typical operating characteristics of fluid bed gasifiers using bituminous coals and blown with steam and oxygen or air

Characteristic	Steam/Oxygen		Steam/Air	
	Hygas	Winkler	U-Gas	Westinghouse
Pressure, MPa	8	0.1	0.5–2.5	1.5
Temperature, °C				
Gasifier	1010*	980	1040	1000
Gas exit	940*	790	840–1040	650
Steam/Oxidant, kg/kg	4.8	0.4–0.6	0.14–0.18	0.18
Oxidant, kg/GJ gas	31	28	123–145	105
Coal residence time, h	1.3	~3	0.8–1	~0.3†
Cold gas efficiency, %	72†	72	79	~84

* Steam/oxygen gasifier; hydrogasifier, 940°C in stage II, 690°C in stage I.
† Estimated.

Fluidized Bed (Indirectly Heated) and Molten Media

We choose to discuss the indirectly heated fluidized bed gasifier and the molten media gasifier together, despite the fact that the reactors for both types fall into one or the other of the basic classes already considered. This is done because both gasifier types employ an added solid or fluid heat transfer or dispersing medium, over and above the usual gasifying agents. We have already mentioned the COGAS and CO_2 Acceptor processes as employing indirect heat. In the COGAS process heat is supplied by hot char, and in the CO_2 Acceptor process by heated dolomite or limestone. Both of these processes employ fluidized beds in their gasification stages for the reason that this type of reactor is the most efficient in transferring heat from a solid to a gas phase. The operating pressure and temperature characteristics are comparable with directly heated fluidized bed gasifiers.

Table 4.8 Typical product gas composition* in mole percent for directly heated fluidized bed gasifiers blown with steam and oxygen or air†

Constituent	Steam/Oxygen		Steam/Air	
	Hygas	Winkler	U-Gas	Westinghouse
CO	24	48	20	19
CO_2	25	14	10	9
H_2	30	35	18	15
CH_4	19	2	3	3
N_2	—	1	49	54
Other	2	—	—	—
Total	100	100	100	100
GCV, MJ/m³	13.8	10.7	5.7	5.0

* Dry basis after scrubbing and cooling.
† Lignite for Winkler, all others bituminous coals.

The advantages of indirect heating to produce a medium-CV gas without the use of pure oxygen have previously been discussed. A disadvantage is the increased solids handling, which when coupled with fluidized bed operation tends to increase the process complexity.

In molten media gasifiers, the coal is dispersed into a molten carrier, which generally takes part in the gasification and acts as a heat source. One such gasifier that has been demonstrated on commercial scale is the Rummel-Otto single-shaft gasifier. The gasifier is actually an entrained flow type, in that pulverized coal is injected together with steam and oxygen or air from a ring of jets into a slag bath in a direction that keeps the bath stirred. The coal particles are, for the most part, entrained in the slag. The slag promotes the gasification reactions and certain constituents, such as iron oxide, act as oxygen-transfer agents through oxidation and reduction reactions. In the Atomics International Molten Salt gasifier, the coal is fed together with air or oxygen alone into a bath of molten sodium carbonate where it is gasified. The sodium carbonate acts as a dispersing medium and as a heat sink with high heat transfer rates for absorbing the heat generated in the exothermic oxidative reactions and for distributing this heat for the endothermic gasification reactions. Most importantly, it also reacts with and adsorbs the sulfur in the coal, as well as retains the ash.

The temperature in molten media gasifiers depends upon the melt, but as a group they have the advantage that they can accept all types of coals and provide an efficient heat sink for the promotion of gasification reactions. However, a limitation is that the processes must handle the molten media and, for example, in the case of molten salt, regenerate and recycle an extremely corrosive material.

Tables 4.9 and 4.10 illustrate typical characteristics and product gas compositions for the gasifiers mentioned. The characteristics and products cannot be compared directly, since they are largely determined by the melt employed.

Table 4.9 Typical operating characteristics of indirectly heated fluidized bed gasifiers and molten media gasifiers*

	Fluidized Bed		Molten Media	
	Indirect Heat (Air)		Steam/Oxygen	Air
Characteristic	CO_2 Acceptor	COGAS	Rummel-Otto	AI Molten Salt
Pressure, MPa	1.1	0.2–0.5	0.1–2.5	0.1–2
Temperature,* °C				
Gasifier	815	870	1480–1700	980
Gas exit	N/A	N/A	815–980	925
Steam/Oxidant, kg/kg	0.48†	0.57	0.4	0
Oxidant, kg/GJ gas	116	N/A	40	123
Cold gas efficiency, %	77	50‡	72	78

* Lignite for CO_2 Acceptor, all others bituminous coals.

† Steam to gasifier, air to regenerator.

‡ Based on pilot plant; 70–75% estimated for commercial plant.

N/A Not available.

Table 4.10 Typical product gas composition* in mole percent for indirectly heated fluidized bed gasifiers and molten media gasifiers†

| Constituent | Fluidized Bed Indirect Heat (Air) | | Molten Media | |
	CO₂ Acceptor	COGAS	Steam/Oxygen Rummel-Otto	Air AI Molten Salt
CO	15	31	54	30
CO₂	9	7	14	4
H₂	59	58	31	13
CH₄	14	4	—	2
N₂	3	—	1	48
Other	—	—	—	3
Total	100	100	100	100
GCV, MJ/m³	14.2	12.5	10.4	5.9

* Dry basis after scrubbing and cooling.

† Lignite for CO_2 Acceptor; all others bituminous coals.

4.2 STEAM/OXYGEN AND STEAM/AIR GASIFICATION

Before discussing the performance of commercial or prototype gasifiers, it is useful to provide a basis for their operating characteristics by briefly reviewing some of the properties of ideal gasifiers. We recall that the ideal gasification route can be represented by the overall reaction of steam and carbon to form methane with the rejection of 50 percent of the carbon as carbon dioxide

$$C + H_2O(g) \rightleftharpoons \tfrac{1}{2}CH_4 + \tfrac{1}{2}CO_2 \qquad \Delta H^\circ_{1000} = +5.8 \text{ kJ/mol} \qquad (4.2)$$

We recall further that while this reaction is relatively thermally neutral, suggesting that gasification could be made to proceed with little heat input, methane formation is slow (unless catalyzed) relative to the other principal gasification reactions

$$C + H_2O(g) \rightleftharpoons CO + H_2 \qquad \Delta H^\circ_{1000} = +135.9 \text{ kJ/mol} \qquad (4.3)$$

$$C + 2H_2O(g) \rightleftharpoons CO_2 + 2H_2 \qquad \Delta H^\circ_{1000} = +101.1 \text{ kJ/mol} \qquad (4.4)$$

The reactions (4.3) and (4.4) are highly endothermic and so require a heat supply. Although the required heat is generally obtained by burning carbon in oxygen, the transfer of heat to the gasifier may be by indirect means via electricity or a heat transfer medium. In this section, however, we are concerned with systems in which the oxidant is introduced directly into the gasifier either with air or as relatively pure oxygen. Such systems are said to be directly heated or autothermal. We will exclude from our discussion the cyclic gas producers, briefly mentioned in the previous section, in which the oxidant (air) and the steam are alternately introduced into the gasifier, as these seem less likely to be developed for future use.

The intended use of the product gas and the coal properties determine whether the oxygen is supplied as air or in the pure form, and the relative rates of steam, oxygen, and coal fed to the gasifier. Consider the limiting case in which no steam is used and the oxygen supplied in air is sufficient to form carbon monoxide only (cf. stoichiometric oxygen constraint in Section 3.2). Then, approximately:

$$C + \tfrac{1}{2}(O_2 + 4N_2) \rightleftharpoons CO + 2N_2 \qquad \Delta H^\circ_{1000} = -111.9 \text{ kJ/mol} \qquad (4.5)$$

This gas is called producer gas, contains about 33 percent by volume of carbon monoxide, and has a calorific value of about 4.2 MJ/m^3. If in addition, steam is supplied to the gasifier then the product gas, sometimes called a "semi-water gas," will contain hydrogen and have a gross calorific value of about 7 MJ/m^3. Such gases are suitable for burning on site, for example to produce electricity in gas turbines, or in so-called "combined cycle plants" in which the hot exhaust gases from the gas turbine are used to raise steam for additional electricity generation.

If the gas is intended for chemical synthesis such as Fischer-Tropsch, methanol, or substitute natural gas (SNG) production, then the presence of nitrogen is undesirable. As discussed previously, the presence of inerts increases the volume of gas to be processed or transported and hence increases processing and transmission costs. While it is shown in Section 5.3 that air separation to produce pure oxygen is expensive, and can in fact amount to 25 to 35 percent of the cost of the product gas in SNG production,[7] it is currently the chosen method of overcoming the nitrogen barrier. The use of a pure oxygen feed to the gasifier does, however, have advantages other than the absence of nitrogen in the product and the concomitant decrease in the size of downstream processing equipment. For instance, the smaller quantity of gas produced means a correspondingly smaller sensible heat loss from the gasifier. More important perhaps is that the gasifier can be operated economically at elevated pressures, as the volume of oxygen to be compressed is only one-fifth the equivalent volume of air. Elevated pressure operation in turn means smaller equipment sizes and increased methane production, and the final gas product will require only minimal, if any, compression to pipeline entry pressures of 5.5 to 7 MPa.

The semi-water gas mentioned above becomes a water gas or blue water gas (Section 1.2) if nitrogen is not present, and would typically have a calorific value in the vicinity of 13 MJ/m^3. In accordance with the material in Section 3.3, we shall call a gas which consists mainly of carbon monoxide and hydrogen a "synthesis gas," as it may be processed to yield a variety of synthetic fuels and chemicals. It may be used for production of SNG which has a calorific value of 37 MJ/m^3, in which case it is desirable that the synthesis gas have a significant methane content, or it may be used for synthesis of methanol or gasoline, in which case the presence of methane is undesirable. A review of Figures 3.11 and 3.12 will show how the composition of the synthesis gas may be controlled by the temperature and pressure of the gasifier, as well as by the feed rates of oxidant and steam.

As shown in Eq. (4.3), the heat required for the gasification of carbon with steam to form carbon monoxide and hydrogen at 1000 K is 135.9 kJ/mol. If this heat is provided by combustion of carbon to carbon dioxide, which yields 394.5 kJ/mol (Table 2.15), then it is necessary to burn 135.9/394.5 = 0.34 moles of carbon for each mole gasified. This results in the thermally neutral chemical equation for gasification to carbon dioxide, carbon monoxide, and hydrogen

$$1.34C + 0.34O_2 + H_2O(g) \rightleftharpoons 0.34CO_2 + CO + H_2 \qquad \Delta H^\circ_{1000} \approx 0 \qquad (4.6)$$

The oxygen needed to satisfy the thermal constraint is a minimum for the case shown here, since it forms carbon dioxide. If part of the oxygen is consumed to produce carbon monoxide, which is favored in high temperature gasifiers, then a greater quantity of oxygen is required to satisfy the thermoneutrality constraint. In other words, the minimum oxygen to provide the heat of reaction for Eq. (4.3) is 0.34/1.34, or 0.25, moles of oxygen per mole of carbon consumed. Actual oxygen rates may be from one and a half times to more than double this value, depending on heat losses from the gasifier and the degree of preheating and drying of the coal and other feeds. Heat losses from the gasifier are not only those through the gasifier walls, but also the sensible heat of the gases and ash or slag leaving the gasifier. The amount of oxygen required will be reduced to the extent that the moderately exothermic gas shift reaction and the highly exothermic carbon-hydrogen reaction occur within the gasifier.

In practice, the oxygen rate is further governed by coal properties and by safety considerations. Highly reactive coals, or coals having a high volatile content, normally have a lower oxygen requirement than less reactive or higher-grade coals. This is because more of the carbon content in reactive coals is devolatilized in reactions that are less endothermic than the reaction of fixed carbon with steam. Oxygen rates may also be reduced to avoid pockets of high oxygen concentration occurring in the gasifier which might result in an explosion. Excessive oxygen concentrations can occur due to difficulties in monitoring the coal feed rate, or due to poor mixing in the feed and within the gasifier. There is usually a higher risk factor in gasifiers with a low carbon content or low coal inventory. The degree of carbon conversion will be reduced in gasifiers in which the oxygen rate is decreased below the thermal or stoichiometric requirement for safety reasons.

To estimate the steam requirement, Eq. (4.6) can again be used. The equation shows that 1/1.34, or 0.75, moles of steam are required per mole of carbon consumed. If the oxygen rate is greater than that given in Eq. (4.6), then more carbon will be oxidized by oxygen, so less steam will be required per unit of carbon fed to the gasifier. More than the stoichiometric quantity of steam is, however, required to satisfy the chemical equilibrium constraint and sometimes to control temperatures within the gasifier.

An indication of the effect of gasifier temperature and pressure on the amount of steam in the feed that is decomposed can be obtained from Figure 3.12. Pressure has a small effect, while increasing temperature tends to moderately decrease the amount of steam decomposed at equilibrium. In practice, kinetics and the carbon inventory in the gasifier dominate equilibrium effects. For instance, in moving bed slagging gasifiers good steam consumption is obtained due to the excess carbon present, the relatively long contact times, and the high temperature which enhances reaction rates. Although temperatures are as high in entrained flow gasifiers, steam decomposition is lower because of the significantly lower contact times and because there is little excess carbon.[7] In non-slagging systems, considerable excess steam is used for temperature control, and steam utilization is consequently poor.

The selection of steam and oxidant rates, as well as gasifier temperature and pressure conditions, is in general based on many other considerations in addition to those for thermoneutrality of Eq. (4.6). Were SNG the desired end product, then the formation of methane within the gasifier at low temperature and high pressure might be an overriding consideration. In this respect, the development of multistage gasifiers with the object of both fully converting the reactants and producing the desired product appears appealing. For instance, the Hygas gasifier has a high temperature stage for hydrogen production with efficient steam utilization followed by lower temperature stages for methane production. The resultant gasifier off gas has a maximum methane content of about 20 percent by volume. There is, however, a serious question as to the feasibility and economics of attempting to promote different reactions with their opposing operating conditions within a single unit, in contrast to leaving the gas shift and methanation steps to separate reactors, each controlled at optimum conditions.

If gas for Fischer-Tropsch or methanol synthesis is desired, then methane formation must be suppressed, and selection of gasifier conditions might be expected to be more straightforward. Again, however, other factors need be taken into account. For instance, the selection of a high temperature to achieve a maximum rate of steam decomposition means that operation may be above the ash melting point. Increased oxygen consumption may also be expected at the higher operating temperature because of the higher sensible heat loss with the off gas and with the slag. If on the other hand, temperatures are to be maintained below the range at which the ash fuses, then additional steam may be required to moderate the temperature in the combustion zone. Much of this steam will not be decomposed, and will represent a sensible heat loss from the gasifier and a latent heat loss from the system. It has been mentioned, however, that in fluidized bed gasifiers there is good heat transfer between exothermic combustion reactions and endothermic gasification reactions. Additional steam feed for temperature moderation will, therefore, be less than in moving bed gasifiers.

The most economical operating conditions will depend on gasifier costs and reliability as well as on the relative costs of steam and oxygen. Capital costs for boilers and air separation plants have been estimated to be 16 percent, and 10 percent, respectively, of the total cost of a plant producing SNG from coal.[8] The heat required to vaporize water, assuming a boiler efficiency of 85 percent, is about 2.7 MJ/kg steam, while energy consumption at an air separation plant, expressed in terms of the heat required, is about 4.0 MJ/kg oxygen product (Section 5.3). There will be additional requirements for oxygen compression and steam superheating. The relative feed rates of steam and oxygen to a gasifier in which 80 percent of the steam is converted and in which there are no heat losses would, for the simplified reaction scheme of Eq. (4.6), be about 3.7 moles of steam per mole of oxygen.

The oxygen and steam rates for the ideal reaction schemes discussed above are summarized in Table 4.11, and form a basis for assessing the rates in the gasifiers discussed below.

Lurgi Dry Ash Gasifier

The Lurgi dry ash gasifier, Figure 4.2, is a relatively high pressure, moderate temperature moving bed gasifier.[4] It was the first high pressure gasifier and was introduced

Table 4.11 Steam and oxygen feed rates in hypothetical gasifiers

Oxygen/Carbon	Steam/Carbon	Remarks
0.50	0	Stoichiometric requirement for producer gas production, Eq. (4.5).
0.25	0.75	Stoichiometric/thermoneutral requirement for synthesis gas, Eq. (4.6). Oxygen rate to supply heat of reaction; no heat losses.
0.50	1.50	Assuming 50% of the steam is not decomposed. Extra steam may be for temperature control and/or for equilibrium requirements. Oxygen assumed to be about double the theoretical rate.

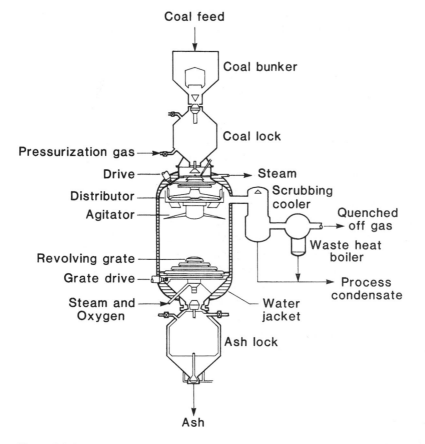

Figure 4.2 Lurgi dry ash gasifier.

in commercial operation in Germany in 1936. The nominal operating pressures of commercial units are now about 3 MPa and the coal throughput is up to 800 t/d. Pilot units have been run at 5.3 MPa and development is targeted for 10 MPa. Gasifiers with diameters of 4 m are in use, and 5 m diameter units are under test. Figure 4.3 is a photograph of a Lurgi gasifier vessel during erection.

Figure 4.3 Photograph of Lurgi gasifier vessel during erection. The gasifier is to be connected to one of the two waste heat boilers shown, and to coal and ash lock hoppers that are not shown in the photograph. *(Courtesy of Lurgi Kohle und Mineralöltechnik GmbH.)*

There are several reaction zones in the gasifier similar to those shown in Figure 3.8. Devolatilization, including the vaporization of coal moisture, occurs at the top where the incoming coal is heated by the rising hot gases. Further down where the coal char reaches temperatures of 650 to 815°C, the gasification reactions proceed. Temperatures in the bottom combustion zone, where the coal burns to supply heat for the endothermic gasification reactions, are kept below the ash melting point, which may vary from about 980 to 1370°C depending on the coal.

As discussed for moving bed gasifiers, the Lurgi gasifier does not readily handle fines, and caking or swelling coals generally require an oxidative pretreatment or internal agitation to prevent plugging and resultant excessive pressure drops. The coal is crushed and screened to between 3 and 40 mm, and the fines are removed and used for raising steam in pulverized coal boilers. If the coal is very friable, the fines content may exceed plant fuel requirements and must be disposed of or sold. Predrying the coal is not necessary; lignites with moisture contents up to 35 percent may be used directly. Ash contents up to 35 percent or more are also not a problem, although the poorer heat transfer characteristics of ash may require increased steam rates to prevent clinkering. The sized coal is introduced through the coal lock hopper into the reactor, which is maintained at a pressure of up to 3.2 MPa. Coal residence time within the gasifier is approximately one hour. The ash is withdrawn through the ash lock hopper at the bottom. Heat recovery in this counterflow system is good; the ash is cooled to about 400 to 500°C by the incoming steam/oxygen mixture, while the off gases are cooled to 200 to 540°C depending on the moisture in the coal.

A typical gas composition based on bituminous coal gasification is given in Table 4.4. The H_2/CO ratio given in the table is 2.2, although ratios less than 2 are possible depending on steam rates and coal type. Much of the steam introduced into the gasifier is used to control the temperature in the combustion zone, with up to 50 percent of the steam passing through the system unreacted. High steam concentrations do, however, serve to shift the equilibrium of the carbon-steam and the gas shift reactions to promote hydrogen formation. The intermediate gasification zone temperature results in some methane formation, although much of the methane in the off gas probably results from coal pyrolysis in the devolatilization zone. Tars and oils are also released in this zone and exit with the off gases from which they are later removed by washing with water in the scrubber/cooler.

Material and Thermal Balance Calculations

Some material and thermal balance calculations for a specific plant design will now be presented. Their purpose is to demonstrate how actual oxygen and steam rates can differ from the theoretical ranges given in Table 4.11, as well as to indicate the relative energy contributions to the cold gas efficiencies given in the previous section. The calculations are presented here in some detail to demonstrate the procedure followed and so to allow balances presented later in the book to be checked using the raw data provided. Data used here have been adopted from the Great Plains Gasification Project,[9] which plans to produce 7×10^6 m³/d (250×10^6 ft³/day) SNG from lignite with Lurgi gasifiers.

Table 4.12 Lignite analysis for Lurgi dry ash gasifier balances

Proximate		Ultimate (daf)	
Constituent	Mass %	Element	Mass %
Moisture	36.0	Carbon	71.5
Ash	7.4	Hydrogen	4.8
Volatile matter	27.2	Nitrogen	1.4
Fixed carbon	29.4	Sulfur	1.3
Total	100.0	Oxygen	21.0
		Total	100.0
GCV, MJ/kg	16.8		

   The reported proximate and ultimate lignite analyses are given in Table 4.12, and show the high moisture, volatile, and oxygen content typical of these coals. As the lignite feed rate to the gasifier is given on an as-received basis, it is convenient to convert the daf ultimate analysis to include the moisture and ash. This is done by multiplying each concentration by the factor (mass daf lignite)/(total mass), or (100 − 36.0 − 7.4)/100 = 0.57. The results are shown in Table 4.13. Note that the hydrogen and oxygen values in the main analysis are for these elements combined in the dry coal only. The values given for "total" hydrogen and oxygen include the hydrogen and oxygen in the moisture, namely 2/18 and 16/18 respectively of the 36 percent H_2O present. The reported total lignite feed rate for the Great Plains Project with 23 gasifiers is 281.8 kg/s, from which it is now a simple matter to determine the feed rate of each element, as well as the moisture and the ash entering the gasifiers with the coal.

   The steam and pure oxygen feeds to the gasifiers are also reported, as is the production rate of the byproduct tars and oils. These values are listed in Table 4.14. The off gas rate is not given, nor is the net amount of condensate collected in the

Table 4.13 Calculated lignite analysis and corresponding feed rate for Lurgi dry ash gasifiers

Constituent	Mass %	Feed Rate, kg/s
Moisture	36.0	101.4
Carbon	40.5	114.1
Hydrogen	2.7	7.6
Nitrogen	0.8	2.3
Sulfur	0.7	2.0
Oxygen	11.9	33.5
Ash	7.4	20.9
Total	100.0	281.8
Total hydrogen	6.7	18.9
Total oxygen	43.9	123.7

Table 4.14 Overall material balance for Lurgi dry ash gasifiers

	kg/s	mol/mol C in Lignite
IN		
Lignite (as-received)	281.8	—
Oxygen	57.0	0.19
Steam	297.8	1.74
Total	636.6	
OUT		
Ash (7.4% of lignite)	20.9	
Tars and oils	13.5	
Off gas* plus scrubber condensate† (by difference)	602.2	
Total	636.6	

* Off gas rate from Table 4.16 = 573.2 kg/s or 3.03 mol (wet)/mol C in lignite.

† Scrubber condensate = 602.2 − 573.2 = 29.0 kg/s (see text).

crude gas scrubber (see Figure 4.2). However, the sum of these two quantities may be estimated by closing the material balance, and the off gas rate determined by a carbon balance. The rate so calculated includes the moisture, ammonia, hydrogen sulfide, and other gases, but excludes the tars and oils which are listed separately, and the fly ash which is included in the total ash. The carbon feed rate in Table 4.13 may be used to calculate the oxygen- and steam-to-carbon ratios which are included in Table 4.14. The fact that the oxygen rate of 0.19 mol/mol C is significantly less than the theoretical range of 0.25 to 0.5 is probably because about half the carbon is volatile and gasified in thermally neutral pyrolysis reactions, rather than in the highly endothermic gasification reactions. The steam rate, on the other hand, is high and this is reflected in the H_2/CO ratio of nearly 2.5 as seen in Table 4.15.

The raw off gas leaves the gasifier at 273°C and 3 MPa. Its composition is reported on a dry molar basis, and the moisture content of the tar-, oil-, and dust-free gas is given as 46.7 percent by mass. The mean molar mass of the dry gas can be calculated using Eq. (8) in Table 2.13, following which the mass concentrations on a dry and wet basis may be determined using Eqs. (10) and (11) in that table. The results are summarized in Table 4.15. Gross calorific values were estimated by summing the products of the heat of combustion (Table 2.12) and the fractional molar concentration for each compound. Total carbon, hydrogen, and oxygen concentrations were determined by summing the products of the mass fraction of the element in the compound and the compound concentration. For example, the contribution to the total carbon by the 10.6 percent carbon monoxide is $(12/28) \times 10.6 = 4.5$ percent.

A carbon material balance may now be completed to determine the off gas rate. To do this it has been assumed that the tars and oils contain 85 percent carbon and 15 percent hydrogen by mass. The carbon balance shown in Table 4.16 suggests that the off gas rate is 573.2 kg/s, and that the net condensate rate from the crude gas scrubber is, by difference, 29.0 kg/s. Table 4.16 includes hydrogen and oxygen

Table 4.15 Off gas composition for Lurgi dry ash gasification of lignite

Constituent	Mol wt	Mole %, Dry	Mass %, Dry	Mass %, Wet
H_2	2	38.8	3.5	1.9
CO	28	15.6	19.9	10.6
CO_2	44	32.5	65.2	34.7
CH_4	16	10.8	7.9	4.2
$C_{2+}(C_3H_8)$	44	0.8	1.6	0.9
Other (N_2)	28	1.5	1.9	1.0
H_2O	18	—	—	46.7
Total		100.0	100.0	100.0
Total carbon				17.9
Total hydrogen				8.3
Total oxygen				72.8
Mean molar mass		21.9 (dry)		19.9 (wet)
GCV, MJ/kg		12.2		6.5
GCV, MJ/m³		11.9		5.8

balances which close within 2 percent, and so give extra confidence to the off gas rate as determined by the carbon balance. Mass balances for other elements such as nitrogen and sulfur could be included as well, although the results for the minor elements tend to be less reliable. In cases where several flow rates and/or concentrations are unknown, they can sometimes be estimated by solving the set of equations made up from the overall material balance and as many elemental material balances as needed. In these and other balances in the text, rounded molar masses are used (Table 2.1).

In addition to elemental material balances, material balances for certain compounds may be set up to determine the amount of the compound that is generated or consumed in the process. For example, the total water entering the gasifier is the coal moisture plus the steam, or $101.4 + 297.8 = 399.2$ kg/s. The steam leaving with the off gases is $0.467 \times 573.2 = 267.7$ kg/s, or some 67 percent of the total water entering the gasifier. However, the moisture in the coal is driven off at the top of the gasifier and does not get a chance to react. Of the steam fed in at the bottom of the gasifier ($267.7 - 101.4$)/297.8, or 56 percent, is not decomposed, and represents a significant heat loss. In the plant design, heat is recovered from the off gas in waste heat boilers, and some of the steam is further converted to hydrogen in the gas shift reactor.

An approximate thermal balance for the gasifier is shown in Table 4.17. Sensible heats and energy for compression and feed preheating are generally insignificant and have therefore been neglected. The sensible heat of the off gas is, of course, not used in calculating the cold gas efficiency (Eq. 2.60a) which is here found to be 79 percent. It is planned to burn the tars and oils on site to raise steam and, as mentioned, some of the (latent) energy content of the off gases is to be recovered in waste heat boilers. If these energy sources are used to reduce the input energy for steam and oxygen production, a modified cold gas efficiency (Eq. 2.60b) of 77 percent is obtained. The overall thermal efficiency for the SNG product in the Great Plains Project is reported to be 67 percent.[9]

Table 4.16 Elemental material balances for Lurgi dry ash gasifiers

Element	kg/s
Carbon	
IN	
Carbon with lignite (Table 4.13)	114.1
Total	114.1
OUT	
Carbon in tars and oils, 0.85×13.5	11.5
Carbon in off gas (by difference)*	102.6
Total	114.1
Hydrogen	
IN	
Hydrogen in lignite (Table 4.13)	18.9
Hydrogen in steam $(2/18) \times 297.8$	33.1
Total	52.0
OUT	
Hydrogen in tars and oils, 0.15×13.5	2.0
Hydrogen in off gas, 0.083×573.2	47.6
Hydrogen in scrubber condensate, $(2/18) \times 29.0$	3.2
Total	52.8 ($+1.5\%$ error)
Oxygen	
IN	
Oxygen with lignite (Table 4.13)	123.7
Oxygen for combustion (Table 4.14)	57.0
Oxygen in steam feed, $(16/18) \times 297.8$	264.7
Total	445.4
OUT	
Oxygen in off gas, 0.728×573.2	417.3
Oxygen in scrubber condensate, $(16/18) \times 29$	25.8
Total	443.1 (-0.5% error)

* Off gas contains 17.9% C, so off gas rate $= 102.6/0.179 = 573.2$ kg/s.

The Slagging Lurgi Gasifier

The major disadvantage of the dry ash Lurgi gasifier is that temperatures have to be maintained below the ash melting point to prevent clinkering. As a consequence steam utilization is poor and efficiencies are relatively low. Moreover, low reactivity coals are of concern since they require higher steam/oxygen ratios to hold down the temperature. To improve the performance of the gasifier—and in line with the current leaning towards high temperature operation for increased reaction rates, higher steam utilization, and reduced methane formation—a slagging version of the gasifier has been developed.[5,10] This work was initiated in Germany in 1954-55, and since then was

Table 4.17 Approximate thermal balance for Lurgi dry ash gasifiers

	Heat content, MJ/kg	Mass flow, kg/s	Heat flow, MJ/s	Heat flow, % of Total
IN				
Lignite (as-received)	16.8	281.8	4734	82.1
Oxygen separation	4.0*	57.0	228	4.0
Latent heat of steam	2.7	297.8	804	13.9
Total			5766	100.0
OUT				
Off gas	6.5	573.2	3726	64.6
Tars and oils	42.0	13.5	567	9.8
Latent heat of steam in off gas	2.7	267.7	723	12.6
Unaccounted (by difference)	—	—	750	13.0
Total			5766	100.0

Cold gas efficiency $= \dfrac{3726}{4734} \times 100 = 79\%$

Modified cold gas efficiency† $= \dfrac{3726}{[5766 - (0.5 \times 723) - 567]} \times 100 = 77\%$

* Heat requirement for oxygen manufacture.

† Assuming 50% of the latent heat of the steam in the off gas can be recovered, and the tars and oils are burned to reduce plant energy input.

continued sporadically in Britain—notably in Westfield, Scotland, in the mid-1970s—in collaboration with the British Gas Council.

The top of the slagging gasifier is identical to the dry ash unit, while the bottom part has been modified as may be seen from Figure 4.4. The ash melts at the high temperatures reached in the combustion zone, up to 2000°C, and forms a slag which runs into the quench chamber—a water bath in which the slag forms small granules of solid ash. Temperatures, and therefore reaction rates, are high in the gasification zone so that the coal residence time is about 20 minutes, or one-third of that in the dry ash Lurgi. Steam decomposition approaches 100 percent, and because the off gas now has less heat capacity, top temperatures are very much dependent on the moisture content of the coal, as is the steam content of the off gas.

In addition to improved steam decomposition, the higher temperatures also result in lower carbon dioxide concentrations due to the formation of carbon monoxide via the Boudouard reaction. The H_2/CO ratio in the off gas is typically 0.5 in comparison with 2 for the dry ash Lurgi. The off gas still contains methane, but this is probably derived from devolatilization at the top of the gasifier rather than by carbon hydrogenation. The off gases in fact contain more tars and oils than in the dry ash system, apparently because of the lower heat capacity of the off gases and consequently lower top temperatures.

To provide a comparison with the dry ash Lurgi, a material and thermal balance for a slagging Lurgi gasifier is summarized in Tables 4.18 to 4.21. Data have been adopted from the Conoco Pipeline Gas Demonstration Plant design, in which a bituminous coal is converted to 0.54×10^6 m³/d (19×10^6 ft³/day) of SNG.[11]

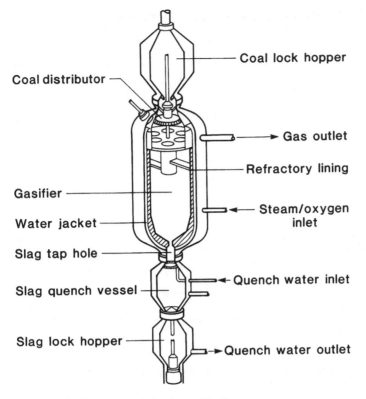

Figure 4.4 British Gas/Lurgi slagging gasifier.[11]

Table 4.18 Bituminous coal analysis for Lurgi slagger balances

Proximate		Ultimate (as-received)	
Constituent	Mass %	Constituent	Mass %
Moisture	2.5	Moisture	2.5
Ash	22.5	Carbon	58.5
Volatile matter	35.0	Hydrogen	4.2
Fixed carbon	40.0	Nitrogen	0.9
Total	100.0	Sulfur	4.7
		Oxygen	6.7
GCV, MJ/kg	24.4	Ash	22.5
		Total	100.0

Table 4.19 Overall material balance for Lurgi slagger gasifying bituminous coal

	kg/s	mol/mol C in Coal
IN		
Coal (as-received)	11.3	—
Fluxing agent	1.1	—
Oxygen	4.8	0.27
Steam	5.8	0.58
Total	23.0	
OUT		
Slag (22.5% of coal) & flux	3.6	—
Tars and oils (net)	0.5	—
Off gas (given)	16.0	1.38
Scrubber condensate (net, by difference)	2.9	—
Total	23.0	

Although this coal is very different from the lignite of the dry ash case, a comparison of the relative flow rates and energies clearly illustrates the essential characteristics of the slagger. The steam rate, for instance, is about one-third that of the dry ash gasifier, and as can be seen from the composition of the off gas in Table 4.20, it is virtually all decomposed. The H_2/CO ratio is 0.47, while the carbon dioxide concentration is only 4.3 percent. Although the calorific value of the gas is similar to that in the dry ash case, the cold gas efficiency is significantly higher, at about 91 percent in comparison to 79 percent. Note from the overall material balance in Table 4.19 that a fluxing agent is used to assist in slag removal by decreasing the slag viscosity. The cost of fluxing will, of course, detract from the gain in thermal efficiency.

Table 4.20 Off gas composition for Lurgi slagger gasifying bituminous coal

Constituent	Mole %	Mass %
H_2	27.4	2.6
CO	58.0	77.4
CO_2	4.3	9.0
CH_4	6.7	5.1
C_{2+}	0.8	1.7
H_2S	2.0	3.2
N_2	0.6	0.8
Moisture	0.2	0.2
Total	100.0	100.0
Mean molar mass	21.0 (wet)	
GCV, MJ/kg	15.7	
MJ/m³	14.7	

Table 4.21 Approximate thermal balance for Lurgi slagger gasifying bituminous coal

	Heat Content, MJ/kg	Mass Flow, kg/s	Heat Flow, MJ/s	Heat Flow, % of Total
IN				
Coal (as-received)	24.4	11.3	275.7	88.8
Oxygen separation	4.0*	4.8	19.2	6.2
Latent heat of steam	2.7	5.8	15.7	5.0
Total			310.6	100.0
OUT				
Off gas	15.7	16.0	251.2	80.9
Tars and oils	40.2	0.5	20.1	6.5
Latent heat of steam in off gas	2.7	0.5	1.4	0.4
Unaccounted (by difference)	—	—	37.9	12.2
Total			310.6	100.0

$$\text{Cold gas efficiency} = \frac{251.2}{275.7} \times 100 = 91\%$$

$$\text{Modified cold gas efficiency} = \frac{251.2}{[310.6 - (0.5 \times 1.4) - 20.1]} \times 100 = 87\%$$

* Heat requirement for oxygen manufacture.

Texaco Gasifier

The Texaco gasifier is an entrained flow, high pressure, and high temperature slagging system that is based on a commercially proven crude oil gasifier. Pilot plants operating with coal are presently being tested, including one in Germany with a nominal coal feed rate of 150 t/d and two 15 t/d units in California. Figure 4.5 is a photograph of the Texaco-Ruhrchemie-Ruhrkohle pilot plant at Oberhausen-Holten, Germany. Development work has been concerned mainly with methods for feeding the coal into the high pressure reactor as well as for removing the slag from the system—problems not encountered when gasifying oil. The gasifier is a tall cylindrical refractory-lined carbon steel shell that may be divided into a gasification and a quench zone as shown in Figure 4.6.

Coal is slurried with water and introduced together with oxygen at the top of the gasifier where the coal is partially oxidized to provide the heat for the gasification reactions. The dominating gasification reactions are the carbon-steam reaction to form carbon monoxide and hydrogen, and the Boudouard reaction in which carbon dioxide is reduced to more carbon monoxide. The higher the temperature the higher the hydrogen and carbon monoxide concentrations in the off gas. Increasing the pressure, however, reduces these concentrations in favor of carbon dioxide and methane. As shown in Table 4.6, methane is not a significant constituent of the Texaco off gas. Tars, oils, and phenols are also not present, as the entire gasification process occurs in the high temperature combustion region so that any pyrolysis products formed are immediately decomposed. The entrained slag remaining is separated from the gas stream in the quench bath and is discharged through the ash lock. Residence times are of the order of a few seconds.

To feed the coal into the reactor, it is pulverized and mixed with 40 to 80 percent water to form a slurry which is then pumped up to pressure. In one version of the

Figure 4.5 Photograph of Texaco-Ruhrchemie-Ruhrkohle 150 t/d pilot plant at Oberhausen-Holten, Germany. *(Courtesy of Ruhrchemie Ag, Ruhrkohle GmbH.)*

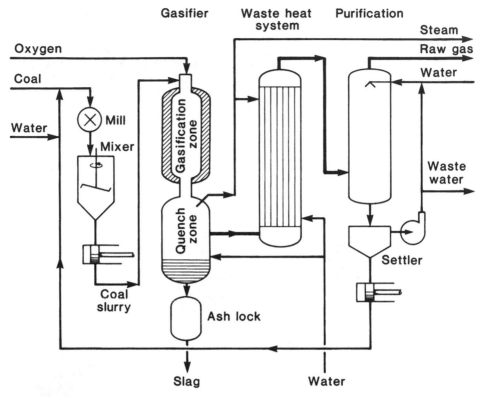

Figure 4.6 Texaco process.[12]

design, the slurry is preheated to vaporize part or all of the water before being blown into the gasifier via a cyclone in which some steam is separated to achieve the desired steam-to-coal feed ratio.[5] The more recent literature suggests that the prevaporization stage has been eliminated. In this version, the steam-to-coal ratio is kept to the desired low level by using a "slurry" of up to 70 percent solids content. The slurry is forced directly into the gasifier using a piston pump.[12,13] Reactor pressures of up to 8 MPa have been used. Temperatures have to be maintained below about 1400°C to avoid damage to the refractory lining. Extra water is used to control the temperature, but unlike the dry ash Lurgi, steam rates are not excessive and it is essentially all converted to hydrogen at the high reactor temperature. The hot slag and gases entering the quench system do, however, produce considerable quantities of steam, so that the moisture content of the off gases may be from 20 percent to as high as 60 percent by volume. Part of this steam may be later converted to hydrogen in a shift reactor to increase the H_2/CO ratio from typically less than 1 to the value required for the downstream synthesis. An alternative gas withdrawal system is possible in cases where steam is not required in the off gases. In this design, the hot gases are removed directly from the gasification zone and cooled outside the gasifier in waste heat boilers.

Heat recovery within the cocurrent Texaco gasifier is poor. Preheaters for the coal slurry and oxygen are required, while the slag and gas cooling produce considerable steam that cannot be directly used for gasification. However, the gasifier itself is of a simpler design than that of the Lurgi, and the higher pressures can mean smaller reactors for the same throughput. In addition, caking and swelling coals can be handled without the need for pretreatment.

Two simplified material and thermal balances are presented here to emphasize the wide range in operating conditions reported in the literature for the Texaco gasifier. Case A, given in Tables 4.22 and 4.23, is based on gasification of Illinois No. 6 bituminous coal to produce a fuel gas for a combined cycle power plant.[13] Case B, given in Tables 4.24 and 4.25, is based on gasification of Montana Rosebud subbituminous coal to form a synthesis gas for indirect liquefaction.[14] The estimated cold gas efficiencies are about 79 percent for Case A and 67 percent for Case B. The difference in efficiency is largely due to the very different CO/CO_2 ratios in the gases produced and consequently in their calorific values. In Case A most of the carbon ends up as carbon monoxide, whereas in Case B some 42 percent of the carbon is completely oxidized to carbon dioxide. In both cases it has been assumed that the coal slurry is fed directly to the gasifier so that no additional energy is required to raise steam.

The difference in the gas composition apparently stems from the very different feed conditions.[15] Although oxygen feed rates are similar and in line with the theoretical rates of Table 4.11, the relative steam rates differ by more than a factor of 2. In fact, the steam rate in Case A is even lower than the stoichiometric requirement given in Table 4.11 for gasification according to Eq. (4.6), suggesting that a significant portion of the coal is gasified by the Boudouard reaction. The high steam rate in Case B may reflect an attempt to increase the H_2/CO ratio in the off gas, and so reduce the amount of gas shift required prior to synthesis. The quantity of steam leaving with the gas is also high and promotes the gas shift reaction. It is not clear, however, that operating the gasifier to suit the downstream shift reactor is justified in view of the resulting low gasifier efficiency.

Table 4.22 Overall material balance around Texaco gasifiers for bituminous coal: Case A

	kg/s	mol/mol C in Coal
IN		
Coal (as-received; 66.6% C, 4.2% H_2O)	105.1	—
Oxygen	88.0	0.5
Slurry water (coal slurry is 66.5% solids)	46.2	0.4
Total	239.3	
OUT		
Quenched slag (coal is 9.6% ash)	10.1	—
Dry off gas*	192.5	1.6
Moisture in off gas (16% by mass of wet gas)	36.7	0.4
Total	239.3	

* 35.1% H_2; 51.7% CO; 10.6% CO_2 by volume, dry.

Table 4.23 Approximate thermal balance around Texaco gasifiers for bituminous coal: Case A

	Heat Content, MJ/kg	Mass Flow, kg/s	Heat Flow, MJ/s	Heat Flow, % of Total
IN				
Coal (as-received)	28.5	105.1	2995	89.5
Oxygen separation	4.0*	88.0	352	10.5
Total			3347	100.0
OUT				
Dry off gas	12.3	192.5	2368	70.8
Latent heat of steam in off gas	2.7	36.7	99	3.0
Unaccounted (by difference)	—	—	880	26.2
Total			3347	100.0

$$\text{Cold gas efficiency} = \frac{2368}{2995} \times 100 = 79\%$$

$$\text{Modified cold gas efficiency} = \frac{2368}{[3347 - (0.5 \times 99)]} \times 100 = 72\%$$

* Heat requirement for oxygen manufacture.

Table 4.24 Overall material balance around Texaco gasifiers for subbituminous coal: Case B

	kg/s	mol/mol C in Coal
IN		
Coal (as-received; 50.6% C, 24.7% H$_2$O)	115.2	—
Oxygen	88.2	0.6
Slurry water (coal slurry is 44.8% solids)	78.7	0.9
Makeup quench water (by difference)	20.4	—
Total	302.5	
OUT		
Quenched slag (coal is 9.7% ash)	11.2	—
Water with quenched slag	1.2	—
Dry off gas*	177.7	1.6
Moisture in off gas (38.8% by mass of wet gas)	112.4	1.3
Total	302.5	

* 33.6% H$_2$; 37.3% CO; 27.2% CO$_2$ by volume, dry.

Table 4.25 Approximate thermal balance around Texaco gasifiers for subbituminous coal: Case B

	Heat Content, MJ/kg	Mass Flow, kg/s	Heat Flow, MJ/s	Heat Flow, % of Total
IN				
Coal (as-received)	20.0	115.2	2304	86.7
Oxygen separation	4.0*	88.2	353	13.3
Total			2657	100.0
OUT				
Dry off gas	8.7	177.7	1546	58.2
Latent heat of steam in off gas	2.7	112.4	303	11.4
Unaccounted (by difference)	—	—	808	30.4
Total			2657	100.0

$$\text{Cold gas efficiency} = \frac{1546}{2304} \times 100 = 67\%$$

$$\text{Modified cold gas efficiency} = \frac{1546}{[2657 - (0.5 \times 303)]} \times 100 = 62\%$$

* Heat requirement for oxygen manufacture.

In either case, the efficiencies obtained for the Texaco gasifier are less than those for the Lurgi slagger. The main difference between the two is the larger losses in the case of the Texaco system as a consequence of the higher temperatures of the gas and slag leaving the system, as well as the higher steam content of the off gas.

Koppers-Totzek and Winkler Gasifiers

The Koppers-Totzek gasifier is a commercially established, high temperature, entrained flow gasifier capable of partially oxidizing a wide variety of feed stocks at pressures slightly above atmospheric.[4,5] It is in many ways similar to the Texaco gasifier. The coal is dried, pulverized, and conveyed with nitrogen from storage to the gasifier service bin and feed bins. Variable speed coal screw feeders then continuously discharge the coal into a mixing nozzle, where a mixture of steam and oxygen entrains the pulverized coal. Moderate temperature and high burner velocity prevent oxidation of the coal.

The gasifier (Figure 4.7) is a refractory-lined steel shell equipped with a steam jacket for producing low pressure process steam. A two-headed burner has heads 180° apart and can handle about 350 t/d of coal. The gasifier is about 7.5 m high and has an inside diameter of up to 3.5 m. A four-headed gasifier, with burners 90° apart has about twice the volume and can handle about 750 t/d of coal. The burner heads are opposed to ensure that particles escaping from one burner will be burnt in the opposite burner. Carbon is oxidized in the gasifier producing a high temperature flame zone of about 1900°C. Heat losses and the endothermic carbon-steam reactions reduce the exit temperature to around 1500°C. The coal is gasified nearly to completion in times on the order of 1 s. Carbon conversion, dependent on the reactivity of the coal, is about

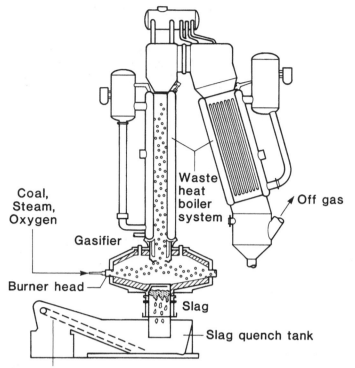

Coal,
Steam,
Oxygen

Gasifier

Burner head

Waste
heat
boiler
system

Off gas

Slag

Slag quench tank

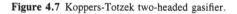

Slag extractor

Figure 4.7 Koppers-Totzek two-headed gasifier.

96 to 98 percent. At the prevailing high operating temperatures, gaseous and vaporous hydrocarbons emanating from the coal decompose so rapidly that coagulation of coal particles during the plastic stage does not occur. Thus any coal can be gasified regardless of its caking property, ash content, or ash fusion temperature. Also, only gaseous products are produced; no tars, condensable hydrocarbons, or phenols are formed. Approximately 50 percent of the coal ash drops out as slag. The remainder of the ash leaves the gasifier as fine slag particles entrained in the exit gas. Gas leaving the gasifier may be directly water quenched to solidify entrained slag droplets, if necessary, before passing through a waste heat boiler where high pressure steam is produced. The gas exits the waste heat boiler at 175 to 260°C.

Development work with Shell is aimed at producing an improved system operating at up to 3 MPa.[16] Shell-Koppers, like Texaco, has a 150 t/d unit under test, and a comparison of the performance characteristics of these competitive units would be most interesting.

Another gasifier that has had wide commercial use but is now mainly of historical interest is the Winkler.[4,5] Unlike the other gasifiers discussed so far, the Winkler, shown in Figure 4.8, is a fluidized bed system. It was the first commercial application of fluidized bed technology, and was developed in the late 1920s with the aim of gasifying coal fines, which were cheaper than lump coal. It operates at atmospheric pressure or slightly above, with a bed temperature of around 950°C. Gasification reactions are similar to the Lurgi except that there is less methane formation due to the lower pressure, and pyrolysis products including tars and oils are decomposed within the gasifier. Being both a low pressure and low temperature process, gasification rates are low and residence times of up to 3 hours are required to achieve acceptable conversions. Carbon conversions are, in fact, as low as 60 percent, partly due to the difficulty in measuring the coal feed rate, and the need to maintain an excess of coal relative to oxygen for safety considerations. In addition, the fluidized bed system cannot handle caking and swelling coals, and the Winkler is really only suitable for use with the highly reactive lignites for which it was originally intended. Consequently, Koppers-Totzek and Lurgi units are replacing the Winkler in commercial applications.

Steam/Air Gasifiers

Most of the gasifiers discussed so far in this section can be operated on air rather than oxygen if a fuel rather than a synthesis gas is to be produced. In many cases, gasifiers

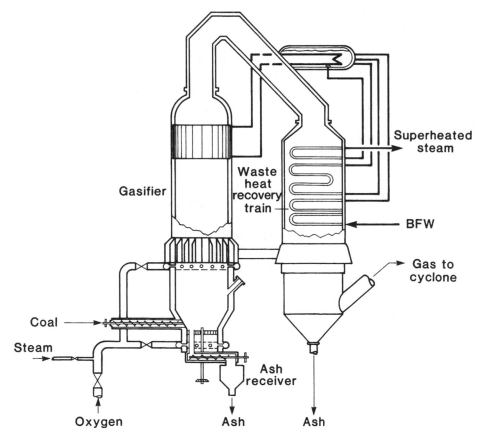

Figure 4.8 Winkler fluidized bed gasifier.

developed specifically for gasification with air are being further developed for relatively pure oxygen feeds. The Westinghouse gasifier is an example.[4,5] A 15 t/d process development unit for production of a low-CV fuel gas intended for on-site combined cycle electricity generation has been under test since 1975. A gas turbine capable of efficiently burning this type of fuel was being developed concurrently by Westinghouse. More recently, the process development unit has been tested on oxygen to examine its potential for producing a medium-CV gas for both the petrochemical industry and for power production via fuel cells.

The gasifier itself is a pressurized fluidized bed system that operates at about 1.5 MPa and 1000°C. Other operating characteristics are given in Table 4.7. In one version of the gasifier, the crushed and dried coal is first devolatized in a separate fluidized bed reactor. One purpose of the devolatizer is to pretreat caking coals to prevent agglomeration and plugging in the gasifier fluidized bed. Simultaneous removal of sulfur gases by absorption into dolomite has also been tested. Some success has been had in operating the gasifier as a single stage unit without initial coal devolatilization, although bridging problems can arise in this mode. Advantages claimed for the Westinghouse gasifier include the capability of handling all ranks of coals with high carbon conversion, and the absence of tars and oils in the product gas.

Another elevated pressure, fluidized bed gasifier designed originally for steam/air operation is the U-Gas process developed by the Institute of Gas Technology.[4,5] Operating temperatures are around 1000°C, at pressures from 0.5 to 2.5 MPa. Additional characteristics are given in Table 4.7. The gasifier is a vertical vessel with external cyclones for returning elutriated fines to the bed. A sloped grid at the bottom containing one or more inverted cones serves as the oxidant and steam distributor and an agglomerated ash outlet. Part of the fluidizing steam and oxidant flows through nozzles in the grid and the remaining gas flows upward at high velocity through a discharge at the cone apex. High temperatures accompany the high velocities, and the ash particles soften and agglomerate, eventually becoming heavy enough to fall out of the bed. The fluidizing velocities are from 0.6 to 1.2 m/s, and the bed is approximately 45 percent ash. This system is said to produce high carbon conversion efficiencies comparable to slagging gasifiers. Caking coals may require oxidative pretreatment with steam at about 400°C in a fluidized bed. A demonstration plant feeding steam and oxygen to produce 5×10^6 m³/d of 10.4 MJ/m³ medium-CV gas is being designed under sponsorship of the U.S. Department of Energy by Memphis Light, Gas and Water, a municipal utility.

The Foster Wheeler (FW) gasifier is a pressurized, entrained flow, slagging, two stage system blown with air and steam.[4,6] It is a hypothetical model based on an O_2-blown gasifier developed by Bituminous Coal Research. Dried pulverized coal is fed into the upper stage reactor shown in Figure 4.9, using gas recycled from the off gas quench or purification section as a transport medium. Steam is injected into this upper stage and heat is provided by the hot gases from the lower combustion/gasification stage. The unreacted char that exits with the off gases at about 980°C is separated in cyclones and returned to the lower stage for combustion in air. The char is transported using steam which also serves to control the temperature in the lower stage to avoid refractory damage. Slag is water quenched before being removed from the reactor. System pressure is about 2.5 MPa.

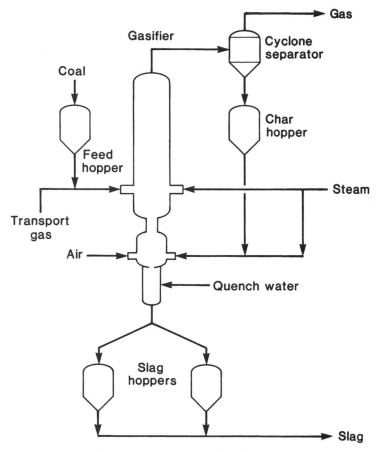

Figure 4.9 Foster Wheeler entrained flow gasifier (idealized design).

An overall material balance for the FW gasifier is shown in Table 4.26. Of particular interest is the low steam rate of 0.14 moles per mole of carbon in the coal. If the moisture entering with the coal and air, and the steam with the char is included, this value increases to 0.21, but is still significantly below the theoretical steam required for gasification according to Eq. (4.6). Clearly the gasifier behaves more like a producer operating according to Eq. (4.5), than a steam gasifier of the Lurgi type. The energy balance for the gasifier, given in Table 4.27, shows a cold gas efficiency of 90 percent and a modified cold gas efficiency of 87 percent. The energy required to compress the air feed to the gasifier has been neglected in the calculation of the modified cold gas efficiency; similarly the energy for oxygen compression was neglected in the previous examples.

Data for the material and energy balances were adopted from a conceptual design for a combined cycle power plant rated at 1200 MW net and which consumes 8800 t/d of bituminous coal.[6] A simplified block flow diagram for such a plant is shown in Figure 4.10. The overall efficiency for electricity generation is 40.5 percent, as given in Table 4.28. Included in the table is the thermal balance for a similar sized plant

Table 4.26 Overall material balance around air-blown Foster Wheeler gasifiers for bituminous coal

	kg/s	mol/mol C in Coal
IN		
Coal (dried; 68.1% C, 2.0% H_2O)	102.6	—
Air: oxygen	66.9	0.36
nitrogen	220.4	—
moisture	4.5	—
Steam	15.1	0.14
Recycle gas*	9.3	—
Char/steam (40% char)	1.1	—
Total	419.9	
OUT		
Slag (9.8% of coal) + unconverted		
carbon	10.5	—
Off gas*	409.0	2.89
Char	0.4	—
Total	419.9	

* 3.4% CH_4, 13.7% H_2, 28.7% CO, 3.0% CO_2, 48.1% N_2, 2.0% H_2O, 0.7% H_2S, 0.4% NH_3 by volume.

Table 4.27 Approximate thermal balance around air-blown Foster Wheeler gasifiers for bituminous coal

	Heat Content, MJ/kg	Mass Flow, kg/s	Heat Flow, MJ/s	Heat Flow, % of Total
IN				
Coal (dried)	28.5	102.6	2924	96.3
Latent heat of steam	2.7	15.1	41	1.3
Recycle gas	6.4	9.3	60	2.0
Char	32.8	0.4	13	0.4
Total			3038	100.0
OUT				
Off gas	6.4	409.0	2618	86.2
Latent heat of steam in off gas	2.7	6.0	16	0.5
Char	32.8	0.4	13	0.4
Unaccounted (by difference)	—	—	391	12.9
Total			3038	100.0

$$\text{Cold gas efficiency} = \frac{2618}{2924} \times 100 = 90\%$$

$$\text{Modified gas efficiency} = \frac{2618}{[3038 - (0.5 \times 16) - 13]} \times 100 = 87\%$$

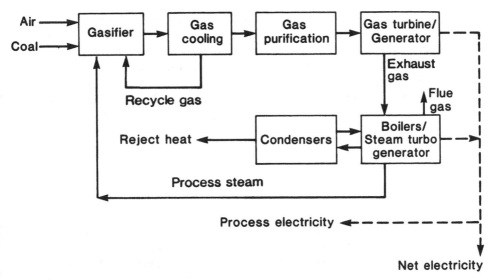

Figure 4.10 Simplified combined cycle plant block-flow diagram.

based on oxygen-blown gasifiers. Differences in the energy consumption for individual processes are not directly comparable for integrated plants that are correctly designed for optimum overall performance as opposed to optimum efficiency of isolated units. What is clear, however, is that the energy expended in oxidant supply for the oxygen-blown case is not recovered by way of improved performance associated with lower gas volumes and higher calorific values, and consequently, significantly higher efficiencies are generally obtained for air-blown gasifiers. An important difference in gasifier operation is that in the oxygen-blown case, nearly four times as much steam is fed to the gasifiers, about half of which goes through undecomposed. The additional

Table 4.28 Combined cycle efficiencies for plants using Foster Wheeler gasifiers[6]

	% of Total Energy	
	Air-blown	Oxygen-blown
IN		
Bituminous coal	100.0	100.0
OUT		
Net electric power	40.5	38.5
Oxygen manufacture	—	10.8
Air compression	3.2	—
Off gas cleaning	2.1	2.5
Heat rejected, steam condensers	28.9	20.1
Gasifier losses	1.8	1.8
Stack losses	18.2	20.5
Other losses	5.3	5.8
Total	100.0	100.0

steam is presumably required for temperature control; in the air-blown case, the high nitrogen flow rates provide sufficient heat capacity to prevent excessive temperatures. A consequence of the additional steam is that the off gas now contains more carbon dioxide and less carbon monoxide in accordance with the equilibria shown in Figure 3.11. While the calorific value of the gas is increased from 7 MJ/m³ to 11 MJ/m³ due to the absence of diluent nitrogen, the modified cold gas efficiency drops from 87 to 76 percent, reflecting the energy needed for oxygen manufacture and the extra steam requirements.

4.3 INDIRECTLY HEATED AND MOLTEN MEDIA GASIFICATION

As discussed in Section 4.1, both indirectly heated and molten media gasifiers are fluid bed reactors that employ a dispersing medium over and above the usual gasifying agents such as steam and oxidant. In the case of the indirectly heated gasifiers an oxidant is not added directly to the gasifier; instead coal or char is burned in air in a separate unit and the heat so generated is then transferred to the gasifier using a recirculating heat transfer medium. Fluidized bed reactors are invariably selected for these gasifiers because of their good solid-to-solid and solid-to-gas heat transfer characteristics. Some examples of indirectly heated and molten media gasifiers follow.

Indirectly Heated Gasifiers

The COGAS char gasifier[4] is an indirectly heated unit that has been developed for use in combination with a pyrolysis process to gasify the residual char remaining after pyrolysis. The combined system is called the COGAS process and is discussed in Section 6.3 in connection with the description of the pyrolysis section. A pilot plant unit with a nominal throughput of over 50 t/d of char has been operated in England since 1974 by the British Coal Utilization Research Association.

The gasifier is shown in Figure 4.11. Char from the pyrolysis section is fluidized by steam and is gasified to produce mainly carbon monoxide and hydrogen. Heat for the endothermic reactions is provided by burning char in a separate combustion unit. Char fines in the gasifier that are entrained by the synthesis gas product are separated from it in a cyclone and fall into the combustion unit. Here they are entrained and heated by the hot combustion flue gases, which lift them to a separation chamber from where they once again fall back into the gasifier. Smaller fines that are not recovered in the separation chamber are removed from the flue gases in a cyclone and used for fuel in the combustion unit. The combustion flue gases are taken to downstream units for further energy recovery.

The COGAS gasifier operates at slightly elevated pressures of 0.2 to 0.5 MPa, and gasification temperatures are about 870°C. Combustion temperatures are in excess of 1900°C, which is sufficient to melt the ash. Molten ash flows from the combustion unit into a quench chamber where it solidifies, settles, and is then taken to disposal. A typical synthesis gas composition for the COGAS gasifier is given in Table 4.10.

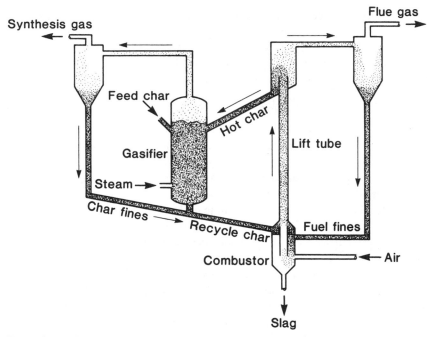

Figure 4.11 COGAS char gasifier.

As shown, the gas has a calorific value of about 12.5 MJ/m³, which is within the range of medium-CV gases. Furthermore, the H_2/CO ratio in the gas is close to 2, so extensive shift is not required prior to downstream synthesis. Much of the sulfur originally in the coal is not removed during pyrolysis and remains in the char, which when gasified is reduced to hydrogen sulfide and has to be removed from the synthesis gas before further processing.

The COGAS process typifies how a circulating heat transfer medium, in this case char, can be heated and then reintroduced into the gasifier to provide the required heat without diluting the product gas with nitrogen and without the need to produce pure oxygen. An obvious drawback of the process compared to directly heated systems is its greater complexity and, consequently, its greater operational and control difficulties. The cold gas efficiency of the gasifier itself is low, at about 50 percent. However, in the integrated plant much of the sensible heat leaving with the synthesis and flue gases is recovered. Additional details on the process and its development are discussed in Section 6.3.

In some indirectly heated systems, the heat transfer medium is selected to serve secondary functions that improve the overall gasifier performance. For example, in the CO_2 Acceptor process,[4,5] dolomite (CaO·MgO) or lime (CaO) is used as a heat transfer medium that simultaneously removes or accepts much of the carbon dioxide and hydrogen sulfide from the synthesis gas:

$$CaO + CO_2 \rightleftharpoons CaCO_3 \qquad \Delta H^\circ_{298} = -183 \text{ kJ/mol} \qquad (4.1)$$

$$CaO + H_2S \rightleftharpoons CaS + H_2O(g) \qquad \Delta H^\circ_{298} = -65 \text{ kJ/mol} \qquad (4.7)$$

In fact, much of the heat required is supplied by the highly exothermic reaction of the dolomite and carbon dioxide, rather than by the sensible heat of the heat transfer medium, as in the case of the COGAS gasifier. Removal of carbon dioxide and hydrogen sulfide from the synthesis gas directly in the gasifier has several important advantages. Firstly, the gas shift reaction

$$CO + H_2O \rightleftharpoons CO_2 + H_2 \qquad (4.8)$$

is promoted so that the gas has a high H_2/CO ratio (Table 4.10) and does not need to be shifted. Secondly, the volume of gas is relatively low, thereby reducing the size of downstream process units. Finally the energy required for synthesis gas purification is considerably reduced. The Conoco Coal Development Company tested the performance of the CO_2 Acceptor process in a pilot plant designed to produce 14 000 m³/d of SNG from 40 t/d of coal and 3 t/d of dolomite. When the pilot plant was shut down in 1977, some problems related to materials handling and corrosion remained unresolved. The process nevertheless remains an interesting one and a short description is therefore included here.

Raw lignite is crushed to approximately 3 to 6 mm in a hot-gas-swept impact mill and lifted with hot flue gas to the preheater. The lignite, originally about 30 percent moisture, enters the preheater with about 5 to 15 percent moisture. The preheater in a large plant would be fluidized with flue gas for energy conservation. The preheater operates at atmospheric pressure, and a temperature of 200 to 260°C which is insufficient to pyrolyze the coal. The preheated lignite is then fed to the bottom of the char phase of the gasifier, shown in Figure 4.12, via lock hoppers and is fluidized by steam with which it reacts. The gasifier and regenerator pressures in the pilot plant were 1.1 MPa. Higher pressures would increase gasifier throughput, but pressures above 2 to 3 MPa would unfavorably affect the chemical equilibrium in the acceptor regeneration. Hot dolomite or acceptor from the regenerator showers through the lignite bed absorbing carbon dioxide and supplying heat to maintain the gasifier at about 800°C. If higher gasification temperatures were used carbon dioxide would be reduced to carbon monoxide by the Boudouard reaction, and carbon dioxide partial pressures would be insufficient to react with the acceptor. The CO_2 Acceptor process is therefore restricted to the lower gasification temperature range and consequently to highly reactive coals such as lignite or subbituminous coal. Methane production is, however, significant (see Table 4.10), while phenols, tars, and oils released in pyrolysis reactions are decomposed in the fluidized bed and are not present in the product gas. Due to the high hydrogen and methane contents, the gas is most suitable for SNG production.

Acceptor accumulates at the bottom of the gasifier from where it is removed. Char, amounting to about 33 percent of the carbon in the feed lignite, is allowed to flow from the top of the gasifier. Withdrawal rates of char and acceptor are controlled so that the residence time of acceptor in the gasifier is less than the residence time of char. Both the char and acceptor are taken to the regenerator, where the char is burned to provide the heat for regenerating or calcining the dolomite.

In the pilot plant design, acceptor is lifted to the regenerator by flue gas. Air may be used if the lift line is made resistant to corrosion. Fresh acceptor is added and used acceptor is withdrawn at a rate required to maintain the activity of the acceptor. Char

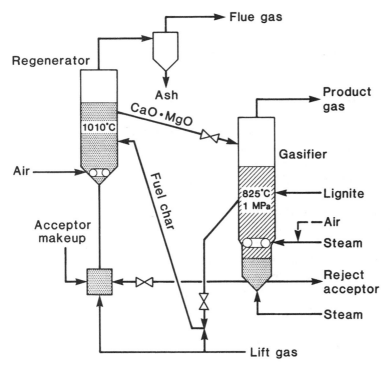

Figure 4.12 Simplified schematic of the CO_2 Acceptor gasifier.

is lifted to the regenerator with nitrogen in the pilot plant. Flue gas may also be used. In the regenerator, air is added to burn the char and raise the temperature to about 1000°C to reverse the acceptor reaction and drive off the carbon dioxide and sulfur. Ash is elutriated from the regenerator and recovered in cyclones from which it is released via lock hoppers. The ash contains less than 10 percent carbon.

The acceptor contains some 80 to 90 percent of the sulfur in the lignite, but only about 10 percent of this is driven off in the regenerator. The regenerator flue gas contains about 350 mg/m³ sulfur, primarily as sulfur dioxide, and requires scrubbing before being released to the atmosphere. About 70 percent of the sulfur in the coal is rejected as calcium sulfide in the spent acceptor, along with the ash from the regenerator. The cold gas efficiency of the CO_2 Acceptor is stated to be as high as 77 percent, while the overall efficiency for an integrated plant would be expected to reflect the effects of the reduced amount of acid gas removal required and the elimination of a gas shift step. On the other hand, additional costs for supply, recirculation, and regeneration of the acceptor, as well as flue gas scrubbing, must be included in comparisons with conventional processes.

Molten Bath Gasifiers

In the CO_2 Acceptor process the added medium, namely calcined limestone or dolomite, not only provides the heat for gasification, but also improves the quality of the synthesis gas by virtue of its chemical activity. In molten bath gasifiers, the added media is selected mainly for its chemical functions, while the heat requirements are provided autothermally by introduction of oxygen or air directly into the gasifier. Molten iron is frequently advocated as a suitable medium due to the affinity of iron for sulfur:

$$Fe + S \rightarrow FeS \qquad (4.9)$$

Although the prime purpose of the added medium is to provide a sulfur-free fuel or synthesis gas, it also serves ancillary functions. It acts as a dispersing medium for the coal; it acts as a heat sink with high heat transfer rates for distributing the heat of combustion; it also physically retains the ash in the coal. In some cases the medium may also catalyze the gasification reactions.

The Atomics International (AI) and Kellogg gasifiers are examples of reactors that employ sodium carbonate as the molten medium.[4,5] The sodium carbonate reacts with the sulfur to form compounds that catalyze the partial oxidation of the coal. The sulfur in the molten salt is removed as hydrogen sulfide in the regeneration system and elemental sulfur may be recovered as a byproduct. Some operating data for the AI Molten Salt process are given in Tables 4.9 and 4.10; a simplified schematic of a molten salt gasifier is shown in Figure 4.13.

Another molten bath gasification process is the Rummel-Otto gasifier, one version of which has been proven commercially in sizes up to 250 t/d of coal. The molten medium in this gasifier is slag. Unlike other molten media gasifiers, the intent is not to remove pollutants from the product gases; instead the slag acts more as a heat shield and as a "reservoir" for oxygen. Iron oxide in the slag accepts oxygen:

$$2FeO + \tfrac{1}{2}O_2 \rightarrow Fe_2O_3 \qquad (4.10)$$

and can then gasify carbon by partial oxidation:

$$Fe_2O_3 + C \rightarrow 2FeO + CO \qquad (4.11)$$

The slag further acts as a promoter of the gasification reactions.

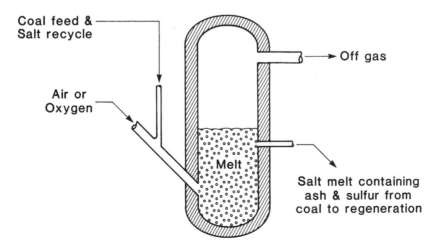

Figure 4.13 Simplified schematic of a molten salt gasifier.

The Rummel-Otto gasifier,[4,17] shown schematically in Figure 4.14, consists essentially of three zones. In the top zone the product gases are cooled by direct water quench, the middle zone is an entrained flow combustion/gasification zone of the Texaco and Koppers-Totzek type, while the bottom zone is the slag bath. Dried and pulverized coal is entrained in a recycle gas and injected along with steam and oxygen into the gasifier. The injection nozzles are oriented tangentially above the slag bath and the entering feed imparts a slow vortex motion to the slag. The turbulence induced in the slag bath by the injected feed enhances mass and heat transfer between the two phases. Lime may be added as a fluxing agent to decrease the viscosity of the slag. Feed rates for a bituminous coal are shown in the overall material balance[17] for the gasifier in Table 4.29. The steam rate is low but sufficient to moderate the flame temperature to about 1650°C. The reactor vessel in the combustion/gasification zone is both refractory lined and water jacketed. Additional heat shielding is provided by the slag bath.

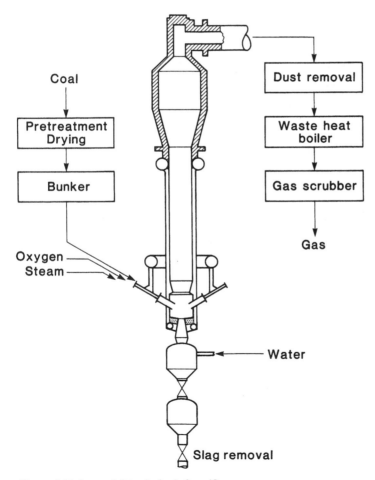

Figure 4.14 Rummel-Otto single shaft gasifier.

Table 4.29 Overall material balance around Rummel-Otto gasifiers for bituminous coal

	kg/s	mol/mol C in Coal
IN		
Coal (dried; 67.6% C, 2.0% H$_2$O)	110.3	—
Oxygen	93.2	0.5
Steam	22.0	0.2
Quench water	149.8	1.3
Recycle gas	6.7	—
Total	382.0	
OUT		
Slag (coal is 11.8% ash)	10.9	—
Fly ash in off gas	2.5	—
Off gas	368.1	3.1
Total	382.0	

The gases leaving the gasification zone have high carbon monoxide concentrations as is typical for the high oxygen-to-steam feed ratio and high reaction temperature. On entering the quench zone they react further with water to increase the H$_2$/CO ratio to about 1 as shown in Table 4.30. However, about 75 percent of the total water entering the gasifier remains undecomposed, leaving a gas containing more than 40 percent moisture by volume. The water quench serves to cool the gases to 980°C before they enter the cyclones where unreacted char is separated and reinjected into the gasifier. The cold gas efficiency for this plant is about 80 percent as derived by the approximate thermal balance given in Table 4.31. This calculated efficiency for the advanced conceptual design is significantly higher than the value of 72 percent given in Table 4.9, based on operating pilot plant data.

The material and thermal balances presented here are based on a conceptual design for a gasifier operating at about 3.5 MPa, whereas the units used commercially were

Table 4.30 Off gas composition for Rummel-Otto gasifier feeding bituminous coal

Constituent	Mole %, Wet	Mass %, Wet
H$_2$	22.1	2.3
CO	23.5	33.7
CO$_2$	9.9	22.3
CH$_4$	0.0	0.0
H$_2$S	0.7	1.2
N$_2$	0.3	0.4
H$_2$O	43.5	40.1
Total	100.0	100.0
Mean molar mass	19.5	
GCV	6.0 MJ/m^3	6.8 MJ/kg

Table 4.31 Approximate thermal balance around Rummel-Otto gasifiers for bituminous coal

	Heat Content, MJ/kg	Mass Flow, kg/s	Heat Flow, MJ/s	Heat Flow, % of Total
IN				
Coal (dried)	28.4	110.3	3133	85.8
Oxygen separation	4.0*	93.2	373	10.2
Latent heat of steam	2.7	22.0	59	1.6
Recycle gas	12.3	6.7	86	2.4
Total			3651	100.0
OUT				
Off gas	6.8	368.1	2503	68.6
Latent heat of steam in off gas	2.7	147.6	399	10.9
Unaccounted (by difference)	—	—	749	20.5
Total			3651	100.0

$$\text{Cold gas efficiency} = \frac{2503}{3133} \times 100 = 80\%$$

$$\text{Modified cold gas efficiency} = \frac{2503}{[3651 - (0.5 \times 399)]} \times 100 = 73\%$$

*Heat requirement for oxygen manufacture.

atmospheric pressure gasifiers. The feasibility of operating at high pressure is dependent, as in the Texaco case, on the successful development of high pressure, solid feed systems. In addition the conceptual design calls for a 9000-t/d, 7-m-diameter unit in contrast to the 250-t/d, 1.8-m-diameter units which have been used commercially. Both the commercial and conceptual units described are the so-called "single shaft" gasifiers in which heat is provided directly by oxygen combustion. A double shaft version has been tested in which the gasifier contains two concentric reactors with a common slag bath. Combustion occurs in one shaft and supplies heat indirectly by conduction through the shaft walls to the gasification shaft. As the two reactors are not directly connected, air, instead of oxygen, can be used as oxidant. This version of the gasifier has so far proved unsuccessful.

4.4 HYDROGASIFICATION AND CATALYTIC GASIFICATION

We have seen that the trend in steam-and-air- or oxygen-blown gasifiers is towards high temperatures and pressures with the purpose of decreasing gasification costs by increasing reaction rates, increasing throughput, and increasing steam and carbon conversion. It was also shown that at high temperatures gasification can be typified by the reaction

$$C + H_2O \rightleftharpoons CO + H_2 \tag{4.3}$$

with some additional carbon being oxidized to carbon dioxide to provide the necessary heat. The off gas characteristically contains little or no methane and has a low H_2/CO ratio. While the absence of methane is ideal for synthesizing liquid products, current synthesis reactors require that the H_2/CO ratio be increased to at least 2 by gas shift. It has been argued that the need to have an additional gas shift process step is more than compensated for by the higher overall cost effectiveness of having the gasifier operate at optimum conditions.[18]

There have nevertheless been some efforts to develop gasifiers yielding off gases having both high methane concentrations and high H_2/CO ratios. Such gasifiers would be used specifically for the production of methane or SNG. Before describing some of these gasifiers, it is useful to briefly review the operating conditions theoretically required to enhance methane formation. An examination of Figures 3.11 and 3.12 shows that low temperatures, high pressures, and a high hydrogen-to-oxidant ratio in the feed increase the concentration of methane in the equilibrium gas mixture. Increasing the hydrogen in the gasifier by increasing the steam (H_2O) rate is not expected to significantly improve methane yields as illustrated by the following example.

Consider steam/oxygen gasifiers operating at a temperature of 1000 K and a pressure of 3.0 MPa; that is, at conditions favorable for methane formation. Let the feed in one case be 0.25 moles of pure oxygen and 0.75 moles of steam per mole of carbon, as suggested in Table 4.11 for synthesis gas production. The composition of the equilibrium gas mixture for these operating conditions, as calculated using Eq. (3.33), is shown in Table 4.32 under "Gasifier A." It can be seen that the low temperature, high pressure conditions do result in a significant methane concentration in the off gas, although in practice the value of 19.3 percent at equilibrium will

Table 4.32 Equilibrium off gas composition for three gasifier feed conditions

Gasifier temperature 1000 K; gasifier pressure 3 MPa (30 atm).

	Gasifier A	Gasifier B	Gasifier C
Gasifier feed, mol/mol C			
Pure O_2	0.25	0	0.25
Pure H_2	0	0	2.00
Steam	0.75	2.0	0
Off gas composition, mol %			
H_2	11.8	23.4	23.7
CO	29.8	9.8	10.8
CO_2	30.1	19.4	6.1
CH_4	19.3	12.6	49.3
H_2O	9.0	34.8	10.1
Total	100.0	100.0	100.0
H_2/CO ratio	0.4	2.4	2.2

Table 4.33 Operating characteristics for gasifiers in Table 4.32

	Gasifier A	Gasifier B	Gasifier C
Percent carbon converted to:			
CH_4	24.4	30.1	74.5
CO_2	38.0	46.5	9.3
CO	37.6	23.4	16.2
Total carbon converted, %	100.0	100.0	100.0
Total steam decomposed, %	85	58	—
Overall heat of reaction, kJ/mol C	− 56	+ 52	− 159

probably not be obtained due to the slow rate of methane formation relative to other gasification reactions. As pointed out in Section 3.2, the concentrations of components in the off gas mixture can be misleading when comparing gasifier products, as the total off gas rate may be different. The concentrations in Table 4.32 have therefore been converted to carbon yields in Table 4.33, where it is seen that for the case just considered some 24 percent of the carbon is converted to methane.

Suppose we wish to further increase the methane formation by increasing the hydrogen in the feed to 2 moles of hydrogen per mole of carbon. This is the stoichiometric quantity of hydrogen required to convert all the carbon to methane. If pure hydrogen is not available, then the necessary hydrogen can be provided by feeding steam at a rate of 2 moles per mole of carbon. Since the steam contains combined oxygen, the pure oxygen feed may be eliminated in the interest of maintaining a high hydrogen-to-oxygen ratio in the gasifier. The heat required for gasification will now of course have to be supplied indirectly as discussed in the previous section. The resulting equilibrium gas mixture is shown under "Gasifier B" in Tables 4.32 and 4.33. It is seen from the latter table that the increased steam rate to the gasifier has not resulted in a large increase in methane production. This is because the combined oxygen in the steam has tied up nearly as much of the carbon in the form of carbon oxides as occurred in gasifier A. The lower methane concentration shown in Table 4.32 is a consequence of the large amount of undecomposed steam remaining in the off gas. While this excess steam does represent an energy loss from the system, the large steam rates enhance the gas shift reaction within the gasifier so that the amount of downstream shifting required is considerably reduced.

If, however, pure hydrogen were available, the situation would change dramatically as shown under "Gasifier C" in Tables 4.32 and 4.33. Nearly 75 percent of the carbon is converted directly to methane in the gasifier; carbon dioxide and steam concentrations in the off gas are low, and further, only a light methanation step is required to produce a pipeline quality product. It is not surprising, therefore, that considerable interest has been shown in hydrogasification, or gasification with a pure hydrogen feed, in spite of the high cost of hydrogen (see Section 5.4).

Hydrane Gasifier

The Hydrane gasifier is an example of a hydrogasifier which is fed with hydrogen and coal only.[4,19] The hydrogen is produced in a separate unit by steam/oxygen gasification of unreacted char from the hydrogasifier. The Hydrane process was investigated by the U.S. Bureau of Mines (now incorporated in the Department of Energy) from 1955 up to the mid-1970s. An experimental reactor, 7.5 cm in diameter and 9 m total height, with a coal throughput of about 130 kg/d was used in the later studies. The development work was pursued on the basis that hydrogenating coal to methane with hydrogen produced by steam/oxygen gasification of char is more thermally efficient than steam/oxygen gasification followed by methanation.[19]

The Hydrane gasifier, shown in Figure 4.15, consists of two separate reactors. The top or first reactor is a free-fall, dilute-phase gasifier in which coal particles are partially hydrogenated in a hydrogen/methane atmosphere. This reactor is essentially a pretreatment stage in which the agglomerating properties of the coal are destroyed. Most American coals have been found to exhibit severe caking properties in high pressure hydrogen atmospheres at elevated temperatures, and coal agglomeration with concomitant plugging of reactors has consequently proved to be a major problem in the development of hydrogasifiers.

This problem is overcome in the Hydrane gasifier by partial gasification of the

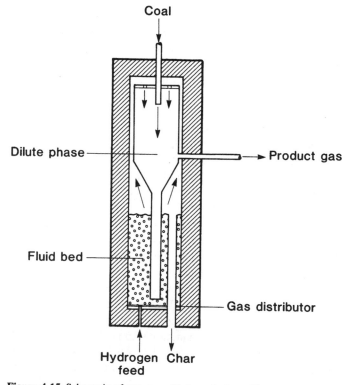

Figure 4.15 Schematic of two-stage Hydrane hydrogasifier.

coal in the dilute-phase stage, where the particles react out of contact with each other. Wall temperatures have to be maintained above about 750°C to prevent the coal particles from adhering to them. Although the experimental gasifier was operated at pressures as high as 20 MPa, a commercial system would probably not operate at pressures above 7 MPa, which is sufficient for pipeline entry without recompression. Pulverized coal, dried to a moisture content of about 2 percent, is fed into the dilute-phase reactor via a series of lock hoppers. The coal is rapidly heated, in part by the hot gases from the second reactor, but mainly by the highly exothermic heat of reaction

$$C + 2H_2 \rightleftharpoons CH_4 \qquad \Delta H^{\circ}_{1000} = -89.5 \, \text{kJ/mol} \qquad (4.12)$$

The gas/coal flow in the dilute-phase reactor is cocurrent, which allows devolatilization products to be cracked, and leaves the product gas relatively free of tars and oils. About 20 percent of the carbon, or essentially the volatile fraction of the coal, is gasified in the first stage.

A fluidized bed reactor was selected for the second stage of the original unit, although a moving bed version was also tested. The second stage operated at temperatures up to 1000 K. Heat evolved in this section may be used for raising part of the steam required for hydrogen production. About 50 percent of the carbon in the coal is required for hydrogen production in the steam/oxygen gasification system, so only some 35 percent of the unconverted carbon leaving the dilute-phase reactor need be gasified in the second reactor. More recently the possibility of simplifying the Hydrane reactor to a single stage process suitable for commercial exploitation has been investigated. The single stage would be an entrained flow reactor similar to the dilute-phase stage of the original unit, but incorporating a Rocketdyne coal injection system. A flash hydropyrolysis Rocketdyne system is discussed in Section 6.3.

To enable the reader to compare operating conditions for hydrogasification with those for conventional steam/oxygen gasification, a material and thermal balance for the Hydrane gasifier is given in Table 4.34. It is emphasized that these data are based on a small experimental unit operating under ideal conditions, and should not be taken

Table 4.34 Approximate material and thermal balances for Hydrane gasification

	Material Balance, kg	Thermal Balance		
		MJ/kg	MJ	% of Total
IN				
Bituminous coal (76% C, 6.0% ash)	100.0	32.2	3220	100.0
Hydrogen	10.0	—	0*	—
Total	110.0		3220	100.0
OUT				
Char and ash	52.5	—	0*	—
Dry off gas	51.1	54.7	2796	86.8
Moisture in off gas	6.4	2.4	15	0.5
Unaccounted (by difference)	0.0	—	409	12.7
Total	110.0		3220	100.0

* The char is used to produce hydrogen.

as being indicative of commercial-scale systems. Further, the inclusion of these data is not meant to imply that hydrogasifiers of the Hydrane type are technically proven. The total energy input to the Hydrane process is the coal fed to the gasifier. Some 47.5 percent of this coal is hydrogasified, with the remainder being withdrawn as char and used for hydrogen production. A typical gas composition, based on data collected from the experimental unit operating at 6.8 MPa, is given in Table 4.35 and confirms that high methane yields are attainable in practice by direct hydrogasification. The off gas has a calorific value of 29 to 30 MJ/m³ and contains about 94 percent of the methane required for a pipeline quality gas. Consequently only a light methanation, probably without shift, would be required to produce the final product.

Hygas Gasifier

The Hygas gasifier, developed by the Institute of Gas Technology,[4.20] was originally designed to receive a relatively pure hydrogen stream produced in a separate unit. Three systems were considered: a steam/oxygen gasification, as in the Hydrane process; an electrothermal gasification system in which the heat for steam gasification is provided electrically, thus reducing the amount of carbon dioxide to be removed; and a steam/iron process. The steam/oxygen process has been the method chosen, and in fact has been incorporated directly into the Hygas gasifier as shown in Figure 4.16. The carbon oxides cannot be removed from the steam/oxygen-gasification off gas in the integrated reactor, so the Hygas system is no longer a true hydrogasification process. Operating conditions of high pressure, 6.8 to 10.2 MPa, and moderate temperature in the hydrogasification stage, 930 to 980°C, are nevertheless selected to enhance methane formation, and the off gases typically have a methane concentration of 15 to 20 percent (Table 4.8), which represents up to 60 percent of the methane in the final SNG product.

The very high methane concentrations obtained in the Hydrane process cannot be realized in the integrated Hygas unit due to the lower hydrogen concentrations prevailing. As with the Hydrane system, the Hygas unit has a first dilute-phase stage followed by a second dense-phase fluidized bed stage. Again, about 20 percent of the coal is converted in the first stage, with a total of 45 to 50 percent of the coal being

Table 4.35 Off gas composition for Hydrane hydrogasifier

Constituent	Mole %	Mass %
H_2	21.0	3.1
CO	3.6	7.4
CO_2	—	—
CH_4	67.0	78.4
H_2O	8.4	11.1
	100.0	100.0
Mean molar mass	13.7	
GCV, MJ/kg	48.7	
MJ/m³	29.7	

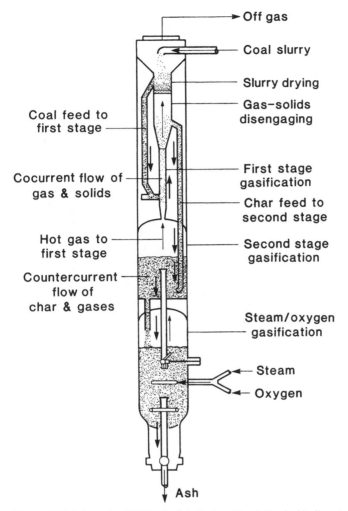

Off gas

Coal slurry

Slurry drying

Gas–solids
disengaging

Coal feed to
first stage

First stage
gasification

Cocurrent flow of
gas & solids

Char feed to
second stage

Hot gas to
first stage

Second stage
gasification

Countercurrent
flow of
char & gases

Steam/oxygen
gasification

Steam

Oxygen

Ash

Figure 4.16 Schematic of HYGAS pilot plant gasifier; 1.7 m inside diameter and 40 m tall (after Ref. 20).

hydrogasified. The remaining char descends to the bottom stage where the hydrogen-rich gas is produced in the presence of steam and oxygen. Caking coals normally require pretreatment to prevent agglomeration in the Hygas gasifier. Pretreatment involves partial oxidation using air in a fluidized bed at 400 to 450°C; heat evolved is used for steam generation.

Catalytic Gasification

The concept of hydrogasification, while superficially seeming to effectively yield a gas ideal for SNG production, in fact departs considerably from the idealized gasification reaction first discussed in Section 2.2:

$$2C + 2H_2O(g) \rightleftharpoons CO_2 + CH_4 \qquad \Delta H^\circ_{1000} = +11.6 \text{ kJ/mol} \qquad (4.13)$$

The ideal gasification reaction is essentially thermally neutral, so temperature control is not a problem. Further, the supply of a separate oxidant is not required, so the cost of producing pure oxygen or separating nitrogen from the off gas is avoided. Steam is the only gasification agent used, and downstream gas shift and methanation steps are eliminated.

By contrast, hydrogasification requires a hydrogen supply which is normally produced by the highly endothermic high temperature carbon-steam reaction

$$C + H_2O(g) \rightleftharpoons CO + H_2 \qquad \Delta H^{\circ}_{1500} = +135.8 \text{ kJ/mol} \qquad (4.14)$$

Pure oxygen must be supplied to provide the heat of reaction by combustion with some of the coal. In addition, a gas shift step is required to convert the carbon monoxide to additional hydrogen, and the carbon dioxide produced is normally removed upstream of the hydrogasifier using one of the processes discussed in Section 5.1. The relatively pure hydrogen stream is then reacted with carbon at low to moderate temperatures according to the carbon-hydrogen reaction, Eq. (4.12). Because this reaction is highly exothermic, some means of temperature control is required. In addition, the previously discussed problems of coal agglomeration in hydrogen atmospheres have to be countered. Consequently, both the hydrogasifier itself and the hydrogasification process in general are more complex than, say, steam/oxygen gasification followed by shift and methanation.

The integration of the hydrogen production step into the Hygas gasifier does simplify the overall process, but has resulted in a very tall and complex gasifier. In addition, the gas produced by the Hygas process, Table 4.8, is not that much better for SNG production than the off gas from the established Lurgi gasifier, Table 4.4.

Table 4.36 Comparison of Lurgi and Hygas gasifiers

	Lurgi*	Hygas†
Type of reactor	Moving bed	Multistage fluidized bed
Coal pretreatment	No‡	Yes§
Coal feed method	Lock hopper	Oil slurry
Relative material rates¶		
Oxygen feed	1	0.62
Steam feed	1	0.69
Dry gas	1	0.98
Methane	1	1.46
Carbon dioxide	1	1.24
H_2/CO volume ratio	2.2	2.1
Tars and oils in off gas	Yes	Yes
GCV of dry off gas, MJ/m³	12.4	12.8
Energy for O_2 production, % of off gas GCV	28.2	19.5
Latent heat of steam, % of off gas GCV	2.9	1.8

* Based on data in Ref. 14 for Illinois No. 6 coal.
† Based on pilot plant data reported in Ref. 20 for Illinois No. 6 coal.
‡ Feed is crushed wet coal; fines may be used for steam generation.
§ Pulverized coal is pretreated by air oxidation; heat evolved may be used for steam generation.
¶ Based on equal energy rates in the off gas.

A more detailed comparison of the two gasifiers, given in Table 4.36, does show that the Hygas process is more effective in oxygen and steam usage, but this advantage may be largely offset by the possibly higher cost of the gasifier itself, and associated practical problems of operation and control.[21]

The major problem with the ideal gasification route of Eq. (4.13) is that the rate of methane formation at low temperatures is slow relative to other gasification reactions. At higher temperatures methane formation is not favored thermodynamically. We recall from the discussion in Section 2.3 that one means of enhancing the rate of a specific reaction is by use of catalysts. It has been known for some time that alkali metals catalyze the reaction of carbon with steam to form mainly carbon monoxide and hydrogen. More recent studies of these catalysts, in particular by Exxon, have shown that potassium catalysts promote the gas phase shift and methanation reactions. This means that the products of carbon/steam gasification, essentially carbon monoxide and hydrogen, can be converted to methane within the gasifier. A further important advantage of alkali metal catalysts is that they reduce agglomeration of caking coals.

Exxon Research and Engineering Co., under sponsorship of the U.S. Department of Energy, has engaged in a program to develop a coal gasification process based on the ideal gasificaton route and potassium catalysts.[22] A 15-cm-diameter, 9-m-tall fluidized bed gasifier operating at pressures up to 0.8 MPa was used for initial assessment of the gasification step. A larger process development unit, 25 cm in diameter and 24 m tall, capable of handling coal at a rate of 1 t/d and at pressures up to 3.5 MPa was subsequently built. The process development unit was designed to operate in an integrated mode with recycle of both synthesis gas and catalyst. A block flow diagram of the process is shown in Figure 4.17.

Earlier catalyst tests showed that the potassium must be available for combination with coal acids to be effective. Potassium hydroxide (KOH), potassium carbonate (K_2CO_3), and potassium sulfide (K_2S) are suitable catalysts, whereas the salts of strong

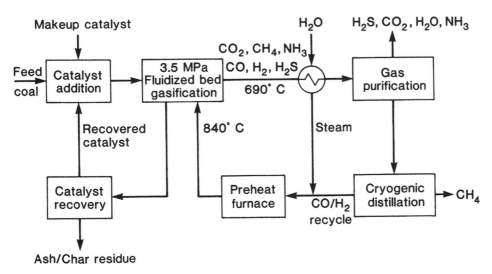

Figure 4.17 Exxon Catalytic Coal Gasification process.

acids such as potassium chloride (KCl) and potassium sulfate (K_2SO_4) do not have an affinity for the coal and are ineffective. It was further found that concentrations of potassium carbonate of from 10 to 20 mass percent of the coal result in commercially acceptable gasification rates at 700°C, in comparison to about 925°C without catalyst. The effect of this decrease in temperature on equilibrium methane concentrations can be calculated using Eq. (3.33). For a steam rate of 1.5 mol/mol of carbon fed to a gasifier operating at 3 MPa, the carbon converted to methane is calculated to increase from about 14 percent at 925°C to 40 percent at 700°C.

The overall reaction scheme for the catalytic gasification can be presented as follows:

$$\text{Gasification} \quad 2C + 2H_2O \rightleftharpoons 2H_2 + 2CO \qquad \Delta H^\circ_{1000} = +271.8 \text{ kJ} \qquad (4.15)$$

$$\text{Gas shift} \quad CO + H_2O \rightleftharpoons H_2 + CO_2 \qquad \Delta H^\circ_{1000} = -34.8 \text{ kJ} \qquad (4.16)$$

$$\text{Methanation } CO + 3H_2 \rightleftharpoons H_2O + CH_4 \qquad \Delta H^\circ_{1000} = -225.4 \text{ kJ} \qquad (4.17)$$

$$\text{Overall} \quad 2C + 2H_2O \rightleftharpoons CO_2 + CH_4 \qquad \Delta H^\circ_{1000} = +11.6 \text{ kJ/mol } CH_4 \qquad (4.13)$$

While catalytic methane production at low temperature is significantly greater than that obtained with conventional gasifiers, the equilibrium gas compositions given in Table 4.37 show that unlike hydrogasification, the off gas still contains significant amounts of carbon monoxide, carbon dioxide, and hydrogen, in addition to undecomposed steam. As in the conventional production of pipeline gas, these substances have to be removed or converted to methane. Instead of using a gas shift and a methanation step, in catalytic gasification the carbon monoxide and hydrogen are separated and

Table 4.37 Equilibrium gas produced by catalytic gasification with recycle*

	Moles per Mole Total Carbon in Feed†	
Steam feed rate	1.00	2.00
Total H_2 in feed	1.30	2.85
Total O_2 in feed	0.64	1.10
Off Gas		
H_2	0.30	0.85
CO	0.28	0.21
CO_2	0.36	0.40
CH_4	0.36	0.40
H_2O	0.28	1.20
Total off gas	1.58	3.06
Total recycle gas (H_2 + CO)	0.58	1.06
Recycle gas, mol/mol CH_4	1.6	2.7
Steam decomposed, %	72.0	40.0

* Calculated using Eq. (3.33) for 1000 K and 3 MPa.
† Includes carbon in recycle gas.

recycled to the gasifier. This is feasible, since the methanation reaction is close to equilibrium and there is consequently no buildup of hydrogen and carbon monoxide as would occur in an uncatalyzed system.

The method selected for separation of carbon monoxide and hydrogen from the methane product is cryogenic distillation. Distillative removal of methane from the synthesis gas is more readily accomplished than oxygen separation from air, due to the larger differences in boiling points, or larger relative volatilities, of the gases in the former case (see Section 5.3). In addition, it has been estimated that the size of the plant required for methane separation is about 20 percent smaller than an oxygen plant serving a conventional gasifier of equivalent output. Both capital and operating costs for the cryogenic step in a catalytic process are expected to be lower than the cost of the oxygen supply to a conventional gasifier. Prior to cryogenic methane separation, it is, however, necessary that the gas be cleaned and purified. In particular, ammonia, hydrogen sulfide, and carbon dioxide have to be removed, as they would otherwise freeze out and plug the low temperature process units. In addition to the usual gas purification steps discussed in Section 5.1, polishing steps with molecular sieves and activated carbon may be required. Preliminary estimates have indicated that the cryogenic gas separation and recycle loop represent about 10 percent of the product gas cost.

Another recycle loop in catalytic coal gasification is that of the catalyst itself. Catalyst rates of 10 to 20 percent of the coal rate represent a relatively high cost for the potassium salt. The catalyst is not retained in the gasifier but is carried out with the ash. Economics dictate that the catalyst be recovered and recycled rather than dumped, which would be preferable from a process point of view. Although much of the catalyst can be recovered by washing the ash with water, some catalyst reacts with alumina in the ash to form potassium aluminosilicates, from which the potassium cannot be recovered by simple water wash. However, the addition of lime releases the potassium by formation of insoluble calcium aluminosilicate, and lime-based catalyst recovery is stated to be economically justified, especially when gasifying high-alumina coals. If recycled, catalyst costs can be reduced to about 4 percent of the product gas cost.

Table 4.38 is a simplified material balance for the Exxon Catalytic Coal Gasification process, based on some limited data provided for a conceptual commercial-scale plant producing 7×10^6 m³/d (250×10^6 ft³/day) of SNG.[22] The plant will consume a total of 16 500 t/d of bituminous coal, of which 13 150 t/d will be gasified, with the remainder being used for coal drying and steam raising. Electric power will be purchased. A flow sheet for the conceptual plant is shown in Figure 4.17.

Coal to be gasified is crushed and dried and then mixed with a solution of catalyst. The coal, now impregnated with catalyst, is dried once more before being fed via a lock hopper to the four single-stage fluidized bed gasifiers. Each gasifier is a relatively simple vessel not requiring complex internals, and is designed to operate at 3.5 MPa and 690°C. The beds are fluidized by a preheated mixture of steam fed at a rate of 1.6 kg per kg of coal plus recycle gas. Lower steam rates would require less steam-raising energy and also result in a lower volume of recycle gas as shown in Table 4.37. However, reaction rates would be lowered too, and a larger gasifier volume would

Table 4.38 Estimated material balance for Exxon catalytic gasification

	kg/s	mol/mol C in Coal
IN		
Bituminous coal (68.3% C, 2% H₂O, 9.5% ash)	152.2	—
Steam (1.6 kg/kg coal)	243.5	1.56
Catalyst (K₂CO₃, 15% of coal)	22.8*	—
Recycle gas (estimated)	70.0	0.82
Total	488.5	
OUT		
Ash	14.5	—
Unconverted carbon (90% C conversion)	10.4	—
Steam (40% steam decomposed)	146.1	—
SNG (270 × 10¹² J/d)	59.0	0.41
CO₂, NH₃, H₂S, etc. (by difference)	175.6	—
Catalyst (K)	12.9	—
Recycle gas	70.0	—
Total	488.5	

* Makeup catalyst is about 10% of the recycle rate.

consequently be required to achieve the equivalent production and carbon conversion. Most designs call for a steam rate of between 1 and 2 moles per mole carbon, and so would have off gas rates falling within the range shown in Table 4.37. It is estimated that carbon conversions of 90 percent and at least 40 percent steam decomposition will be achieved in the Exxon gasifier. As is typical of fluidized bed operation, pyrolysis products are cracked and the off gas is essentially free of hydrocarbons heavier than methane.

Undecomposed steam in the off gas is condensed in a waste heat boiler which produces some of the steam for gasification. The gas is then scrubbed to remove particles and ammonia, following which the acid gases, carbon dioxide and hydrogen sulfide, are removed primarily by physical absorption. The product methane is then separated cryogenically and the remaining synthesis gas recycled to the gasifier. Char and ash removed from the gasifier are digested with lime for catalyst recovery and then discharged.

An approximate thermal balance for the demonstration process, given in Table 4.39, shows the overall thermal efficiency for SNG production to be about 55 percent. This is significantly lower than the value of 84 percent obtained in the example for an idealized process presented in Section 2.2. Much of the unaccounted energy in Table 4.39 is for gas cleaning, purification, and compression, as well as for sensible heat losses from the system. Further development to improve conversion efficiencies

Table 4.39 Approximate thermal balance for proposed demonstration of Exxon Catalytic Gasification process using bituminous coal

	Heat Content, MJ/kg	Mass Flow, kg/s	Heat Flow, MJ/s	Heat Flow, % of Total
IN				
Coal (dried)	29.0	152.2	4413	77.6
Energy to raise steam and to dry coal, etc.*	—	243.5†	1125	19.8
Purchased electricity	—	—	147	2.6
Total			5685	100.0
OUT				
Unconverted carbon	32.0	10.4	333	5.9
Latent heat of steam in off gas	2.7	146.1	394	6.9
SNG	53.2	59.0	3137	55.2
Unaccounted (by difference)	—	—	1821	32.0
Total			5685	100.0

Overall thermal efficiency $= \dfrac{3137}{5685} \times 100 = 55\%$

* Based on a coal rate of 38.8 kg/s.
† Steam rate.

is aimed at reducing or eliminating the recycle of the catalyst and synthesis gas. Both these improvements are dependent on advances in catalyst technology. The availability of a cheap catalyst that could be discarded without significant economic penalty would, of course, reduce the need for catalyst recycle. More important, perhaps, would be the development of catalysts active at even lower temperatures to further favor methane equilibrium. For instance, if gasification rates were economical at 500°C, then methane conversion would be in excess of 90 percent per pass and so the need for gas recycle would be virtually eliminated. It can be assumed that in a commercial operation, appropriate equipment would be installed to reduce the large unaccounted losses of 32 percent shown in Table 4.39 for the demonstration plant, thereby increasing the overall thermal efficiency.

In addition to the fundamental advantages of gasification according to the essentially thermoneutral ideal gasification reaction, Eq. 4.13, catalytic gasification offers many practical advantages over conventional first- and second-generation gasifiers. One advantage already mentioned is that the catalyst reduces agglomeration of caking coals, so fluidized bed gasifiers may be used with a wide range of coal feedstocks without pretreatment. Further, there is no need to stage the gasification process, so gasifier construction and operation is relatively straightforward. The low temperatures and moderate pressures serve to reduce material and mechanical problems. Finally there is no need to handle a slag, and tars and oils are cracked in the well-mixed fluidized bed and are absent from the off gas. Although not suitable for indirect liquefaction processes requiring a synthesis gas low in methane, catalytic gasification does appear to offer an attractive and practical method for SNG production.

4.5 UNDERGROUND GASIFICATION

We have so far considered a wide variety of schemes for the gasification of coal, all of which first involve the mining and preparation of the coal prior to its gasification in a reactor vessel. If, however, the coal could be gasified in place (*in situ*), then not only would it be possible to eliminate the need for underground mining and its attendant costs and problems, but it might also be possible to recover the fuel value from identified resources considered uneconomical to mine or which have been left in place from previous mining operations.

Underground coal gasification, or UCG, is not a new idea, it having been first suggested by Siemens in 1868 for the gasification of coal left in place after mining. Mendeleev independently proposed the true underground gasification of coal in 1888. It is interesting that Lenin recommended that work be carried out in this area as a way to liberate workers from the drudgery of mining, and in 1933 the Soviet government began its first UCG field experiments. Up to the present the most extensive developments of this method have been in the Soviet Union, where several commercial plants have been operated with the product gas used on site, principally for the generation of electric power.[23,24] The Soviet program peaked about 1966, but at the time of writing indications are that the program has all but ceased, possibly due to a combination of unfavorable economics and geology.

The basic concepts of UCG operation may be illustrated with reference to Figure 4.18. Our discussion here and in what follows draws heavily on the excellent review of the subject by Gregg and Edgar.[24] Two boreholes, or wells, are drilled to the bottom

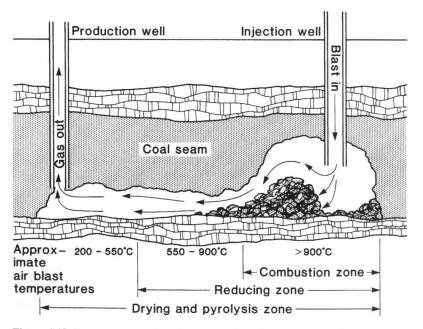

Figure 4.18 Side view of linked-vertical-well method of underground coal gasification.[24]

of the coal seam to be gasified. Normally the natural permeability of the coal bed is not high enough nor the fissures large enough to permit gas percolation through the seam, so it is necessary to enhance the permeability. This is called "linking" of the wells, in which narrow linear channels of high permeability are formed without increasing the permeability of the bulk of the coal seam. The formation of channels is accomplished by several means discussed below, including directional drilling and the burning of a hole in the coal seam, resulting in a highly permeable narrow channel of char or ash. Air or oxygen, either alone or with steam, is blown down the injection well. The coal is ignited at the base of the well and the combustion maintained by continuous injection of the oxidant. The combustion which is pushed along the seam by the injected gases is preceded by zones of gasification and pyrolysis, while the product gases that are formed percolate through the link and the seam, and are withdrawn from the production well.

With the linking channel at the bottom of the seam, the coal is initially consumed there. As indicated in Figure 4.18, as the void grows, the unburned coal falls into it creating a bed of coal rubble that is relatively reactive because of the large surface area it presents. This self-rubbling effect is greater with lower-rank coals. Although the process is shown in only two dimensions, it is in fact three dimensional, with the channel widening at the same time that it lengthens.

The seam between the wells, though it may be a coherent material, can be thought of as a fixed packed bed with the reaction chemistry essentially the same as for any packed bed gasifier (see Figure 3.8). Unlike the moving bed gasifier in which the different reaction zones remain fixed in space, here the bed is fixed and the different zones propagate through it. Moreover, as we observed above, there is a replenishment of coal. In UCG terminology, the gasification zone is referred to as the reducing zone, probably because most developments have been with air injection alone. In that case the principal gasification reaction is the endothermic reduction of carbon dioxide to carbon monoxide by carbon, following the Boudouard reaction (3.12). Of course, steam may be injected and in any case water is often present in the seam as well as in the coal, so that if the temperature is high enough as a result of sufficient oxidant injection, then the carbon-steam reaction (3.14) will also be important. Whether the pyrolysis gases are drawn off as tar or are thermally cracked depends on the extent to which they interact with the hot gasification products.

The composition and calorific value of the product gas is dependent on whether air, oxygen, or a steam/oxygen mixture is injected. The product gas is similar to that obtained in moving bed surface gasifiers for the same feed gases. With air injection alone the gross calorific value of the dry product gas has ranged between about 4 and 7 MJ/m³ for U.S. tests on Wyoming subbituminous seams. Data on burning with pure oxygen or with a steam/oxygen mixture are more limited, but gas with a calorific value of about 10 MJ/m³ has been produced.[25]

One of the most important parts of UCG technology is the means of providing the permeable link between the injection and production wells, along with the permeable horizontal links that may manifold the injection wells or that may manifold the production wells. One method developed in the Soviet Union for UCG is that of directional drilling, which allows drilling holes with radii of curvature as small as

600 m.[24] The drill is guided in its course by information transmitted to the surface from a downhill compass and pendulum. A layout for the underground gasification of a steeply pitching coal seam using directional drilling is shown in Figure 4.19. Note that in this layout the production wells, the second stage injection wells, and the horizontal linkage path making up the bottom manifold are directionally or slant drilled. Holes with diameters of 0.3 m and as long as 100 m have been drilled by the directional drilling technique at rates from 70 to 250 m/d. In the steeply pitching seam shown in Figure 4.19 combustion is initiated at the base of the production well and advances up the seam toward the surface. The orientation of the seam is such that as the gasification proceeds, the coal easily falls into the void created by the advancing combustion front.

A second procedure for linking is that of reverse or countercurrent combustion. The gasification procedure previously described for moving and fixed packed beds in which the combustion front and gas flow propagate in the same direction represents forward or cocurrent combustion. In forward combustion, the rate at which the coal is consumed defines the front speed. In reverse combustion, the front moves opposite to the direction of gas flow, that is, it moves towards the source of oxidant. Here, the front speed is determined by the rate at which heat is conducted upstream against the gas flow. In reverse combustion, the coal upstream is heated to its ignition temperature thereby drawing oxygen away from the coal downstream which therefore does not burn completely. On the other hand, in forward combustion, the coal is for the most part burned completely. Even more importantly, forward combustion propagates on a broad cross-section compared to reverse combustion, which tends to burn narrow

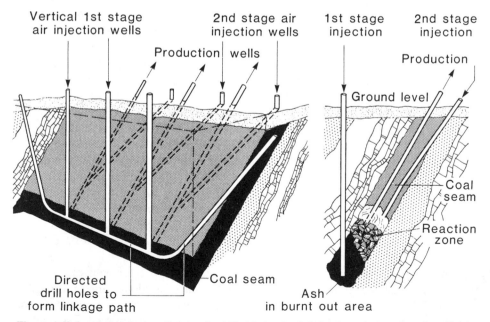

Figure 4.19 Layout employing directionally drilled holes for linkages in underground coal gasification system at Yuzhno Abinsk, U.S.S.R. (after Ref. 24).

channels. This is a consequence of an insufficiency of oxygen behind the moving combustion front, which inhibits any additional burning that would otherwise expand the permeable channel formed by the advancing front.[24] Reverse combustion linking-rates using air injection are on the order of 1 to 3 m/d, compared to forward gasification rates of 0.5 to 1 m/d.

The procedure by which reverse combustion linking of vertical wells is established and production gasification initiated is shown schematically in Figure 4.20. The linked-vertical-well process is usually layed out in a rectangular array as illustrated in the plan view shown in Figure 4.21 of an underground gasification plant operated near Moscow. The dotted lines indicate linkage paths established by reverse combustion and production paths. In the layout illustrated, production by forward combustion is taking place in the direction from row 1 to row 2.

Other procedures used for linking include hydraulic fracturing where high pressure water is injected to open up natural breaks in the coal; pneumatic linking by high pressure air (~7 MPa) wherein the air percolates from one borehole to the other, forming an enlarged passage; and implanting shaped charges in the borehole to blow long holes parallel to the bedding planes, then using these as pilot holes for reverse combustion. Mining may also be used for preparation of the underground manifolds.

The design of a UCG system and the efficiency with which it operates are strongly site specific. Physical and chemical models can at best serve to indicate trends, not

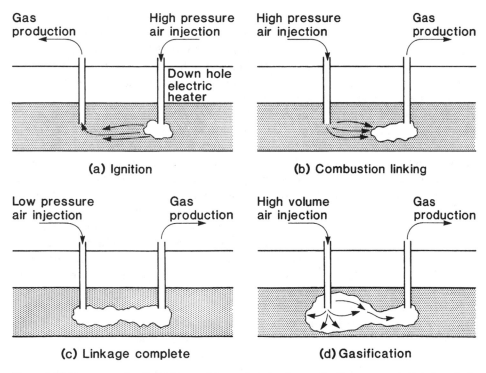

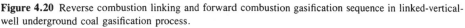

Figure 4.20 Reverse combustion linking and forward combustion gasification sequence in linked-vertical-well underground coal gasification process.

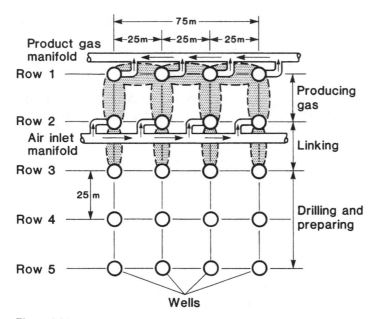

Figure 4.21 Plan view of linked-vertical-well underground gasification plant operated near Moscow.[24]

only because of the heterogeneous character of the coal seam, but also because of numerous other factors which enter that are not always predictable, such as gas leakage, water intrusion, and channeling to name but a few. Most UCG systems operate below the water table, so prior to gasification the coal has to be dewatered. Intrusion of some water during the burn is beneficial to the gasification reaction but excess water results in a decrease in the calorific value of the gas.

The importance of water intrusion lies not only in the amount, but also in where it intrudes. If the water enters in the combustion zone, it will participate in the gasification reaction. If it enters downstream after the principal reactions, there may be no change in gas composition but a loss in sensible heat of the product gas because the temperature is reduced. There is clearly an optimum amount of water intrusion in the gasification zone. As water is added carbon monoxide and hydrogen are produced by the gasificaton reaction (3.14), but, as in the dry ash Lurgi gasifier, if too much is added the temperature is dropped and more carbon dioxide and less carbon monoxide are produced, giving a gas of lower calorific value and a lower cold gas efficiency. Water intrusion can be controlled by high pressure operation, which reduces the influx but which also increases any gas leakage. Low pressure operation has the opposite effect.

Seam thickness is also an important parameter. Thick seams not only allow access to a large amount of coal from one well, but also have relatively low heat losses to the surrounding formation for the volume of coal gasified. The cooling effect by heat conduction to the surroundings may be seen by considering a rectangular seam of thickness H, width W, and length L between the injection and production wells.[24] The conductive heat loss to the surroundings from the hot gases will be proportional to the surface area $2L(W + H)$, while the heat generated will be proportional to the burning area $W \cdot H$. It follows that

$$\frac{\text{Heat Loss}}{\text{Heat Generation}} \sim \left(\frac{L}{W}\right)\left(1 + \frac{W}{H}\right) \qquad (4.18)$$

Generally $W \gg H$, whence the rate of heat loss is inversely proportional to the seam thickness H. It also follows that a larger water intrusion can be tolerated for a thicker seam to obtain a product gas of the same calorific value. This has been shown in Soviet tests. Heat losses can eventually become rate limiting and for this reason gasification of seams of less than about 1 m thickness is not practical.

Various modeling procedures have been developed to predict the chemical composition of the product gas as a function of the injected gas composition and rate, coal type, seam thickness, and water intrusion rate. We have already observed that the UCG system behaves much like a packed bed gasifier. This is borne out by the fact that material- and energy-balance models employed for moving bed surface gasifiers, when applied to properly operating UCG systems, are reasonably successful in predicting product gas compositions. Inasmuch as we have previously discussed these models and illustrated their application in connection with moving bed gasifiers, we shall not consider them further here.

One important design question related to UCG systems concerns the fraction of coal recoverable with a particular well pattern. Channel enlargement and subsequent coal consumption depend to a large extent on heat transfer considerations such as discussed above in connection with Eq. (4.18). However, the problem is very much site specific since it is a function of how the coal will fracture and how the shape, size, and composition of the underground reactor will vary spatially and temporally.

The most extensive UCG tests in the United States have been linked-vertical-well experiments carried out by the Laramie Energy Technology Center of the Department of Energy at a site near Hanna, Wyoming, and by the Lawrence Livermore Laboratory of the University of California at Hoe Creek, Wyoming. These sites are technically most favorable because the coal is a subbituminous one and the seams are thick. In Tables 4.40 and 4.41 we have shown typical operating characteristics and product gas

Table 4.40 Typical operating characteristics of linked-vertical-well underground coal gasification systems for subbituminous coal* with injection of air or steam and oxygen

	Air	Steam/Oxygen
Pressure, system, MPa	0.3	0.3
Temperature, gas exit, °C	340	340
Steam/Oxygen, kg/kg	—	0.84
Oxidant, kg/GJ gas	145	27
Combustion front speed, m/d	0.7	0.7
Cold gas efficiency, %	80	80

* Seam thickness 10 m.

Table 4.41 Typical product gas composition* in volume percent from underground gasification of subbituminous coal

Constituent	Air (Hanna II)	Steam/Oxygen (Hoe Creek 2)
CO	15	23.5
CO_2	12	32.5
H_2	17	37.5
CH_4	3	5.5
N_2	51	—
Other	2	1
Total	100	100
GCV, MJ/m^3	5.7	10.1

* Dry basis after scrubbing and cooling.

compositions for air and steam/oxygen injection that have been derived from these tests. A summary of information on these tests may be found in a number of papers contained in the collections of Refs. 25 and 26.

From what has been discussed it is evident that there are a number of technical problems which require solution before large-scale commercialization. These include the determination of the best methods of seam preparation, operation without excessive water influx or excessive gas leakage, and the optimum layout of wells. There is, however, no reason to believe that these problems are not soluble, the best proof of which is the large-scale commercialization of UCG that has already taken place in the Soviet Union.

Some of the advantages of UCG that have been noted include the ability to increase mineable reserves and the elimination of underground mining. Environmental reasons that have also been cited in its favor are that problems of air pollution and solid waste disposal are reduced. Among environmental problems that mitigate against UCG are groundwater contamination following the burn, and subsidence effects. Because aquifers are often found either above or within coal seams, influx of water after gasification results in groundwater pollution by organic and inorganic materials. Moreover, the burning of thick seams may result not only in subsidence but also could lead to gas leakages to the surface. An understanding of the control of these problems will be required.

REFERENCES

1. Hottel, H. C., and Howard, J. B., *New Energy Technology*. MIT Press, Cambridge, Mass., 1971.
2. Hebdon, D., "High Pressure Gasification Under Slagging Conditions," *Proc. Seventh Synthetic Pipeline Gas Symposium*, Chicago, Ill., October 1975.
3. Yoon, H., Wei, J., and Denn, M. M., "A Model for Moving-Bed Coal Gasification Reactors," *AIChE J.* **24**, 885-903, 1978.

4. Dravo Corp., "Handbook of Gasifiers and Gas Treatment Systems," Report No. FE-1772-11, U.S. Energy Research & Development Administration, Washington, D.C., February 1976.
5. Baughman, G. L., *Synthetic Fuels Data Handbook*. 2nd Edition, Cameron Engineers, Inc., Denver, Colorado, 1978.
6. Fluor Engineers and Constructors, Inc., "Economic Studies of Coal Gasification Combined Cycle Systems for Electric Power Generation," Report No. EPRI AF-642, Electric Power Research Institute, Palo Alto, Calif., January 1978.
7. von Fredersdorff, C.G., and Elliott, M. A., "Coal Gasification," in *Chemistry of Coal Utilization*, Supplementary Volume (H. H. Lowry, ed.), pp. 892-1022. Wiley, New York, 1963.
8. Penner, S. S. (ed.), "Assessment of Long Term Research Needs for Coal Gasification Technologies," Fossil Energy Research Working Group, Mitre Technical Report No. MTR-79W00160, The Mitre Corp., McLean, Va., April, 1979.
9. U.S. Department of Energy, "Great Plains Gasification Project, Final Environmental Impact Statement, Vol. 1," Report No. DOE EIS-0072 F, U.S. Department of Energy, Washington, D.C., August, 1980.
10. Massey, L. G., "Coal Gasification for High and Low Btu Fuels," in *Coal Conversion Technology* (C. Y. Wen and E. S. Lee, eds.), pp. 313-427. Addison-Wesley, Reading, Mass., 1979.
11. Conoco, Inc., "Phase 1: The Pipeline Gas Demonstration Plant—Environmental Report. Vol. I Project Description," Report No. FE-2542-25 (Vol. 1), U.S. Department of Energy, Washington, D.C., January, 1980.
12. Cornilis, B., *et al.*, "The Ruhrchemie-Ruhrkohle Technical Version of the Texaco Coal Gasification Operating at Oberhausen-Holten," in *Coal Processing Technology*, Vol. VI, pp. 209-216, American Institute of Chemical Engineers, New York, 1980.
13. Schlinger, W. G., Falbe, J., and Specks, R., "Coal Gasification for Hydrogen Manufacturing," in *Hydrogen: Production and Marketing* (W. N. Smith and J. G. Santangelo, eds.), pp. 177-190. ACS Symposium Series No. 116, American Chemical Society, Washington, D. C., 1980.
14. U.S. Environmental Protection Agency, "Pollution Control Guidance Document for Indirect Coal Liquefaction," Coal Gasification and Indirect Liquefaction Working Group. Industrial Environmental Research Laboratory, U.S. Environmental Protection Agency, Research Triangle Park, N. C., 1981 (to appear).
15. Wen, C.Y., and Chaung, T. Z., "Entrainment Coal Gasification Modeling," *Ind. Eng. Chem. Process Des. Dev.* **18**, 684-695, 1979.
16. Vogt, E. V., and van der Burgt, M. J., "Status of the Shell-Koppers Process," *Chem. Eng. Prog.* **76**(3), 65-72, 1980.
17. Badger Plants, Inc., "Conceptual Design of a Coal-to-Methanol-to-Gasoline Commercial Plant. Vol. 1 Technical," Report No. FE-2416-43 (Vol. 1), U.S. Department of Energy, Washington, D.C., March, 1979.
18. Shinnar, R., "Gasoline from Coal: A Differential Economic Analysis," *ChemTech* **8**, 686-693, 1978.
19. Channabasappa, K. C., and Linden, H. K., "Hydrogenolysis of Bituminous Coal," *Ind. Eng. Chem.* **48**, 900-905, 1956.
20. Bair, W. G., and Lau, F. S., "Status of the Hygas Process Development," in *Coal Processing Technology*, Vol. VI, pp. 166-169, American Institute of Chemical Engineers, New York, 1980.
21. Rogers, K. A., and Hill, R. F., "Coal Conversion Comparisons," Report No. FE-2468-51, U.S. Department of Energy, Washington, D.C., July 1979.
22. Gallagher, J. E., and Euker, C. A., "Catalytic Coal Gasification for SNG Manufacture," Sixth Annual Intn'l. Conf. on Coal Gasification, Liquefaction and Conversion to Electricity, Pittsburgh, Pennsylvania, July/August, 1979.
23. Elder, J. L., "The Underground Gasification of Coal," in *Chemistry of Coal Utilization*, Supplementary Volume (H. H. Lowry, ed.), pp. 1023-1040. Wiley, New York, 1963.
24. Gregg, D. W., and Edgard, T. F., "Underground Coal Gasification," *AIChE J.* **24**, 753-781, 1978.
25. U.S. Department of Energy, *Proceedings 5th Underground Coal Conversion Symposium*, June 1979. Report CONF. No. 790630, U.S. Department of Energy, Washington, D.C., May 1979. (Available from NTIS).
26. U.S. Department of Energy, *Proceedings 4th Underground Coal Conversion Symposium*, July 1978. Report No. SAND 78-0941, Sandia Laboratories, U.S. Department of Energy, Sandia, New Mexico, June 1978. (Available from NTIS).

FIVE

GAS UPGRADING

5.1 GAS CLEANING AND PURIFICATION

Gases leaving a synthetic fuel reactor, whether they are product gases from a gasifier or byproduct gases from an oil shale retort or direct liquefaction reactor, contain components that may be categorized as desirable, neutral, or undesirable. A "component" is here defined as desirable or undesirable primarily from a process viewpoint. A desirable component should be present in the end product, an undesirable one should be absent. A component may also be undesirable because its presence in a processing stage could be detrimental. For example, hydrogen sulfide is undesirable in SNG because its level specified for pipeline gas contracts is often as low as 5.5 mg/m³. This is many orders of magnitude less than what is present in raw coal-derived off gas used to manufacture SNG. In addition, however, the hydrogen sulfide present in the raw gas to be used for SNG manufacture generally has to be removed just because of the detrimental effect of sulfur on shift and methanation catalysts. It should be noted that in our definition we distinguish the cleaning and purification necessitated by process considerations from that required by environmental considerations, although sometimes they overlap. The removal of environmental contaminants from gases discharged into the atmosphere is discussed in Section 9.2.

Desirable components generally include carbon monoxide, hydrogen, methane, and other fuel vapors. In low-CV gas for combustion without long distance transportation carbon dioxide and water vapor are neutral. However, these same components present in a low- or medium-CV gas that is to be further processed or upgraded are undesirable simply because the processing units would have to be oversized to handle gases that make no contribution to calorific value. Moreover, these inert diluents could consume heat and compression energy. In addition to hydrogen sulfide, undesirable components include particulate matter, ammonia, hydrogen chloride, and tars. A good single reference on the topic of gas purification is the book by Kohl and Riesenfeld,[1] to which the reader is referred for a detailed discussion.

The removal of particulate matter larger than about 5 μm from off gas streams is generally accomplished with cyclone separators. For smaller size particles, venturi scrubbers, electrostatic precipitators, and fabric filters are used. In cyclone separators, the particle laden gas enters a cylindrical or conical chamber tangentially. Centrifugal forces accelerate the particles toward the outside wall and they are removed from the bottom of the separator. The cleaned gas is drawn off overhead from an outlet at the core of the separator. Cyclone separators are operated at temperatures up to about 1000°C and pressures as high as 50 MPa.

The gases exiting synthetic fuel reactors are hot, with temperatures ranging from 400 to 1000°C for most reactors, except for steam/oxygen-blown, entrained flow gasifiers where the gas exit temperatures can reach 1500°C. One of the simplest cleaning procedures is to cool the gas to a temperature low enough that the water and higher boiling hydrocarbons condense. The temperature to which the gas is cooled will depend on its pressure. As a result of the cooling, tars that are present may be partially solidified. The condensed liquids will tend to capture the solidified tar particles as well as any solid particles not removed by the cyclone separators.

The sensible heat of the process gas is normally removed by generating process steam, although not all of the heat can be recovered. Below about 130 to 150°C, useful heat cannot be economically extracted from the gas stream and further cooling is accomplished by air or water cooling, with the heat dissipated to the atmosphere. Even in the steam generation process the sensible heat is degraded and there is a loss in available energy, thereby reducing the overall conversion efficiency. For this reason, efforts have been made to develop high temperature cleaning and purification procedures, most especially for low-CV gas which is to be used on site for heat or electricity generation. We shall speak more of this later on in the section.

The gas cooling may be accomplished by indirect heat exchange or by quenching and scrubbing with water. In any case, washing of the gas with water will normally be used as one of the means of gas cleaning and purification. A wide variety of procedures are available to wash the gas. The most commonly used device is the spray tower in which liquid droplets produced by spray nozzles are allowed to settle down through the rising gas stream. The water will bring down with it soluble gases and particulate matter. If a main aim is to remove fine particulate matter, the collection efficiency can be increased by increasing the relative velocity between the dirty gas and the water droplets. One method for accomplishing this is the venturi scrubber, in which the gas flows through a convergent-divergent venturi nozzle and the water is sprayed into the throat where the gas velocity is highest and the pressure lowest. Venturi scrubbers remove particulates down to 0.5 to 1 μm.

Of the contaminant gases, ammonia is very soluble in water and so can be readily removed by washing. Ammonia is a valuable commodity and may be subsequently recovered from the water by fractional distillation (see Section 9.3).

The extent to which a gas dissolves in a solvent at equilibrium is given by Henry's law, Eq. (13), Table 2.13, which we repeat here:

$$p_i = h_i \cdot x_i \tag{5.1}$$

In a mixture of gases, p_i is the partial pressure of the gas species i, x_i its concentration, and h_i is Henry's law constant. The lower the value of h_i, the more soluble is the gas. The constant h_i depends on temperature; and values for dilute solutions of ammonia, carbon dioxide, and hydrogen sulfide are given in Table 5.1 for 25 and 100°C. Ammonia is seen to be very much more soluble than either carbon dioxide or hydrogen sulfide, neither of which by themselves are particularly soluble in water.

Unfortunately it is not a straightforward matter to apply Henry's law to ammonia dissolution because it is only applicable to the concentration of unionized dissolved gas in solution. Ammonia solubilizes in water in the form of unionized ammonia molecules (NH_3) and ionized ammonium ions (NH_4^+). However, what is of interest is the total concentration of the gas in solution. The ratio of the ammonium to ammonia will depend on the solution pH, that is, the hydrogen ion concentration, since hydrogen ions will combine with ammonia to form ammonium ions

$$H^+ + NH_3 \rightleftharpoons NH_4^+ \qquad (5.2)$$

The NH_3/NH_4^+ ratio can be calculated as a function of pH from a knowledge of the equilibrium constant for ammonia dissociation.

Concentrations of ammonia from 6000 to 20 000 mg/L are obtained in wash waters before the solution is wasted. The limitation on the extent to which the ammonia is absorbed is rarely a solubility limit, but rather one imposed by the economics of the tower design and operation. As the concentration of ammonia increases in the scrubber water, it takes a longer length of scrubber to push in the same amount of ammonia because the driving force, as measured by the concentration difference between the gas and the solution, has been reduced. A tradeoff must therefore be made between using fresh water, recycling water containing the dissolved gas, or, alternatively, increasing the scrubber length.

Even though the acid gases carbon dioxide and hydrogen sulfide by themselves are not very soluble in water, if they are in a mixture with ammonia, which is a weak alkali, they will be solubilized to a considerably greater extent than indicated by Henry's law. This is because carbon dioxide dissociates into HCO_3^- and CO_3^{2-} ions, and hydrogen sulfide into HS^- and S^{2-} ions. As a result, for example, the bicarbonate ions will combine with the ammonium ions to form ammonium bicarbonate, and the bisulfide ions will combine to form ammonium bisulfide. We may estimate that the concentration in solution of the carbon dioxide plus the hydrogen sulfide will about equal the ammonia concentration, so long as one or both of the other components is in excess in the gas.

In gasifier off gases, carbon dioxide concentrations are very much greater than

Table 5.1 Henry's law constants for unionized gas solubility in dilute solutions
(h = gas partial pressure/dissolved gas concentration)

Gas	h, kPa/(mol/kg soln.)	
	25°C	100°C
Ammonia (NH_3)	1.7	25
Hydrogen sulfide (H_2S)	1090	3030
Carbon dioxide (CO_2)	3040	8400

hydrogen sulfide concentrations and the amount to which they are absorbed in an aqueous ammonia solution will be about in proportion to their ratios in the gas. On the other hand, in the off gases from direct liquefaction reactors, the hydrogen sulfide is formed in excess of carbon dioxide and the concentrations in solution would be reversed from those resulting from washing gasifier off gas. In either case, the water wash does not usually purify the gas adequately of carbon dioxide and hydrogen sulfide. The additional removal of these acid gases must therefore be accomplished by other means.

Acid Gas Removal

Of the two acid gases it is the hydrogen sulfide which presents the most difficult removal problem since it cannot be vented to the atmosphere and must be collected, with sulfur recovery generally the end step. In the particular case where the gas containing the hydrogen sulfide is to be used as a fuel gas, an alternative procedure would be to burn the gas, converting the hydrogen sulfide to sulfur dioxide, and then to remove the sulfur dioxide by stack gas scrubbing procedures (see Section 9.2). However, because the flue gas volumes after combustion are greater than the fuel gas volumes prior to combustion, the equipment for sulfur dioxide removal is larger and the capital expenditures higher, so it generally proves to be more economical to remove hydrogen sulfide from the fuel gas.

Acid gas removal processes generally fall into one of the following categories:

Absorption into a liquid. This is the most important gas purification technique. It involves the transfer of a substance from the gaseous to the liquid phase through the phase boundary. The absorbed material may dissolve physically in the liquid or react chemically with it. Desorption, or stripping, represents a special case of the same operation in which the material moves from the liquid to the gaseous phase.

Absorption into a solid. This is a procedure of limited use and is discussed later in connection with hot gas purification. It involves the transfer of a substance from the gaseous to the solid phase. The absorbed material diffuses throughout the absorbent and may react chemically with it.

Adsorption on a solid. In this procedure the impurities are removed from the gas stream by concentration on the surface of a solid material. The quantity of material adsorbed is proportional to the surface area so that the adsorbents generally are granular solids with a large surface area per unit mass.

Chemical conversion to another compound. Here the impure gaseous contaminant is converted to a compound which is not objectionable or which can be subsequently removed with greater ease than the original compound. This is most often carried out in the presence of a solid catalyst.

A variety of processes are commercially available or under development for acid gas removal. Most of these processes are proprietary and are known by trade names.

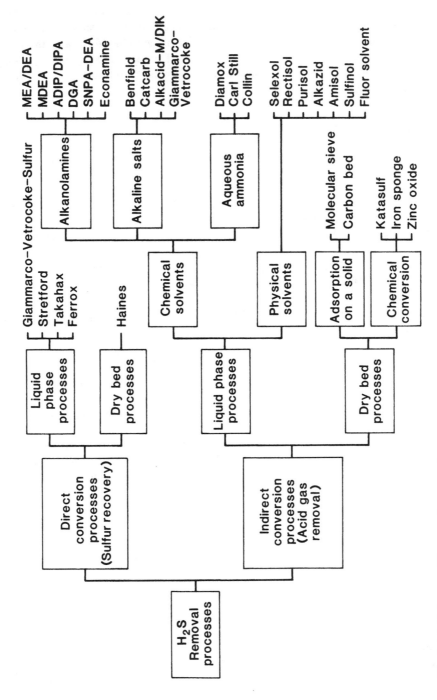

Figure 5.1 Hydrogen sulfide removal processes.[2]

They may be classified in various ways. One classification for hydrogen sulfide removal is shown in Figure 5.1. Direct conversion processes remove hydrogen sulfide by chemically converting it to sulfur. Indirect conversion processes simply remove the hydrogen sulfide molecules from the gas. The most common procedure is to absorb the hydrogen sulfide (and carbon dioxide) in a liquid. The liquid absorbents are regenerated for reuse, yielding a concentrated stream of acid gases which are then treated for sulfur recovery.

Among the factors which control the choice of acid gas removal process are the gas flow rate, the concentration of acid gases, and the need to remove carbon dioxide as well as hydrogen sulfide. For large gas volumes containing high concentrations of carbon dioxide mixed with hydrogen sulfide, a likely but by no means unique sequence of treatments is that shown in Figure 5.2. Most of the carbon dioxide and hydrogen sulfide are removed in a regenerable liquid absorbent which is continuously circulated. Final traces of hydrogen sulfide which, for example, might poison downstream catalysts are removed in a solid adsorbent. The solid adsorbent is regenerated intermittently or discarded. Concentrated gas from the regeneration of the liquid absorbent is treated for bulk recovery of elemental sulfur by the Claus process discussed later in the section. The final cleanup of the Claus plant off gas, usually referred to as "tail gas," can be by a direct conversion process.

The actual choice of treatment scheme will depend on a number of factors other than those mentioned, including the presence of other impurities, gas pressure, solvent selectivity, and energy requirements. This last factor is a particularly important one, especially in plants for the manufacture of SNG, since gas purification consumes a large amount of energy.

Bulk Acid Gas Removal by Liquid Absorption

Most currently practiced, large-scale bulk-acid-gas removal procedures are indirect ones in which the acid gases are selectively dissolved in a liquid, passed countercurrent to the gas. In a separate vessel the absorbing liquid is stripped of its gas content and thereby regenerated. It is then recycled back to the absorber. A simple flow scheme

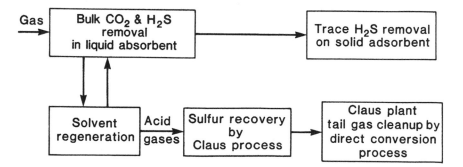

Figure 5.2 Possible treatment sequence for removal and recovery of acid gases from a high volume gas stream.

is shown in Figure 5.3, although in an actual design multiple absorption columns and multiple regeneration columns are always used.

As shown in Figure 5.1, liquid absorption processes may be broadly classified as physical solvent processes or chemical solvent processes. Physical solvent processes employ an organic solvent, with absorption enhanced by low temperature or high pressure or both. Absorption generally takes place at temperatures below 50°C. In chemical solvent processes, absorption of the acid gases is caused principally by using alkaline solutions such as amines or carbonates. Of course, the temperature must be sufficiently low and the pressure sufficiently high, but the operating requirements are generally less severe than for physical solvent processes. Regeneration, or desorption, is caused by reduced pressure, high temperature, and boiling. This causes the solvent vapor to rise through the liquid, stripping out the acid gases. Most of the energy for gas purification goes into boiling the solvent, which all the regeneration schemes involve, and this energy is ultimately dissipated by condensing the solvent.

Physical solvent processes are the probable choice for the purification of high pressure gases. Chemical solvent carbonate systems can be cheaper for medium pressure gases, but they do leave the gas saturated in water vapor and have low removal of higher-molar-mass organics; amine systems are best used for low pressure gases.

Among processes that have been applied to the purification of coal-derived gases, which may be mentioned as illustrative of physical solvent procedures, are the Rectisol

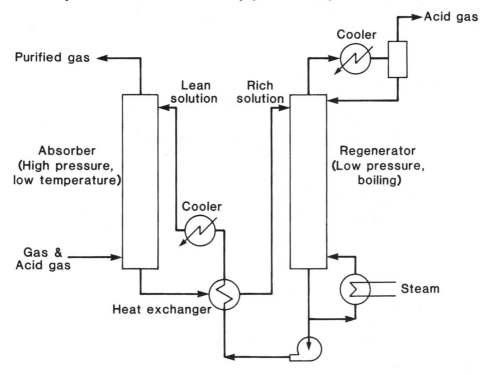

Figure 5.3 A simple gas purification arrangement.

and Selexol processes. The Benfield hot potassium carbonate process has been used extensively to clean up gases derived from the partial oxidation of petroleum and coal, and may be cited as illustrative of the chemical solvent procedures.

The Rectisol process is a Lurgi licensed process which employs the fact that hydrogen sulfide, carbon dioxide, and other gases usually considered to be impurities are especially soluble in cold methanol, while desired product gases such as hydrogen, carbon monoxide, and methane are relatively insoluble. Moreover, as shown in Table 5.2, the solubilities of the impurities increase with decreasing temperature, while the solubilities of the desired product gases are not much changed by temperature. The process is therefore most efficiently operated at low temperatures, generally $-40°C$ or lower, with precooling of the gas feed and refrigeration of the methanol.

An advantage claimed for the process is a comparatively low energy requirement, despite the large refrigeration load, because the solution is cooled by pressure reduction in the regeneration step, and the gas to be purified is cooled by heat exchange with the purified gas stream. Another advantage may be seen from Table 5.2, which shows hydrogen sulfide to be considerably more soluble than carbon dioxide. This permits the partial separation of hydrogen sulfide from gas streams containing both gases. Other advantages are high levels of impurity removal, and production of a gas of low water content. A main disadvantage is relatively high solvent loss because of the high vapor pressure of the methanol even at low temperatures.

The Selexol process is a proprietary process developed by the Allied Chemical Corporation and uses an organic solvent (the dimethylether of polyethylene glycol). The solvent does not have to be extremely cold, as does the methanol in the Rectisol process, and the absorption can be designed for temperatures around 20°C. The solvent is about 10 times more selective for hydrogen sulfide over carbon dioxide, so as with the Rectisol process partial separation of the acid gases can be accomplished. Although the process does not require refrigeration and the solvent losses are low, the solvent is comparatively expensive.

The Benfield process uses a 25 to 35 percent solution of hot potassium carbonate in water. Absorption requires pressures above about 2 MPa; but absorption and

Table 5.2 Henry's law constants for gas solubility in methanol* (h = gas partial pressure/dissolved gas concentration)

	h, MPa/(mol/kg Methanol)		
Gas	$-40°C$	$-55°C$	$-70°C$
Hydrogen sulfide (H_2S)	0.28	0.10	0.026
Carbonyl sulfide (COS)	0.45	0.22	0.067
Carbon dioxide (CO_2)	1.4	0.71	0.31
Methane (CH_4)	70	40	30
Carbon monoxide (CO)	170	160	110
Hydrogen (H_2)	670	590	930

* Interpolated from graphical data in Ref. 3.

regeneration take place at about the same temperature, in the range of 80 to 150°C, so strong cooling is not required. The difference in relative absorption rates between hydrogen sulfide and carbon dioxide of about 4 is sufficient to allow for a selective absorption design. The process has the advantage of being quite simple. Its main disadvantages are that it must handle a corrosive solution, the product gas is saturated with moisture, and the absorption of hydrocarbons is not as high as with physical solvents.

As mentioned, essentially all of the energy requirement for gas purification processes is for regenerating the solvent. The energy requirements of gas purification systems, particularly those employing organic solvents, are for the most part proprietary and partly controlled by the capital investment made. For estimating purposes, the regenerator steam requirement may be taken to be about 68 MJ/kmol of acid gas removed. The solvent pump work requirement is about 2 MJ/kmol of acid gas removed, with the equivalent heat requirement about 3 times this amount when account is taken of the conversion efficiency of heat to work. In integrated plants for the production of SNG, gas purification represents some 6 percent of the plant energy requirements and some 15 percent of the plant capital cost (see Section 5.5).

Claus Process

The Claus process is the principal method by which sulfur is recovered from separated hydrogen sulfide. It is not a gas purification process in the usual sense, since its purpose is to recover sulfur from a gas stream which itself is a product of a previous gas purification process. However, in most synthetic fuel plants employing indirect liquid absorption systems, Claus plants can be expected to be mandatory adjuncts and therefore it is appropriate to discuss the process in this connection.

The Claus process is usually carried out in two stages—a thermal stage where part of the hydrogen sulfide is burned to elemental sulfur and sulfur dioxide, and a catalytic stage where the remaining hydrogen sulfide is reacted with sulfur dioxide in the presence of a catalyst to form additional sulfur. The stoichiometry for the formation of the elemental sulfur and sulfur dioxide in the thermal stage can be represented by the exothermic reactions

$$H_2S + \tfrac{3}{2}O_2 \rightarrow H_2O + SO_2 \qquad \Delta H° \approx -520 \text{ kJ/mol} \qquad (5.3)$$

$$2H_2S + SO_2 \rightleftharpoons 2H_2O + 3S(g) \qquad \Delta H° \approx -88 \text{ kJ/mol} \qquad (5.4)$$

The heats of reaction[1] are only approximate, since the exact values will depend on the type and state of sulfur that is formed. The reaction characterizing the overall stoichiometry of the sulfur formation is given by the sum of the two equations

$$3H_2S + \tfrac{3}{2}O_2 \rightarrow 3H_2O + 3S(g) \qquad \Delta H° \approx -608 \text{ kJ/mol} \qquad (5.5)$$

In one form of the process—where the hydrogen sulfide content of the gas stream is sufficiently large to supply the process heat, normally greater than about 35 volume percent—a stoichiometric amount of air sufficient to convert one-third of the hydrogen sulfide to sulfur dioxide is fed to the reaction furnace. The hydrogen sulfide is burned near atmospheric pressure at temperatures between 1100 and 1600°C. About 60 to 70 percent of the sulfur is evolved in the thermal stage by reaction (5.4). The elemental sulfur which is formed is condensed out by cooling effluent gases from the furnace by heat recovery.

The cooled mixture—consisting ideally of two-thirds hydrogen sulfide and one-third sulfur dioxide—is reheated above the sulfur dewpoint, a temperature usually less than 400°C, and then fed into the catalytic reaction stage where additional sulfur is produced in the presence of an aluminum oxide catalyst following the reaction (5.4). It can be seen that the reaction (5.4) is only mildly exothermic and that only about 15 percent of the total heat of reaction is evolved in the catalytic reactor. The temperature rise is therefore relatively small and the presence of the catalyst allows the conversion in this stage to be carried out at a high rate at the lower temperature. This results in higher equilibrium conversions since the reaction is exothermic and the equilibrium constant is increased by decreasing temperature (Eq. 2.49).

The temperature in the catalytic stage must be kept somewhat above the sulfur dewpoint so that the sulfur does not condense on the alumina catalyst and deactivate it. To increase the conversion, the gas is passed through a series of catalytic reactors at successively lower temperatures. This is possible since in each reactor more sulfur is removed, which lowers the dewpoint and thereby enables operation at a lower temperature in the next reactor. Before the gas enters each reactor it is cooled to condense out the sulfur formed in the preceding reactor and then heated again above the new sulfur dewpoint. The temperature in the last reactor is usually somewhat higher than 200°C.

The tail gas leaving the last catalytic reactor after the sulfur has been condensed will still contain small amounts of sulfur compounds. Complete conversion is precluded by chemical equilibrium considerations. Cleanup of this tail gas may be required by environmental considerations and a number of methods for doing this have been developed. Included among the approaches are the extension of the Claus reaction itself, but carried out in the liquid phase, and the hydrogenation of the sulfur compounds in the gas to hydrogen sulfide with subsequent removal in an indirect amine liquid absorption process or subsequent conversion directly to sulfur by the liquid phase Stretford process discussed below.

Direct Conversion of Hydrogen Sulfide by Liquid Absorption

A number of processes have been developed that absorb hydrogen sulfide in an alkali solution with subsequent chemical conversion to elemental sulfur. One such process is the *Giammarco-Vetrocoke process*, which uses aqueous sodium or potassium carbonate solutions containing arsenic compounds for absorption. The hydrogen sulfide is then oxidized to elemental sulfur by the exchange of one oxygen atom for one sulfur atom in the arsenic compound which is pentavalent. The overall stoichiometry is the same as (5.5) except that the sulfur produced is in solid form. The process is stated to be capable of producing a purified gas containing less than 1 mg/L of hydrogen sulfide at absorption temperatures up to 150°C. It has been mainly used for the desulfurization of coke oven and synthesis gases.

Another direct conversion liquid phase process developed initially for the removal of hydrogen sulfide from coal gases and widely used in Europe for the treatment of refinery gas, synthesis gas, and natural gas is the *Stretford process*. This process is proposed for use on several projected commercial-scale coal and oil shale conversion plants. It has a high removal efficiency, with values up to 99.9 percent reported. It is, however, a low temperature process which operates most economically below 40°C. In the process, hydrogen sulfide is absorbed in an alkali solution to form sodium bisulfide, which is in turn oxidized to elemental sulfur using a vanadium oxidant (sodium vanadate, $NaVO_3$). The reaction that takes place may be represented by the equation

$$2V^{5+} + H_2S \rightarrow 2V^{4+} + 2H^+ + S \tag{5.6}$$

The sulfur formed by this reaction is recovered by flotation. To regenerate the reduced vanadate, a relatively expensive organic acid (ADA = anthraquinone disulfonic acid) is added, which in the presence of air leads to the oxidation of the reduced vanadate. The overall oxidation is represented by the reaction

$$2V^{4+} + 2H^+ + \tfrac{1}{2}O_2 \rightarrow 2V^{5+} + H_2O \tag{5.7}$$

This reaction, which is promoted by blowing in air, takes place in a tank separate from the flotation tank. The reduced ADA is regenerated by contact with air.

Hot Gas Purification

A disadvantage of present commercial processes for hydrogen sulfide removal is that they are generally designed for operation at low temperatures. On the other hand, the raw gases exiting synthetic fuel reactors have temperatures ranging from 400 to 1000°C, and for high temperature gasifiers they reach 1500°C. This means that the gas must always first be cooled, thereby degrading its sensible heat. Alternative high temperature purification procedures under development use a solid absorbent that is not volatile in the temperature range of interest. The hydrogen sulfide is absorbed into the absorbent, reacts with it, and forms a stable sulfur compound which is retained in the absorbent.

Hot gas purification methods appear particularly attractive for combined cycle power systems, where the low-CV gas is not upgraded but is burned directly in turbines for the generation of electric power.

The sensible heat that must be removed to cool the gas that is at 1000 to 1500°C amounts to about 0.75 to 1.1 MJ/m³. In addition, depending on the coal and the gasification temperature, condensable tars representing up to 0.82 MJ/m³ would also be removed from the gas. These two energy quantities amount to about 30 percent of the 5.6 MJ/m³ calorific value of a low-CV gas derived from an air-blown coal gasifier. Of course, the sensible heat will in part be recovered by heat exchange with the purified gas, and the tars will be recovered and might be burned within the plant.

The absorbents for high temperature purification processes must be regenerable without creating atmospheric or process pollution problems. Two absorbents meeting these criteria are iron oxide and cuprous oxide. Iron oxide supported on a fly ash matrix for use as a high temperature sorbent, has been under development at the Morgantown Energy Research Center of the U.S. Department of Energy.[4] The chemistry involved in absorption and regeneration shows that iron sulfides are produced when hydrogen sulfide reacts with iron oxide, with the empirical composition approaching $FeS_{1.5}$:

$$Fe_2O_3 + 3H_2S \rightarrow 2FeS_{1.5} + 3H_2O \qquad (5.8)$$

During regeneration, air oxidizes the iron sulfide and the reactions can be represented by

$$6FeS_{1.5} + 13O_2 \rightarrow 2Fe_3O_4 + 9SO_2 \qquad (5.9)$$

$$2Fe_3O_4 + \tfrac{1}{2}O_2 \rightarrow 3Fe_2O_3 \qquad (5.10)$$

The sulfur dioxide can subsequently be reduced to elemental sulfur by reduction over carbon or by capture by calcium (see Section 9.2).

Problems with hot gas purification are that dust and other vapor phase contaminants are not removed as when water washing is practiced, nor is ammonia removed, which could lead to excess nitric oxide emissions when the gas is burned. Moreover, the overall thermal efficiency gains claimed for hot purification systems may not be as large because of the energy penalty incurred in reducing sulfur dioxide and because effective heat recuperation can be practiced with cold absorption systems. Despite these caveats, development of economic hot absorption systems would appear worthwhile.

5.2 SHIFT AND METHANATION

Coal-derived synthesis gas composed of carbon monoxide and hydrogen is the basic ''feedstock'' in a variety of processes for the manufacture of synthetic fuels. These include methanation to produce substitute natural gas (SNG) and indirect liquefaction to produce methanol and other liquid hydrocarbons. Synthesis gas is also a principal source of hydrogen needed in direct liquefaction processes (see Section 5.4).

The first step in the production of synthesis gas is the generation of a medium-CV gas either by the gasification of coal (or char) with steam and oxygen, or with steam and air if an indirect heat transfer procedure is used to remove the nitrogen barrier. In hydrogen manufacture in direct liquefaction plants it may also be appropriate to gasify the residuum fraction from the product fuel.

If the desired end product is SNG, then moving and fluidized bed intermediate temperature gasifiers will be appropriate to generate the medium-CV gas because of the relatively high methane concentrations produced in the gasifier. On the other hand, if liquid hydrocarbons or hydrogen are the end products, then high temperature gasifiers are appropriate because they generate a gas consisting mainly of carbon monoxide, hydrogen, and carbon dioxide with little methane.

Shift

In lower temperature gasifiers, the H_2/CO ratio is generally in the range of 1 to 2, whereas in higher temperature gasifiers it is between 0.5 and 1. In addition to carbon monoxide and hydrogen, the gas will also contain a number of undesirable impurities including hydrogen sulfide, ammonia, and carbon dioxide. The means of their removal is discussed in the previous section. However, even with the impurities removed the gas is still not a satisfactory feedstock since the H_2/CO ratio is normally not that which is needed. We recall that for conversion to methane the stoichiometric H_2/CO feed requirement is 3, while for indirect liquefaction by Fischer-Tropsch synthesis it is 2 (see Section 3.3). Depending upon the reactors used and the desired end products, optimum H_2/CO ratios may be somewhat lower or higher than indicated by the stoichiometry. Of course, if hydrogen is to be the end product, then all of the carbon monoxide must be removed.

The means by which the H_2/CO ratio is adjusted to the desired level is through the catalyzed, moderately exothermic gas shift reaction, Eq. (3.41). Assuming the carbon monoxide is essentially completely converted at equilibrium, we can use the stoichiometry to determine the fraction of the gas that must be shifted to achieve the desired H_2/CO ratio. For example, if the required H_2/CO ratio is x and we start with a gas having a value of x_0 for this ratio, then using Eq. (3.41) we can write

$$CO + H_2O \rightarrow CO_2 + H_2 \qquad (5.11)$$

$$\text{No. of} \begin{cases} \text{Initial} \\ \text{moles} \end{cases} \begin{matrix} \text{Initial} & 1 & & x_0 \\ \text{Final} & 1 - f & & f + x_0 \end{matrix}$$

where f is the fraction of gas passed through the shift reactor. The H_2/CO ratio in the mixture of shifted and unshifted gas is then

$$x = \frac{f + x_0}{1 - f} \qquad (5.12a)$$

which gives the fraction of gas shifted as

$$f = \frac{x - x_0}{x + 1} \qquad (5.12b)$$

The carbon dioxide that is formed is removed following the reaction. Because the reaction is moderately exothermic, the equilibrium constant is increased slowly by decreasing temperature (Eq. 2.49). Pressure has no effect on the equilibrium fraction of reactants converted because of the equal number of moles of reactants and products. The variation of the equilibrium constant with temperature is given in Figure 2.2.

The shift reaction is an old and commercially well-developed process for increasing the hydrogen fraction of synthesis gas and converting carbon monoxide. It is particularly important in ammonia manufacture to reduce the carbon monoxide which strongly deactivates (poisons) ammonia synthesis catalysts.

Both high and low temperature commercial catalysts are available to promote shift conversion.[5,6] The basic high temperature catalysts in use for many years are iron oxide-chromium oxide catalysts, which operate at temperatures typically in the range

350 to 475°C. The reaction is usually carried out in uncooled fixed bed reactors (adiabatic reactors) in which the temperature increases from the inlet to the outlet. The catalyst activity and reaction rate increase with higher temperature, but the equilibrium conversion of the carbon monoxide is reduced. At a reactor exit temperature of 425°C, the equilibrium constant is approximately 10 (Figure 2.2b) and since

$$K = \frac{p_{CO_2} p_{H_2}}{p_{CO} p_{H_2O}}$$

we may calculate that with 50 percent excess steam, the equilibrium carbon monoxide concentration is about 5 mole percent. The reader may determine the effect of both a variation in the equilibrium constant and the amount of steam on the stoichiometrically determined values for the fraction of gas that must be shifted (Eq. 5.12).

The carbon monoxide concentration can be further reduced in a second reactor in which more active low temperature zinc oxide-copper oxide catalysts are used. The temperature range for this type of catalyst is usually between 200 and 250°C, for which the equilibrium constant is increased by about a factor of 10 over the value for the example of the high temperature catalyst. The equilibrium carbon monoxide concentration is decreased correspondingly. These low temperature catalysts are prone to deactivation by sulfur compounds not removed by gas purification. Sulfided cobalt-molybdate catalysts, for which high sulfur levels are not a problem, have been developed. They are, however, somewhat higher temperature catalysts that are active above about 230°C. These catalysts convert the sulfur compounds to hydrogen sulfide, which can then be removed together with the carbon dioxide subsequent to the shift reaction. Such a catalyst has been successfully employed in the shift section of a partial oxidation plant where the feed gas is not first desulfurized.[7]

The pressure in the shift reactor is largely determined by requirements in other stages of the process. Higher system pressures increase the reaction rate and decrease the reactor volume, so that if high pressures are required downstream, it is appropriate to operate the shift reactor at these pressures. At high pressure operation of 3 MPa or greater, gas-catalyst contact times are on the order of 1 s and catalyst life is typically 1 to 3 years.

Shift catalysts must be periodically regenerated to remove accumulated carbon deposits, the formation of which is discussed in Section 3.3. As noted there, carbon formation is generally not a problem with shift catalysts because of the presence of excess water; the water inhibits carbon deposition by gasification of the deposited carbon and by formation of hydrogen at the catalyst surface. Increased hydrogen partial pressures shift the equilibrium away from carbon formation, so in general carbon deposition is hindered at high H_2/CO ratios.[5,6]

Methanation

Methanation is the final major processing step in upgrading cleaned and shifted medium-CV gas to SNG.[5,6,8] The synthesis of methane from carbon monoxide and hydrogen has been studied since the turn of the century. It has been employed for some time as a gas purification step in ammonia production to convert low levels of carbon monoxide, which as we have already noted is a strong poison to ammonia synthesis catalysts, whereas methane is inert to the catalysts. Methanation has not as yet been applied in full-scale commercial operations.[8,9]

In Section 3.3 it was pointed out that the basis for commercial methanation is the formation of methane with the rejection of steam

$$CO + 3H_2 \rightleftharpoons CH_4 + H_2O \tag{5.13}$$

The thermodynamic parameters for methane production through this reaction are given in Table 2.20. The methanation reaction is highly exothermic with a very strong dependence of the equilibrium constant on temperature (Figure 2.2b). Low temperatures and high pressures favor methane yields, although as noted in Section 3.3, satisfactory yields can still be obtained at temperatures up to 525°C. The reaction must be promoted by a catalyst. The active component in almost all commercial methanation catalysts is nickel supported on an alumina or other oxide support. The reaction rate for the nickel catalyzed methanation reaction is discussed in Section 2.3 and the methane formation rates at various pressures and temperatures are summarized in Table 2.21. Nickel catalysts operate at temperatures generally from 300 to 400°C. They have a very low tolerance to sulfur with the result that hydrogen sulfide levels must be reduced to less than 1 to 0.1 part per million to achieve practical catalyst lifetimes. Kinetic effects generally tend to limit carbon formation because the reaction rate for carbon monoxide methanation is much larger than that of the Boudouard reaction (3.12).

Of importance in coal gasification processes is that the methanation reaction is strongly exothermic. Previous industrial applications of methanation, as in gas purification for ammonia synthesis, have all been with relatively low levels of carbon monoxide. With low carbon monoxide concentrations the temperature rise was generally not large, usually less than 75°C, and simple uncooled packed bed reactors could be used. However, the heat of reaction causes the temperature to increase by about 60°C for each one percent of carbon monoxide converted to methane in an uncooled adiabatic reactor. This may be seen from a simple energy balance around a reactor with no heat removal which gives

$$f_{CO} = \frac{C_p \Delta T}{-\Delta H^\circ} \tag{5.14}$$

Here, f_{CO} is the fractional conversion of carbon monoxide in the stoichiometric mixture, ΔH° is the heat of reaction, and $C_p \Delta T$ is the heat to raise the gas temperature from the initial to final value. For an average reactor temperature of 350°C the mixture specific heat is about 36 J/(mol·K), and the heat of reaction is -218 kJ/mol. With $f_{CO} = 0.01$, the temperature rise from Eq. (5.14) is $\Delta T = 60°C$. Because of the high concentrations of carbon monoxide present in the synthesis gas, this imposes the need for some form of cooling or method to hold down the temperature.

Numerous possibilities exist for limiting the temperature rise in the highly exothermic methanation reaction, with the choice generally based on economic or operational considerations. Two of the principal methods used are dilution of the feed with recycled product gas or cooling by recovery of the reaction heat. The maximum temperature which the reactor is allowed to reach is imposed by the equilibrium conversion requirement and the catalyst thermal stability.

In the dilution procedure, the reaction temperature is kept low by reducing the carbon monoxide concentration so that the heat capacity relative to the quantity of heat released is increased. The recycle gas may or may not be cooled. By this procedure or a number of variants, with an inlet temperature of, say, 300°C, the exit temperature can be held below 400°C without the use of cooling.

In the cooling procedure, the reactor is operated with the reaction heat removed, for example, by water in the form of high temperature, high pressure steam. The heat may be removed either externally or internally. One reactor type used for external heat removal is a fixed bed reactor consisting of a bundle of catalyst-filled tubes in a vessel. The gas flows inside the tubes and the heat generated by the exothermic reaction is removed through the tube walls by the generation of steam outside the tubes. In an alternate design, called the "tube wall reactor," the nickel catalyst is applied to the inside walls of the tubes making up the heat exchanger bundle.

Cooling may also be effected in fluid bed reactors. In an entrained flow design employed in the Sasol Fischer-Tropsch plant, circulating catalyst is entrained by the gas feed. The large mass of catalyst absorbs the heat of reaction which is removed by oil coolers and used for steam raising. The temperature distribution is very uniform in this type of reactor. Heat removal can be accomplished by internal cooling, by absorption of the reaction heat in the oil with subsequent external cooling in a heat exchanger, and by evaporation of a portion of the oil. Combinations of two or all three of these approaches can also be used.

At least two processes have been demonstrated which combine the shift and methanation reactions. In one process tested on pilot scale by the R. M. Parsons Co., shift and methanation are accomplished simultaneously in a series of adiabatic fixed bed reactors operating at successively lower temperatures.[8] Heat is removed from the gas between the reactors by the generation of high pressure steam. The system operates at quite high temperatures with sufficient steam addition to cause the shift reaction to occur over a nickel catalyst, while at the same time avoiding carbon formation. The gas is fed to the first reactor at about 480°C and the exit temperature is about 770°C. Six reactors have been employed in series with exit temperatures in the last reactor as high as 470°C. The process is designed to operate at pressures from atmospheric to 7 MPa.

Conoco, Inc., has also demonstrated on semi-commercial scale a combined shift-methanation process in connection with the British Gas Corp. demonstration of the slagging Lurgi gasifier at Westfield, Scotland.[9] The process employs two stages. The first stage consists of a number of adiabatic fixed bed reactors in a series-parallel configuration, each containing the same nickel catalysts. The gas exiting each first stage reactor is cooled by generating steam. Temperature control is also maintained by recycling a portion of the product gas from the final first stage reactor. Carbon dioxide is removed from the first stage product and then fed to a single adiabatic fixed bed reactor for conversion of unreacted carbon monoxide and hydrogen. The process is operated at temperatures from 260 to 520°C and a pressure of about 2 MPa.

In our discussion of methanation we have so far spoken only of carbon monoxide methanation. However, carbon dioxide is present in both the unshifted and shifted gas, and it would be useful if the carbon loss in "waste" carbon dioxide could be minimized by also methanating the carbon dioxide through the reaction

$$CO_2 + 4H_2 \rightleftharpoons CH_4 + 2H_2O \tag{5.15}$$

Laboratory experiments with composite catalysts which appear to simultaneously methanate carbon monoxide and carbon dioxide have been reported.[10] The active constituents of the catalysts are nickel, lanthanide oxide (La_2O_3), and small amounts of ruthenium. Rapid and complete conversion with high selectivity was obtained at temperatures of 270 to 300°C. This would appear to be a most promising advance.

5.3 OXYGEN PRODUCTION

The principal route to the production of a medium- or high-CV gas is the endothermic gasification of coal with steam, the needed heat being supplied by burning the coal or char with relatively pure oxygen. When air is used instead of oxygen, the nitrogen in the air acts as a diluent and the product is a low-CV gas, unless the nitrogen is removed afterwards from the gasifier product stream or during the gasification process. In Section 4.1 it is shown that by means of an indirect heat transfer procedure, using a circulating solid or liquid heat carrier heated in a separate furnace, the nitrogen can be rejected in a stream of combusted gas. At the time of writing, no process has been developed commercially for removing the nitrogen from the product stream. It remains that by far the most straightforward procedure is to use oxygen directly.

In pipeline gas manufacture, depending on the process and the coal, between 0.5 and 1 mole of oxygen is consumed in the gasifier for each mole of methane produced. Because of the large quantities required, the oxygen plant will be of comparable magnitude to the gasification plant. As a consequence, it will have to be an integral part of any synthetic fuel complex producing a medium-CV gas either as the product or for upgrading to pipeline gas, or for use in liquids synthesis. Capital cost breakdowns for pipeline gas facilities show that between 10 and 20 percent of the total investment is for the oxygen plant. This investment range is the same as that required for the gasification section.

The efficiency of gasification is to a large extent also dominated by the oxygen requirement. The heat generated by the complete conversion of the 0.5 to 1 mole of oxygen needed to produce pipeline gas is 394 kJ/mol. If this is taken to be the heat required to drive the methane forming reactions, since methane has a calorific value of 885 kJ/mol, a thermal efficiency of about 70 to 80 percent would be indicated. However, energy is also lost in the production and compression of the oxygen. This energy requirement can represent an additional 10 percent loss in efficiency for each mole of oxygen supplied to the gasifier. The overall thermal efficiencies derived by this argument are representative of the range attainable in practice (see Table 1.16 and Section 5.5).

Because the oxygen plant represents such a large component of the capital and energy required to produce a synthesis gas, this suggests that improvements in the technology and efficiency of oxygen production could have an important effect in reducing product cost. In order to understand the constraints to improvement, as well as those directions which might be pursued, we next discuss the ideal and actual requirements to produce oxygen by various known methods.

The two main sources of oxygen in nature are air and water. Recovery of oxygen from air involves a physical separation of gases, whereas its recovery from water involves a chemical dissociation. In Section 2.2 we showed that the ideal minimum work to separate oxygen from air is equal to the increase in the Gibbs free energy. This minimum occurs when the system starts and ends at atmospheric pressure and temperature and is allowed to exchange heat freely with the atmosphere [see Eq. (2.52) and discussion]. The minimum work for the separation of air as a tertiary mixture into pure oxygen, nitrogen, and argon was given as 6 kJ/mol of O_2. At standard conditions, the increase in Gibbs free energy to dissociate water into one mole of hydrogen and one-half mole of oxygen is 474 kJ/mol of O_2. Comparison of the two energy requirements shows why the electrolysis of water is not used commercially for the production of oxygen. The electrolysis route is, however, not as bad as might appear. First, the electrolysis of water can be carried out in practice with energy consumptions much closer to the ideal minimum than can air separation. Secondly, the electrolytic dissociation of water also yields two moles of hydrogen for each mole of oxygen. Hydrogen by itself is an important product for liquefaction plants, whereas the nitrogen recovered in air separation is not needed in synthetic fuel processing.

In this section we shall discuss the production of oxygen by cryogenic liquefaction and membrane separation techniques. Both of these procedures involve air separation, so their efficiency can be measured against the ideal minimum value of 6 kJ/mol of O_2. Oxygen production by electrolysis is treated as a byproduct of hydrogen manufacture by this method, and as such it is looked at in the next section in connection with hydrogen production.

The principal method by which oxygen is produced is the low temperature distillation of air, wherein air is first liquefied by refrigeration and the liquid is then distilled, also termed rectified, to separate the more volatile nitrogen from the less volatile oxygen. Air liquefaction falls within the science of cryogenics, which depending on the source is taken to begin at anywhere from -100 to $-185°C$, although generally temperatures of $-100°C$ and below are considered "cryogenic." Table 5.3 lists the critical temperatures and boiling points of cryogenic gases and Table 5.4 the major components of dry air at standard conditions.

In all refrigeration cycles low temperatures are produced by absorbing heat at a lower temperature and dissipating it at a higher one. The sources of refrigeration used to liquefy air are the Joule-Thomson effect, the expansion in an engine doing external work, and the vaporization of a liquid. Actual cycles commonly employ combinations of two and sometimes all three cooling methods.

Joule-Thomson cooling is produced by expanding compressed air at constant enthalpy through a throttle. For a perfect gas allowed to reach equilibrium there would be no temperature change; however, the Joule-Thomson effect is an irreversible one resulting from the gas imperfection. The extent of the temperature change that occurs

Table 5.3 Critical temperatures and boiling points of cryogenic gases[11]

Gas	Critical Temperature, °C	Boiling Point at 101 kPa
Methane	− 82	− 156
Oxygen	− 119	− 183
Argon	− 122	− 186
Fluorine	− 129	− 188
Air	− 140	− 195
Nitrogen	− 147	− 196
Neon	− 229	− 246
Hydrogen	− 240	− 253
Helium	− 268	− 269

depends on the slope of the constant enthalpy curve with respect to pressure at the initial pressure and temperature conditions of the expansion. For cooling, that is, temperature decreasing on expansion, the slope must be positive.

The expander method utilizes a nearly isentropic expansion with external work instead of an isenthalpic expansion. The temperature is reduced as the energy in the gas is removed by using an expansion engine or turbine. Work or heat is removed from the gas at the same time the pressure is reduced. For a given pressure difference the cooling is greater from the nearly isentropic expansion than from throttling, as illustrated in the temperature-entropy diagram for air of Figure 5.4.

From Figure 5.4 it can be seen that temperature changes from a Joule-Thomson expansion are generally not large. Even for an isentropic expansion, a high initial pressure would be required if liquefaction were to be achieved by expanding to atmospheric pressure in a single step. The compression work for such an initial to final pressure would be large (see Table 2.16) and lead to a low overall process efficiency. As a result, the temperature changes obtained from either expansion procedure are relatively small. However, the refrigeration effect can be made accumulative by first cooling the compressed air by heat exchange with the cold product gases. This is illustrated in Figure 5.5, which shows the cycle first developed by Carl von Linde for the commercial production of oxygen using Joule-Thomson cooling. In

Table 5.4 Major components of dry air at standard conditions[11]

Gas	Volume %	Mass %
Nitrogen	78.03	75.47
Oxygen	20.99	23.19
Argon	0.94	1.30
Carbon dioxide	0.03	0.04
Hydrogen	0.01	Trace

place of the throttling expansion shown, an expansion engine or turbine could be substituted. In practice, the expander cycle will also employ a Joule-Thomson expansion for the final refrigeration step, in order to avoid liquid formation in the engine.

The common features of the two expansion methods are:[11]

Compression of the air to pressures which in modern plants range from 0.5 to 1 MPa.

Cooling of the air after compression to near ambient in an aftercooler, followed by further cooling by countercurrent heat exchange with the cold product gases, and

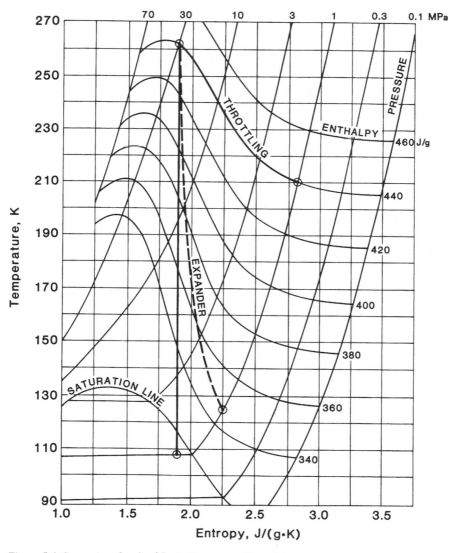

Figure 5.4 Comparison for air of Joule-Thomson cooling with cooling by a mechanical expander.

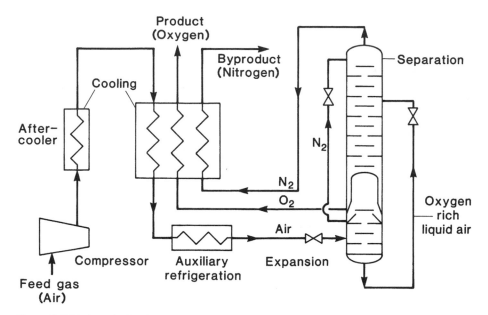

Figure 5.5 Linde cycle for air separation.

by auxiliary refrigeration. Large plants also use "regenerators" in which the incoming air is passed through beds of cold packing to remove contaminants, including carbon dioxide and water.

Expansion to further cool the air. This may be by throttling the gas through a reducer valve or by using an expansion engine or turbine. The development of efficient expansion turbines operating at relatively low pressures has made the expander method the most widely used procedure for the large-scale production of oxygen, as would be required in a synthetic fuel plant. The turbine efficiency is important, not because of useful work obtained, but because the greater the expander efficiency, the greater the amount of heat removed; that is, the expander is a refrigeration device and the mechanical work developed is normally wasted.

Separation of the oxygen and nitrogen in a distillation column to which the liquid air is sent. The higher-boiling-point oxygen collects at the bottom as a liquid and the lower-boiling-point nitrogen leaves from the top of the column as a gas. This is normally done in two steps, as shown in Figure 5.6, except where 90 to 95 percent oxygen purity is acceptable in which case one step is sufficient.

The last procedure used for cryogenic liquefaction employs a liquid vaporization cycle in which a refrigerant gas is compressed to a high pressure, and the heat of compression is removed and the refrigerant condensed by water or air cooling. The condensed refrigerant is then allowed to evaporate at a lower pressure—and by heat exchange with the fluid to be cooled removes heat in an amount equivalent to the heat of vaporization. The refrigerant is then compressed to close the cycle. Of course, the fluid to be liquefied can itself be the refrigerant. In cryogenic liquefaction, this procedure is usually applied in a "cascade system" in which a series of refrigerants are used, each boiling at a successively lower temperature. Cascade processes have about the same efficiency but are more complex than expansion procedures and are not generally used in large air separation plants.

As we have noted, the minimum work for the separation of air as a tertiary mixture into pure oxygen, nitrogen, and argon is 6 kJ/mol of O_2. The energy required to separate oxygen from air in modern large-scale oxygen plants is typically around 40 kJ/mol O_2. This would indicate a thermodynamic efficiency of about 15 percent. All of the energy consumed in an oxygen plant goes into compressing the incoming air. The plant energy requirement for oxygen production is therefore the work to drive the compressors. Using an overall efficiency for conversion of heat to work of about 30 percent, the heat requirement for oxygen manufacture becomes about 128 kJ/mol or 4.0 MJ/kg.

The pressure to which the air must be compressed in a single stage is given from Eq. (14) in Table 2.16 by

$$\frac{p_2}{p_1} = \left[1 + \frac{\gamma - 1}{\gamma} \frac{\eta}{RT_1} \left(\frac{w}{n} \right) \right]^{\gamma/\gamma - 1} \tag{5.16}$$

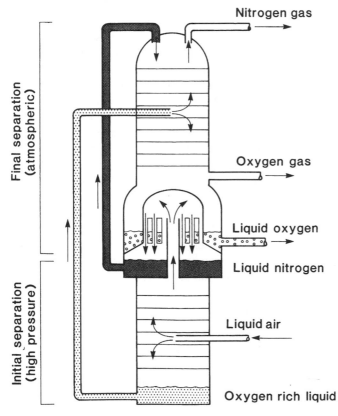

Nitrogen gas

Final separation (atmospheric)

Oxygen gas

Liquid oxygen

Liquid nitrogen

Liquid air

Initial separation (high pressure)

Oxygen rich liquid

Figure 5.6 Distillation column for fractional separation of liquid air (after Ref. 11).

Taking 5 mol air/mol O_2, a typical work input is then $40/5 = 8$ kJ/mol air. For a compressor efficiency $\eta = 0.8$ with $T_1 = 298$ K, $R = 8.31 \times 10^{-3}$ kJ/(mol·K), and $\gamma = 1.4$, we find $p_2 \approx 7p_1$ or 0.7 MPa for compression from atmospheric. This is about the middle of the pressure range to which the air is compressed in modern oxygen plants.

It should be noted that the energy requirement discussed is only for the separation of the oxygen from the air. In most synthetic fuel plants, the oxygen will in turn be compressed to a high pressure before being fed to the gasifier. The work to do this can also be calculated from Eq. (5.16). However, this operation is normally done in stages with interstage cooling to hold down the temperature rise. For this case, $\gamma/(\gamma - 1)$ is replaced by $s\gamma/(\gamma - 1)$, where s is the number of stages.

The thermodynamic efficiency of the air separation procedures discussed has been seen to be quite low, 15 percent being a typical figure for modern plants, with an upper limit of about 20 percent. Although there are many losses in the system, the largest single contributing factor is the distillation column itself.[12] Ideal column efficiency is 67 percent, but in practice, actual efficiencies are on the order of 35 to 40 percent. Although the thermal efficiencies of the liquefaction procedures could undoubtedly be improved, this could be done only at the expense of increased equipment size in order to reduce the driving forces.

An approach that has been investigated for air separation, which avoids the phase changes and associated losses inherent in liquefaction procedures, is the use of membrane separation techniques. It has been known for some time that polymers in the form of flat membranes or hollow fiber membranes can be used to separate gases by virtue of the difference in permeation rates through the membrane under the action of a hydrostatic pressure difference across the membrane. One limitation to the development of these processes has been the manufacture of membranes with fast enough permeation rates to make them economically viable. Of the polymeric materials, silicone rubber is known to be among the most permeable to common gases. Recently, composite sheet and hollow fiber membranes consisting of a porous polymeric substrate coated with silicone rubber have been commercially prepared.[13,14] Through the use of composites, the membrane resistance to permeation can be reduced, since the resistance is proportional to the thickness of the portion of the membrane where separation takes place.

The driving force for membrane gas separations is the pressure difference across the membrane. The degree of oxygen enrichment with a single stage membrane system, to which a large excess of feed air is supplied, can be derived by writing the flux for each component through the membrane and then solving for the fraction of oxygen in the product gas. It can be shown from Fick's law that the flux of any component i through the membrane is given by

$$J_i = \frac{\Delta p_i}{R_i} \tag{5.17}$$

where J is the flux in m³/(s·m²), Δp is the partial pressure difference across the membrane, and R is the resistance to the gas permeation. Here,

$$R_i = l/\kappa_i \tag{5.18}$$

where l is the membrane thickness, and κ_i the permeability. It may be recognized that the permeability is simply the product of the gas diffusion coefficient and Henry's law constant.

Assuming, for simplicity, air to consist of 21 percent oxygen and 79 percent nitrogen, the volume fraction of oxygen in the product gas can be derived as a function of the pressure ratio across the membrane and the ratio of the oxygen-to-nitrogen permeabilities.[13] For a very large pressure ratio, the volume fraction of oxygen becomes dependent only on the permeability ratio and can be shown to be given by the relation

$$X_{O_2} = \frac{J_{O_2}}{J_{O_2} + J_{N_2}} = \frac{0.21\alpha}{1 + 0.21(\alpha - 1)} \qquad (5.19)$$

Here, α is the ratio of the permeabilities of oxygen to nitrogen and is termed the separation factor. For silicone rubber α is 2.2. From Eq. (5.19) this shows that the maximum enrichment in a single stage silicone rubber permeation cell is only 37 percent. Although the efficiency and economics for this level of enrichment have been shown to be attractive in comparison with oxygen production by air liquefaction procedures, the level of enrichment is too low to be useful in synthetic fuel plants. Recently, however, it has been shown[15] that much higher levels of enrichment can be achieved by using a continuous membrane column, which is a column where the feed is separated from the product by a membrane and flows countercurrent to the product. By recycling the product gas, after compressing it, the more permeable gas can be enriched to any degree desired. It still remains to be seen whether by further membrane and separation system development, membrane procedures can succeed in improving the efficiency and reducing the cost of oxygen production. The outlook for large-scale production is not favorable at this time, but continued work is very much warranted.

5.4 HYDROGEN PRODUCTION

The manufacture of hydrogen is an integral part of all pyrolysis and direct liquefaction plants producing liquid fuels from coal. It is used directly in the process or for upgrading of the primary coal liquids. Hydrogen is also needed for the upgrading of crude shale oil and the bitumen separated from tar sands, as well as in the hydrogasification of coal. The cost of manufacturing hydrogen is a major component of the cost of direct liquid synthetic fuel production, just as is the cost of manufacturing oxygen in gas production.

The hydrogen consumed in direct liquefaction plants will generally range between 2 and 6 percent by mass of the dry and ash-free coal feed, depending on the extent to which the final product is hydrogenated. The lower value is representative of a solvent refined product and the upper one of a synthetic crude. Any additional on-site refining, to remove contaminants or to produce gasoline or mid-distillate products, would increase the requirements correspondingly. For example, the theoretical amount of hydrogen needed to go from a coal represented by $CH_{0.8}$ to a crude oil represented by $CH_{1.6}$ would be about 6 percent by mass of the original coal. To further upgrade to gasoline, CH_2, would require an additional 3 percent. Therefore, to produce a high-quality motor fuel from bituminous coal would consume hydrogen in an amount equal to about 10 percent by mass of the original coal, or around 1 m^3/kg of coal.

As we have seen in Section 3.4, this large consumption of hydrogen is the principal reason for the relatively low thermal efficiencies of direct liquefaction processes. The capital requirements for hydrogen production are correspondingly large. Indeed, the fractional cost of the hydrogen plant, including the cost of any associated oxygen plant, ranges between 35 and 50 percent of the total capital cost of direct liquefaction facilities now projected for demonstration on large scale.[16] It is interesting to note that the liquefaction reactor sections for the projected demonstration facilities on average constitute only about 25 percent of the total capital cost.

Hydrogen, unlike oxygen, is not found free in nature in any significant concentration. Its source must therefore be either hydrocarbons or water. Hydrocarbons are, however, the products being manufactured in synthetic fuel plants, so water will most generally be the main source of any hydrogen, although a byproduct hydrocarbon that is not desired in the product mix may be used. The principal methods for the manufacture of hydrogen involve reactions of coal, char, or hydrocarbons with steam at high temperatures.

The most widely used method of producing hydrogen is by the catalytic steam reforming of natural gas, represented by the reaction (Table 2.22)

$$CH_4 + H_2O \rightarrow CO + 3H_2 \tag{5.20}$$

This is the reverse of the methanation reaction Eq. (5.13). It is endothermic and proceeds in the direction indicated only with the addition of heat. The reaction is carried out over a nickel catalyst at a high temperature in a direct-fired furnace fueled by the methane. The catalyst is poisoned by sulfur, so any sulfur present in the feed must be removed. The synthesis gas is in turn passed through a catalytic shift converter, where the carbon monoxide is reacted exothermically with steam to produce hydrogen and a carbon dioxide byproduct which is removed from the system. In practice, the efficiency of producing hydrogen from methane is about 70 percent.

Feedstocks for hydrogen production by reforming can be any desulfurized, light hydrocarbon, including methane, liquefied petroleum gas, and naphtha. As we noted, these are generally the products desired in synthetic fuel manufacture. For this reason, hydrogen production by reforming will not normally be employed in integrated plants for synthetic liquids production by direct hydrogenation procedures. Reforming may, however, be used in indirect liquefaction plants where light hydrocarbon gases other than the desired product are formed, as in Fischer-Tropsch synthesis. It may also be used where refinery operations produce sufficient light hydrocarbons that can be diverted to manufacture hydrogen.

Gasification, or so-called partial oxidation, is the procedure that is likely to be the most widely used for hydrogen production in synthetic fuel plants. The technologies available are discussed at length in the preceding chapter. Coal and char are obvious feedstocks, though in most direct liquefaction plants the heavy bottoms fractionated from the liquid product generally will also be used. Of significance in the production of hydrogen by partial oxidation is that oxygen is required since the endothermic heat of the gasification and "reforming" reactions is supplied directly by combustion rather than by heat transfer across a tube wall. The capital investment for the oxygen plant constitutes about one-third of the hydrogen plant cost for the direct liquefaction facilities now projected for demonstration that employ partial oxidation.

The characteristics that the gasifiers should have when used for hydrogen production depend to some extent on the feedstock. With hydrogen the desired product, high temperature gasifiers will be used in order to minimize methane formation. We recall from Section 4.2 that the oxygen requirement for these gasifiers is relatively high, and that the carbon conversion and the CO/CO_2 ratio are relatively high. This leads to a low H_2/CO ratio, usually between 0.5 and 1.0, so that any excess steam raised by quenching the gasifier product gases is needed for the shift reaction to produce hydrogen. Apart from throughput considerations, gasifier pressures for hydrogen manufacture will be high since the hydrogen that is produced is needed at high pressures for the subsequent liquefaction.

A simplified flow scheme for hydrogen manufacture by gasification of char is shown in Figure 5.7. A high temperature gasifier has been assumed with an oxygen feed rate of 0.3 mol/mol C and a steam rate of 0.53 mol/mol C, and producing an off gas of composition typical for these second generation gasifiers. The values given in Figure 5.7 are based on those for the Lurgi slagger and Texaco gasifiers, Tables 4.19, 4.20, and 4.22, but with some modification to eliminate the pyrolysis-derived methane produced in the case of the Lurgi. Apart from the gasifier and steam boilers, gas shift and purification systems are required as well. As it is required to convert essentially all the carbon monoxide to hydrogen, the entire off gas stream must be taken to the shift reactor, which has been assumed to operate on 50 percent excess steam.

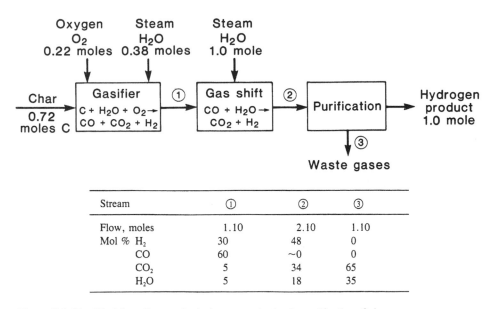

Stream	①	②	③
Flow, moles	1.10	2.10	1.10
Mol % H_2	30	48	0
CO	60	~0	0
CO_2	5	34	65
H_2O	5	18	35

Figure 5.7 Simplified flow diagram for hydrogen production by gasification of char.

The flow rates and compositions in Figure 5.7 have been used to construct the thermal balance of Table 5.5, which suggests a total heat input of 376/2, or 188 MJ/kg of hydrogen produced. The unaccounted heat in the table is made up of sensible heat losses and the exothermic heats of the shift and overall gasification reactions. In an integrated plant some of this heat would be recovered, for example by raising steam, and so reduce the quoted value for hydrogen production. On the other hand, gas purification requirements were not included, and can be expected to amount to a further 6 percent of the char feed calorific value. If on balance some 60 percent of the unaccounted heat is recovered, the net requirement for hydrogen production is 130 MJ/kg, which is equivalent to a thermal efficiency for the gasification-based process of 62 percent.

The residuum fraction from the product fuel in direct liquefaction plants is too heavy for catalytic reforming and will normally be used for the manufacture of hydrogen. Two of the gasifiers discussed in Chapter 4, which are presently being used to produce hydrogen from liquid hydrocarbons, are the Shell-Koppers and Texaco gasifiers.[17] They are now also being applied to the manufacture of hydrogen from solid fuels. These gasifiers require an oxygen plant. One indirect gasification procedure that employs air as the oxidant, and which is presently used for hydrogen manufacture from residuum oils, is the Exxon Flexicoke system. In the Flexicoke system the fuel is circulated, being air oxidized in one reactor and steam reformed and cooled in another, with streams of hot and cold particles circulating between the reactors. In this way the combustion gases, including the carbon oxides and hydrogen, are separated from the nitrogen in the air.

Table 5.5 Approximate thermal balance for hydrogen production by gasification of char

	Heat Content, MJ/kg	Mass Flow,* kg	Heat Flow, MJ	Heat Flow, % of Total
IN				
Char	32.8	8.6	282	74.8
Oxygen	4.0	7.0	28	7.4
Steam to gasifier	2.7	6.8	18	4.8
Steam to shift	2.7	18.0	48	13.0
Total		40.4	376	100.0
OUT				
Hydrogen	80.1	2.0	160	42.4
Steam in waste gases	2.7	7.0	19	5.0
CO_2 in waste gases	—	31.4	—	—
Unaccounted (by difference)	—		197	52.6
Total		40.4	376	100.0

Energy to produce hydrogen† $= 376 - 0.6 \times 197 \approx 260$ MJ/kmol or 130 MJ/kg

Thermal efficiency† $= \dfrac{160}{376 - (0.6 \times 197)} \times 100 = 62\%$

* Based on idealized flow scheme of Figure 5.7.

† Assuming 60% of the unaccounted energy is recovered.

One quite different method for hydrogen manufacture, which is a fully commercial and reliable technology, is water electrolysis.[18] The overall chemical reaction is

$$H_2O(l) \rightarrow H_2(g) + \tfrac{1}{2}O_2(g) \tag{5.21}$$

In conventional operation, a dc potential is applied across electrodes bounding a cell containing an alkaline electrolyte solution. Current flows through the cell, and the electrochemical dissociation of the water leads to hydrogen gas evolved at the cathode and oxygen evolved at the anode. The electrode reactions may be written

$$2H_2O + 2e^- \rightarrow 2H + 2OH^- \qquad \text{(cathode)} \tag{5.22}$$

$$2H \rightarrow H_2 \tag{5.23}$$

$$2OH^- \rightarrow 2OH + 2e^- \qquad \text{(anode)} \tag{5.24}$$

$$2OH \rightarrow H_2O + \tfrac{1}{2}O_2 \tag{5.25}$$

Among the advantages of electrolysis are that it is operable at room temperature, pure hydrogen and pure oxygen are produced, both gases are generated separately, and the equipment is relatively compact. The principal problem is that the energy requirement is large. The minimum value of the energy to dissociate water is equal to the increase in Gibbs free energy at standard conditions, which is 237 kJ/mol H_2, equivalent to 1.23 volts. Despite the fact that electrolysis efficiencies are quite high when measured against the theoretical minimum energy, with voltages of less than 2 volts common, the difficulty lies in the need to provide electrical energy. For a typical conversion efficiency of heat to electricity of about 33 percent, this means that for a 2 volt potential the overall conversion efficiency would be about 20 percent, and for a 1.7 volt potential about 24 percent. This is about the range of conventional electrolysis conversion efficiencies and may be compared with the 60 to 65 percent range attained with partial oxidation systems, and the 70 percent for reforming procedures. The large difference in efficiencies results from the need to supply high availability electrical energy.

Because of the high energy requirement, conventional electrolysis is not likely to be applied to the manufacture of hydrogen in synthetic fuel plants. Moreover, most plants must have a major processing step built into them to convert heavy residuum and other byproduct hydrocarbons, the most logical choice of which is to hydrogen. Even the simultaneous production of oxygen together with hydrogen is not advantageous in synthetic fuel plants, since when oxygen is required hydrogen is generally not, and when hydrogen is needed any oxygen requirement is to produce it.

Innovative electrolysis processes are under development,[17] such as one employing a solid ion exchange membrane as the "electrolyte." The membrane is highly conductive to hydrogen ions and the electrodes are attached to each side of it. Water is supplied to the oxygen evolution electrode where it is decomposed to oxygen, hydrogen ions, and electrons. The hydrogen ions move through the membrane to the hydrogen evolving electrode, while the electrons pass through the external circuit. At the hydrogen electrode, the hydrogen ions and electrons recombine to produce hydrogen. Operating voltages as low as 1.37 volts, corresponding to a 90 percent thermodynamic efficiency, have been achieved. The overall conversion efficiency is, however, still only 30 percent and does not appear competitive at the scales of hydrogen production required for synthetic fuel plants.

Because of the high cost of producing hydrogen, recovery from byproduct and various off gas streams may be economically employed. Among commercially available processes for doing this are cryogenic separation and membrane separation, both of which were discussed in connection with oxygen production, in addition to adsorption systems.[17,19]

5.5 INTEGRATED PLANTS

In broad terms we can state that coal gasifiers are designed to produce one of only two prime products. These are a *low-calorific value gas* and a *medium-calorific value gas*. Indeed, a medium-CV gas may simply be considered to be a low-CV gas without the diluent nitrogen. As we have seen, nitrogen dilution can be avoided either by blowing the gasifier with pure oxygen, or by using an indirectly heated gasifier.

A closer look at off gas properties reveals that there is a wide range of compositional differences in the gases falling into each category, as well as large differences in their temperature and pressure as they leave the gasifier. Such differences are dependent essentially on the operating characteristics of the gasifier, and these are discussed in some detail in Chapter 4. It is possible, however, to further manipulate the properties of the off gas to suit its end use, by the incorporation of various downstream processing units. Many of these process steps are discussed in preceding sections of this chapter. In this section we show how these processes are combined to form the integrated plants that produce the desired synthetic fuel product. While discussing factors relating to the selection of gasifier type as well as the selection and sequencing of downstream processing steps, no definite statement as to the *best* integrated plant for a specific end product will be made. The word *best* implies the cheapest dollar cost per unit of product, and this can really be determined only from detailed engineering and cost studies of the entire integrated plant for each alternative arrangement considered. Apart from requiring several man years of effort, such studies are at best time-dependent estimates in view of the rapid technological advances being made, and the limited information available on commercial-scale performance. In addition, the immense quantity of data generated often serves to hide relevant information that would otherwise provide insight as to the real source of improved process performance. One technique worthy of mention in this connection is that of differential economic analysis, in which similar process units are eliminated from consideration and the cost evaluation made for only those units which perform differently.[20]

Numerous contributing factors are imbedded in a detailed cost analysis. These include capital costs, overall operating costs, energy costs, energy efficiency, labor requirements, environmental factors, and as always, process control and reliability (see Section 10.1). The manner in which a particular process choice may affect one or more of these factors can often be assessed on the basis of more fundamental aspects than detailed, but routine, cost estimates, and this is done wherever meaningful. The main purpose of this section is to emphasize the variety of plant configurations available for a given product, and to provide some insight into which operating characteristics most significantly determine the effectiveness of the overall energy conversion.

To simplify the discussion, only two basic integrated plants have been selected for presentation here. The variations possible within these two plants do, however, cover most of the process principles of relevance. One plant is based on a low-CV gas, the other on a medium-CV gas. Low-CV gas is invariably consumed on site to provide heat, and this will often be for electricity generation. *Combined cycle electricity generation* appears to be the most efficient process currently under consideration and was therefore selected for discussion. A medium-CV gas may also be burned to produce heat, but there is no real advantage in manufacturing this higher quality product for on-site consumption. The extra processing steps—and concomitant lower conversion efficiency—associated with producing a medium-CV gas provide a product that can be transported over moderate distances and sold as a clean-burning industrial fuel. More important, however, is that the medium-CV gas may be upgraded to produce a high-CV *substitute natural gas* (SNG) for wider distribution to both domestic and industrial consumers. SNG production is consequently the second process selected for presentation. Coal-based gas processes not discussed here include the production of hydrogen, which is reviewed in the previous section, and the synthesis of liquid fuels, which is discussed in Section 6.2. A brief description of an integrated plant based on the important catalytic gasification concept is given in Section 4.4. The synthesis of petrochemicals, another important aspect of coal gasification, is beyond the scope of the book. Finally, it is stressed again that it is not our intention to give a mechanistic type description of integrated plants; rather, we wish to highlight the selection of equipment and operating variables in terms of the process principles discussed in previous chapters.

Combined Cycle Electricity Generation

It is convenient to discuss conventional coal-fired boiler, steam-electric power plants before progressing to the combined cycle systems. A simplified block flow diagram for a conventional process is shown in Figure 5.8. It can be seen that the only gas stream of consequence in this system is the flue gas produced by complete combustion of the coal in excess air. A typical flue gas composition based on a modern pulverized-coal boiler burning a high sulfur coal is shown in Table 5.6. Apart from carbon dioxide, nitrogen, and oxygen, the flue gas contains oxides of sulfur and nitrogen, and also particulate matter, called "fly ash," derived from the ash content of the coal. To comply with emission limits, it is necessary to reduce the sulfur and nitrogen oxides as well as the particulate levels in the flue gas (see Sections 9.1 and 9.2). Particulates and sulfur oxides emissions are usually controlled by filtering and/or scrubbing the flue gas, while nitrogen oxides (NO_x) control will most likely be achieved by modifying combustion conditions to prevent its formation. The incorporation of a flue gas desulfurization (FGD) system into a power plant significantly increases the cost of the product electricity. A wet lime scrubber, for example, can reduce the overall conversion efficiency by 2 to 3 percent, and increase the capital cost of the plant by some 15

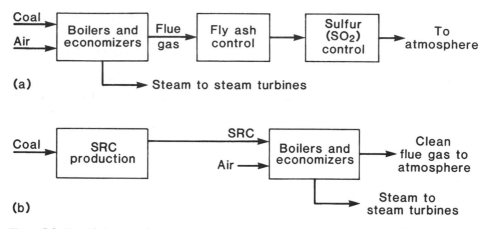

Figure 5.8 Simplified process flow sheet for conventional steam turbine electricity generation. (*a*) Burning coal and cleaning flue gas. (*b*) Burning clean coal produced by solvent refining (see Section 6.4).

percent.[21] Capital costs can be even higher if the FGD system is retrofitted to existing plants. In view of the substantial costs involved in FGD, alternative means of satisfying emission criteria are constantly being sought.

One such method, sometimes advocated as being economical for existing plants, is to switch to a low-sulfur feedstock. In the limit, a clean solid fuel containing neither sulfur nor ash, for example solvent refined coal (SRC) or one of the other products discussed in Chapter 6, could be used. Both the FGD and the fly ash systems can be eliminated when burning a clean fuel, resulting in the considerably simplified power plant flow sheet shown in Figure 5.8b. Unfortunately the production of a clean solid fuel is in itself complex, and it is not clear whether this route will in fact be more economical than add-on FGD.

Detailed designs of grassroots power plants have shown that a definite improvement in conversion efficiency results if plant energy is derived from a clean gaseous fuel. Much of the advantage is due to the use of inherently more efficient modern gas turbines available to gas-fueled power plants. To demonstrate this we briefly compare

Table 5.6 Typical coal-fired boiler flue gas analysis*

Constituent	Mole %
Nitrogen	73.8
Oxygen	4.8
Carbon dioxide	12.3
Sulfur dioxide	0.24
Sulfur trioxide	0.0024
Nitrogen oxides, NO_x	0.06
Hydrogen chloride	0.01
Moisture	8.8

* 3.9% sulfur in the coal.

the efficiencies theoretically achievable by steam and gas turbine generating systems. The analysis given here is necessarily superficial and does not account for the many refinements available for enhancing performance in both systems; it is nevertheless adequate for our purpose.

The efficiency of electricity production as related to the turbine section can be defined as

$$\eta = \frac{\text{Net work done by the turbine}}{\text{Energy input}} \tag{5.26}$$

If it is assumed that the turbine operates ideally without losses, then the work done is given by the isentropic enthalpy change (see Table 2.16) of the gases entering and leaving the turbine

$$w = \Delta H \tag{5.27}$$

The energy input into the system is normally under constant pressure conditions and so can also be measured as the enthalpy change either across the boiler in the steam system, or the combustion chamber in the gas turbine system. For example, the efficiency of the closed loop steam cycle shown in Figure 5.9 becomes

$$\eta_{\text{steam cycle}} = \frac{H_2 - H_3}{H_2 - H_1} \tag{5.28}$$

Typical conditions for a simple steam turbine of 8.6 MPa and 538°C at the turbine entrance, and 6.8 kPa and 48°C in the condenser, result in enthalpy values of $H_1 = 0.20$ MJ/kg, $H_2 = 3.48$ MJ/kg, and $H_3 = 2.11$ MJ/kg. The quantities H_1 and H_2 are

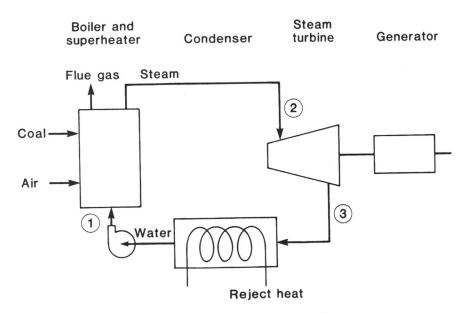

Figure 5.9 Simplified closed cycle steam turbine electricity generation.

tabulated values for water and superheated steam respectively, while the value for H_3 is for wet steam at 6.8 kPa having the same entropy as the superheated steam at 8.6 MPa. The value for H_2 is most easily read from the Mollier diagram supplied with most steam tables. Substituting the indicated values in Eq. (5.28) yields

$$\eta_{\text{steam cycle}} = \frac{3.48 - 2.11}{3.48 - 0.20} \times 100 = 41.8\% \tag{5.29}$$

The efficiency given in Eq. (5.29) can be improved upon by techniques such as reheat (withdrawing intermediate pressure steam from the turbine, reheating in the boiler, and returning to the turbine) and regeneration (withdrawing a small part of the steam from the turbine for preheating the condensate being returned to the boiler). Increasing the value of $(H_2 - H_3)$ will clearly increase the efficiency, and this can be done by increasing the pressure and temperature of the superheated steam. Increasing the pressure requires more turbine stages to accommodate the higher pressure ratio, and will also require reheat to prevent the moisture content of the wet steam from becoming excessive near the turbine exit. Increases in temperature much above 500°C are limited by materials of construction and problems of slagging and heat exchanger fouling in the boiler. In practice, ideal steam turbine efficiencies as calculated above are probably limited to about 50 percent. When internal turbine inefficiencies, boiler inefficiencies and losses associated with auxiliaries are accounted for, the overall conversion efficiency is reduced to 30 to 33 percent.

Modern-day gas turbines can operate at much higher temperatures of up to about 1100°C and, as we shall see, consequently achieve significantly higher conversion efficiencies. Because adiabatic flame temperatures of even low-CV gases are considerably higher than the permissible temperature of 1100°C, considerable excess air has to be supplied to the combustion chamber. Air-to-fuel ratios may be about 50 for a high-CV natural gas or SNG, or about 5 for a low-CV gas such as from the air-blown Foster Wheeler gasifier. The air supplied to the combustion chamber must be compressed and, as shown in Figure 5.10, this work is provided by the gas turbine itself.

For large excess air rates, the quantity of air passing through the compressor is approximately equal to the quantity of combustion products expanding in the turbine, so the net work available from the turbine can be expressed as $(H_3 - H_4) - (H_2 - H_1)$. The efficiency equation therefore becomes

$$\eta_{\text{gas turbine}} = \frac{(H_3 - H_4) - (H_2 - H_1)}{H_3 - H_2} \tag{5.30}$$

Because the pressure ratio of the compressor and turbine are the same, $H_2/H_1 = H_3/H_4$, and Eq. (5.30) simplifies to

$$\eta_{\text{gas turbine}} = 1 - \left(\frac{H_4}{H_3}\right) \approx 1 - \left(\frac{T_4}{T_3}\right) \tag{5.31}$$

Using Eq. (15) in Table 2.16, this can be written

$$\eta_{\text{gas turbine}} = 1 - \left(\frac{p_4}{p_3}\right)^{0.286} \tag{5.32}$$

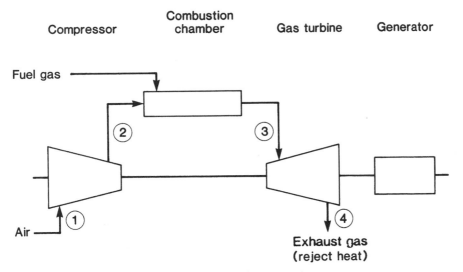

Figure 5.10 Simplified open cycle gas turbine electricity generation.

which for a typical pressure ratio of 14, results in a gas turbine conversion efficiency of nearly 53 percent. Again, efficiency can be enhanced by regeneration—utilizing the exhaust gases to preheat the compressed combustion air—but when inefficiencies of the turbine, compressor and auxiliaries are accounted for, overall conversions are reduced to about 37 percent.

Although Eq. (5.32) for a simple non-regenerative turbine shows the efficiency to be independent of temperature, a more rigorous development accounting for system losses shows that there is an optimum temperature which increases with increasing pressure ratio. As the efficiency also increases with increasing pressure ratio according to Eq. (5.32), maximum efficiencies are determined by the maximum possible temperature the turbine can withstand. Developments in gas turbines are expected to lead to inlet temperatures of 1300°C and more, with an associated ideal turbine efficiency of 55 percent, and actual overall conversion efficiencies to electricity approaching 40 percent.

The exit gas from a gas turbine operating under conditions of optimum efficiency is still very hot. For example, for an inlet temperature of 1100°C and a pressure ratio of 14, the exit gas temperature according to Eq. (15) in Table 2.16 is 373°C. Higher inlet temperatures, when permissible, will result in even higher exit temperatures. Significant improvements in conversion efficiency would result if the sensible heat in the exit gas could be recovered. In the combined cycle concept, the hot gases from the gas turbine are used to raise steam, and the steam is in turn used for electricity generation by a steam turbine. If 30 percent of the energy in the gas turbine exhaust gases is recovered in the steam cycle, overall conversion efficiencies will be increased from the projected 40 percent for an advanced gas turbine, to more than 50 percent for the combined gas-steam process. If now the clean gaseous fuel were raised from coal in gasifiers operating at efficiencies of 70 to 80 percent, then overall thermal efficiencies, coal to electricity, can be expected to range from 35 to 40 percent.

A comparison of Figures 5.8 and 5.11 shows that the combined cycle plant is considerably more complex than the conventional steam-electric power plant. Nevertheless, engineering studies[21,22] have indicated that the capital costs of the two types of plant are not significantly different, and may in fact be slightly lower for the combined cycle plant. This is partly attributable to the high cost of the steam generators and flue gas desulfurization systems in conventional plants. Further, the cost of electricity produced in the more efficient combined cycle plants was estimated to be about 20 percent lower than in conventional power plants. It is stressed, however, that the costs for the combined cycle systems are based on conceptual designs and may require modification as field experience is gained.

Although the gasifier itself generally represents only a small portion of the total cost of a combined cycle plant,[22] the type of gasifier selected largely determines the overall conversion efficiency attainable. Clearly, the higher the gasifier efficiency, the higher will be the overall conversion efficiency. Note, however, that cold gas efficiencies are particularly misleading in gauging gasifier effectiveness in combined cycle plants as the sensible heat in the off gases can normally be effectively recovered. Also, factors such as high H_2/CO ratios and methane formation which are important in synthesis and SNG processes, are not of consequence in electricity production.

The temperature and pressure of gasification do not appear to significantly affect overall conversion efficiencies other than by the degree to which it affects the efficiency of the gasification process itself. The sensible heat of high temperature off gases can be recovered in waste heat boilers as shown in Figure 5.11, and the steam so produced may be used either for gasification or in the steam power cycle. One problem with the waste heat boiler recovery that must be mentioned is the fouling of heat exchanger surfaces by dirty off gases. Particulate matter can to some extent be controlled by high temperature cyclones as shown for the Foster Wheeler gasifier in Figure 4.9, while other gasifiers incorporate a water quench. The main purpose of the scrubber quench in the Lurgi system, Figure 4.2, is to remove tars and oils that would otherwise

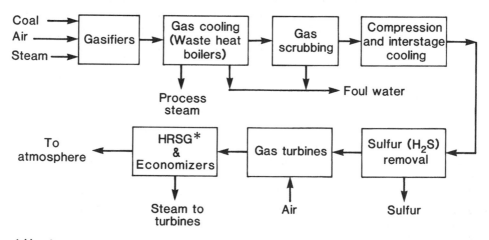

*Heat recovery steam generators

Figure 5.11 Simplified process flow sheet for combined cycle electricity generation.

condense in downstream process units such as the waste heat boiler and low temperature desulfurization units. Removal of tars and oils if present decreases the calorific value of the fuel gas and, because energy is expended in their recovery, represents a small process inefficiency. While low temperature gasification is associated with lower irreversible sensible heat losses, particularly in the cooling and reheating of the gases for desulfurization, the poor steam utilization and the often high tar and oil formation are normally overriding factors in overall performance.

The ideal gasifier pressure is probably that which is just sufficient to supply the gas to the turbine at the desired entry pressure. Gasification pressures of 2 to 3 MPa appear to be adequate. Increasing the gas turbine pressure ratio will decrease its exit temperature and so reduce the level of high pressure steam raised. The efficiency of the steam power cycle is reduced as steam temperatures fall below 540°C, but is not importantly increased by producing steam above this temperature. Consequently, as practical turbine inlet temperatures increase, so will the pressure, but not by more than that necessary to provide an exit gas sufficient to produce 540°C superheated steam. Other alternatives do exist but may not be practical due to control or equipment constraints. For instance, final heating of the steam could be accomplished in the waste heat boilers, or even in the gas turbine combustion system itself.

Gasification at pressures higher than needed for turbine conditions is by no means excluded, and may in fact improve overall efficiency if the excess pressure energy is recovered by first passing the fuel gas through an expansion system. The increased efficiency will be obtained at the expense of increased plant complexity and need not necessarily result in a lower-cost product. Selection of atmospheric pressure gasifiers will require that the fuel gas be compressed. Theoretically the energy of compression is not lost but will largely be recovered in the gas turbine. However, the energy required for compression increases with increasing temperature—see Eq. (14) in Table 2.16—and it is customary to reduce costs by compressing in several stages with interstage cooling. The energy of compression is then lost to the extent that the heat content of the compressed gases is dissipated to the atmosphere in cooling.

There appears to be some advantage in gasifying with air and so avoiding the expense of producing oxygen. Although the presence of nitrogen increases the volume of gas passing through the gasification and gas purification systems, it is not otherwise detrimental, and would in any case be added with excess air in the turbine combustor for temperature control. For example, the mass ratio of air to fuel is less than 6 for the low-CV Foster Wheeler air-blown gas, GCV = 6.4 MJ/kg, while it is 12 for a medium-CV gas from an oxygen-blown Texaco gasifier, GCV = 12.3 MJ/kg. The compression energy for the excess air is not lost other than by compression inefficiencies as, unlike fuel gas compression, it is not customary to incorporate interstage cooling for air compression. Air required for a pressurized gasifier can be bled off from the compressed combustion air. If additional compression is required for a high pressure gasifier, the bleed air may first be cooled by raising intermediate pressure steam.

An important step in the gas processing scheme is that of sulfur removal. The currently available desulfurization processes, discussed in Section 5.1, are low temperature processes that require the gas to be cooled below the temperatures attained in the waste heat boilers. Wet scrubbing is typically used as a first step which cleans the gas in addition to reducing its temperature. Any tars, oils, and particulates still present are removed, as is ammonia, which is important for NO_x control. In the case of a low pressure gasifier it is possible that the interstage cooling for fuel gas compression can be combined with the ammonia scrubbing function. Compression of the fuel gas prior to desulfurization has an added advantage in that the consequent higher hydrogen sulfide partial pressure facilitates its removal and can bring a wider selection of desulfurization processes into contention. While many of the hydrogen sulfide removal processes are not affected by the presence of ammonia, control of tars and oils will be required to prevent condensation and fouling. Were a high temperature desulfurization process available, condensation would not be a problem, removal of tars and oils would not be required, and their energy content could be recovered in the gas turbine. The combination of the elimination of gas cooling for desulfurization and the retention of tars and oils in the fuel gas can result in significant improvements in overall conversion efficiency.

The advantages of high temperature desulfurization will not be realized unless it is possible to dispense with the low temperature ammonia removal step. Even though the amount of nitrogen in the combustion air is several orders of magnitude larger than that in the fuel, the fuel-derived nitrogen, as for example in the ammonia, may contribute most significantly to NO_x formation. This is because nitrogen-hydrogen and nitrogen-carbon bond strengths are weaker than for molecular nitrogen. The oxidation of ammonia, for example, proceeds at rates equivalent to the combustion of the hydrocarbon fuel and at temperatures lower than required for the thermal oxidation of molecular nitrogen.[23] Combustion techniques appropriate for reducing NO_x formation from nitrogen may consequently not be adequate for meeting emission limits if the fuel contains ammonia. Methods of removing NO_x from flue gases are available, for example by scrubbing with ammonia solutions:

$$NO + NO_2 + 2NH_3 \xrightarrow{\text{catalyst}} 2N_2 + 3H_2O \qquad (5.33)$$

As we have seen, however, the volume of flue gas is at least six times that of the fuel gas, so flue gas scrubbing does not appear at all attractive for combined cycle plants. The decomposition of ammonia to nitrogen and hydrogen is another possibility, and as shown in Figure 2.2c, is thermodynamically favored at temperatures above about 500 K. Decomposition rates appears to be slow, however, and would require a catalyst for this ammonia control method to be effective. One catalyst, an iron oxide/alumina formulation, is of interest as it simultaneously adsorbs hydrogen sulfide, but the process has yet to be developed commercially. Other suitable catalysts are poisoned by even trace quantities of sulfur compounds. Because of the thermodynamic equilibrium discussed in connection with ammonia decomposition, high temperature gasifiers tend to produce less ammonia and so might be preferred for use in conjunction with high temperature desulfurization. Methods for estimating sulfur and nitrogen oxides emissions are given in Section 9.2.

Apart from the environmental problems of NO_x emissions, there are turbine-related problems associated with hot gas purification. For example, trace elements such as chlorides and alkali metal vapors are not removed at high temperature and will be harmful to turbine machinery. In addition, particulate removal which is accomplished incidentally in the ammonia scrubbers, may not be adequate in the high temperature schemes and could lead to early turbine failure.

The combined cycle configuration nevertheless provides the opportunity of de-sulfurizing a small volume of relatively high-sulfur-content fuel gas before its combustion instead of scrubbing flue gas which has a low sulfur concentration. Also, processes for hydrogen sulfide scrubbing are in general more proven and developed than processes for sulfur dioxide scrubbing. In addition, hydrogen sulfide processes produce elemental sulfur as a byproduct, whereas processes considered for sulfur dioxide scrubbing produce a sulfate-containing solid waste for disposal. A possible exception to the implied rule for sulfur control may arise at oil shale plants where the retort off gas is used to raise electricity. Spent oil shale has sulfur dioxide adsorption properties, is freely available at these plants, and has to be disposed of in any case. Whether or not it proves more economical than, say, an upstream Stretford process will depend on site and process specific factors. A determining factor in deciding between pre- and post-sulfur control will be the ability of the turbine to handle corrosive sulfur-containing gases.

An indication of the distribution of energy in combined cycle plants is given in Table 5.7 for both an air-blown and an oxygen-blown gasifier. Nearly half the unrecovered energy is lost in condensing steam in the steam-electric power cycle, while much of the remaining energy is dissipated in stack gases and by cooling. Energy for low temperature gas purification is larger for the air-blown gasifier due to the larger

Table 5.7 Energy allocation in combined cycle plants*

	Percent of Total Energy	
	Air-blown Foster Wheeler Gasifier†	Oxygen-blown Texaco Gasifier
IN		
Coal to gasifier	100.0	100.0
OUT		
Gas turbine power‡	24.2	24.2
Steam turbine power	16.3	14.5
Total electric power	40.5	38.7
CONSUMED AND LOSSES		
Oxygen production	—	15.1
Air compression	3.2	—
Steam condensers	28.9	24.9
Gas purification and cooling	2.1	1.4
Stack gases, cooling, etc.	23.4	18.9
Byproducts	1.9	1.0
Total	59.5	61.3

* Values for bituminous coal adapted from conceptual designs in Ref. 22.
† See also Table 4.28.
‡ 1315°C inlet temperature.

gas volume, but in both cases amounts to a relatively small fraction of total energy losses. In fact, conceptual designs[24] have indicated that only marginal improvements in performance can be expected on switching to hot gas purification with the types of gasifiers considered in Table 5.7. The situation changes in the case of dry ash Lurgi gasifiers, which show efficiencies of around 33 percent with cold gas purification, and nearly 40 percent for hot gas purification. We recall that the Lurgi off gases contain tars and a high steam concentration and that the energy associated with these components is lost if condensed out prior to low temperature gas purification. In view of the problems of corrosion and NO_x emissions that are anticipated, it is concluded that the use of gasifiers other than the dry ash Lurgi be considered for combined cycle application and that the need for further development of hot gas purification processes be carefully reviewed.

It is stressed that the overall conversions of 40.5 percent and 38.7 percent shown in Table 5.7 are for high-grade electrical energy that can be converted to work at efficiencies exceeding 90 percent. Conversion to heat is possible at even higher efficiencies, but electrical heating is considered a wasteful use of this valuable commodity. In countries having an existing pipeline gas distribution system it is far more economical to satisfy domestic heating requirements by conversion of coal to substitute natural gas.

Table 5.8 Gas properties

| Composition, mole % | Gasifier Off Gases | | High Quality Louisiana Natural Gas | Typical SNG Specification[25] |
	Koppers-Totzek*	Lurgi Dry Ash†		
H_2	27.6	18.8	—	3.00
CO	49.3	7.5	—	0.05
CO_2	5.9	15.7	0.2	0.40
CH_4	0.1	5.2	94.7	95.95
C_2H_6	—	0.2	2.8	—
C_2H_4	—	<0.1	—	—
C_{3+}	—	0.1	—	—
NH_3	0.1	0.5	—	—
$H_2S + COS$	0.4	0.2	—	—
$N_2 + Ar$	1.0	<0.1	2.3	0.60
Tars and oils	—	0.5	—	—
H_2O	15.6	51.1	—	—
Total	100.0	100.0	100.0	100.0
Molar mass	20.2	20.3	16.7	15.8
GCV, MJ/m³	9.9	6.1‡	39.6	>36.1
Pressure, MPa	0.1	3.0	—	5.5§

* Gasification of a subbituminous coal.
† Gasification of lignite.
‡ Including tars and oils.
§ Typical pressure for pipeline entry.

Substitute Natural Gas Production

Compositions of off gases produced by several gasifiers are presented in the previous chapter. While there is a wide variation in off gas compositions depending on gasifier type and operating conditions, there is an even wider difference between a typical off gas and natural gas or SNG. Table 5.8 lists compositional data for two commercially proven gasifiers, a natural gas, and a published SNG specification for a coal gasification process.[25] Clearly, extensive gas processing is required to convert an off gas to SNG, and it is desired to effect this conversion as efficiently as possible. A simplified block flow diagram showing the major gas processing equipment involved is shown in Figure 5.12. The equipment consists essentially of gas purification units to remove tars, ammonia, and the acid gases, a gas shift unit to increase the H_2/CO ratio to above 3, and a methanator to convert the hydrogen and carbon monoxide to methane.

As in the combined cycle plants, the gasifier itself is not normally the major cost item in the plant. It does, however, set the stage for the number and size of the other process units required for the conversion, and these units do significantly impact plant costs. A capital cost breakdown for a SNG plant of the type shown in Figure 5.12, based on dry ash Lurgi gasifiers, is given in Table 5.9 and shows the cost of equipment for gas upgrading, steam, and oxygen supply to be nearly four times that of the gasifier. Apart from capital investment, it is also necessary to consider the energy cost associated with the process steps. In this regard we know that both oxygen and steam production are highly energy consumptive, whereas the gas shift, and in particular methanation, are exothermic processes.

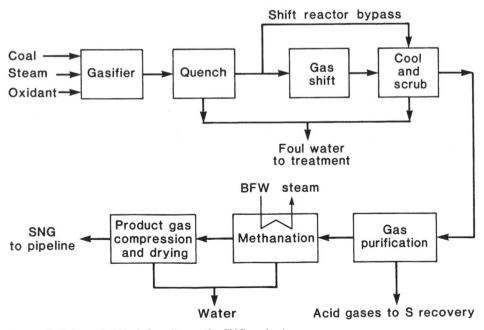

Figure 5.12 Simplified block flow diagram for SNG production.

Table 5.9 Cost breakdown for a Lurgi gasification plant[26]

Process		Percent of Total Plant Investment
Gasification		14.8
Steam and oxygen plants		
Steam plant	16.6	
Oxygen plant	10.0	
Total		26.6
Gas purification and upgrading		
Shift conversion	2.5	
Gas cooling	3.0	
Acid gas removal	15.6	
Methanation	6.2	
SNG compression	1.0	
Total		28.3
Effluent treatment		
Effluent water treatment	1.7	
Phenol recovery	3.2	
Tar and oil recovery	2.6	
Sulfur recovery	3.3	
Total		10.8
General facilities and utility systems		7.8
Miscellaneous		11.7
Total		100.0

Pure oxygen is used in directly heated gasifiers so as to produce a medium-CV gas that is not diluted with nitrogen. In the previous chapter we saw that both indirectly heated gasifiers and catalytic gasifiers produce off gases free of nitrogen, and without the need for air separation. Indirectly heated gasifiers are more complex than conventional gasifiers and can therefore be expected to cost more. In addition, they generally show greater sensible heat losses in both the transport of the heat transfer medium and in the discharged combustion flue gases, and so can be expected to operate at lower efficiencies. Indirectly heated gasifiers operate at relatively low temperatures where steam decomposition and reaction rates are low, although direct methane formation, if at high pressure, can be significant. Direct methane formation in catalytic gasifiers is, of course, excellent, and as we saw in Figure 4.17, catalytic systems for SNG production are considerably less complex than the system shown in Figure 5.12. While catalytic gasification holds great promise, further development is required prior to its commercialization.

An alternative method for nitrogen control is to separate it from the off gases by cryogenic distillation, similar to what is planned for hydrogen and carbon monoxide separation in catalytic gasification. This approach would eliminate the need for an oxygen plant but replace it with an off gas distillation system. In view of the higher relative volatilities in the off gas case, off gas separation is expected to be less costly than air separation. This savings must be balanced against the increased cost of the gasifier and gas purification units, which must now handle larger quantities of nitrogen-containing gas. However, in some gasifiers such as the dry ash Lurgi, gas volumes are in any case high due to the large amounts of steam used for controlling temperature in the combustion zone. Gasification with air instead of oxygen would conceivably reduce the need for part of this high-cost steam.

Gasification with oxygen is invariably used in the case of high temperature slagging gasifiers. High temperature entrained gasifiers have low residence times and are consequently smaller and less expensive than low temperature systems of the same throughput. For example, the cost of a Texaco gasifier is estimated to be about one-quarter of that of a dry ash Lurgi. However, as shown in Table 5.9, the cost of the gasifier itself is only a small part of the overall gasification plant, and can be misleading when comparing overall capital costs. In fact the cost of the Lurgi plant, including steam, oxygen, and ancillary systems, is reportedly only 12 to 15 percent more than a Texaco-based plant.[22] Further, comparison of these two systems shows that while the Texaco gasifier requires about one-quarter or less of the steam required by the Lurgi, it consumes three times as much oxygen. Although much of the oxygen is consumed for raising steam from the slurry feed within the gasifier and results in reduced boiler sizes, this is considered an expensive use of pure oxygen. In this respect the Lurgi slagger offers a significant advantage as its oxygen consumption is about one-half that of the Texaco, while it has a similar steam requirement. A comparison of capital costs for Lurgi slagger and Texaco gasification systems including oxygen, steam raising, and ancillary equipment, shows the Lurgi slagger costs to be about 30 percent less than those for the Texaco.[22,27] The lower cost for the Lurgi system is due largely to its lower off gas exit temperature and the associated lower costs for gas cooling equipment.

A factor stated to favor low temperature and high-steam-rate systems such as the dry ash Lurgi is the high H_2/CO ratios and high methane concentrations in the off gas. This is well illustrated by a comparison of the high temperature Koppers-Totzek and low temperature Lurgi gas compositions shown in Table 5.8. The degree of shift and methanation required in the Lurgi case is significantly less than for the Koppers-Totzek gas, which will also require an additional steam input to the shift converter. It is possible using simple stoichiometry to roughly estimate the relative costs of shift and methanation for high and low temperature gasification. The relative gas rates and compositions for two selected cases are shown in Figure 5.13. The calculated flows are based on 100 moles of product methane formed, and consider only the CH_4, H_2, and CO components, and the steam requirements for the gas shift reaction. It is assumed that 40 percent of the final methane product is formed in the low temperature gasifier as is typical for a dry ash Lurgi, while the high temperature gasifier produces no methane. The stoichiometric equations for the shift reaction can be used to show the following relations for the carbon monoxide and hydrogen concentrations (see Eqs. 5.11 and 5.12):

$$CO_{\text{before shift}} = \frac{CO_{\text{after shift}}}{1 - f} \qquad (5.34)$$

$$H_{2\ \text{before shift}} = \frac{x_0 H_{2\ \text{after shift}}}{x_0 + f} \qquad (5.35)$$

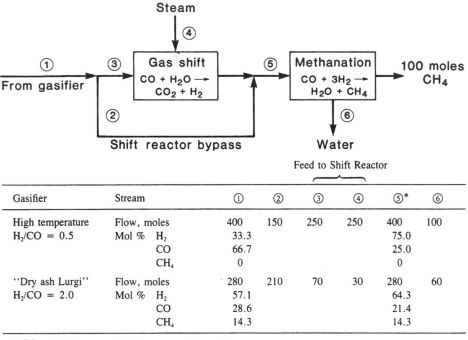

Gasifier	Stream		①	②	③	④	⑤*	⑥
High temperature	Flow, moles		400	150	250	250	400	100
$H_2/CO = 0.5$	Mol %	H_2	33.3				75.0	
		CO	66.7				25.0	
		CH_4	0				0	
"Dry ash Lurgi"	Flow, moles		280	210	70	30	280	60
$H_2/CO = 2.0$	Mol %	H_2	57.1				64.3	
		CO	28.6				21.4	
		CH_4	14.3				14.3	

* CO_2 and H_2O assumed to be removed from shifted gas.

Figure 5.13 Stoichiometric comparison of shift and methanation requirements for two representative off gases.

where x_0 is the initial H_2/CO ratio in the gas. The fraction of gas f that must be passed through the gas shift reactor to achieve a H_2/CO ratio of 3 is

$$f = \frac{3 - x_0}{4} \tag{5.36}$$

Figure 5.13 shows that the gas shift reactor for the high temperature off gas is some five times the size required for the low temperature off gas. It is assumed that 50 percent excess steam is fed to the gas shift reactor. Figure 5.13 shows that the methanator, too, is larger in the high temperature case. According to this simplified analysis, overall capital costs cannot be expected to be much, if at all, lower when using cheaper high temperature gasifiers. The reader is reminded, however, that the size of the shift reactor for the high temperature case would be significantly reduced if methanation could be accomplished by rejection of carbon dioxide (see Section 3.3):

$$2CO + 2H_2 \rightleftharpoons CO_2 + CH_4 \tag{5.37}$$

Gasification efficiencies are invariably higher for the slagging systems, as exemplified by the cold gas efficiencies of 80 percent for the dry ash Lurgi and 90 percent for its slagging counterpart given in Table 4.3. We must, however, also account for the energy consumed in producing the steam and oxygen feeds, factors not included in cold gas efficiency data. An exact determination of overall plant efficiencies requires a detailed evaluation of all the process units in the plant. We can nevertheless obtain an approximate indication of plant efficiency by considering the major process steps only, and neglecting sensible heat effects as well as energy for compression, pumping, gas purification, etc. The results of such a simplified analysis have been summarized in Table 5.10 for a low temperature and a high temperature gasifier. Gasifier feed rates for the low temperature case are based on the dry ash Lurgi material balance in Table 4.14, and for the high temperature case on the Texaco gasifier material balance in Table 4.22. The gas rates in Figure 5.13, together with data for the moles of gas per mole of carbon gasified, given in the material balances, are used to estimate the carbon feed rate in each case. We conclude from Table 5.10 that while capital costs may possibly be higher for slagging systems, they do appear to have a significantly higher thermal efficiency.

According to the capital cost breakdown of Table 5.9, the cost of gas purification equipment is a significant factor in SNG production. In fact, if we sum the cost for gas purification equipment with the cost of equipment for the associated liquid effluent treatment, a total of nearly 26 percent of the overall plant investment is obtained. This value is particularly high for the Lurgi gasifiers, in which the off gas contains high loadings of phenols, tars, and oils. Unlike the combined cycle plants, high temperature desulfurization is not expected to significantly decrease purification costs, as these contaminants must be recovered in the production of SNG. Cocurrent slagging gasifiers and fluidized bed gasifiers producing a relatively clean gas might be preferred for this reason in spite of the advantages of high sensible heat recovery in countercurrent moving bed gasifiers. It is pointed out that while the energy for gas purification in

Table 5.10 Approximate determination of efficiency of SNG production for flow scheme of Figure 5.13

			Low Temperature Gasifier*		High Temperature Slagging Gasifier†	
	Stream	Heat Content, kJ/mol				
No.	Description		moles	MJ	moles	MJ
1	Carbon to gasifier	394	315	124	280	110
2	Oxygen to gasifier	128‡	60	8	140	18
3	Steam to gasifier	48	548	26	112	5
4	Steam to shift§	48	0	0	140	7
5	Heat from methanator	225¶	60	13	100	22
6	Product methane	890	100	89	100	89
Thermal efficiency $= \dfrac{6}{1 + 2 + 3 + 4 - 5}$			61%		75%	

* Based on dry ash Lurgi, Table 4.14.
† Based on Texaco gasifier, Table 4.22.
‡ Heat requirement to manufacture oxygen.
§ Allowing for moisture content of off gas.
¶ Heat of reaction.

combined cycle plants was found to be about 2 percent of the total (see Table 5.7) it will be closer to 6 percent in SNG plants. This is because overall energy flows in the production of SNG (heat) are about one-third of those in the production of an equivalent amount of electricity (work).

Finally we recall that considerable effort is being devoted to the development of high pressure gasifiers. In relatively low temperature gasifiers this might be justified in view of the increased methane produced directly, as is shown in Figure 3.12. In high temperature processes, methane is not produced in the gasifier and the only advantage to be had is in the reduced size of all gas processing equipment and the reduction or elimination of the amount of product gas compression required to attain pipeline entry pressures of at least 5.5 MPa. These factors will be offset by the cost of compressing the oxidant to gasifier pressures. Note that high pressure steam is obtained by pumping boiler feedwater, or the coal slurry in the case of the Texaco gasifier, and the associated energy requirement is small. Referring to Table 5.10 it is seen that in the Lurgi case, 60 moles of oxygen are required per 100 moles of product SNG, while the equivalent number for the Texaco gasifier is 140. Consequently there is a 40 percent energy saving in the Lurgi case, but a 40 percent energy *penalty* in the Texaco case on comparing gasification at atmospheric pressure to that at pressures sufficient for pipeline entry. The situation is very different for production of a medium-CV gas, where the oxygen feed rate to the Texaco gasifier is about 30 to 40 percent of the product gas rate. The energy required to compress methane from atmospheric pressure to pipeline entry conditions is, from Eq. (14) in Table 2.16, about 1 percent of the calorific value of the gas. We note also from Table 5.9 that the capital cost of compressing the Lurgi product gas, estimated to be at about 3 MPa, to pipeline entry pressures is 1 percent of total plant investment.

Summary

On the basis of the limited data and simplified analyses considered here we conclude that air-blown, moderate pressure gasifiers producing little or no tars appear most suitable for gasification-combined cycle power plants. The air-blown Foster Wheeler unit is one example of such a gasifier. For SNG production, the high temperature slagging gasifiers appear most efficient in spite of their rather unfavorable off gas composition. Low oxygen consumption and high pressure characteristics are advantageous, while countercurrent operation resulting in good gasifier heat interchange and a reduced need for downstream waste heat recovery equipment is also attractive. The slagging Lurgi has most of these characteristics. Having made these conclusions, we now point out that they are not inviolate. For example, an estimated energy balance for a commercial-scale SNG plant based on dry ash Lurgi gasification, reproduced in Table 5.11, shows surprisingly good agreement with the analysis of Table 5.10. However, a projected balance for a demonstration-scale SNG plant based on the slagging Lurgi[9] and, included in Table 5.11, shows a lower efficiency. It could be argued on the basis of the large heat loss shown in the table that low equipment costs rather than efficient energy recovery had priority in the demonstration plant design. This may be true. The real test of the numerous theoretical evaluations being made must, however, await actual operating data from commercial-scale plants.

Table 5.11 Estimates of projected efficiencies of SNG production

(a) Lignite gasification in dry ash Lurgi gasifiers[25] (see Table 4.14)

	10^6 MJ/d
IN	
Coal to gasifier	416
Electric power: to plant	12
to mine	1
Vehicles	3
Total	432
OUT	
SNG	283
Other and Losses	149
Total	432

SNG Thermal efficiency = $(283/432) \times 100 = 65.5\%$

(b) Bituminous coal gasification in slagging Lurgi gasifiers[9] (see Table 4.19)

	10^3 MJ/d
IN	
Coal to gasifier	1036
Electricity	90
Total	1126
OUT	
SNG	688
Sulfur and Ammonia	24
Tars, etc.	72
Heat losses	342
Total	1126

SNG Thermal efficiency = $(688/1126) \times 100 = 61.1\%$

REFERENCES

1. Kohl, A., and Riesenfeld, F., *Gas Purification*. 2nd Edition, Gulf Publishing Co., Houston, Texas, 1974.
2. Lovell, R. J., Dylewski, S. W., and Peterson, C. A., "Control of Sulfur Emissions from Oil Shale Retorts," Report under EPA Contract No. 68-03-2568-T 7012, U.S. Environmental Protection Agency, Cincinnati, Ohio. 1981.
3. U.S. Environmental Protection Agency, "Pollution Control Guidance Document for Indirect Coal Liquefaction," Coal Gasification and Indirect Liquefaction Working Group. Industrial Environmental Research Laboratory, U.S. Environmental Protection Agency, Research Triangle Park, N.C., 1981 (to appear).
4. MERC Hot Gas Cleanup Task Force, "Chemistry of Hot Gas Cleanup in Coal Gasification and Combustion," Report No. MERC/SP-78/2, Morgantown Energy Research Center, U.S. Department of Energy, Morgantown, W. Va., February, 1978.
5. Satterfield, C. N., *Heterogeneous Catalysis in Practice*. McGraw-Hill, New York, 1980.

6. Cusumano, J. A., Dalla Betta, R. A., and Levy, R. B., *Catalysis in Coal Conversion*. Academic Press, New York, 1978.

7. Auer, W., Lorenz, E., and Grundler, K. H., "A New Catalyst for the CO-Shift Conversion of Sulfur-Containing Gases," AIChE 68th National Meeting, Houston, Texas, February 28–March 4, 1971.

8. Seglin, L. (ed.), *Methanation of Synthesis Gas*. Advances in Chemistry Series No. 146. American Chemical Society, Washington, D.C., 1975.

9. Conoco, Inc., "Phase I: The Pipeline Gas Demonstration Plant—Environmental Report, Vol. 1., Project Description," Report No. FE-2542-25 (Vol. 1), U.S. Department of Energy, Washington, D.C., January 1980.

10. Inui, T., Funabiki, M., and Takegami, Y., "Simultaneous Methanation of CO and CO_2 on Supported Ni-Based Composite Catalysts," *Ind. Eng. Chem. Product Res. Dev.* **19**, 385-388, 1980.

11. Gibbs, C. W. (ed.), *Compressed Air and Gas Data*. 2nd Edition, Ingersoll-Rand Co., Woodcliff Lake, N. J., 1971.

12. Johnson, V. J., "Cryogenic Processes," *Chemical Engineers' Handbook* (R. H. Perry and C. H. Chilton, eds.), pp. 12-49 to 12-53. 5th Edition, McGraw-Hill, New York, 1973.

13. Ward, W. J., III, Browall, W. R., and Salemme, R. M., "Ultrathin Silicone/Polycarbonate Membranes for Gas Separation Process," *J. Membrane Sci.* **1**, 99-108, 1976.

14. Henis, J. M. S., and Tripodi, M. K., "A Novel Approach to Gas Separations Using Composite Hollow Fiber Membranes," *Separation Sci. and Tech.* **15**, 1059-1068, 1980.

15. Hwang, S-T., and Thorman, J. M., "The Continuous Membrane Column," *AIChE J.* **26**, 558-566, 1980.

16. Rogers, K. A., and Hill, R. F., "Coal Conversion Comparison," Report No. FE-2468-51, U.S. Department of Energy, Washington, D.C., July, 1979.

17. Smith, W. N., and Santangelo, J. G. (eds.), *Hydrogen: Production and Marketing*. ACS Symposium Series No. 116. American Chemical Society, Washington, D.C., 1980.

18. Bockris, J. O'M., *Energy, The Solar-Hydrogen Alternative*. Wiley, New York, 1975.

19. Knierem, H., Jr., "Membrane Separation Saves Energy," *Hydrocarbon Processing* **59**(7), 65-68, 1980.

20. Shinnar, R., "Gasoline from Coal: A Differential Economic Analysis," *ChemTech* **8**, 686-693, 1978.

21. Bechtel Power Corporation, "Coal-fired Power Plant Capital Cost Estimates," Report No. EPRI AF-342, Electric Power Research Institute, Palo Alto, Calif., January 1977.

22. Fluor Engineers and Constructors, Inc., "Economic Studies of Coal Gasification Combined Cycle Systems for Electric Power Generation," Report No. EPRI AF-642, Electric Power Research Institute, Palo Alto, Calif., January 1978.

23. Robson, F. L., Blecher, W. A., and Colton, C. B., "Fuel Gas Environmental Impact," Report No. EPA-600/2-76-153, U.S. Environmental Protection Agency, Washington, D.C., June 1976.

24. Stone and Webster Engineering Corporation, "Comparative Evaluation of High and Low Temperature Gas Cleaning for Coal Gasification—Combined Cycle Power Systems," Report No. E/N AF-416, Electric Power Research Institute, Palo Alto, Calif., April 1977.

25. U.S. Department of Energy, "Great Plains Gasification Project, Final Environmental Impact Statement, Vol. 1," Report No. DOE EIS-0072 F, U.S. Department of Energy, Washington, D.C., August, 1980.

26. Penner, S. S. (ed.), "Assessment of Long Term Research Needs for Coal Gasification Technologies," Fossil Energy Research Working Group. Mitre Technical Report No. MTR-79W00160, The Mitre Corp., McLean, Va., April, 1979.

27. Shinnar, R., and Kuo, J. C. W., "Gasifier Study for Mobil Coal to Gasoline Processes," Report No. FE-2766-13, U.S. Department of Energy, Washington, D.C., October 1978.

LIQUIDS AND CLEAN SOLIDS FROM COAL

6.1 LIQUEFACTION AND COAL REFINING TECHNOLOGIES

The three principal routes by which liquid fuels can be produced from coal have been shown to be pyrolysis, direct liquefaction, and indirect liquefaction. A clean fuel that is a solid at room temperature can also be produced by direct liquefaction. Here, as in Chapter 4 on the production of gas from coal, emphasis is placed on developed and developing technologies as viewed from an American perspective. As with gasification, however, where a technology is felt to have potential, we shall discuss it even though it may not have progressed beyond bench scale.

Pyrolysis of coal, as described in some detail in Section 3.1, yields through destructive distillation condensable tar, oil and water vapor, and noncondensable gases together with rejected char. The condensed pyrolysis product must be further hydrogenated to remove the sulfur and nitrogen and to improve the liquid fuel quality. Pyrolysis technologies differ principally in their methods and rates of applying the heat, their ultimate temperatures, and their gas atmospheres.

Pyrolysis processes are developed and commercially available. In Table 6.1 are listed the more important ones and their status, together with representative operating conditions and yields (see, e.g., Ref. 1). We have chosen to illustrate the process severity and yields with a representative operating condition that has been indicated by the developer as being generally appropriate to maximizing the liquid product. It must be emphasized, however, that a large number of factors, including coal rank and type, together with the desired product, determine the process operating conditions and yields.

Examination of Table 6.1 shows that the shorter-residence-time processes give higher liquid yields, as does pyrolysis in a hydrogen atmosphere. The reasons for this are discussed in Section 3.1. Supercritical gas extraction, which also gives high tar yields, is a special low temperature pyrolysis procedure in the atmosphere of a compressed hydrocarbon gas that makes use of the ability of the gas near its critical temperature to extract relatively involatile substances. The main limitation of all of the pyrolysis processes is that the principal product is char. The effectiveness of any

Table 6.1 Representative operating conditions and yields for coal pyrolysis processes

Process	Status	Reactor Temperature, °C	Reactor Pressure, MPa	Residence Time, s	Coal	Yield, mass % dry coal			
						Char	Tar/Oil	Gas	Water
COED, FMC Corp.	Developed	290–565*	0.12–0.19	1200*	Bit.	62	21	14	3
TOSCOAL	Developed	520	0.1	300–600	Subbit.	69	13	9	9
Lurgi-Ruhrgas	Commercial	595	0.1	~3	Subbit.	50	~32†	11	~7†
Occidental Flash Pyrolysis	Developing	610	0.3	1.5	Bit.	56	35	7	2
Rockwell/Cities Service Flash Hydropyrolysis	Developing	845	3.5	~0.1	Bit.	46	38	16	—
Supercritical gas extraction, NCB	Developing	400	10	1800	Bit.	63	33‡	2	2

* In pyrolysis stages.
† Water plus tar/oil sum 39%, individual values estimated.
‡ Tar extract, softening point 70°C.

of these technologies must therefore rest on the ability to utilize the char, for example, to produce gas or electricity.

The COED (Char-Oil-Energy-Development) process developed by the FMC Corporation is a fluidized bed process which is carried out in successive stages at progressively higher temperatures. By partially devolatilizing the coal at a lower temperature, the coal can be heated to a higher temperature in the next reactor without agglomerating and plugging the fluid bed. In one design, the process embodies three pyrolysis stages with the stage temperatures ranging from about 290°C in the first stage pyrolyzer, to 565°C in the third stage pyrolyzer. Heat for the process is generated by burning the product char with a steam-oxygen mixture and using the hot gases and the hot char to heat the other vessels. In another configuration, the char is gasified indirectly using air.

In the TOSCOAL process, crushed coal is fed to a horizontal rotating kiln where it is heated by hot ceramic balls to between 425 and 540°C. The hydrocarbons, water vapor, and gases are drawn off, and the char is separated from the ceramic balls in a revolving drum with holes in it. The ceramic balls are reheated in a separate furnace by burning some of the product gas.

The Lurgi-Ruhrgas process feeds finely ground coal and hot product char to a chamber containing a variable-speed mixer with two parallel screws rotating in the same direction. Temperature equalization and devolatilization are very rapid due to the uniform mixing and high heat transfer rates. The pyrolysis liquids and gases are removed overhead from the end of the chamber. Some of the new product char is burned in a transfer line and recycled to the reactor to provide the heat for the pyrolysis.

In the Occidental Flash Pyrolysis process, hot recycle char provides the heat for the flash pyrolysis of pulverized, dried coal in an entrained flow reactor at a temperature of about 600°C. The pyrolysis vapors are separated from the recycle and product char and condensed by scrubbing with a cooled recycle oil which stabilizes the free radicals. A portion of the product char is partially burned with air in an entrained bed furnace for recycle to the reactor.

Rockwell International is developing a flash hydropyrolysis process in which the coal is pyrolyzed rapidly in a hot high pressure hydrogen atmosphere. Pulverized coal is fed from a high pressure feeder in a dense-phase flow into the reactor where it is mixed with hydrogen at temperatures from 760 to 1040°C. Reactor pressures are in the range of 3.5 to 10 MPa, and the reactions are allowed to take place for about 20 to 200 ms before rapidly quenching with water to prevent the liquids formed initially from being hydrogenated to gaseous products.

The last process listed in Table 6.1 is that of supercritical gas extraction, which is being developed by the National Coal Board in England. We consider this process to be a special case of pyrolysis, although it may also be thought of as a special case of solvent extraction, the distinction being unimportant for our purposes. In this process, the coal is pyrolyzed at a temperature around 400°C in the presence of a compressed "supercritical gas;" that is, a gas whose temperature is above the critical temperature at which it can be liquefied. At a temperature around 400°C, the coal liquids that are formed would normally be too involatile to distill. However, the process takes advantage of the fact that substances volatilize more readily in the presence of a compressed gas, the more so the greater the gas density. At elevated pressures the gas density is a maximum at its critical temperature, approaching that of a liquid. Therefore, if the critical temperature of the extractant gas is close to but below the pyrolysis temperature, then the gas acts as a strong solvent that causes the liquids to volatilize and be taken up by the vapor. Gases that may be used are suitable coal tar or petroleum naphtha fractions with critical temperatures below but not too far from 400°C as, for example, toluene, whose critical temperature is 319°C. The extracted tar is recovered by reducing the pressure, which causes it to precipitate. The product is a glassy solid that softens at about 50 to 70°C, depending on the parent coal. The extract is fairly hydrogen rich and can be readily hydrocracked to distillable oils. The unconverted coal and ash are discharged as char.

In the discussion of direct liquefaction in Section 3.4, a distinction is made between hydroliquefaction, involving the catalytic addition of the hydrogen to the coal, and solvent extraction, in which the hydrogen is added through the intermediary of a solvent that exchanges hydrogen with the coal. Despite this differentiation, the major elements of both processes are similar and a generalized direct liquefaction process train may be laid out as in Figure 6.1. Coal is slurried with recycled oil or a coal-derived solvent, mixed with hydrogen, and liquefied at high pressures—in the case of hydroliquefaction in the presence of an externally added catalyst. The resulting mixture is separated into gas and liquid products and a heavy bottoms slurry containing mineral matter and unconverted coal. Generally, a large fraction of the carbon in the feed ends up in the bottoms and a major processing step is the means that must be incorporated to convert this material. Most processes are designed to gasify the bottoms slurry to produce fuel gas and hydrogen by, for example, partial oxidation as discussed in Section 5.4.

Listed in Table 6.2 are those U.S. direct liquefaction processes which are now operating on large pilot plant scale or have operated on large pilot scale and are being developed for commercial demonstration. Included in the table is the Conoco Zinc Chloride process, which although it has only been operated at a process development unit (PDU) scale, is deemed to have sufficient potential to warrant inclusion.

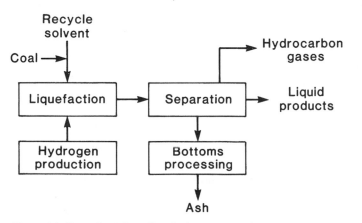

Figure 6.1 Generalized direct liquefaction process train.

Table 6.2 Representative operating conditions and yields for direct liquefaction processes*

Process	Current Project, t/d†	Reactor Pressure, MPa	Reactor Temperature, °C	Residence Time,† h	Consumed Hydrogen	Yield, mass % daf coal		
						Liquids‡	HC Gas	SRC/ Residuum
Solvent Extraction								
Solvent Refined Coal-I	5400	12	440	0.4	2.3	17	7	60
Exxon Donor Solvent	225	10	450	0.7	4.3	39	7	42
Solvent Refined Coal-II	5400	13	460	1.0	4.8	41	18	28
Hydroliquefaction								
H-Coal, fuel oil mode	530	19	455	0.5	3.8	34	12	35
H-Coal, syncrude mode	190	19	455	1.2	5.6	48	15	18
Conoco Zinc Chloride	1	21	415	1.6	8.4	60	15	8

* All coals bituminous, except subbituminous for Conoco process.
† Nominal coal feed rate and coal residence time, respectively.
‡ Principally fuel oil and naphtha, except syncrude in one H-Coal mode and gasoline in Conoco process.

In the Solvent Refined Coal-I (SRC-I) process, hydrogen is mixed with the slurry prior to the dissolution of the coal in an upflow non-catalytic reactor. Typical reactor operating conditions are 440°C and 12 MPa total pressure, with residence times about 30 minutes. In one procedure, the hot reactor effluent is filtered to remove unconverted coal and ash, which are sent to partial oxidation to make hydrogen and fuel gas for the process. The filtrate is vacuum distilled and the recycle solvent and some distillate are separated from the heavy bottoms product, which is cooled and solidifies to produce a low-sulfur, deashed, pitch-like material known as solvent refined coal (SRC). The SRC has a melting point between 175 and 200°C, depending on its hydrogen content. In another version of the process, designated SRC-II, a portion of the coal solution from the reactor is recycled and used in slurrying the feed. This increases the reaction time and together with a higher concentration of mineral matter, which catalyzes the reaction, increases the hydrogen consumption. The result is that more higher-molar-mass species are converted to lower ones.

The principal differentiating feature of the Exxon Donor Solvent (EDS) process is that the recycle solvent is a distillate product that has been externally hydrogenated. An upflow non-catalytic reactor is used as in the SRC processes.

The H-Coal process incorporates an ebullated bed reactor operating at about 20 MPa and 450°C. The upward passage of slurry and gases in the reactor maintains millimeter-size cobalt-molybdenum catalyst pellets in a bubbling, fluidized bed. The principal product can be a synthetic crude or a low-sulfur fuel oil, depending largely on the residence time of the mixture in the reactor. As shown in Table 6.2, the low-sulfur fuel oil mode consumes less hydrogen than does the syncrude mode.

All four of the major direct liquefaction procedures being developed in the United States are single-step systems operating at about the same temperature and pressure. Although the pressures are much reduced from those of the earlier German technology, the processes are constrained by the requirement of handling heavy bottoms as a major unit operation, in contrast to the German systems, where much higher distillable product yields were obtained. It is of interest that direct liquefaction technology is once more being commercialized in Germany based on the Bergius-I.G. Farben technology used during World War II. As with the older German processes, the new IG process uses disposable, finely divided iron oxide catalyst, employs quite severe operating conditions with reactor pressures of about 30 MPa, and has large hydrogen treatment requirements.

One interesting hydroliquefaction approach still in the process development stage is the Conoco Zinc Chloride process listed in Table 6.2. In this system, molten zinc chloride is used as the reagent/catalyst. The catalyst has the property of causing a highly selective conversion of the coal to a high-quality motor fuel in a single step and with essentially no residual fuel oil to dispose of. Moreover, there is an apparent suppression of carcinogens in the product in comparison with the relatively high levels found in the liquids from the other processes cited. The zinc chloride must, however, be continuously regenerated because of its reaction with the heteroatoms in the coal. In addition, it is a highly corrosive material that is difficult to handle.

We compare the processes mentioned, along with other less developed direct liquefaction procedures, in Section 6.4. However, we may observe from Table 6.2

that, as expected, higher hydrogen consumption, coupled with greater residence times, lead to higher liquid yields, as do higher pressures together with external catalyst addition.

The last major category for the production of liquid fuels is that of indirect liquefaction. As discussed in Section 3.3, indirect liquefaction includes those processes that synthesize liquid hydrocarbon fuels from a synthesis gas, composed mainly of carbon monoxide and hydrogen, that has been produced by the gasification of coal. It also includes those processes, still under development, wherein carbon monoxide is reacted with steam to synthesize hydrocarbons. Also included in the category of indirect liquefaction are processes where a liquid synthesized from a coal-derived gas is transformed to a liquid fuel. The specific process to be discussed is the Mobil M-gasoline process, in which methanol is converted to high octane gasoline.

In Table 6.3 are listed representative processes, operating conditions, and products for the three major indirect liquefaction procedures. The methanol and Fischer-Tropsch processes are commercial, while the Mobil M-gasoline process is under development, with a 2×10^6 L/d plant planned for New Zealand with natural gas as the feedstock.

The Fischer-Tropsch process, as described in Section 3.3, starts with a medium-CV gas that has been purified and shifted if necessary, to produce a synthesis gas with a H_2/CO ratio of 2 or greater. In the 8×10^6 L/d commercial-scale Sasol plants in South Africa the gas is reacted in entrained flow reactors using promoted iron oxide catalysts. The reactor pressures are typically in the range of 2.2 to 2.8 MPa and the temperatures 315 to 330°C. The hydrocarbon products are principally in the gasoline and mid-distillate range, with olefins about 60 to 70 percent and paraffins about 15 percent.

Methanol is synthesized over a range of pressures from synthesis gas, which is at present produced mainly by the steam reforming of natural gas. For the conversion of coal-derived synthesis gas having a H_2/CO ratio of 2 or somewhat greater, the newer lower pressure processes such as the ones developed by Imperial Chemical Industries (ICI) and Lurgi are preferred to minimize gas compression. These processes use copper/zinc catalysts containing chromia or alumina or both, and generally operate in the range of 5 to 10 MPa at temperatures from about 250 to 270°C.

Methanol can be used to make gasoline in the Mobil M-gasoline process. The key

Table 6.3 Representative operating conditions and products for major indirect liquefaction processes

	Low Pressure Methanol	Sasol Fischer-Tropsch	Mobil M-Gasoline
Pressure, MPa	5	2.2	0.4
Temperature, °C	250	320	400
Reactor	Fixed bed	Entrained flow	Fluidized bed
Catalyst	Cu/Zn	Iron oxide	Zeolite
Feed	$H_2/CO \geq 2$	$H_2/CO = 2$–3	Methanol
Product	Methanol	Gasoline, mid-distillates	High octane gasoline

to the process is the previously discussed synthetic zeolite catalyst ZSM-5 which results in a selective penetrability by molecules of intermediate size. A consequence is that negligible amounts of hydrocarbons are produced that boil above 210°C, the end point of conventional gasoline. In this reaction, water is removed from the methanol and the remaining hydrocarbons are rearranged and recombined to form high octane gasoline. For fluidized bed operation the process conditions are mild, as may be seen in Table 6.3, although the reaction is highly exothermic, so the heat release is substantial. The hydrocarbon product is about 80 percent in the gasoline range, with the remaining 20 percent light hydrocarbon gases in the C_1 to C_4 range.

An advantage of indirect liquefaction is that the heteroatoms in the coal are removed in the gasification step and the synthesized liquids are clean. The liquids like the ones produced in the Conoco Zinc Chloride process are less carcinogenic than those from the other pyrolysis and direct liquefaction procedures. However, it is important to emphasize again that indirect liquefaction technology starting with coal tends to be inherently inefficient, since it involves first producing a synthesis gas from the coal endothermically at a high temperature and then forming the gas into a liquid in highly exothermic reactions at lower temperatures. There are additional losses when there is further conversion of the liquids, as in the Mobil M process.

6.2 INDIRECT LIQUEFACTION

In the last section the three major indirect liquefaction procedures cited as commercial or being readied for commercialization are methanol synthesis, oxygenated hydrocarbon production by Fischer-Tropsch, and the Mobil methanol-to-gasoline process. All of these technologies have the common processing step of producing a cleaned and purified synthesis gas with a H_2/CO ratio of 2 or greater. All the technologies then catalytically synthesize the gas either to methanol or other liquid organics. In the case of Mobil M, the methanol is subsequently converted to gasoline in a further processing step. Overall plant configurations are not that different from those to produce SNG. Indeed, if a methane-producing gasifier such as Lurgi is used, SNG could be a coproduct. In any case, in Fischer-Tropsch synthesis the product selectivity is such that a large fraction of the product could be methane (Section 3.3). The methane can, of course, always be recycled and reformed to carbon monoxide and hydrogen if SNG is not desired as a coproduct, but the efficiency of this procedure is only about 70 percent (Section 5.4) and this would significantly reduce the overall conversion efficiency.

In Section 3.5 we compared the above three indirect liquefaction procedures based on ideal reactions and hypothetical heat recovery schemes. In this section we describe and compare proposed synthesis procedures based on conceptual designs for large-scale plants that employ different gasifier types and that may have more than one product.

According to the discussion in Section 5.5, highest process efficiency could be achieved by gasifying coal to produce a low H_2/CO synthesis gas and then in a separate step shifting the gas to the desired ratio for synthesis. In this way each of the steps could be optimized for maximum efficiency and the combined efficiency would be greater than if both gasification and shift were carried out in the gasifier alone. Ideally, then, indirect liquefaction plant designs would incorporate high temperature, high pressure entrained flow gasifiers which produce low H_2/CO, methane-free gas. Two gasifiers meeting these criteria are the Texaco and Shell-Koppers gasifiers; both, however, are second generation gasifiers still under development. The Koppers-Totzek gasifier is commercially available but, as we have seen, operates at low pressure and has quite a high oxygen consumption. Weighed against the second generation gasifiers are fully developed and reliable dry ash Lurgi units which produce a high H_2/CO product together with methane. These gasifiers have been successfully employed in the Sasol Fischer-Tropsch plants on commercial scale for many years. Their use, however, would almost certainly require that SNG be coproduced if a low overall conversion efficiency is to be avoided.

The feed to a methanol or Fischer-Tropsch reactor is a cleaned and purified synthesis gas with H_2/CO and H_2/CO_2 ratios specified by the empirical relation[2,3]

$$\frac{H_2}{2CO + 3CO_2} = 1.03 \tag{6.1}$$

The H_2/CO and H_2/CO_2 ratios are somewhat greater than the stoichiometric values because some hydrogen is generally purged in the processes and some is lost.

For purposes of comparing the different indirect liquefaction processes, we consider the two different feed gases shown in Table 6.4. One is assumed to be the product of Lurgi gasification with the carbon dioxide removed down to a level that Eq. (6.1) is satisfied. The other is assumed to be a gas from a Koppers-Totzek or Texaco gasifier that has been shifted to obtain a composition which satisfies Eq. (6.1). The carbon dioxide level can have any value consistent with Eq. (6.1), and a value of 3 percent is taken. Lower levels would incur larger acid gas removal expense. The synthesis

Table 6.4 Feed gas compositions for methanol and Fischer-Tropsch synthesis[3]

Constituent	M	Dry Ash Lurgi, mol %	Koppers-Totzek/ Texaco, mol %
H_2	2	57.6	67.5
CO	28	20.9	28.2
CO_2	44	4.7	3.0
CH_4	16	15.4	0.1
$C_{2+}(C_3H_8)$	44	0.2	—
N_2	28	1.2	1.2
Total		100.0	100.0
Mean molar mass		12.0	10.9
GCV, MJ/kg		30.5	25.1
MJ/m³		16.3	12.2

feed gas compositions are not much dependent on the particular coal type once they have been purified and the compositions adjusted. The compositions shown in Table 6.4 are representative of what might be obtained from gasifying bituminous or sub-bituminous coals.[3]

Methanol

Of the indirect liquefaction procedures, methanol synthesis is the most straightforward and well-developed and for this reason we discuss it first. The fundamentals of methanol synthesis have already been described in Section 3.3, so we shall concentrate here on the process description and illustrative material and energy balances. Although methanol synthesis plants today use natural gas as the feedstock, we are here concerned only with coal as the feedstock from which, as discussed above, the necessary synthesis gas is obtained by gasification, purification and, if necessary, shifting. Methanol manufacture from coal has been practiced in the past, and at present one large commercial methanol-from-coal plant is in operation in South Africa, with the synthesis gas generated by Koppers-Totzek gasifiers.

A simplified picture of low pressure methanol synthesis, representative of the ICI process, is shown in Figure 6.2. The synthesis gas is compressed to a pressure of from 5 to 10 MPa, with the extent of the compression dependent on the pressure at which the coal gasifier is operated. The gas is mixed with recycled gas and heated by exchange

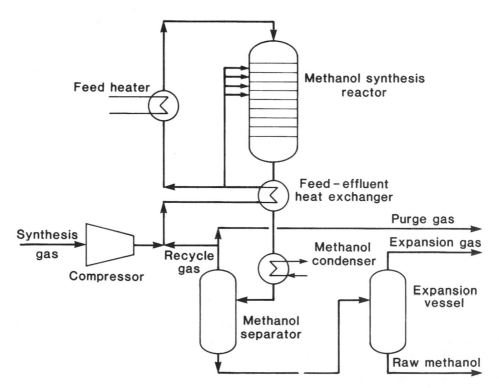

Figure 6.2 Simplified picture of low pressure methanol synthesis.

with the methanol product, as well as by a separate feed heater, after which it is fed to the top of an adiabatic fixed bed reactor. Temperature control, normally between 250 and 270°C, is maintained by injecting a part of the relatively cool gas into the reactor at various levels. The methanol vapors leaving the bottom of the reactor are partially condensed and cooled by the heat exchange with the feed. The reactor effluent then enters a high pressure separator where the methanol is separated. Gas is purged from the separator to prevent the accumulation of inert gases in the system. The gases dissolved in the methanol are removed by a subsequent reduction of pressure in an expansion vessel.

In the Lurgi low pressure methanol synthesis process, the fixed bed reactor is a vessel containing tubes filled with Cu/Zn catalyst, inside each of which the gas flows downward, with boiling water on the outside. The reactor temperature is normally maintained between 235 and 250°C, with operating pressures from 4.8 to 7.7 MPa.

A simplified but representative material balance for methanol synthesis can be made on the basis that 98 percent of the carbon monoxide and 50 percent of the carbon dioxide in the feed gas are converted to methanol stoichiometrically following the reactions

$$CO + 2H_2 \rightarrow CH_3OH \tag{6.2}$$

$$CO_2 + 3H_2 \rightarrow CH_3OH + H_2O \tag{6.3}$$

The water formed by reaction (6.3) is assumed to be condensed out with the methanol. All of the unconverted carbon monoxide and carbon dioxide, together with the methane, higher hydrocarbons, and nitrogen in the feed gas, are taken to appear in the purge and expansion gases. The unconverted hydrogen also ends up in the purge and expansion gases.

Tables 6.5 and 6.6 show material balances for two plants, each with a nominal 10^8 MJ/d output of products. This is about one-quarter the size of the standard plants

Table 6.5 Material balance for a plant producing a nominal 10^8 MJ/d of methanol and methane product from synthesis gas derived by Lurgi gasification

	IN	OUT	
Constituent, mol/s	Feed Gas	Raw Methanol	Purge and Expansion Gases
H_2	2105	—	351
CO	763	—	15
CO_2	172	—	86
CH_4	564	—	564
$C_{2+}(C_3H_8)$	8	—	8
N_2	44	—	44
CH_3OH	—	~~826~~ 834	8
H_2O	—	86	—
Total, mol/s	3656	912	1076
kg/s	43.8 75	28.02 8.24	15.8 5.1

Table 6.6 Material balance for a plant producing a nominal 10^8 MJ/d of methanol from synthesis gas derived by Koppers-Totzek or Texaco gasification

	IN	OUT	
Constituent, mol/s	Feed Gas	Raw Methanol	Purge and Expansion Gases
H₂	3731	—	425
CO	1558	—	31
CO₂	168	—	84
CH₄	6	—	6
N₂	65	—	65
CH₃OH	—	1595	16
H₂O	—	84	—
Total, mol/s	5528	1679	627
kg/s	60.4	52.6	7.8

discussed in Section 1.2. It can be seen that the plant using gas derived from a high temperature gasifier has about twice the methanol output as the one using Lurgi-derived gas. A more revealing picture, however, emerges from the thermal balances given in Table 6.7. These balances, like those in the previous chapters, are all based on gross calorific values. It may be noted that for both plants, about one-third of the losses shown in Table 6.7 represent heat of condensation of the methanol, heat that is generally dissipated by air cooling.

For the high temperature gasifier plant with a gasifier efficiency of 70 percent, the overall thermal efficiency to liquid methanol is 54 percent, assuming that the purge and expansion gases are used to drive the plant and supply the energy necessary to

Table 6.7 Thermal balances for methanol plants with nominal outputs of 10^8 MJ/d

	Lurgi		Koppers-Totzek/ Texaco	
	MJ/s	% of Total	MJ/s	% of Total
IN				
Feed gas	1337	100.0	1513	100.0
OUT				
Methanol (liquid)	600	44.9	1159	76.6
Purge and expansion gases	630	47.1	148	9.8
Losses (by difference)	107	8.0	206	13.6
Total	1337	100.0	1513	100.0

purify and shift the gas in preparing the feed. This is about the middle of the range of 51 to 59 percent reported for projected coal-to-methanol designs.[4]

An alternative design is to use Lurgi gasification and to recover the purge and expansion gases as an SNG coproduct after converting the residual hydrogen and carbon monoxide to methane. In this case, with about the same use of purge gas for plant driving energy and feedstock preparation as for the high temperature gasifier plant, as much as a 10 point higher overall thermal efficiency is obtainable. This is principally because of the relatively high efficiency of the Lurgi gasifier (compare Tables 4.17 and 4.25). This advantage would be dissipated, however, were the methane in the purge gas to be reformed to provide additional feedstock for methanol synthesis.

From what has been said it can be seen that product considerations may have an effect not revealed when assessing processes on the basis of idealized conversion schemes. In addition, it must be emphasized that no consideration has been given here to capital cost requirements which play a major role in any actual design, as discussed more completely in Section 10.1.

Fischer-Tropsch

The most extensive production of synthetic liquid fuels today is that being carried out by Fischer-Tropsch reactions at the South African Sasol complexes. Sasol I, in operation since 1955, produces motor fuels, liquid byproducts, and industrial gas from coal. Dry ash, oxygen-blown Lurgi gasifiers are used to generate some 10^7 m³/d of raw gas from which about 850 t/d of liquid fuels are manufactured, equivalent to 1.25 $\times$ 10^6 L/d of gasoline, together with about 1.7 $\times$ 10^6 m³/d of 19 MJ/m³ medium-CV gas. Two second generation plants, Sasol II and Sasol III, each with an output of approximately 8 $\times$ 10^6 L/d (50 000 bbl/day) of motor fuels are being built at the time of writing. The first of these, Sasol II, is scheduled for initial production of motor fuels about mid-1981, with full capacity achieved by the end of 1982. Sasol III, a carbon copy of Sasol II and located adjacent to it, is scheduled for startup in 1984. Figure 6.3 is a photograph of the Sasol II plant from which the magnitude and scope of the engineering undertaking is evident.

At Sasol I there are 13 Lurgi gasifiers each about 4 m in diameter which operate at a pressure of 2.9 MPa and produce a total of about 10^7 m³/d of raw gas. This gas is quenched and scrubbed for removal of particulate matter, tar, and ammonia, following which hydrogen sulfide and carbon dioxide are removed by the Lurgi Rectisol system (see Section 5.1). At Sasol I there are two kinds of Fischer-Tropsch synthesizers operating in parallel: the Arge system, which uses a pelletized iron catalyst in a fixed bed tubular reactor, and the Synthol system, which uses a promoted iron catalyst powder in a circulating entrained flow design (cf. Section 5.2). The fluid bed system operates at higher temperatures, has a higher output per reactor volume, and is useful for the production of lighter hydrocarbons, while the fixed bed system is suitable for a significant contribution of higher-molar-mass hydrocarbons.

The Sasol II plant has a design output of about 60 kg/s of 48 MJ/kg fuel, principally gasoline. This is equivalent to 2.5 $\times$ 10^8 MJ/d. The plant has 36 Lurgi gasifiers and incorporates many improvements gained from its predecessor. It differs from Sasol I

Figure 6.3 Photograph of Sasol II plant for the production of 8×10^6 L/d of motor fuels from coal. Seven Synthol reactors are to left of cooling towers, thirty-six Lurgi gasifiers are in right background. *(Courtesy of Sasol, Ltd.)*

mainly in that only the Synthol circulating fluid bed reactor is incorporated, because the main purpose is to produce gasoline. The plant contains 7 Synthol reactors, each with a design capacity of about 3.5 times those of Sasol I or about 1.1×10^6 L/d capacity compared with 0.3×10^6 L/d. An additional difference is that all of the methane and lighter gaseous hydrocarbons are steam reformed to synthesis gas. This considerably reduces the overall thermal efficiency, but is done because gas is not used in South Africa and a distribution system is not in place to pipeline it. In the United States, on the other hand, as we have noted, both methane and gasoline are desired products, so their coproduction would be appropriate. A simplified flow diagram for the Sasol II plant is given in Figure 6.4.

Figure 6.5 is a schematic of the Synthol reactor and Figure 6.6 a photograph of three of the seven reactors at the Sasol II plant. Fresh feed gas mixed with 2 to 3 volumes of recycle gas enters at 160 to 200°C and about 2.2 MPa. The gas entrains the catalyst which is added through slide valves at 340 to 350°C. The suspension enters the fluidized bed reaction section where the Fischer-Tropsch and the gas shift reactions proceed at a temperature of from 315 to 330°C. Heat released by the exothermic reactions is removed by circulating coolant within tubes inside the reactor. In Sasol I, cooling oil is circulated and subsequently used to raise steam. In Sasol II, water is the coolant and high pressure steam is generated directly. Some of the steam is used to preheat the gas fed to the reactor. The mixture of product, reactants, and catalyst enters the catalyst hopper where the decreased gas velocity causes most of the catalyst to settle out. The gas then passes through two banks of internal cyclones in series which separate out the remaining catalyst from the off gas. The off gas is scrubbed with cooled recycle oil to condense out the heavier hydrocarbon products. The vapors are then further cooled and washed with water in a tower where the light oil and aqueous chemicals are condensed and separated from the remaining light gas. Part of the gas is recycled, and in the Sasol II plant the remaining "tail gas" goes to a gas reforming unit for conversion back to carbon monoxide and hydrogen. Alternatively, this tail gas could be converted to SNG or used directly.

In our discussion of the Fischer-Tropsch process in Section 3.3, it is pointed out that a wide spectrum of hydrocarbons are produced. Most of the more recent information on product selectivity has come from the Sasol experiences, which indicate that kinetic

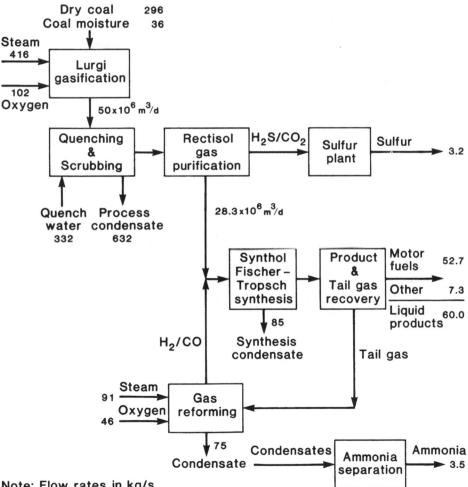

Figure 6.4 Simplified flow diagram for Sasol II Fischer-Tropsch plant with a nominal output of 8×10^6 L/d of motor fuels.[5]

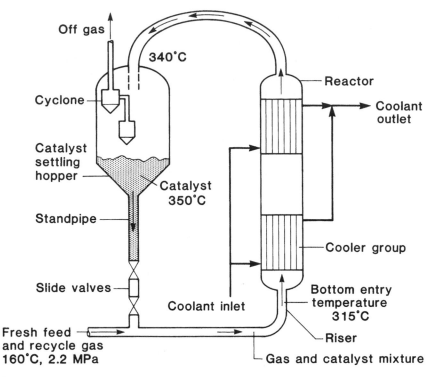

Figure 6.5 Synthol Fischer-Tropsch reactor.

rather than equilibrium considerations are controlling. Although the primary products are olefins and alcohols, they are unstable relative to paraffins. The extent to which the products initially formed exit the reactor depends on their residence time within the catalyst particles and other reactor characteristics.[6] For the high temperature, fluid bed Synthol reactor and a specific catalyst, the importance of the carbon dioxide concentration on product selectivity is also noted in Section 3.3. With the caveat that a wide range of product distributions is possible, we have shown in Table 6.8 a "typical" Synthol reactor product slate for the general range of operating conditions discussed.[2] The molar masses given in the table for the liquid products are mean values and the molar formulas are representative of the average product. The heats of combustion are for the molar compositions listed.

The Fischer-Tropsch reactions have till now been discussed in terms of the ideal stoichiometry

$$CO + 2H_2 \rightarrow (-CH_2-) + H_2O \tag{6.4}$$

Consistent with the yield breakdown of Table 6.8, the overall stoichiometry may be characterized by the empirical relation[2]

$$CO + 2.12H_2 \rightarrow (HC, \text{Alcohols, Acids}) + 0.95H_2O \tag{6.5}$$

The stoichiometric coefficients of 2.12 and 0.95 instead of 2 and 1, respectively, are

Figure 6.6 Photograph of Synthol reactors at Sasol II plant. *(Courtesy of Sasol, Ltd.)*

Table 6.8 Representative selectivities for Fischer-Tropsch reactions in Synthol entrained flow reactor

Constituent	Molar Formula	Mass %	M	$-\Delta H°$, MJ/kg
Gases				
Methane	CH_4	11	16	55.5
Ethene	C_2H_4	4	28	50.3
Ethane	C_2H_6	6	30	51.9
Propene	C_3H_6	11	42	49.0
Propane	C_3H_8	2	44	50.4
Butene	C_4H_8	8	56	48.5
Liquids				
$C_5 - C_7$	$C_{5.5}H_{11}$	8	77	46.9
Light oils	C_8H_{16}	33	112	47.4
Heavy oils	$C_{20}H_{42}$	6	282	47.1
Alcohols	$C_{2.4}H_{6.8}O$	9	52	71.1
Acids	$C_{2.4}H_{4.8}O_2$	2	66	57.6
Total		100		
Mean molar mass			47.5	

a consequence of the fact that for every 100 moles of CO reacted there are found combined in the products 100 C atoms, 224 H atoms, and 5 H_2O groups.

The H_2/CO and H_2/CO_2 ratios in the feed are the same as for methanol synthesis and are specified by Eq. (6.1). Unlike methanol synthesis, in the Fischer-Tropsch reactions little carbon dioxide is hydrogenated. On the contrary, because of the large amount of water formed, carbon dioxide is produced by the gas shift reaction

$$H_2O + CO \rightleftharpoons CO_2 + H_2 \qquad (6.6)$$

At the relatively high temperatures in the Synthol reactor the reaction is in equilibrium. Although carbon dioxide is produced, the amount is relatively small and for our purposes here may be neglected in making material and thermal balances.

From the criteria given, material and thermal balances can be carried out for the same feed gas compositions used for the methanol synthesis calculations (Table 6.4). The carbon monoxide conversion may be taken to be the same as for the methanol synthesis (98%), a value which is probably somewhat high, but chosen for convenience of comparison since small differences in this value will have little effect on the balance.

In Tables 6.9 and 6.10 are given molar material balances based on the stoichiometry of Eq. (6.5). From these balances, the combined mass rate of hydrocarbon, alcohol, and acid production can be calculated by difference, and the separate component mass fractions then obtained from the product slate distribution given in Table 6.8. With the mass of each component produced known, the thermal balance can be derived using the calorific values in Table 6.8. The thermal balances so calculated are shown in Table 6.11.

Comparison of the Fischer-Tropsch and methanol synthesis thermal balances, Tables 6.11 and 6.7, shows an interesting result. When the sum of the calorific values

Table 6.9 Material balance for a plant producing a nominal 10^8 MJ/d of Fischer-Tropsch products from synthesis gas derived by Lurgi gasification

Constituent, mol/s	IN Feed Gas	OUT From Reaction	From Feed
H_2	2105	—	519
CO	763	—	15
CO_2	172	—	172
CH_4	564	*	564
$C_{2+}(C_3H_8)$	8	*	8
N_2	44	—	44
HC, Alcohols, Acids	—	238†	—
H_2O	—	711	—
Total, mol/s	3656	949	1322
kg/s	43.8	24.1	19.7

  * Included in hydrocarbons.
  † Based on mean molar mass of 47.5 and 11.3 kg/s calculated by difference.

Table 6.10 Material balance for a plant producing a nominal 10^8 MJ/d of Fischer-Tropsch products from synthesis gas derived by Koppers-Totzek or Texaco gasification

Constituent, mol/s	IN Feed Gas	OUT From Reaction	From Feed
H_2	3731	—	494
CO	1558	—	31
CO_2	168	—	168
CH_4	6	*	6
N_2	65	—	65
HC, Alcohols, Acids	—	486†	—
H_2O	—	1451	—
Total, mol/s	5528	1937	764
kg/s	60.4	49.2	11.2

* Included in hydrocarbons.

† Based on mean molar mass of 47.5 and 23.1 kg/s calculated by difference.

of the Fischer-Tropsch reaction products is compared with the methanol value, and when the calorific value of the Fischer-Tropsch gases from the feed is compared with the value for the purge and expansion gases in methanol synthesis, the results of the two balances are almost the same.

Because most of the product is gas, a simple estimate for an all-liquid thermal efficiency is somewhat arbitrary. However, an estimate can be made for the overall thermal efficiency for the combined gas and liquid products. This may be done by taking an 80 percent and 70 percent gasification efficiency, respectively, for generating the Lurgi and Koppers-Totzek/Texaco feed gases and then assuming that all the

Table 6.11 Thermal balances for Fischer-Tropsch plants with nominal outputs of 10^8 MJ/d

	Lurgi MJ/s	Lurgi % of Total	Koppers-Totzek/ Texaco MJ/s	Koppers-Totzek/ Texaco % of Total
IN				
Feed gas	1337	100.0	1513	100.0
OUT				
Liquids from reaction	337	25.2	688	45.5
Gases from reaction	243	18.2	497	32.9
Gases from feed	672	50.3	155	10.2
Losses (by difference)	85	6.3	173	11.4
Total	1337	100.0	1513	100.0

unconsumed gases from the feed are recycled and reduce the net feed gas requirement by this amount. For the Lurgi case the thermal efficiency is then $(337 + 243)/(1671 - 672) = 58$ percent, and for the Koppers-Totzek/Texaco case $(688 + 497)/(2161 - 155) = 59$ percent. Surprisingly, despite the quite different gasification efficiencies, there is seen to be little to choose between the processes on overall thermal efficiency grounds. This is the more so because the liquid-to-gas product ratio for the two gas feeds is the same. It should be noted that 58 percent is the overall conversion efficiency calculated for an engineering conceptual design for a Fischer-Tropsch gas/liquid product slate using Lurgi gasification.[2] Though we have not estimated an efficiency for the conversion to an all-liquid product, it would, of course, be lower than for a gas/liquid slate.

Before concluding our discussion of Fischer-Tropsch synthesis, note should be made of the Kölbel slurry column reactor, in which low-H_2/CO-ratio gas is synthesized directly to Fischer-Tropsch products. The Kölbel reactor is a three-phase reactor in which synthesis gas is passed upward through a slurry of finely divided iron-based catalyst in a relatively non-volatile heavy oil medium, generally a molten wax. It is similar in nature to the direct liquefaction reactors mentioned in the previous section.

Kölbel and his associates ran a large pilot plant in the early 1950s with a 10-m³ reactor, and achieved 90 percent conversion to typical Fischer-Tropsch products with a feed having an H_2/CO ratio of 0.67.[7] The operating pressure was 1.2 MPa and the product output was about 100 g/s at 268°C and 350 g/s at 280°C. The catalyst lifetime of the relatively cheap powdered iron catalyst, of average particle size less than 30 μm, was claimed to be much greater than for operation in a conventional two-phase reactor.

Unlike the conventional Fischer-Tropsch reaction (6.4), in which water is rejected, in the Kölbel reaction carbon dioxide and water are rejected. In the limit of an H_2/CO ratio of 0.5, only carbon dioxide is rejected following the stoichiometry

$$2CO + H_2 \rightarrow (\text{---}CH_2\text{---}) + CO_2 \qquad \Delta H^\circ_{700} = -187.7 \text{ kJ/mol} \qquad (6.7)$$

The advantage of the Kölbel reactor is that second generation, low-H_2/CO-ratio, high temperature gasifiers can be employed without necessitating a separate gas shift step and without extensive gas recycle. An assessment of the Kölbel reactor has been made in which are discussed other studies where the results were somewhat less successful.[8] The reader is referred to this report, both for its excellent description of the status of the process and for an extended bibliography. Deckwer et al.[9] present a good summary of the hydrodynamics of the Kölbel reactor. This paper also discusses the work of the Kölbel group and has an extensive bibliography.

In conclusion, we mention the limit in which molecular hydrogen is altogether absent from the feed and carbon monoxide is reacted directly with steam in accordance with the reaction

$$3CO + H_2O \rightarrow (\text{---}CH_2\text{---}) + 2CO_2 \qquad (6.8)$$

This reaction is called the Kölbel-Engelhardt reaction and for iron-based catalysts takes place at 250 to 300°C. It has so far not been exploited commercially, but is a limit of the Kölbel reaction described above.[8]

Mobil M

The last indirect liquefaction process we shall discuss is the Mobil M (methanol-to-gasoline) process, the basic features of which are outlined in Section 3.3. As discussed, the key to the process was the development by Mobil of the ZSM-5 shape-selective zeolite catalyst. Zeolite is crystalline alumino-silicate, and by shape selective is meant a catalyst whose cavity geometry and pore dimensions have been tailored so that it selectively produces hydrocarbon molecules within a desired size range. In particular, the ZSM-5 catalyst has pore dimensions such that intermediate-size molecules in the C_5 to C_{10} groups can get out of the catalyst, whereas molecules of a larger size cannot. The C_9 to C_{10} aromatic hydrocarbons are essentially the largest molecules that can exit the catalyst and therefore the largest to show up in the product. This selectivity enables the conversion of methanol into a mixture of hydrocarbons with the composition, octane number, boiling range, and other specifications of high-quality gasoline.[10,11]

Unlike the processes so far considered, the conversion of methanol to gasoline is one of dehydrogenation in which the material is first dehydrated to dimethylether (Eq. 3.50) and then in a second step the ether is converted to gasoline product (Eq. 3.51). The overall reaction may be represented by the relation

$$CH_3OH \rightarrow (—CH_2—) + H_2O \qquad (6.9)$$

As we have pointed out, this reaction is strongly exothermic with a heat of reaction for gaseous products of about 50 kJ/mol. A major problem in any plant design is therefore the reactor system to effect the necessary heat removal. Two systems are being developed by Mobil, a fixed bed and a fluidized bed system.

In the fixed bed reactor system the control of the reaction heat is achieved by splitting the reaction into two stages. The first stage is essentially the formation of the dimethylether (DME), and in it some 20 percent of the heat is released. The product is actually an equilibrium mixture of dimethylether, water, and methanol, corresponding to the reaction temperature. The temperatures are about 300°C at the DME reactor inlet and 410°C at the outlet. The operating pressure is typically 2.3 MPa. In the second stage where hydrocarbon conversion takes place, a large recycle gas stream of 6 to 9 moles of recycle gas per mole of makeup is used to limit the temperature rise to 70°C. A typical design value is 7.5 moles per mole. The temperatures are about 330°C at the M-gasoline reactor inlet and 400°C at the outlet, and the pressure is 2.2 MPa. In one suggested design to produce 3×10^8 MJ/d of product, equivalent to about 50 000 bbl/day of fuel oil, seven fixed bed reactors are proposed. Two are DME reactors and five are M-gasoline reactors, four of which are in service at any time with the fifth off cycle for catalyst regeneration. This is necessitated by the fact that the M-gasoline catalyst requires regeneration by burning with air about every two weeks to remove accumulated coke.

The fluidized bed design enables the replacement of the seven fixed bed reactors with one large fluidized bed reactor at a projected operating temperature of 400°C and 0.4 MPa. As for the Fischer-Tropsch Synthol reactor, temperature control can be achieved by internal coolant coils. Studies indicate a somewhat improved efficiency with the fluidized bed system; however, it has not been developed to the point that the fixed bed system has. For this reason and because of simpler scaleup problems, the first large-scale 2×10^6 L/d Mobil M plant, which is to be constructed in New Zealand, will employ fixed bed reactors.

Table 6.12 presents product yields projected for fixed bed operation, as adapted from Ref. 2. The very small amounts of alcohols and acids that are formed are neglected along with other trace components, and the wide range of hydrocarbons in the gasoline fraction are lumped together into elementary paraffinic, olefinic, and aromatic groups. The liquid component distribution is typical of a conventional high octane gasoline. Also shown in Table 6.12 is the GCV for combustion of each of the products.

With the product distribution known, the calculation of material and thermal balances is a straightforward matter. For purposes of simplification it may be assumed that Eq. (6.9) holds, so that for every mole of methanol converted one mole of water is produced. Moreover, any water present in the raw methanol can be taken to pass through without participating in the reaction. Table 6.13 presents the material and thermal balances for a 100 kmol of methanol feed equivalent to 3200 kg of methanol. The products are 1400 kg of hydrocarbons and 1800 kg of water. From the product

Table 6.12 Product yields for Mobil M fixed bed operation

Constituent	Molar Formula	Mass %	M	$-\Delta H°$, MJ/kg
Gases				
Methane	CH_4	1	16	55.5
Propane	C_3H_8	5	44	50.4
Butane	C_4H_{10}	12	58	49.6
Butene	C_4H_8	1	56	48.5
Mean gas molar mass			47	
Gasoline Fraction (liquids)				
Pentane	C_5H_{12}	14	72	49.0
Pentene	C_5H_{10}	2	70	47.0
Hexane	C_6H_{14}	14	86	48.3
Hexene	C_6H_{12}	2	84	46.6
Heptane	C_7H_{16}	6	100	48.1
Heptene	C_7H_{14}	4	98	46.5
Octane	C_8H_{18}	2	114	47.9
Octene	C_8H_{16}	5	112	46.6
Nonane	C_9H_{20}	1	128	47.8
Nonene	C_9H_{18}	2	126	46.8
Toluene	C_7H_8	2	92	42.5
Xylene	C_8H_{10}	9	106	43.0
Trimethyl-benzene	C_9H_{12}	10	120	43.3
Tetramethyl-benzene	$C_{10}H_{14}$	8	134	43.8
Total		100		
Mean liquid molar mass			96	

Table 6.13 Material and thermal balances for conversion of 100 kilomoles of methanol in Mobil M process

	Material Balance		Thermal Balance	
	kmol	kg	MJ	% of Total
IN				
Methanol (gas)	100	3200	76 400	100.0
OUT				
Gases	5.7*	266	13 300	17.4
Gasoline (liquid)	11.8†	1134	52 500	68.7
Losses (by difference)	—	—	10 600	13.9
Water	100	1800	—	—
Total	117.5	3200	76 400	100.0

$$\text{Overall thermal efficiency} = \frac{65\,800}{76\,400} \times 100 = 86\%$$

* Based on 266 kg and mean molar mass of 47 for gas products.
† Based on 1134 kg and mean molar mass of 96 for gasoline fraction.

breakdown of Table 6.12 the mass of each component is specified so that the heat of combustion of the components is easily calculated. Since the heat of the methanol feed is known, which is here taken to be gaseous, the losses are determined by difference.

From Table 6.13 it can be seen that 80 percent of the product heating value is in gasoline and that the overall thermal efficiency based on gross calorific values is 86 percent. If our earlier estimate from Table 6.7 of the efficiency to convert coal to methanol liquid is adjusted for conversion to methanol gas, then the thermal efficiency for conversion to methanol becomes 56 percent instead of 54 percent. The product of the two process efficiencies gives an overall thermal efficiency of 48 percent for conversion of coal to gasoline and hydrocarbon gas products. This value is what has been estimated from detailed conceptual design studies.[2] The relatively low value again points up the importance of co-producing SNG together with liquid products.

In concluding our discussion of the Mobil M process, we would again point out that methanol derived from coal need not necessarily be the feed to the process. The methanol could be the product of biomass conversion or could be produced from synthesis gas obtained by steam reforming of methane, as is being done in the Mobil M plant being developed for New Zealand. One other point of note is that there are indications that the ZSM-5 shape-selective zeolite catalyst also yields a high octane gasoline with Fischer-Tropsch products as the feed in place of methanol. This suggests the possibility of an advanced indirect liquefaction scheme for the conversion of coal to gasoline which would consist of a second generation, high temperature gasifier, followed by a Kölbel slurry Fischer-Tropsch reactor, followed by a Mobil M unit.[8]

6.3 PYROLYSIS

The liquid yields obtained from coal pyrolysis are generally small and the major product is char which can, however, be gasified or used as a combustion feedstock. As we have seen, the tar yield is very dependent on the reactor configuration and the process variables. Packed bed, entrained flow, and fluidized bed reactors have been used, although feed and agglomeration problems are encountered with bituminous coals at pyrolysis temperatures.

In our discussion of pyrolysis a basic distinction was made between slow and fast pyrolysis of coal. We recall that all else being equal, fast pyrolysis produces higher liquid yields, though not necessarily of the same quality. The yields shown in Table 6.1 for pyrolysis processes that are developed or being developed bear out the distinction. According to the categorization of Figure 3.5, of the processes listed in Table 6.1, the COED and TOSCOAL processes may be classified as slow pyrolysis processes, the Lurgi-Ruhrgas and Occidental processes as intermediate rate ones, and the Rockwell hydropyrolysis system as a fast process. Supercritical gas extraction is a special case of slow pyrolysis. Both the TOSCOAL and Lurgi-Ruhrgas processes are being applied to the retorting of oil shale and we shall discuss them in that connection.

Pyrolysis tars and oils are neither stable nor suitable final products. They are naphthenic and aromatic, and tend to be high boiling. Because they are more aromatic than petroleum crudes, they are correspondingly more hydrogen deficient. Low temperature-pyrolysis tar is composed roughly of 70 percent pitch boiling above 360°C, about 10 percent tar acids, mainly cresols and phenols, and about 20 percent neutral oils. The impurity levels in the liquids are determined by the impurity levels in the coal. For example, the percentage of sulfur in the liquids is usually about half the percentage of the sulfur in the coal. When the pyrolysis processes are carried out in fluidized bed or entrained flow reactors, the raw pyrolysis product is contaminated, typically with about 10 mass percent fine char that must be filtered out. Because of the high boiling character of the pyrolysis liquids, their most likely use after upgrading is for fuel oil.

COGAS and COED

Since our interest lies with synthetic fuels and not char production, we begin our discussion of pyrolysis technologies with a combined process that has one part designed to produce both char and a liquid fuel and a companion part to gasify the char as a step in producing SNG. The part of the process to produce char and synthetic liquid fuels, the latter by hydrotreating the pyrolysis liquids, is based on the technology of the multistage, fluidized bed COED process, an acronym for the Char-Oil-Energy-Development process.[1] The gasifier that is combined with the pyrolysis process to gasify the char is the COGAS char gasifier described in Section 4.3. The combined process to produce SNG and synthetic liquid fuels is called the COGAS process and is licensed by the COGAS Development Co., a consortium of natural-gas-industry firms and manufacturing companies.[12]

The Illinois Coal Gasification Group, a joint venture of a number of gas and electric utilities, under the sponsorship of the U.S. Department of Energy has prepared the conceptual design of a commercial three-train COGAS plant and the detailed design of a demonstration plant. A very simplified diagram for the ICGG conceptual commercial plant is shown in Figure 6.7.[12,13] The overall thermal efficiency is about 64 percent, as seen from the input and output calorific values. The proposed demonstration plant has an output about one-tenth that of the commercial plant and the units are about one-fourth scale. The plant will have a coal feed rate of 2000 t/d of Illinois No.5-6 bituminous coal, and produce about 650 000 m³/d of SNG and 275 000 L/d of liquid fuels. The liquid fuel mix is projected to consist of 178 000 L/d of No. 2 fuel oil, 39 000 L/d of No. 6 fuel oil, and 57 000 L/d of gasoline grade naphtha. Yields from the projected commercial plant are shown in Figure 6.7.

The COED process was conceived with the intention of converting coals to more useful fuels using the simplest and most developed technology. Its development was begun in 1962 by the FMC Corporation, which from 1970 to 1975 ran a 33 t/d pilot plant in Princeton, New Jersey. The core of the plant consisted of a fluidized bed dryer, followed by three indirectly heated, fluidized bed pyrolysis vessels in series that were staged to operate at progressively higher temperatures. The last pyrolyzer in turn was in series with a fourth fluidized bed reactor, in which a portion of the char was burned with a steam/oxygen mixture. The hot gases and the hot char from the combustion provided the heat for the pyrolyzers. The temperature was lowest in the first pyrolysis stage, where the feed coal entered. It increased progressively in each

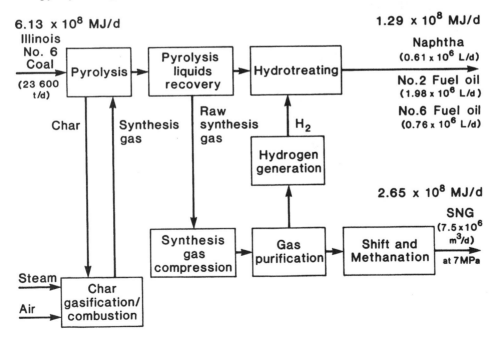

Figure 6.7 Simplified flow diagram for ICGG proposed commercial plant employing the COGAS process of pyrolysis and gasification.

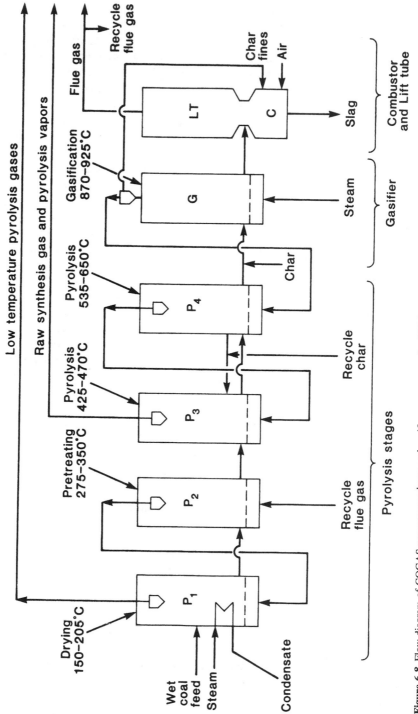

Figure 6.8 Flow diagram of COGAS process pyrolyzer and gasifier arrangement for ICGG demonstration plant.

succeeding stage, with the consequence that only partial devolatilization took place in each stage. The key to the process is the technique of partial devolatilization in one stage to enable higher temperature heating in the next stage without agglomeration and plugging of the fluid bed pyrolyzers.

Figure 6.8 is a flow diagram of the COGAS process combination of pyrolysis, gasification, and combustion that has been reported in U.S. Department of Energy progress reports, as projected for the ICGG demonstration plant. Each vessel represents a stage. The configuration is quite similar to that employed in the FMC pilot plant. In this regard, we note that the number of stages, the operating temperatures, and the residence times within the reactors vary with the agglomerating properties of the coal. The char combustion stage in the original FMC design is here replaced by the COGAS char gasifier and combustor which provide the heat for the sequence. The gasifier operates at a relatively low pressure of about 0.5 MPa, with the pyrolyzers run at about the same pressure. In the original FMC design, the operating pressure was closer to atmospheric, with a value around 0.15 MPa.

With reference to Figure 6.8, crushed coal of about 3 mm is transported to the first stage using a suitable carrier gas. The dried coal from the first stage flows to the second stage, where it is fluidized and further devolatilized by recycle flue gas from an auxiliary furnace. The temperatures in the first two stages are held below 350°C, which is the temperature at which active thermal decomposition begins (Section 3.1). The principal purpose of the first stage is to dry the coal, and of the second stage to provide a partial devolatilization to prevent subsequent agglomeration. The third and fourth stages are the true pyrolysis stages, where the dried and treated coal is fluidized and pyrolyzed by hot synthesis gas at temperatures high enough that active thermal decomposition takes place and char is produced. The char from the fourth stage feeds the gasifier, where the synthesis gas is produced. The interstage char recycle shown between the fourth and third stages is for purposes of transferring the sensible heat of the char, and temperature control. Cyclones inside the pyrolyzers separate out char particles that become entrained in the off gas.

In the FMC pyrolysis pilot plant operation, typical residence times ranged from 12 minutes in the drying stage to 10 minutes in the first stage pretreating pyrolyzer, to 7 minutes and 3 minutes, respectively, in the second and third stage pyrolyzers. The principal variables affecting the liquid yields are volatile matter content of the feed coal and the severity of oxidative pretreatment to prevent agglomeration.

Table 6.14 lists the properties of a typical COED pyrolysis liquid derived from an Illinois No. 6 bituminous coal in the FMC pilot plant runs. Also shown are the composition and properties of a synthetic crude oil obtained by hydrotreating the pyrolysis product. The synthetic crude is highly naphthenic, with the hydrocarbons composed of 10 percent paraffins, 40 percent naphthenes, and 50 percent aromatics.[14]

In Table 6.1 is given a typical product slate from the COED pyrolysis stages, as reported for the early FMC experiments with an Illinois No. 6 coal. The conversion efficiency of the pyrolysis process to these products is about 90 percent, when measured as a percent of the gross calorific value of the feed coal, adjusted for the energy input in the utilities.[1] From the product slate shown in Figure 6.7 and the overall thermal efficiency of 64 percent, we have constructed in Table 6.15 a simplified thermal

Table 6.14 Typical properties of COED liquids[14]

Mass %	Raw Pyrolysis Liquid	Synthetic Crude Oil
Carbon	79.6	87.9*
Hydrogen	7.1	11.3
Nitrogen	1.1	0.032
Sulfur	2.8	0.068
Oxygen	8.5	0.64
Ash	0.9	—
H/C atom ratio	1.07	1.54
Specific gravity, 15°C	1.11	0.88
GCV, MJ/kg	35.9	45.2

* Adjusted in accordance with FMC data.

balance for conversion of the pyrolysis products to SNG and refined liquids. The lumped thermal efficiency of 53.6 percent listed for converting the char to SNG absorbs all of the plant requirements, including gas cleaning and purification, hydrogen production and hydrotreating. This "efficiency" is derived from forcing a balance. It is nevertheless of interest that this efficiency is at the upper end of the range of cold gas efficiencies reported for the COGAS pilot plant.[12] Again, however, we emphasize that the criterion of efficiency alone is insufficient to justify this or any other process and that process complexity, reliability, and economics must also be considered.

Table 6.15 Thermal balance for ICGG demonstration plant normalized to 100-kg output from pyrolysis section

	MJ/kg	kg	MJ
FROM PYROLYSIS			
Char	25.7	62	1591
Raw pyrolysis liquid	34.7	21	729
Raw pyrolysis gas	39.6	14	555
Water	—	3	—
Pyrolysis losses (at 90% eff.)	—	—	319
Total		100	3194
FROM PLANT			
Refined liquids	45.1	14.9	672
SNG from char (at 53.6% eff.*)	53.1	16.1	853
Pyrolysis gas as SNG†	53.1	9.9	526
Losses (by difference)	—	—	1143
Total		40.9	3194

Overall thermal efficiency $= \dfrac{2051}{3194} \times 100 = 64\%$

* Lumped "efficiency" derived from forcing balance.
† Raw pyrolysis gas calorific value minus calorific value of H_2S.

Occidental Flash Pyrolysis

The Occidental Flash Pyrolysis process is an indirectly heated process which uses a stream of hot product char to provide the heat for the pyrolysis of the coal.[15,16] Pulverized and dried coal of less than 250 μm is transported pneumatically in an inert gas such as nitrogen or carbon dioxide to an entrained flow pyrolysis reactor sketched in Figure 6.9. There it is mixed in a ratio of about 10:1 with recycled char that has been heated by partial combustion with air in an entrained bed furnace at temperatures from 650 to 980°C. The combustion gases are separated from the char in cyclones and the hot char is fed to the reactor in a dense-phase fluidized flow, where the mixing with the coal takes place in a turbulent jet at 0.3 MPa. The coal is rapidly heated to the reaction temperature of 510 to 730°C. Residence times are from 1 to 3 s.

The pyrolysis products are passed through a series of cyclones where the char is separated from the vapors. The vapors are then quenched to about 100°C by direct contact with recycle oil. Condensed tar products are subsequently recovered from the recycle oil in a vacuum flash unit, and the effluent gases from the cyclones are cooled by water scrubbing for recovery of light oils. As in the COED process, the pyrolysis products are undiluted by air, so the product gases have a high calorific value.

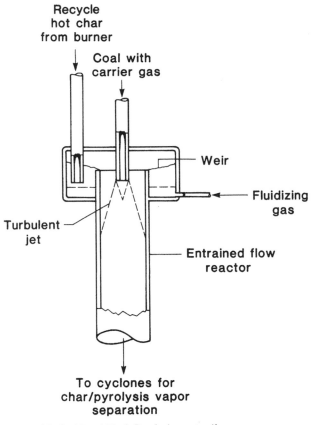

Figure 6.9 Occidental Flash Pyrolysis reactor.[16]

Agglomeration of caking bituminous coals inside the reactor is prevented by sizing the reactor for high heating rates and by using small particles, so that when they travel downward in the mixing jet they pass through the plastic regime before striking the wall.

More importantly, however, is that a tar yield of as much as 35 percent by mass of a caking bituminous coal (daf) has been reported for operation in a nominal 3 t/d process development unit at feed rates of about 1.7 t/d. This yield was obtained for reaction times of 1 to 2 s at temperatures of 590 to 650°C. The yield is almost twice the Fischer assay value, and is a consequence of the rapid pyrolysis and subsequent quenching. Char and gas yields were not reported for the tests, but values obtained for corresponding liquid yields in small-scale bench tests are given in Table 6.1.

Data and kinetic studies indicate that the liquid yield is consistent with a fast pyrolysis into a primary tar, on the order of a fraction of a second, and subsequent cracking into smaller gaseous molecules until quenching is complete. Little loss in total yield occurs during the cracking phase. The liquid products have been hydrotreated, and analysis shows that both the raw pyrolysis liquids and the finished products are comparable with those from the COED process.

Rockwell Flash Hydropyrolysis

It is pointed out in Section 3.1 that coal is highly reactive with hydrogen during pyrolysis, and that pyrolysis in the presence of hydrogen can lead to an increase in volatile yield which may be attributed to the hydrogenation of free radical fragments. To achieve high liquid yields, Rockwell International is developing a flash hydropyrolysis process in which pulverized coal is rapidly mixed with hot gaseous hydrogen, resulting in fast pyrolysis.[17] Rapid mixing is achieved through the use of a rocket engine-type injector for the coal and hydrogen. The pyrolysis products are quenched to avoid secondary reactions, as in the Occidental process. In a commercial coal liquefaction plant using such a reactor, the product char and coal would be used to generate the hydrogen by gasification. Cities Service Co., together with Rockwell, is working to commercialize this process.

The Rockwell reactor has been demonstrated in a 0.25 kg/s (22 t/d) process development unit and been shown to function smoothly with caking bituminous coals. The coal is pulverized to 70 or 90 percent less than 75 μm, and fed into the reactor in a dense-phase flow where it is mixed with high temperature hydrogen at from 3.5 to 10.5 MPa. The reactor is operated typically below 1100°C and, to prevent agglomeration, above 815°C. Because the fast pyrolysis process is close to thermoneutral, the temperature of the feed hydrogen must be above the desired reactor temperature.

Thorough and rapid mixing in the reactor is achieved by impinging several jets of hydrogen on a jet of coal. The reactors tested in the process development unit have been tubes ranging from about 4 to 14 cm in diameter by 1.5 m long, which is small in light of the high throughput. The coal residence times ranged between about 20 and 200 ms. The pyrolysis reactions were quenched by cooling to around 480°C by spraying water into the reaction products. After separation of the char, the pyrolysis vapors were then reduced in pressure to 0.5 MPa and cooled to condense the liquid product. From a thermal efficiency point of view, water quenching is not desirable, and a heat exchange quench unit that recovers the heat and rapidly quenches is under development.

Yields and carbon conversions from a wide range of tests in the 0.25 kg/s unit and in a smaller 0.06 kg/s unit are shown in Figure 6.10. The scatter is attributed to variations in injector mixing. In most cases, however, from 30 to 40 percent of the carbon in the coal was converted to liquid products. Approximately 30 mass percent of the liquid product was a low-sulfur, low-nitrogen content naphtha, which was largely benzene with lesser amounts of toluene and xylene. The liquid yield was found to be insensitive to reactor pressure in the range 3.5 to 7 MPa, although the overall conversion and the gas fraction did decrease somewhat with decreasing gas pressure.

The main attractions process-wise of pyrolysis are that in the preliminary step it avoids the need for expensive hydrogen and the need for high pressure equipment. Although the results of the Rockwell tests are encouraging, these particular advantages are lost. They are replaced, however, by a very high throughput capability in a small reactor and with a higher quality liquid yield. Clearly, detailed process and economic considerations will have to be put forward to properly evaluate the process.

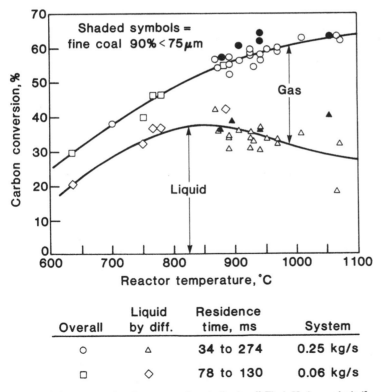

Overall	Liquid by diff.	Residence time, ms	System
○	△	34 to 274	0.25 kg/s
□	◇	78 to 130	0.06 kg/s

Figure 6.10 Yields and carbon conversions in Rockwell Flash Hydropyrolysis.[17]

Supercritical Gas Extraction

In Section 6.1 the concept of supercritical gas extraction is outlined, wherein use is made of the power of a compressed gas that is near but somewhat above its critical temperature, to extract substances that would normally not distill at that temperature.[18] Coal pyrolysis liquids evolved at 400°C do not volatilize to any extent, while at higher temperatures they decompose. However, if the pyrolysis is carried out in the presence of an extractive supercritical gas, the low-molar-mass liquids formed by thermal depolymerization are soluble in the gas phase and can be removed directly as they are formed. Increases in volatility of up to 10 thousand times can be realized. The development of this process has been carried out mainly by the National Coal Board in England, where a 120 kg/d continuous plant has been operational since 1977, with operation of a 25 t/d plant scheduled for the end of 1982.[19,20]

The extraction gas should preferably have a critical temperature at or somewhat below the pyrolysis temperature, for example, in the range 300 to 400°C and consequently be a liquid at ambient temperature. Tests by the National Coal Board have shown that it is not essential for the extracting fluid to be supercritical to be effective; however, the advantages of operating in a supercritical state are control of density and lower viscosity of the fluid compared with a liquid. Extractants that are supercritical in the range 300 to 400°C include common liquids derived from coal tar or petroleum naphtha fractions. One substance that has proved suitable is toluene, whose critical temperature is 319°C and whose critical pressure, that is the saturation pressure corresponding to the critical temperature, is 4.1 MPa. Toluene is stable under extraction conditions and does not cause the coal to cake.

The volatilized liquids can be removed from the extractant in a number of ways, the most straightforward of which is to transfer the gas phase from the pyrolysis reactor, where the pressure is high, to another vessel at lower pressure, in particular atmospheric pressure. This reduces the density of the solvent gas together with its solvent capabilities, and the extracted tar precipitates out.

The product is a low-melting glassy solid, with a softening temperature of 50°C reported for a Wyoming Wyodak subbituminous coal and 70°C for high-volatile U.K. bituminous coal, which is similar to Illinois No. 6 coal. The extract represents the hydrogen-rich fraction of the coal, consistent with our discussion in Section 3.1 of the volatilized material that is evolved on pyrolysis. It is essentially free of mineral matter and solvent, and contains less nitrogen and sulfur than the coal. The mean molar mass for an extract derived from a Wyodak subbituminous coal is reported to be about 330, and for extract derived from a high-volatile U.K. bituminous coal about 500. By hydrotreating, the extract can be converted relatively easily to distillable oils.

The residue, which contains virtually all the mineral matter, is a porous char with a reactivity and calorific value similar to that of the coal from which it is derived. It is different from the char evolved in normal pyrolysis in that it retains much of the volatile content of the original coal.

A typical yield for a high-volatile U.K. bituminous coal, using toluene as the extract, is given in Table 6.1. It is reported that an extract yield as high as 47.5 percent based on daf coal is within the process capabilities.[20] A reported yield for a Wyodak coal, though one that was not optimized, is about 20 percent extract, 70 percent char, and 10 percent gas and water. The yields are little affected for particle sizes below about 2 mm. Because of the relatively large pore sizes of the coal particles, typically 0.5 to 1 μm, the particles are easily penetrated by the supercritical solvent molecules and the extraction rate is found to be insensitive to particle sizes up to about 3 mm.[19]

The effect of pressure and temperature on yield is substantial. For coal/supercritical toluene systems studied by the National Coal Board (NCB), extraction pressures of about 10 MPa, with pyrolysis temperatures in the range 400 to 440°C, were typical. However, both higher pressures and temperatures were used, with the 120 kg/d facility designed for temperatures up to 500°C and pressures to 40 MPa.

Coal residence time was found to have a strong effect on extract yield. In the NCB 120 kg/d facility, a fluidized bed reactor was used with coal residence times in the reactor preferably at least 30 minutes, although the gas residence times were kept as short as possible, generally less than 2 minutes. Some studies were carried out using an entrained flow reactor in which short residence times of less than or around 10 minutes were employed with encouraging results.

Figure 6.11 is a simplified flow diagram of the supercritical gas extraction process as proposed for a commercial-scale design.[19] The coal is pulverized, dried, and heated to a temperature of 315°C and then charged to the feed lock hopper. Hot pressurized recycle toluene is used to pressurize the feed lock hopper before the coal is charged to the top of the moving bed reactor where the extraction takes place. For this type of reactor, high-volatile, low-caking coals are preferred to avoid equipment blockage caused by agglomerating particles. The feed is assumed to be a Wyodak subbituminous coal, and for a design input of about 9000 t/d of dried coal, eight moving bed reactors operating in parallel are used. The toluene, heated to 440°C and pressurized to 10.4 MPa, enters the bottom of the reactor and exits at the top with the extracted product. The moving bed of residue exits the bottom of the reactor and enters a lock hopper from which it is discharged after it is filled. The need to withdraw a stream of residual solids from the system across a high pressure differential poses a number of mechanical engineering problems. Toluene which is above its boiling point comes off overhead from the hopper with residual toluene purged by steam stripping.

The effluent from the extractor is condensed and the pressure is then reduced, causing precipitation of the extract as a liquid in toluene. The hydrocarbon gases are drawn off overhead. The toluene, extract, and product water are all liquid and are separated by heating in a flash tank and then further separated as shown in Figure 6.11. Although not shown, part of the char is gasified with oxygen in a Koppers-Totzek gasifier to produce fuel gas for firing the toluene heaters and other process needs. This gas is supplemented by the hydrocarbon gases evolved in the pyrolysis. The tar extract is the primary product and the char that has not been gasified is a byproduct.

The design is based on a supercritical gas extraction yield as a mass percent of dry coal of 21.2 percent extract, 67.5 percent char, and 5.8 percent gas. A simplified overall thermal balance for this plant is given in Table 6.16. The heat requirement to manufacture the oxygen is about 4 MJ/kg of oxygen distilled (Section 5.3). The overall conversion efficiency for the process is seen to be quite high, at about 94 percent. The

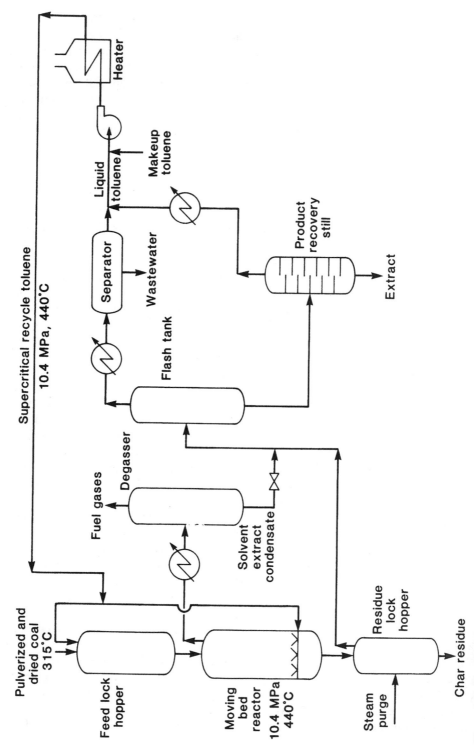

Figure 6.11 Simplified flow diagram of supercritical gas extraction process for commercial-scale design.[19]

290

products are, however, still not final products, with most of the energy contained in the char, as is characteristic of pyrolysis outputs. However, mention may be made of an NCB design in which an extract yield of 47.5 percent is said to be feasible.[20] Here the char residue is just sufficient to satisfy the power needs to hydrotreat the extract and turn it into a finished product slate of high-quality liquids and a small amount of SNG. An overall thermal efficiency of 68 percent is projected, which is interesting because it is quite high.

Of the pyrolysis processes, the supercritical gas extraction process appears to have a number of advantages. It maintains the main advantages of simple pyrolysis, despite the high pressures, since the extractant is compressed as a liquid and not as a gas. Moreover, the solvent is virtually all recoverable, the separation of the residue from the tar extract is not difficult because of the low viscosity of the supercritical phase, and the hydrogen-rich character of the extract structure makes it a material that is easily hydrocracked to substitute petroleum products at a relatively high efficiency.

6.4 DIRECT LIQUEFACTION

The major elements in processes for the direct liquefaction of coal have been shown to be similar both from the point of view of fundamentals, as discussed in Section 3.4, and from a process viewpoint, as outlined in Section 6.1. For a given coal rank and type, among the more important factors affecting the level of hydrogenation and the related yield and product distribution are the catalyst and the operating variables—pressure, temperature, and residence time. The interaction among some of these variables can be seen from the illustrative yields shown in Table 6.2.

Table 6.16 Thermal balance for supercritical gas extraction plant feeding 10 660 t/d of as-received Wyodak subbituminous coal[19]

	Heat Content, MJ/kg	Mass Flow, kg/s	Heat Flow, MJ/s
IN			
Coal (as-received)	22.1	123.4	2727
Oxygen separation	4.0*	13.6	54
Total		137.0	2781
OUT			
Extract	37.5	22.4	840
Char	34.0	52.1	1771
Losses (by difference)	—	—	170
Total		74.5	2781

Overall thermal efficiency $= \dfrac{2611}{2781} \times 100 = 94\%$

* Heat requirement to manufacture oxygen.

Process-wise, two methods of hydrogen addition have been distinguished—hydroliquefaction and solvent extraction. In hydroliquefaction, the hydrogen is added directly from the gas phase in the presence of a suitable catalyst. In solvent extraction procedures, a coal-derived liquid, which may or may not be separately hydrogenated, transfers the hydrogen to the coal without external catalyst addition.

Direct liquefaction processes may be further categorized by whether the products are produced in one or two stages. In two stage processes, the coal is first hydrogenated in a "liquid phase" stage transforming it into a deashed, liquid product and then in a second "vapor phase" hydrogenation stage, the liquid product is catalytically cracked to produce clean, light distillate fuels using conventional petroleum refining technology.[21] The reasoning behind this approach is much the same as that for making SNG by first producing a low-H_2/CO-ratio gas in a gasifier, and then shifting it and methanating it in separate units, rather than trying to gasify and shift in one reactor. The argument is that each operation can be optimized more closely with a higher overall efficiency than can a combined operation in a single unit. In liquefaction, optimal conditions and intrinsic selectivities are different for the liquid-phase stage, which is principally thermal, and the vapor-phase stage, which is principally catalytic. On this basis, it is argued that greater overall efficiency and product slate flexibility may be achieved by separately optimized stages, in place of carrying out dissolution, deashing, desulfurization, and hydrocracking to gases and distillate fuels under severe hydroprocessing conditions in a single reactor stage.[22,23]

It is not without interest that the Bergius-IG hydroliquefaction process,[21] which was operated so successfully in Germany before and during World War II, was a two stage process. The products of the first stage were middle distillate oils boiling below 325°C. These oils were then catalytically cracked to motor fuels and light hydrocarbons in a vapor-phase hydrogenation stage. For the liquefaction of bituminous coals, quite severe operating conditions were employed in the first stage, with pressures of from 30 to 70 MPa and temperatures from 475 to 485°C. It has been argued recently that the high pressures were necessitated by, among other things, the fact that German coals are much more difficult to liquefy than, for example, U.S. coals. The IG process was carried out using disposable iron oxide catalysts. In the liquid-phase stage from 92 to 97 percent of the coal was liquefied; however, the residence times were quite long, about 1.4 h, and the hydrogen consumptions quite large, about 11 percent by mass of the daf coal. For lignites, operating conditions were somewhat less severe and the hydrogen consumptions lower. The vapor-phase catalytic cracking of the products was carried out typically in downflow reactors over fixed beds of tungsten sulfide catalysts at about 3 MPa and 400°C with residence times of 2 to 4 minutes.

The German liquid-phase technology, although not considered economic today, has nevertheless served as a model for most of the direct liquefaction processes presently under development. Efforts have been directed toward reducing the severity of the operating conditions. This has been achieved principally by utilizing the hydrogen-donor capabilities of the slurry, rather than just relying upon the direct reaction between the coal and molecular hydrogen as in the German plants. The aim has been to maximize the liquid yield, improve the product slate, and minimize the gaseous products with the least consumption of hydrogen since the conversion efficiency correlates most directly with the hydrogen consumed. As we discussed in Section 3.4, the amount of hydrogen required to convert a unit mass of coal to a desired product cannot be specified by straightforward stoichiometry because of the co-production of a large number of byproducts including carbon oxides, gaseous hydrocarbons, water, hydrogen sulfide, and ammonia.

Like pyrolysis liquids, direct liquefaction liquids are also naphthenic and aromatic. The boiling ranges are partly dependent on the coal feed and can be changed by the processing conditions. Direct liquefaction makes a much lighter product slate than pyrolysis, at about one-quarter to one-half the impurity level. The naphtha from direct liquefaction processes reforms easily to high octane levels at high yields, although it must be hydrotreated to remove sulfur and nitrogen. The mid-distillate material that is produced is suitable for home heating oil, while the high-boiling material is useful mainly as fuel oil.

Thermal efficiency or even product slate are not the only criteria. Lower capital and operating costs are important, as is greater reliability. Less severe reaction conditions can help. However, there are still many other problems that require either solution or development, including the maintenance of catalyst activity and improved catalyst performance, solids separation from high viscosity fluids, efficient processing of distillation bottoms and slurry residues, reduction of reactor residence times, and improved high pressure letdown systems.

Solvent Refined Coal

The Solvent Refined Coal or SRC process, where the coal is largely dissolved in a solvent that donates hydrogen to the coal in the absence of any added catalyst, gives a product which may be a solid or liquid at room temperature depending upon the amount of hydrogen donated. There are several versions of the process and the one called SRC-I is a single-step procedure in which the hydrogenation is mild and the principal product is an essentially ash-free, low-sulfur, pitch-like extract that is a solid at room temperature. The "solvent refined coal" can, for example, be used directly as a boiler fuel without the need for pollution control equipment that would otherwise be required were the original coal burned directly. The SRC product can also be upgraded by catalytic hydrocracking to light hydrocarbons in a second stage. There are two procedures under preliminary development that employ this two stage liquefaction process. One of the processes, which uses a short-contact-time dissolution stage plus catalytic upgrading, has been dubbed SRC-$\frac{1}{2}$.

A later version of solvent refining that is in an advanced development stage is the SRC-II process. It is like SRC-I, except that a portion of the product solution is recycled and used as feed slurry with the aim of increasing the conversion to lighter products.

Forerunners of the SRC-I process were the Pott-Broche and Uhde-Pfirrman processes, both of which were developed in Germany at about the same period as the Bergius-IG process.[21] The Pott-Broche process dissolved the coal in a hydrogen donor solvent at 10 to 15 MPa and 415 to 430°C for about an hour. The product was filtered and distilled to yield an extract with a softening point of 210 to 220°C. The Uhde-

Pfirrman process added gaseous hydrogen to the coal-solvent mixture and carried out the dissolution at about 30 MPa and 410°C. The extract contained more hydrogen, and its softening point, 60 to 120°C, was lower than that obtained with solvents alone.

Bench-scale work on the SRC-I process, then just the SRC process, was begun in the United States in 1962 by the Spencer Chemical Co. In 1965 Spencer was acquired by Gulf Oil Corp. and the SRC work was continued by Pittsburg & Midway Coal Mining Co., a subsidiary of Gulf. Under sponsorship of the U.S. Office of Coal Research, now absorbed in the U.S. Department of Energy, a 45 t/d pilot plant was constructed at Fort Lewis, Washington. It began operation in the SRC-I mode in 1974. In 1977 it was converted to the SRC-II mode and subsequently has been operated in both modes. Another pilot plant running in the SRC-I mode was also put into operation in 1974, this one in Wilsonville, Alabama, under management of the Southern Company Services, Inc., with support from the electric utility industry. The nominal coal feed capacity of the plant was 5.4 t/d.

At the time of writing, the U.S. Department of Energy has entered into a cost-sharing agreement for the design of an SRC-I demonstration plant to be constructed in Newman, Kentucky by the International Coal Refining Co. The nominal processing capacity of the plant is 5400 t/d of coal and the primary product will be a clean solid fuel. Startup is scheduled for 1984. Pittsburg & Midway has also been contracted to construct a 5400 t/d SRC-II demonstration plant near Morgantown, West Virginia, with the primary product to be a clean fuel oil. Startup is also scheduled for 1984, but whether either or both of these demonstration plants is built remains dependent on U.S. Government decisions. It is of interest that each of these demonstration plants is designed to process 50 percent more coal than the largest of Germany's bituminous coal processing plants operated during World War II.

Although the SRC-I and SRC-II processes have many common features, there are sufficient distinctions to warrant our discussing them separately. Figure 6.12 is a simplified flow diagram of a commercial-scale SRC-I plant comparable to the configuration for the proposed demonstration plant.[22,24] Raw coal is initially dried to about 1 to 2 percent moisture, pulverized to less than 3 mm, and mixed with filtered recycle solvent from the process, that boils in the general range 290 to 425°C. The recycle solvent-to-coal mass ratio is about 1.5, and this slurry is pumped together with hydrogen at from 10 to 14 MPa to a fired preheater where it is heated to 400 to 450°C, after which it enters the dissolver. The dissolver is a vertical entrained flow reactor in which there are three phases flowing cocurrently upward. The reaction is slightly exothermic and there is about a 15°C temperature rise in the dissolver, with the temperature rise somewhat dependent on the hydrogen consumed. The temperature in the dissolver is governed mainly by the temperature at the exit of the preheater, so that for a 425°C temperature at the preheater exit, the dissolver temperature is about 440°C. The residence time in the dissolver is about 30 minutes and under these conditions most of the carbonaceous material dissolves, with values of 91 to 93 percent typical.

From the dissolver the mixture passes to a vapor/liquid separator where it is cooled to about 290°C and the unreacted hydrogen is separated from the hydrocarbon gases, water, and hydrogen sulfide. The pressure in the slurry of undissolved solids and coal solution is then let down to about 1 MPa with recovery of light oils prior to passing

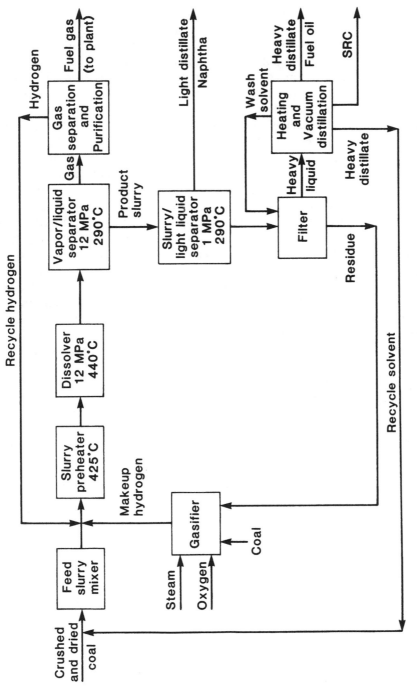

Figure 6.12 Simplified flow diagram of SRC-I process for a commercial-scale design with hydrogen production by gasification.

the slurry through a filter. We would point out that although pressure filtration employing leaf filters has been used with some success, solids separation still remains a problem area in SRC manufacture. This is because the reactor product is very viscous, the solids concentration is high, and separation has to be carried out at elevated temperatures, usually from 250 to 300°C. At these temperatures, pressures up to 1 MPa have to be applied to keep the solvent in a liquid state.

Two alternative procedures to pressure filtration are being pursued, one of which is the use of promoted gravity settling through the addition to the reactor effluent of a solvent that agglomerates the solid dispersed matter. This procedure has been developed by the C-E Lummus Company as a means of removing ash from coal solutions and has been called by them an "antisolvent" deashing technique. It has been applied to SRC solutions.[25] Solvents that cause agglomeration are paraffinic. Ones that are used, and kerosene is most typical, have a boiling range from 215 to 260°C to facilitate recovery. Although the agglomerated particles can be separated by sedimentation, this still must be carried out at 250 to 300°C and under pressure.

The second procedure, developed by the Kerr-McGee Corp., is termed "critical solvent deashing."[41] The principle used is similar to that of supercritical gas extraction (Section 6.3), in that it employs the increased dissolving power of a solvent near its critical temperature and pressure. The solvent is mixed with the ash-containing coal solution, and the mixture flows to a series of settlers. The ash and undissolved coal separate as a heavy fluid phase. Subsequent separation of the coal liquid from the solvent is accomplished by heating the mixture, which decreases the solvent density, causing rejection of the coal liquids as a heavy fluid phase.

As shown in Figure 6.12, with filtration employed, separation of the recycle solvent and the SRC extract is by vacuum distillation. The solvent refined coal has a solidification temperature between 175 and 200°C and is solidified by cooling. It generally contains less than 0.2 mass percent ash and 1 mass percent sulfur, and has a heating value of about 37 MJ/kg. Two representative analyses of solvent refined coals derived from eastern bituminous coals are shown in Table 6.17. Perhaps the most striking feature of these results is that there is actually a *decrease* in the hydrogen

Table 6.17 Analyses in mass percent of solvent refined coals derived from two eastern bituminous coals[5]

	Pittsburgh		Kentucky	
	Coal	SRC	Coal	SRC
C	75.1	88.4	70.6	88.5
H	5.1	5.5	5.0	5.1
N	1.3	1.7	1.4	1.8
O	7.6	3.4	9.2	3.7
S	2.6	0.9	3.4	0.8
Ash	8.3	0.1	10.4	0.1
Total	100.0	100.0	100.0	100.0
Molar Rep.	$CH_{0.82}$	$CH_{0.75}$	$CH_{0.85}$	$CH_{0.69}$
GCV, MJ/kg	31.6	37.2	29.8	36.7

Table 6.18 Typical SRC-I process yield for Kentucky bituminous coal in mass percent of daf coal fed to dissolver[22]

	Mass %	MJ/kg
Consumed		
Hydrogen	2.3	141.8
Total	2.3	
Produced		
Fuel gas (C_1–C_4)	6.9	52.9
Naphtha (C_5–175°C)	4.9	44.6
Fuel oil (175–455°C)	11.7	40.5
SRC	60.1	36.7
Unconverted carbon	7.2	32.8
Hydrogen sulfide	2.0	16.6
Carbon dioxide	1.1	—
Water	6.0	—
Ammonia	0.1	—
Total	100.0	

content in the solvent refined coal compared to that in the original coal, as shown by the molar representation. As we have noted, the main uses of hydrogen in the process are in the removal of oxygen from the coal by conversion to water, the removal of sulfur by conversion to hydrogen sulfide, and the hydrogenation to liquid and gaseous hydrocarbons of the carbon in the coal that is not converted to the solvent refined coal or carbon oxides.

Table 6.18 summarizes the typical process yield structure based on data from the Wilsonville pilot plant using a Kentucky bituminous coal. The nitrogen in the liquid products is about 1 percent, which is quite high compared to comparable petroleum products, but the sulfur content is relatively low, averaging about 0.3 percent. The yield is that expected from the 5400 t/d demonstration plant. In that plant all of the fuel gas produced will be burned for process needs. The unconverted carbon together with additional coal will be gasified in Koppers-Totzek gasifiers to produce hydrogen. The calorific value of the additional coal may be estimated assuming a 65 percent efficiency of conversion to hydrogen from the relation

$$\text{CV of gasifier coal} = \left(\frac{\text{CV of H}_2 \text{ consumed}}{0.65} \right) - \text{CV of unconverted carbon} \quad (6.10)$$

Detailed material and thermal balances for the SRC-I demonstration plant show a thermal efficiency of about 73 percent for conversion to the SRC, fuel oil, naphtha slate of Table 6.18.[22] If an efficiency is calculated on the basis of the yield structure of Table 6.18 and the added coal requirement given by Eq. (6.10), a value of about 80 percent is obtained. The difference is attributable to purchased electric power at a penalty of about 10 MJ/kWh.

SRC-II is a modification of SRC-I to produce a liquid product with byproduct gases. The means by which this is accomplished is to split the product slurry, which

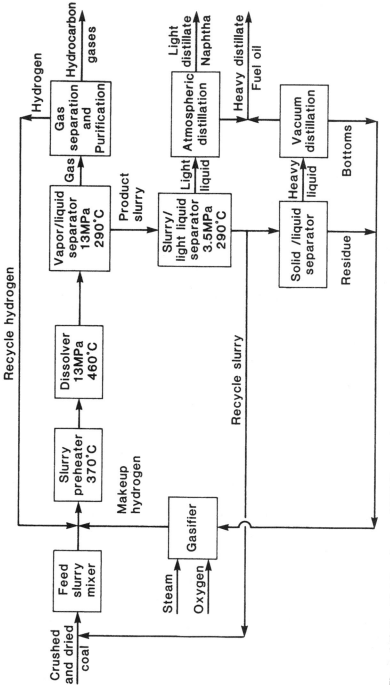

Figure 6.13 Simplified flow diagram of SRC-II process for a commercial-scale design with hydrogen production by gasification.

contains dissolved and undissolved coal, and recycle part of it back to the feed mixer. The splitting may be seen in Figure 6.13, which is a flow diagram of a commercial-scale SRC-II plant comparable to the proposed demonstration plant configuration. The recycle increases the conversion to lower-molar-mass products in two main ways. First it increases the reaction time for conversion, and second the recycling tends to concentrate the coal mineral matter which is believed to catalyze the reaction in the dissolver. Therefore, for the same residence time the degree of hydrogen consumption is increased in the SRC-II mode. In addition, the dissolver is operated at somewhat higher temperatures and pressures, and with residence times somewhat more than twice that for SRC-I, or about an hour.

An important feature of the SRC-II plant is that filtration is not required since the residue slurry liquids from fractionation and distillation are sent to a gasifier where synthesis gas is generated to produce hydrogen. Gasification with Texaco gasifiers is planned for the SRC-II demonstration plant.

In Table 6.19 is shown a typical process yield, based on data from the Fort Lewis pilot plant, for the same Kentucky coal as was used for the tests characterizing the SRC-I yield in Table 6.18. It can be seen that the hydrogen consumption is more than doubled. A detailed material and thermal balance has been carried out for a plant producing only liquids and gas, in which all the SRC is consumed for process needs.[22] In addition to a requirement for imported power in about the same amount as for the SRC-I plant about 10 percent of the product slate must also be burned, leaving about 90 percent of the fuel gas, naphtha, and fuel oil yield shown in Table 6.19 as the exportable product. The overall thermal efficiency for this conversion is reduced to about 64 percent. The lowered efficiency can be correlated with the increased hydrogen consumption. An alternative to the single stage approach is the two stage operation discussed earlier.

Table 6.19 Typical SRC-II process yield for Kentucky bituminous coal in mass percent of daf coal fed to dissolver[22]

	Mass %	MJ/kg
Consumed		
Hydrogen	4.8	141.8
Total	4.8	
Produced		
Fuel gas (C_1–C_4)	17.6	52.6
Naphtha (C_5–175°C)	13.0	44.6
Fuel oil (175–455°C)	25.8	40.5
SRC	26.5	36.7
Unconverted carbon	6.3	32.8
Hydrogen sulfide	2.5	16.6
Carbon oxides	2.0	4.6
Water	5.7	—
Ammonia	0.6	—
Total	100.0	

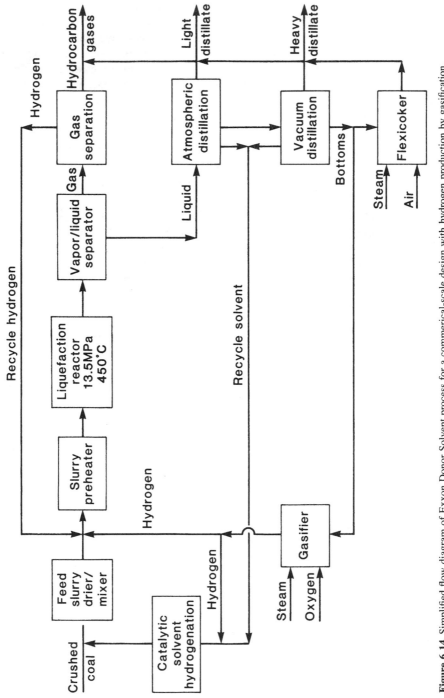

Figure 6.14 Simplified flow diagram of Exxon Donor Solvent process for a commerical-scale design with hydrogen production by gasification.

Exxon Donor Solvent

The key element of the Exxon Donor Solvent or EDS process[26-28] is that the spent solvent is first catalytically hydrogenated in a separate fixed bed reactor prior to its use as a hydrogen donor solvent in the dissolver. A 225 t/d pilot plant was started up in Baytown, Texas, in 1979 by Carter Oil Co., an affiliate of Exxon. It is sponsored by a number of companies and the U.S. Department of Energy under a cooperative agreement.

Figure 6.14 is a simplified flow diagram of a commercial design embodying the EDS process. In this design, it is assumed that the hydrogen is generated by partial oxidation of vacuum bottoms using high temperature gasifiers, the Shell-Koppers having been one considered by Exxon. In the "base" commercial design, hydrogen production by steam reforming of product methane, ethane, and propane was assumed. If it is desirable to export these gases, then hydrogen production by gasification may be employed. For comparison with the other direct liquefaction designs discussed we have indicated the latter option.

With reference to Figure 6.14, feed coal is crushed and then dried by mixing with the hot hydrogenated recycle solvent. The coal-solvent slurry is fed to the dissolver, which as in the SRC designs is simply a vertical, cocurrent upward, entrained flow reactor. Photographs of the reactors in the 225 t/d pilot plant are shown in Figure 6.15. The design temperature is from 425 to 480°C at pressures from 13.5 to 17 MPa. Optimum residence times are about three-quarters of an hour. The reactor product is separated by flashing, and atmospheric and vacuum distillation. The bottoms from the vacuum distillation contain the heavy liquids, unconverted coal, and ash.

Part of the 200 to 425°C fraction of the C_4–540°C distillate is taken as the recycle solvent. The solvent is hydrogenated in conventional fixed bed catalytic hydrotreaters prior to its introduction into the drier/mixer.

In the design shown, part of the vacuum bottoms is used for gasification and the remainder is fed to integrated coking and gasification reactors employing circulating fluidized beds. The integrated coking/gasification units constitute a commercial Exxon petroleum refining process termed "Flexicoking." The purpose of the Flexicoking operation is to produce additional liquid products and a low-CV gas, in order to convert essentially all of the feed carbon. The liquids produced in the coking operation are heavier ones with boiling points above 345°C. The operating pressure of the Flexicoker is 0.35 MPa, with a temperature of 480 to 650°C in the coker and 815 to 980°C in the gasifier.

EDS testing has been carried out on a wide variety of coals, and in our earlier discussion on direct liquefaction we showed in Figure 3.16 the results from a 23 kg/d bench-scale liquefaction unit, to illustrate the variation in liquid yields for different coal types and residence times. Longer residence times at first increase the conversion to coal liquids but then lead to secondary hydrocracking and conversion to gas. The optimal times for liquid production ranged from about 30 to 60 minutes, depending upon the coal. However, quite different liquid yields were obtained for very similar coals and it was indicated that the reason for this was not clear. Table 6.20 summarizes the product-yield distributions reported for the bench-scale liquefaction tests at optimum

(a) (b)

Figure 6.15 EDS 225 t/d pilot plant, Baytown, Texas. (*a*) Four coal liquefaction reactors operated in series; each reactor is 0.61 m inside diameter and 16.8 m long. (*b*) Liquefaction reactors as seen from top with coal-feed hopper in center of photograph and coal storage silo in right background. (*Courtesy of Exxon Research and Engineering Co.*)

Table 6.20 Product distributions from EDS liquefaction reactor in mass percent of daf coal at optimum liquid yields[26]

	Illinois No. 6, Bit.	Wyoming Subbit.	Texas Lignite
Consumed			
Hydrogen	4.3	4.6	3.9
Total	4.3	4.6	3.9
Produced			
Gas (C_1–C_3)	7.0	8.9	8.8
Liquids (C_4–540°C)	37.2	31.8	32.0
Residuum (>540°C)	40.1	37.1	36.7
H_2S + NH_3	4.0	0.9	1.6
CO_x + H_2O	11.7	21.3	20.9
Total	100.0	100.0	100.0

liquid yields for three different coals. The lower-rank coals have lower liquid yields, although the difference is not large.

Of interest in the commercial design is that Exxon has shown in their bench-scale and small pilot plant tests that at optimum liquid yield conditions in the dissolver, the *combined* liquid yield from the dissolver *plus* the Flexicoker is roughly a constant for similar coals. A total liquid yield of about 43 to 45 percent is obtained for bituminous coals, 40 to 42 percent for subbituminous coals, and about 37 percent for lignite.

An interesting result reported from the 900 kg/d pilot plant tests is that younger subbituminous coals and lignite are more difficult to process than bituminous coals. This is related to the higher liquid viscosities of the liquefaction bottoms. Although recycling of the vacuum bottoms produces a lighter product slate, it also increases the residence time and leads to hydrocracking of the light liquids to gas, as we have already noted. Evidently further optimization is required through tests. However, in their "base" design with steam reforming of the product gases to produce the needed hydrogen, Exxon has reported an overall thermal efficiency of 63 percent for a 43 percent conversion to liquids. Higher liquid yields will produce higher conversion efficiencies.

H-Coal

The H-Coal process is a hydroliquefaction procedure in which hydrogen is added to the coal in the presence of catalyst.[24,29] The introduction of a catalyst distinguishes this process from the preceding three we have discussed. It is being developed by Hydrocarbon Research, Inc. At the time of writing, a pilot plant is going into operation at Catlettsburg, Kentucky, that is capable of processing 530 t/d of dry coal to about 215 $\times$ 10^3 L/d of low-sulfur fuel oil, or 190 t/d to a synthetic crude consisting principally of naphtha and middle distillates. The reason for the difference in feed rates for the different products may be seen in Table 6.2, where illustrative liquefaction yields are given. To produce a synthetic crude, more hydrogen must be added and this is brought about by an increase in reactor residence time through a corresponding decrease in feed rate.

Figure 6.16 is a flow diagram of an integrated H-Coal plant. As in all of the processes discussed so far, coal is crushed (in this case to less than 0.2 mm), dried, slurried with recycle oil in a ratio typically between 2 and 3 to 1, and then pumped to a pressure of around 21 MPa but up to 24 MPa. Compressed hydrogen produced by gasification is added to the slurry and the mixture is preheated to 340 to 370°C and fed to an ebullated bed catalytic reactor (see Section 2.4) that is operated at temperatures to 455°C. It may be noted that in the pilot plant operation, hydrogen is not manufactured on site, but is purchased.

A sketch of the pilot plant reactor is shown in Figure 6.17, while Figure 6.18 is a photograph of it in its plant setting. The preheated slurry of coal, recycle oil, and hydrogen is introduced to a plenum chamber at the bottom of the reactor. The three-phase flow passes up through a distributor tray into an active, bubbly bed of catalyst. The bed is kept in a fluidized state by an internal recycle of slurry from above the top

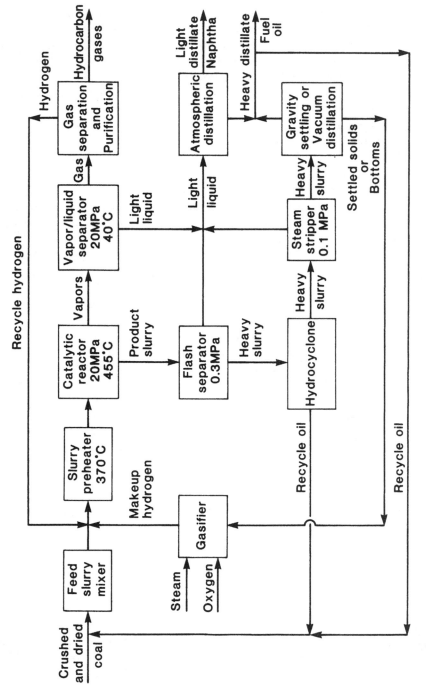

Figure 6.16 Simplified flow diagram of H-Coal process for a commerical-scale design with hydrogen production by gasification.

of the catalyst bed through a downcomer to an ebullating pump. The height of the expanded catalyst bed is controlled by the rate of slurry recycle.

The standard catalyst used is a commercial cobalt-molybdate preparation used in petroleum refining. A portion of the catalyst is withdrawn and fresh catalyst added continuously at a rate sufficient to maintain a constant level of catalyst activity. This is made necessary in part by the fairly rapid rate of catalyst deactivation due to sintering, metal deposition, and carbon deposition. Typically the catalyst makeup is about 0.5 kg/t of coal processed. Typical reactor operating conditions are around 20 MPa and 455°C. The temperature is maintained by adjusting the preheater outlet temperature. Residence times range from about 30 minutes to somewhat more than an hour. Longer residence times and severer operating conditions are used when the desired product is a synthetic crude, and shorter times and milder conditions if a low-sulfur fuel oil and high-CV gas are the desired products.

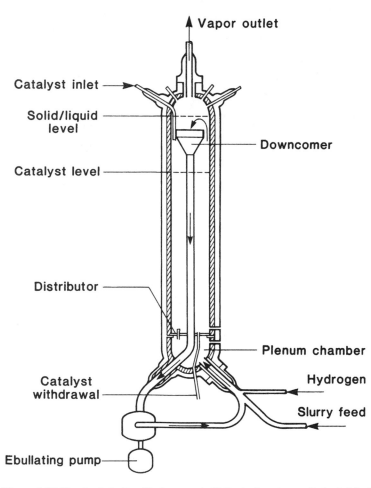

Figure 6.17 Sketch of ebullated bed reactor in H-Coal pilot plant; cylinder height 9.5 m, inside diameter 1.5 m.[29]

Figure 6.18 Photograph of H-Coal pilot plant in Catlettsburg, Kentucky, showing the coal feed hoppers above the ebullated bed reactor. *(Courtesy of Ashland Synthetic Fuels, Inc.)*

Product vapors and oil, ash, and unconverted coal pass out of the top of the reactor. The vapors are cooled to separate the oil from the excess hydrogen and light hydrocarbon gases. The slurry leaving the reactor is flashed in a series of steps and then sent to a bank of hydrocyclones. Oil overflow from the hydrocyclones is mixed with heavy distillate product and recycled to slurry the feed coal. The underflow from the hydrocyclones is sent to an atmospheric steam stripper for recovery of distillate oils. In the fuel oil mode of operation the slurry from the stripper goes to a Lummus "antisolvent" gravity settling, deashing unit described above in connection with the SRC-I process. The solids from this system contain the unconverted coal and ash, and in a commercial plant would probably be gasified to produce hydrogen. The clarified oil is taken as part of the fuel oil product. In the syncrude mode of operation the deashing step is bypassed and the slurry from the steam stripper passes directly to a vacuum distillation unit. Bottoms from the vacuum system then contain all of the unconverted coal and ash. The bottoms are subsequently gasified to produce the hydrogen required in the plant.

Typical process yields in the fuel oil and syncrude mode are shown in Table 6.21. The products are highly aromatic relative to petroleum products and although containing a high level of nitrogen, averaging about 0.5 percent, they have about half that in the SRC liquids. The naphtha typically contains 0.1 to 0.2 percent sulfur and the fuel oil less than 0.5 percent.

Table 6.21 Typical H-Coal liquefaction yields for Illinois No. 6 bituminous coal in mass percent of daf coal fed to reactor[30]

	Fuel Oil Mode, mass %	Syncrude Mode, mass %
Consumed		
Hydrogen	3.8	5.6
Total	3.8	5.6
Produced		
Fuel gas	12.3	15.0
Naphtha (C_4–205°C)	14.2	22.8
Fuel oil (205–525°C)	19.7	25.1
Residuum (>525°C)	34.9	18.2
Unconverted carbon	7.3	4.8
Hydrogen sulfide	2.7	2.9
Carbon oxides	1.1	0.6
Water	7.2	9.6
Ammonia	0.6	1.0
Total	100.0	100.0

Integrated commercial-plant designs with accompanying detailed material and energy balances have been prepared for both modes. The overall thermal efficiency for conversion to the product fuel is calculated to be about 66 percent in the fuel oil mode[31] and about 63 percent in the syncrude mode.[32] These values may be compared with an estimated 61 percent for both modes based on the developer's predicted operational yields for the large-scale pilot plant when feeding an Illinois No. 6 coal.[29]

Other Catalytic Processes

A number of other catalytic processes are under development, although none have progressed to the H-Coal scale. In Section 6.1 the Conoco Zinc Chloride process is mentioned, the goal of which is to produce gasoline directly from coal by hydroliquefaction.[33] The process utilizes the ability of zinc chloride to catalyze the hydrocracking of coal to single ring aromatics. It is of interest that much of the small-scale development has used solvent refined coal as feedstock.

The feed is reacted in roughly a one-to-one ratio with molten zinc chloride. Typical yields from the process are given in Table 6.2. The conversion is very selective, and of the distillate yield approximately two-thirds is C_5—200°C product. Moreover, the sulfur and nitrogen contents are some 100 times lower than those found in H-Coal and SRC products. Because the zinc chloride reacts with the heteroatoms in the coal, it must be continuously regenerated. This leads to difficulties associated with moving molten, corrosive, and viscous zinc chloride solutions around. Nevertheless, the process does have potential because of the high-quality product that is produced.

Another catalytic procedure of interest is the Dow coal liquefaction process, which at the time of writing, has only been operated at a 100 kg/d scale.[34] One main feature of the process is the use of an expendable catalyst made up of a water soluble molybdenum compound probably less than 1 μm in size. The procedure of feeding the catalyst once-through as an emulsion which is mixed with the coal and recycle oil is a unique one that leads to high catalyst activity and a simple design.

The catalyst-coal-recycle oil slurry is preheated and reacted at about 460°C and 14 MPa. As in the H-Coal process, the initial solids separation from the product slurry is carried out using hydrocyclones with the overhead stream recycled as slurry oil. Although the ash concentration in the overflow is much lower than in the underflow, the catalyst concentration is essentially equal to that in the hydrocyclone feed because the hydrocyclone does not separate out the fine dispersed catalyst particles. The recycle results in an increase in catalyst concentration in the reactor by about a factor of 2 over the level due to the addition of fresh emulsion.

A second interesting feature of the process is the use of a countercurrent solvent extraction column to separate out some of the asphaltenic material and essentially all of the ash from the heavy oils in the hydrocyclone underflow. The solvent is a light process-derived aliphatic/naphthenic stream. The unique operating characteristic claimed for the reactor is that the mass transfer of the heavy soluble oils to the solvent-rich phase is enhanced by a surface tension instability of the coal oil droplets which causes them to break up. Agglomerated ash, catalyst, unreacted coal, and heavy asphaltenes are removed from the bottom of the column as a viscous residue which in an integrated plant would be gasified to produce hydrogen. The deashed, deasphalted oil comes out overhead, with a part recycled and the rest taken as product. The ability to separate ash out from heavy oil compounds that cannot be readily vacuum distilled would be important if borne out on large scale.

In light of the basic German contributions to direct liquefaction technology, it is appropriate that mention be made here of the "new" German direct coal liquefaction program.[35] In the earlier Bergius-IG Farben process, bituminous coal was hydrogenated at pressures typically about 70 MPa. Disposable iron oxide catalysts were used. High pressures were employed because bottoms containing asphaltenes were a component in the recycle oil. The asphaltenes tended to build up coke in the reactor and were difficult to liquefy. To achieve liquefaction of the asphaltenes it was necessary to increase the hydrogen partial pressure.

In the new German "IG" process, disposable iron oxide catalysts are again used, but the reactor product is vacuum distilled to separate out the asphaltenes from the oil that is used for recycle. As a result of the asphaltene removal, the reactor operating pressure has been reduced to about 30 MPa, a value still quite high by comparison with the pressures used in U.S. processes. It is reported that the high pressures are necessary because German bituminous coals are more difficult to liquefy than corresponding U.S. coals. In this regard, it may be noted that the carbon content of a Ruhr bituminous coal is about 85 mass percent of the daf coal, compared to about 78 percent for a Kentucky or Illinois coal.

The feed slurry mixture is preheated to 420°C and reacted at 475°C for about an hour with a hydrogen consumption, measured as percent of the daf coal, of from 5 to 6 percent. A typical product yield distribution from the Bergbau-Forschung

Table 6.22 Typical Kohleoel liquefaction yields for German bituminous coal in mass percent of daf coal fed to reactor[35]

	Mass %
Consumed	
Hydrogen	5.8
Total	5.8
Produced	
Gas (C$_1$–C$_4$)	20.3
Naphtha (<200°C)	15.1
Middle distillate (200–325°C)	34.0
Residuum + unconverted carbon	20.1
Water + inorganic gases	10.5
Total	100.0

"Kohleoel" process development unit, which has a nominal maximum coal throughput of 0.5 t/d, is given in Table 6.22. A 200 t/d pilot plant is scheduled for startup in the Ruhr in 1981. The severity of the operating conditions employed does raise questions concerning the economics of the process.

6.5 UPGRADING COAL LIQUIDS

The primary products of the direct coal conversion processes are, depending on the degree of hydrogenation, either clean solid fuels, fuel oils, or synthetic crude oils. Pyrolysis oil is yet another product, but this is often hydrotreated within the overall process scheme to a fuel oil or synthetic crude, as is done in the COED and COGAS processes. Typical compositions of the three primary coal liquefaction products, given in Table 6.23, show that the hydrogen-to-carbon atom ratio, which is the most

Table 6.23 Composition of primary coal liquids and petroleum crude oils

Mass %	Primary Coal Liquids[36]			Petroleum Crudes[37]	
	SRC* (Solid)	Fuel Oil	Synthetic Crude	Heavy Asphaltic	Light Paraffinic
Carbon	87.9	89.0	88.3	86.8	83.9
Hydrogen	5.7	7.9	9.4	11.4	14.0
Nitrogen	1.7	0.8	0.4	1.7	0.1
Sulfur	0.6	0.4	0.1	0.1	2.0
Oxygen	3.5	1.9†	1.8	up to 0.5	
Aromatic carbon	75	65	55	20 to 35	
H/C atom ratio	0.78	1.07	1.28	1.58	2.00

* Not including ash.
† Adjusted to balance.

distinguishing feature, increases from less than about 0.8 for a solvent refined coal, to more than 1.2 for a synthetic crude. Products obtained in practice may fall anywhere in the continuous spectrum of solid to syncrude depending on the coal type, the hydrogenation process, and the selected product fractionation procedures. As discussed in the previous section, gases may be produced as well, but are often consumed as fuel on site rather than exported.

A measure of the quality of a liquid fuel commonly used in the petroleum industry is the API gravity, which is inversely related to the specific gravity by

$$\text{Degrees API} = \frac{141.5}{\text{s.g.}} - 131.5 \tag{6.11}$$

The specific gravity (s.g.) is the ratio of the density of the liquid fuel to the density of pure water, with both fluids at 15.6°C (60°F). Generally the higher the API gravity the higher the fuel quality. Values for common fuels range typically from around 12°API for a heavy (No. 6) fuel oil, to 40°API for a lighter (No. 1) fuel oil, while gasoline has a value of about 60°API. The specific gravity of the fuel can be related to its hydrogen content by

$$H = 26 - 15 \text{ s.g.} \tag{6.12}$$

where H is the mass percent hydrogen. This equation is valid to within about 1 percent for petroleum liquids that contain no sulfur, water, or ash. Using the quoted API gravities, Eqs. (6.11) and (6.12) can be used to show that the H/C atom ratio ranges from 1.51 for No. 6 fuel oil, 1.89 for No. 1 fuel oil, and 2.10 for gasoline, assuming these fuels to be pure hydrocarbons. In general, the larger the API gravity or H/C ratio of the fuel, the better its combustion properties, the lower its viscosity, and of course the lower its density. As in Table 6.23, we shall normally refer to the H/C ratio rather than the API gravity when comparing the quality of various fuels. Not only is the H/C ratio a more fundamental measure, but it is applicable to solid and gaseous fuels as well.

Coal liquefaction products are intended for consumption as boiler fuels in the process and electric power industries, as transportation fuels, as well as for residential and commercial space heating. These uses together with the corresponding fuel categories are summarized in Table 6.24. Non-fuel uses including metallurgical coke and petrochemical feedstocks are not considered in the book. Coal liquefaction is primarily aimed at replacing petroleum resources, but just as with petroleum crude, coal liquids are in general not suitable for direct use in the raw state. A certain amount of upgrading, followed by refining for the case of high-quality fuel products, is required. Whether or not the additional processing is integrated with the coal conversion plant, or done at existing refineries, is a matter of logistics and economics, but is not of immediate concern here. What is of interest is to identify those fuel types listed in Table 6.24 which can most expediently be supplied by coal products. As in the discussion of integrated coal gasification plants, the yardstick is economics. Again, however, it is not our intent to carry out or present detailed cost and efficiency studies, but rather to provide some insight to the key parameters involved by assessing the processes on the basis of the more fundamental concepts developed in the book.

Table 6.24 Utilization of coal conversion liquids and solids

Fuel Name	Distillate Category	Boiling Range, °C	Use
SRC	Solid	>480	A low-sulfur solid fuel replacement for power station and industrial boilers
Heavy fuel oil	Heavy distillate	370–525	A low-sulfur liquid fuel replacement for power station and industrial boilers
Light fuel oil	Mid-distillate	175–370	Residential and commercial space heating
Diesel fuel	Mid-distillate	200–400	Transportation
Jet fuel	Mid-distillate	150–280	Transportation
Gasoline	Naphtha	30–210	Transportation

A comparison of the costs involved in the production of several coal conversion products is given in Table 6.25 to serve as a basis in the selection of the most appropriate process/fuel combination for the applications listed in Table 6.24. The costs in Table 6.25 are for specific flow sheets selected in the evaluation[38] and will vary somewhat with process modifications, selection of alternative gasifiers, etc. The costs are given relative to that of solvent refined coal, the product requiring the least amount of hydrogenation. It is emphasized that in comparing the costs, due recognition should be given to the value of the product and whether or not further upgrading is required for a given use. In this respect it has been estimated[38] that the value of solid SRC is 50 percent, fuel oil 56 percent, and naphtha 82 percent of the value of a premium-grade gasoline. Further, while the major product has been loosely classified as fuel oil or synthetic crude, each process produces a very different product slate, as is described in the previous section, and this must be taken into account when ascribing a value to the various products.

Coal itself, clearly the least expensive of the fuels listed in Table 6.25, is extensively used as a boiler fuel. An important application is in power station electricity generation, but regulations governing the emission of particulates, sulfur dioxide, and nitrogen oxides are now restricting its use. New conventional coal-fired power stations will require a flue gas desulfurization step as well as means for NO_x control. As discussed in Section 5.5, future power stations may well be coal gasification combined cycle systems in which the important pollutant control steps are incorporated within the fuel gas flow sheet. At existing power stations burning moderate- to high-sulfur fuels, sulfur emissions may be reduced either by switching to a low-sulfur fuel, or by adding on a flue gas desulfurization (FGD) step. Selection of a low-sulfur fuel produced by one of the coal conversion processes under consideration will depend on it being economically competitive with coal and add-on FGD. Recent experience in sulfur scrubbing systems seems to suggest that for large base-load power stations, even retrofitted FGD may be more economical than switching to SRC, one of the least

Table 6.25 Relative costs of coal liquefaction processes[38]

Process	Major Product	Relative Cost*	
		Capital	Total
Coal mining	Coal	—	0.30†
SRC-I	Clean solid fuel	1.00	1.00
SRC-II	Fuel oil	1.16	1.07
H-Coal, fuel oil	Fuel oil	0.87	1.10‡
H-Coal, syncrude	Synthetic crude	1.04	1.16§
EDS	Fuel oil	1.16	1.17
Gasification/methanol	Methanol	—	1.29
Gasification/Mobil M	Premium gasoline	1.11	1.43
Gasification/Fischer-Tropsch	Gasoline	1.03	1.48

* For equivalent total energy production.
† Based on coal price of about $1.00/GJ ($1.00/10^6 Btu).
‡ Based on 66% efficiency; originally given as 0.98 for a 74% efficiency.
§ Based on 63% efficiency; originally given as 1.06 for a 69% efficiency.

expensive of the coal hydrogenation products. The situation is different for smaller industrial users and for peak-load power generation. The production of clean coal products in bulk facilities where full economy of scale can be realized, and subsequent distribution of the product to the small user, is inherently more cost effective than provision of a multitude of small flue-gas scrubbing systems. It should in fact be possible to determine a break-even boiler size above which FGD becomes economical. There are, however, factors important to small-scale operations that cannot easily be included in an economic analysis. The convenience of a less complex system with the virtual elimination of solid waste disposal problems, as well as lower workforce, maintenance, and downtime requirements makes clean fuels very attractive even if (marginally) less economical. Operational difficulties associated with FGD systems serving non-steady processes make clean coal liquids additionally attractive for some industrial and peak-load power generation applications.

The solid SRC product can be used as a replacement fuel for coal-fired boilers without incurring significant conversion costs, and would supposedly eliminate the need for fly ash and sulfur dioxide control. However, sulfur dioxide emission limits of 0.3 μg/J (equivalent to about 0.75 lb/10^6 Btu) translate to a sulfur content of 0.6 percent by mass for an SRC heating value of 37 MJ/kg. This is equivalent to the sulfur concentration reported in Table 6.23, but higher sulfur concentrations are not atypical of SRC, as can be seen in Table 6.17. The utility of SRC as a "low-sulfur" fuel substitute is at best marginal, and probably unsatisfactory for new power plants, particularly where local regulations are more stringent than the 0.3 μg/J sulfur dioxide limit quoted above. A further hydrotreating step, or production according to the two stage "SRC-$\frac{1}{2}$" process discussed in the previous section, is consequently of interest for a true low-sulfur solid fuel.

Although many countries have based their electricity generation on coal, in the United States the large majority of both power station and industrial boilers are fired with liquid fuels or natural gas. Conversion to a solid fuel with the attendant increased costs of transportation, storage, and handling is not attractive. An inspection of Table 6.25 suggests that the cost of a fuel oil may be very similar to that of solid SRC, and further, according to Table 6.23, sulfur contents of the liquid products are in general lower than for solid SRC. Conversion to a coal fuel oil may therefore be more attractive than conversion to SRC. However, in evaluating the conversion from a petroleum-based fuel to any of the coal-derived fuels, it is necessary to consider the effects on combustion and emission levels. The data in Table 6.23 show that although sulfur concentrations are comparable, coal liquids have significantly higher nitrogen concentrations than a good-quality light petroleum crude. It can be further seen that coal liquids are deficient in hydrogen and have more aromatic carbon than petroleum crude oils. Some of the consequences of the direct replacement of a petroleum boiler fuel oil by coal liquids are summarized in Table 6.26 taken from Ref. 36. It is concluded in this study that "all . . . [coal] liquids contain unacceptable amounts of nitrogen, oxygen and polynuclear aromatics and require further purification."

Upgrading

Upgrading is essentially a hydrotreatment step with the objective of
(a) removing the oxygen, nitrogen, and sulfur heteroatoms,
(b) increasing the stability of the oil, and
(c) increasing the H/C atom ratio.

Table 6.26 Principal differences between distillate coal liquids and No. 6 fuel oil[36]

Combustion Properties
Coal liquids produce fewer large particulates because fuel can vaporize during combustion.
Coal liquids have a greater tendency to form soot because of their lower hydrogen contents.
Coal liquids produce a more luminous flame due to their lower hydrogen-to-carbon ratio; this produces a different heat transfer balance.

Emissions
Coal liquids contain less sulfur than most unprocessed petroleum fuel oils.
Coal liquids produce more NO_x because of their higher organic nitrogen content (0.5–1.5 percent compared to 0.1–0.5 percent for petroleum-derived fuel oil).
Conventional oil-burning equipment will generally exceed NO_x and particulate emission standards when using coal liquids, but control technology is being developed.

Toxicity
Coal-based oils contain higher concentrations of skin carcinogens than virgin petroleum and will probably require special handling.

Stability
Coal liquids tend to polymerize and increase in viscosity during storage.
Deposits in atomizing nozzles are a greater problem with coal liquids.

Compatibility
Since mixtures of petroleum-based and coal-based fuel oils form precipitates, separate storage and handling facilities will be required.

In the case of subsequent refining to high-quality transportation fuels, the initial upgrading step serves to remove the nitrogen and sulfur components, which would otherwise poison the cracking catalysts.

As a first step in upgrading the primary coal liquids, they may be fractionated into narrower-boiling-range products. Typical distillation curves of some of the primary coal liquids are given in Figure 6.19, in which the volume fraction of the liquid boiling off below the indicated temperature is plotted. For example, only about 15 percent of the solid SRC product boils off at 525°C, while virtually all the Texas petroleum crude given as a reference material has boiled off by 300°C. An indication of the property changes that can be effected by distillation is given in Table 6.27 for a H-Coal fuel oil mode fraction boiling below about 345°C. Not only can cuts be obtained with H/C ratios significantly higher than the value of 1.1 typical for a coal-derived fuel oil,

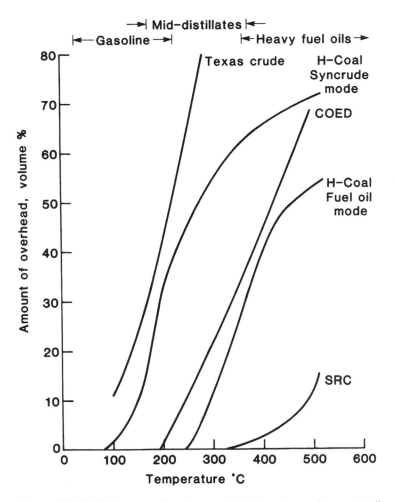

Figure 6.19 Distillation curves for primary coal liquids and a petroleum crude oil.

Table 6.27 Composition of light-end fractions of H-Coal fuel oil products*

Mass %	Temperature Range, °C			
	IBP†–150	150–290	290–345	>345
Carbon	85.9	87.6	89.2	89.0
Hydrogen	13.3	11.2	10.0	9.0
Oxygen	0.60	0.96	0.52	1.30
Nitrogen	0.07	0.17	0.17	0.50
Sulfur	0.13	0.07	0.11	0.20
H/C atom ratio	1.84	1.54	1.35	1.21
% of total monoaromatics	14	45	30	11
% of total polycyclic aromatics	0	6	42	52
% of total liquid	25	62	8	5

* Adapted from Ref. 39.
† Initial boiling point.

but considerable reductions in nitrogen content, particularly for the lighter fractions, can also be obtained. The properties of the remaining fractions are, of course, proportionately less desirable than the typical fuel oil of Table 6.23.

Separating light ends from the primary coal liquids may have some merit if it is desired to produce high-grade transportation fuels as well as heavier boiler fuels. The limited data so far available suggest that upgrading and refining of the lighter fractions is much more readily accomplished, with less catalyst deactivation problems, than when treating the bulk primary products. For example, the composite of the four H-Coal fractions shown in Table 6.27 was catalytically hydrotreated to produce a product that could meet a jet fuel specification.[39] On the negative side, this light end composite represented only 39 percent of the ash-free fuel oil product, and it is estimated that only some half of this quantity would be suitable for hydrotreatment to gasoline.

A simplified flow diagram for a distillation/hydrotreatment upgrading process is shown in Figure 6.20. The hydrotreatment itself may follow the scheme shown in Figure 6.21, in which hydrogen and the oil are passed cocurrently over a fixed or fluidized catalyst bed. Several commercial metal hydrogenation catalysts are available for black oil conversion in petroleum refining. Some of these have been tested with moderate success on the lighter fractions of coal-derived liquids, and it has been stated[36] that the processing of these feedstocks for the production of gasoline and heating oil is relatively straightforward. In Figure 6.20, the distillation is done prior to hydrotreating, although in practice it proved necessary to introduce an initial mild hydrotreatment to improve stability and to avoid coking at elevated distillation temperatures. As further shown in Figure 6.20, the major part of the feed leaves the distillation tower as a bottoms product too heavy for upgrading to high-quality fuels without hydrocracking. This product is best reserved for heavy boiler fuel, although it can be expected to require considerable hydrogenation–both to be compatible with existing firing systems and to comply with emission regulations.

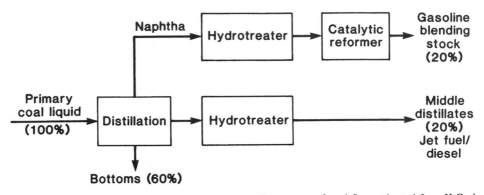

Figure 6.20 Simplified coal liquid upgrading process. (Percentages of total flow estimated from H-Coal data.[36])

The ability to hydrotreat the hydrocyclone overflow portion of the H-Coal product, using existing catalysts has been investigated. As can be seen from Figure 6.16, the hydrocyclone underflow represents the entire heavy-ends product. Its composition is given in Table 6.28 along with the composition after hydrotreatment. In a commercial operation, this upgrading step might be along the lines shown in Figure 6.21, and would probably be integrated as a second stage in the coal liquefaction plant. In the laboratory investigation, some improvement in the H/C ratio was obtained and the oxygen concentration was significantly reduced, while the sulfur concentration was well below that required for a boiler fuel. The value of such a product as a replacement

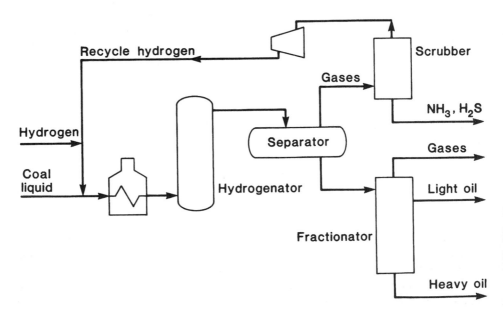

Figure 6.21 Simplified flow diagram of a hydrotreatment process.

Table 6.28 Composition of heavy ends of H-Coal product before and after hydrotreatment[39,40]

Mass %	Feed (Hydrocyclone underflow)	Hours on Stream	
		20–38	166–178
Carbon	86.70	89.58	90.60
Hydrogen	6.96	9.33	8.77
Oxygen	4.38	0.59	—
Nitrogen	1.30	0.47	0.59
Sulfur	0.66	0.03	—
H/C atom ratio	0.96	1.25	1.16

boiler fuel is largely dependent on achieving adequate combustion, in spite of its relatively low hydrogen content, while at the same time limiting NO_x formation. Nitrogen reduction is a major problem area. The levels remaining in the hydrotreated product of Table 6.28 are too high to consider it as a feedstock to a hydrocracker for refining to high quality fuels. In fact, a comparison of the initial and final product compositions in Table 6.28 shows that there was a rapid deactivation of the metal hydrogenation catalyst over the 170 hour test period, so that obtaining a suitable low quality boiler fuel is itself not without problems.

Refining

Refining operations are aimed primarily at producing high-quality products, in particular transportation fuels. While this may be readily done by hydrotreatment of the lighter fractions, the ultimate aim of refining is to process the bulk of the primary coal liquids produced in a liquefaction process. As is suggested in the discussion of the hydrogenation upgrading step, refining of the heavy product fractions is not expected to be straightforward, and there appears to be considerable merit in reserving these heavier ends for boiler fuels. The petroleum products now used as boiler fuel could then be used for refining to transportation fuels. The refining operations of specific interest for coal liquids are the cracking processes for reducing mean molecular size and for increasing the hydrogen content of the product. There are three principal processes, namely hydrocracking, fluid catalytic cracking (FCC), and thermal cracking or coking. The first two employ catalysts, while coking is a purely thermal process. These processes have been well developed in petroleum refineries and it is the intent to apply them to coal liquids. Coking is, however, said to be unsuitable for the highly aromatic coal liquids because the heavy ends would yield little but coke and gas. Further, thermal processes are less selective than the catalytic processes, which in general can be operated to give better yields of the desired liquid product. The major problems in the utilization of catalyst-based processes are the high sulfur and nitrogen levels, particularly in relation to the nitrogen combined in the polynuclear aromatics. Sulfur and nitrogen are strong poisons for both cracking and hydrocracking catalysts. In addition, the presence of aromatics can result in excessive coke yields with consequent catalyst fouling, while the high ash content of some coal liquids can result in plugging of fixed catalyst beds.

The first step in catalytic refining is therefore a hydrotreatment step as shown in Figure 6.21. However, the degree of hydrogenation required is more severe than discussed under upgrading, where a reduction of sulfur and nitrogen to levels acceptable for boiler fuel was satisfactory. Hydrodesulfurization is not a problem in the initial hydrotreatment step, with sulfur concentrations down to 0.02 percent and lower being readily obtained. Hydrodenitrogenation is accomplished at higher temperatures and pressures and at lower reactor space velocities. Even so, final nitrogen concentrations are seldom lower than 0.1 percent, and are often 0.3 percent and higher. Hydrogen consumption to attain these sulfur and nitrogen levels is about 4.6 mass percent for a H-Coal hydrocyclone underflow product, and 6.3 percent for solvent refined coal.[40] Some 7 percent of the hydrogen ends up as ammonia, while more than 70 percent is simultaneously consumed for carbon hydrogenation, which reflects the poor denitrogenation selectivity of available catalysts.

Of the two catalytic cracking processes, hydrocracking more readily handles higher molar mass constituents and has better selectivity characteristics than does fluid catalytic cracking. A dual function catalyst having both hydrogenation and cracking activity is required. Metals such as nickel and platinum for the hydrogenation function, supported on silica alumina or zeolites for the acid cracking function, are typically used. The first step in the hydrocracking sequence is the hydrogenation process, in which the aromatic rings are saturated with hydrogen. The greater the aromatic content the greater the hydrogen consumption. Once saturated, the molecules are cracked to the lower-boiling-point products.

The acid sites of the hydrocracking catalysts are extremely sensitive to nitrogen poisoning. Fluid-catalytic-cracking catalysts are reportedly less sensitive, with nitrogen concentrations of 0.2 to 0.3 percent being acceptable. For this reason FCC may be preferred to hydrocracking for refining coal liquids, although selectivities and yields of gasoline products will be lower. Hydrogen is not consumed in FCC; in fact some hydrogen may be released in the cracking process and may be recovered and recycled to the hydrotreatment step. Some preliminary tests using EDS and COED liquids have indicated that following an appropriate degree of hydrotreatment, coal-derived distillates can be readily processed into high-quality gasolines by application of advanced FCC technology. It is stressed, however, that many of the investigations of catalyst performance have been of a preliminary nature, using selected fractions of the total coal liquid product. Extensive long-term testing is still required to assess the commercial viability of these processes.

Summary

While upgrading and refining of the lighter ends of coal liquids to high quality fuels is readily accomplished using available petroleum-based technology, these lighter products represent only a small fraction of the total liquid product. Upgrading of the heavier fractions by hydrogenation, and possibly fluid catalytic cracking, has met with some success, although problems related to catalyst deactivation, high hydrogen consumption, and the economical processing of the full product slate have yet to be satisfactorily resolved. More severe processing of the heavy liquids by hydrocracking to increase yields of high-quality transportation fuels currently appears to be unrealistic due to rapid catalyst deactivation.

The development of poison- and coke-resistant catalysts as well as related refining technology advances, may in future render the high nitrogen/aromatic coal liquids susceptible to conversion to high-quality fuels. Until then it would seem appropriate to utilize coal liquids, after moderate hydrotreatment, for boiler fuels, while reserving petroleum feedstocks for the high-quality products. Whether or not the lighter coal liquids are separated for conversion to transportation fuels will depend on economics, demand, and the impact of other synthetic fuels, such as shale oil. In countries having a pipeline gas distribution system, there may be good reason to emphasize coal gasification for boiler fuels, but with the co-production of liquid transportation fuels. The type of plant envisaged would gasify coal primarily for SNG production, but with some portion separated for hydrogen/coal liquefaction using one of the processes discussed in the previous section. Instead of attempting to further process the residuum and difficult-to-upgrade heavy liquid fractions, these would be recycled to the gasifier. Production of transportation fuels from the light fractions would, of course, have to be competitive with their production from the gasifier synthesis gas for such a plant to be viable. Costs of the three indirect liquefaction processes relative to direct liquefaction are included in Table 6.25.

REFERENCES

1. Ralph M. Parsons Co., "Coal Liquefaction Process Research Survey, R & D Interim Report No. 2, Data Source Book," Oak Ridge National Laboratory Report No. ORNL/Sub-7186/13, U.S. Department of Energy, Washington, D.C., December, 1977.
2. Schreiner, M., "Research Guidance Studies to Assess Gasoline from Coal by Methanol-to-Gasoline and Sasol-Type Fischer-Tropsch Technologies," Report No. FE-2447-13, U.S. Department of Energy, Washington, D.C., August 1978.
3. U.S. Environmental Protection Agency, "Pollution Control Guidance Document for Indirect Coal Liquefaction," Coal Gasification and Indirect Liquefaction Working Group. Industrial Environmental Research Laboratory, U.S. Environmental Protection Agency, Research Triangle Park, N.C., 1981 (to appear).
4. Salmon, R., Edwards, M. S. and Wham, R. M., "Production of Methanol and Methanol-Related Fuels from Coal," Report No. ORNL-5564, Oak Ridge National Laboratory, Oak Ridge, Tennessee, May 1980.
5. Probstein, R. F., and Gold, H., *Water in Synthetic Fuel Production.* MIT Press, Cambridge, Mass., 1978.
6. Dry, M. E., "Advances in Fischer-Tropsch Chemistry," *Ind. Eng. Chem. Product Res. Dev.* **15**, 282-286, 1976.
7. Kölbel, H., and Ralek, M., "Fischer-Tropsch-Synthese," in *Chemierohstoffe aus Kohle* (J. Falbe, ed.), pp. 219-234, 257-271. G. Thieme Verlag, Stuttgart, 1977.
8. Poutsma, M. L., "Assessment of Advanced Process Concepts for Liquefaction of Low H_2:CO Ratio Synthesis Gas Based on the Kölbel Slurry Reactor and the Mobil-Gasoline process," Report No. ORNL-5635, Oak Ridge National Laboratory, Oak Ridge, Tennessee, February 1980.

9. Deckwer, W.-D., Louisi, Y., Zaldi, A., and Ralek, M., "Hydrodynamic Properties of the Fischer-Tropsch Slurry Process," *Ind. Eng. Chem. Process Des. Dev.* **19**, 699-708, 1980.
10. Meisel, S. L., McCullough, J. P., Lechthaler, C. H., and Weisz, P. B., "Gasoline from Methanol in One Step," *ChemTech* **6**, 86-89, 1976.
11. Chang, C. D., *et al.*, "Process Studies on the Conversion of Methanol to Gasoline," *Ind. Eng. Chem. Process Des. Dev.* **17**, 255-260, 1978.
12. McCray, F. L., McClintock, N., and Bloom, R., Jr., "The ICGG Approach—What Is It?" in *Coal Processing Technology*, Vol. V, pp. 156-165. American Institute of Chemical Engineers, New York, 1979.
13. Bloom, R., Jr., "Coal Dilemma II, COGAS," in *Coal Conversion Technology* (A. H. Pelofsky, ed.), pp. 23-35. ACS Symposium Series No. 110, American Chemical Society, Washington, D.C., 1979.
14. Cusumano, J. A., Dalla Betta, R. A., and Levy, R. B., *Catalysis in Coal Conversion*. Academic Press, New York, 1978.
15. Chang, P. W., Durai-Swamy, K., and Knell, E. W., "Kinetics of Coal Pyrolysis Reactions in a Flash Pyrolysis Process," in *Coal Processing Technology*, Vol. VI, pp. 20-27. American Institute of Chemical Engineers, New York, 1980.
16. Che, S. C., Durai-Swamy, K., Knell, E. W., and Lee, C.-K., "Flash Pyrolysis Coal Liquefaction Process Development," Report No. FE-2244-26, U.S. Department of Energy, Washington, D.C., April 1979.
17. Oberg, C. L., and Falk, A. Y., "Coal Liquefaction by Flash Hydropyrolysis," in *Coal Processing Technology*, Vol. VI, pp. 159–165. American Institute of Chemical Engineers, New York, 1980.
18. Gangoli, N., and Thodos, G., "Liquid Fuels and Chemical Feedstocks from Coal by Supercritical Gas Extraction," *Ind. Eng. Chem. Product. Res. Dev.* **16**, 208-216, 1977.
19. Maddocks, R. R., Gibson, J., and Williams, D. F., "Supercritical Extraction of Coal," *Chem. Eng. Prog.*, **75**(6), 49-55, 1979.
20. Whitehead, J. C., "Development of a Process for the Supercritical Gas Extraction of Coal," *Fuels and Petrochemicals Division Reprints*, 1980. *Vol. I, 88th National Meeting AIChE, Philadelphia, June 8-12, 1980* (E. G. Foster, ed.), pp. 402-413. American Institute of Chemical Engineers, New York, 1980.
21. Donath, E. E., "Hydrogenation of Coal and Tar," in *Chemistry of Coal Utilization*, Supplementary Volume (H. H. Lowry, ed.), pp. 1041-1080. Wiley, New York, 1963.
22. Phillips, E. M., *et al.*, "A Comparative Study of Coal Liquefaction Performance and Economics for Solvent Refined Coal-Based Processes," in *Coal Processing Technology*, Vol. VI, pp. 193-208. American Institute of Chemical Engineers, New York, 1980.
23. Whitehurst, D. D., Mitchell, T. O., and Farcasiu, M., *Coal Liquefaction*. Academic Press, New York, 1980.
24. Nowacki, P., *Coal Liquefaction Processes*. Noyes Data Corp., Park Ridge, N.J., 1979.
25. Peluso, M., and Ogren, D. F., "Antisolvent Deashing," *Chem. Eng. Prog.*, **75**(6), 41-43, 1979.
26. Mitchell, W. N., Trachte, K. L., and Zaczepinski, S., "Performance of Low Rank Coals in the Exxon Donor Solvent Process," *Ind. Eng. Chem. Product Res. Dev.* **18**, 311-314, 1979.
27. Epperly, W. R., and Taunton, J., "Exxon Donor Solvent, Coal Liquefaction Process Development," in *Coal Conversion Technology* (A. H. Pelofsky, ed.). ACS Symposium Series No. 110, American Chemical Society, Washington, D.C., 1979.
28. Epperly, W. R., and Wade, D. T., "Exxon Donor Solvent Coal Liquefaction Process: Development Program Status," Preprint No. AM-80-39, 1980 Nat'l. Petroleum Refiners Assoc. Ann. Meeting, New Orleans, March 1980.
29. Stotler, H. H., and Schutter, R. T., "Status and Plans of H-Coal Pilot Plant," in *Coal Processing Technology*, Vol. V., pp. 73-77. American Institute of Chemical Engineers, New York, 1979.
30. Hydrocarbon Research, Inc., "H-Coal Integrated Pilot Plant," Report No. EPRI AF-681, Vol. 2, Electric Power Research Institute, Palo Alto, Calif., March 1978.
31. Fluor Engineers and Constructors, Inc., "H-Coal Commercial Evaluation," Report No. FE-2002-12, U.S. Department of Energy, Washington, D.C., March 1976.
32. Dickson, E. M., *et al.*, "Impacts of Synthetic Liquid Fuel Development, Vol. II—Analysis," Report No. ERDA 76-129/2, U.S. Government Printing Office, Washington, D.C., 1977.

33. Green, C. R., *et al.*, "Gasoline from Coal via Molten Zinc Chloride Hydrocracking," in *Coal Processing Technology*, Vol. VI, pp. 103-109. American Institute of Chemical Engineers, New York, 1980.
34. Moll, N. G., and Quarderer, G. J., "The Dow Chemical Company Coal Liquefaction Process," *Chem. Eng. Prog.* **75**(11), 46-50, 1979.
35. Friedrich, F., Strobel, B., and Romey, I., "New Coal Hydrogenation Process: Proven Effective by Three-Year Experimental Plant Experience," in *Coal Processing Technology*, Vol. VI, pp. 174-178, American Institute of Chemical Engineers, New York, 1980.
36. Energy Engineering Board, Assembly of Engineering, "Refining Synthetic Liquids from Coal and Shale," National Research Council, National Academy Press, Washington, D.C., 1980.
37. Lom, W. L., and Williams, A. F., *Substitute Natural Gas*. Wiley-Halsted, New York, 1973.
38. Rogers, K. A., and Hill, R. F., "Coal Conversion Comparisons," Report No. FE-2468-51, U.S. Department of Energy, Washington, D.C., July 1979.
39. Grey, D., "Upgrading of Primary Coal Liquids," in *Coal Processing*, Notes from Spring School in Coal Processing, Rand Afrikaans University. South African Coal Processing Society, Johannesburg, 1978.
40. de Rossett, A. J., Tan, G., and Gatsis, J. G., "Upgrading Primary Coal Liquids by Hydrotreatment," in *Refining of Synthetic Crudes* (B. M. Harvey and M. L. Gobatty, eds.). ACS Symposium Series No. 179, American Chemical Society, Washington, D.C., 1979. See also "Characterization of Coal Liquids," Report No. FE-2010-07, U.S. Department of Energy, Washington, D.C., Nov. 1976.
41. Adams, R. M., Knebel, A. H., and Rhodes, D. E., "Critical Solvent Deashing of Liquefied Coal," *Chem. Eng. Prog.* **75**(6), 44-48, 1979.

SEVEN

LIQUIDS FROM OIL SHALE AND TAR SANDS

7.1 OIL SHALE RETORTING

Oil shale deposits occur throughout the world and may, in fact, represent the most abundant form of hydrocarbon on earth.[1] As discussed in Chapter 1, however, the majority of the deposits so far identified are not considered to be viable resources because they are either too deeply buried, occur in very thin layers, or have too low a hydrocarbon content to be economically recovered. By far the largest identified oil shale resource that is suitable for commercial exploitation is that of the Green River Formation in Colorado, Utah, and Wyoming, in the western United States. Other commercially important deposits that are currently being processed or that are being considered for commercial development are in Australia, Brazil, Estonia, Morocco, and Fushun in China. In addition, France, Scotland, Spain, South Africa, and Sweden have supported commercial enterprises that were terminated in the fifties and sixties when natural petroleum became abundantly available at low cost.

We recall that oil shale may be characterized by its grade, that is, its oil yield expressed in liters per ton as determined by the modified Fischer assay.[2] Oil shale is defined as having a yield greater than 42 L/t, and may be found in a range up to 10 times this value. Grades typical of existing and proposed commercial ventures are from 100 to 160 L/t. In the case of Green River oil shales, we may apply Eq. (1.4)

$$\text{Yield (L/t)} = 8.22 \times \text{Organic Matter (mass \% of shale)} - 10.8 \qquad (7.1)$$

to show that the organic content of the commercially important western United States shales ranges from 13.5 percent to 21 percent. By comparison, the organic matter in coal ranges typically from 75 percent to more than 90 percent by mass. Consequently, the amount of oil shale to be processed will range from 3.5 times to 7 times the amount of coal for an equivalent hydrocarbon throughput and process thermal efficiency. This is an important consideration not only in evaluating oil shale recovery processes, but also in relation to the quantity of processed shale that must be disposed of as a solid waste.

Coal can be burned directly as a fuel for, say, electricity production, it can be partially oxidized to produce a gaseous fuel, or it can be hydrogenated to produce liquid fuels. Pyrolysis of coal produces both a gaseous and a liquid product, although typically in excess of 50 percent of the organic content is not converted but remains as a carbonaceous residue or char. Although all of the foregoing techniques can theoretically be applied to oil shale, pyrolysis is by far the most important for oil shale processing. As with coal, when oil shale is pyrolyzed, part of the organic matter is converted to an oil and a gas, with the unconverted organics remaining behind on the shale as a coke-like residue. In a modified Fischer assay of Colorado oil shale, 75 to 80 percent of the organic matter is converted to oil and gas, with the oil yield representing up to 70 percent of the organic content of the shale.[2,3] Pyrolysis consequently offers a relatively straightforward method of recovering a large fraction of the organic content of Green River oil shale in the desirable liquid form.

The high liquid recovery on pyrolysis is not applicable to all oil shale deposits. For example, the shale from Brazil, which represents the second largest of the world's identified resources, yields only 40 to 45 percent of its organic content as oil.[3] Similarly, Fischer assay liquid yields of about 47 percent for Moroccan shale and 30 percent for the Devonian shales of the eastern United States have been reported.[4] Gas yields are generally greater than with Green River shales, but more than half the organic matter remains as a char in the pyrolyzed shale under Fischer assay conditions. One method that has been advocated for more completely utilizing the organic matter is to burn the shale itself and use the heat for electricity production. This practice is in fact the principal use of the relatively high-grade (200 L/t) kukersite shales in Estonia.[1]

Pilot-scale testing of the direct combustion of lower-grade shales in a countercurrent moving bed system has been conducted in Israel.[3] Initiating and sustaining combustion was not a problem, and 80 to 90 percent utilization of the 11 to 14 percent organic matter originally present was obtained. These tests were followed up with trials using an industrial boiler designed for combustion of the residues from washing high-ash coal. The results were stated to be encouraging. Preliminary estimates suggested that the cost of the shale, including mining, crushing, conveying, and disposal, would be about one-seventh that required of fuel oil for an equivalent electric power production. This cost does not include the additional expense incurred for the removal of sulfur and nitrogen oxides from the flue gas, which can be expected to be present at higher concentrations than obtained with oil or coal combustion. It is possible that these pollutants may be at least partially removed by absorption into shale ash, although it is not known how this procedure will impact the environmental problems associated with spent shale disposal.

While direct combustion of oil shale does appear to be feasible, it is not of immediate interest in the United States, where the current energy infrastructure is based largely on liquid fuels. Pyrolysis of oil shale to produce liquids is consequently the only process considered here in detail. Further, much of the discussion is directed towards the Green River deposits and it should be recognized that other shales may have very different characteristics. Indeed, there is considerable variation in characteristics within the Green River deposit itself, and while the discussion will be generally applicable, it may not be valid for shales at a specific location, or at a specific depth within the deposit. Where such variation has a significant bearing on processing characteristics, this will be noted.

Before proceeding to a description of specific pyrolysis processes, we will briefly review some of the properties of oil shale and its pyrolysis products. The reader is reminded that the properties and pyrolysis of oil shale are discussed in Sections 1.3 and 3.1 respectively.

Properties

Kerogen is the major organic constituent of all oil shales. It is a high-molar-mass organic that, like the organics in coal, falls into the category of substances known as pyrobitumens. Pyrobitumens are insoluble in carbon disulfide and common petroleum solvents, so shale oil is not in general amenable to recovery by solvent extraction. However, about 10 percent of the total organic content of Green River oil shale is a relatively low-molar-mass bitumen,[5,6] and up to 25 percent of the total organics can be solubilized under severe conditions of extended extraction times, elevated temperatures, and strongly polar solvents.[7]

Both aliphatic and aromatic carbon are found in the organic matter in oil shale. While the oil yield based on the total organic matter as given by Eq. (7.1) is valid for Green River oil shale only, it has been demonstrated[8] that an equation valid for many oil shales results if the oil yield is expressed in terms of the aliphatic portion of the total carbon

$$\text{Yield (L/t)} = 12.8 \times \text{Aliphatic Carbon (mass \% of shale)} - 35.8 \qquad (7.2)$$

A comparison of Eqs. (7.1) and (7.2) indicates that 70 to 80 percent of the organic matter in Green River oil shale is aliphatic carbon. If the total organic carbon is 80.5 percent of the organic material, then this suggests that about 90 percent of the organic carbon present is bound in aliphatic structures. This value is in fair agreement with analytically determined aliphatic carbon concentrations of 75 to 85 percent of the total organic carbon.[8]

An estimation of the aliphatic carbon in Devonian shale can also be made on the basis of Eq. (7.2). A Fischer assay yield for typical Devonian shale with an organic carbon content of 9.8 percent is about 40 L/t.[4] According to Eq. (7.2) this yield corresponds to an aliphatic carbon content of 5.9 percent, indicating that about 60 percent of the carbon in these shales is aliphatic. Chemical analysis shows there to be about equal portions of aliphatic and aromatic carbon in typical Devonian oil shales.[8]

As mentioned above, an important practical difference between Green River and Devonian shales is the amount of organic matter converted to oil during pyrolysis. This difference can be expressed by the following representative equations for a unit mass of organic matter:

Green River oil shale

Organic matter $\rightarrow$ 0.69 Oil + 0.10 Gas + 0.21 Residue $\qquad (7.3)$

Devonian oil shale

Organic matter $\rightarrow$ 0.26 Oil + 0.19 Gas + 0.55 Residue $\qquad (7.4)$

These equations are based on several reported Fischer assay yields, and the coefficients have been adjusted to exclude water and the estimated quantities of carbon dioxide and hydrogen sulfide present in the gas resulting from decomposition of inorganic substances. They are intended for comparison purposes only and should not be used for prediction of yields for design purposes.

The inorganic content of the Green River oil shales is a mix of carbonates, silicates, and clays, as shown in Table 7.1. Carbonate-containing material accounts for nearly half the inorganic matter, or stated differently, about 17 percent of the shale is inorganic bound carbon dioxide. This carbon dioxide may be evolved on heating, with the formation of either oxides or silicates. The calcination of calcite to lime is an example of oxide formation:

$$\underset{\text{Calcite}}{CaCO_3} \xrightarrow{\text{heat}} \underset{\text{Lime}}{CaO} + CO_2 \qquad \Delta H^{\circ}_{298} = +183 \text{ kJ/mol} \qquad (7.5)$$

The reaction to form silicates may be represented by

$$\underset{\text{Calcite}}{CaCO_3} + \underset{\text{Quartz}}{SiO_2} \xrightarrow{\text{heat}} \underset{\text{Wollastonite}}{CaSiO_3} + CO_2 \qquad \Delta H^{\circ}_{298} = +87 \text{ kJ/mol} \qquad (7.6)$$

The formation of silicates absorbs less process heat, and results in an environmentally inert end product. Some control can be exercised over the relative rates of oxide and silicate formation by the retorting atmosphere. For example, carbon dioxide and steam atmospheres have been found to enhance silicate formation relative to a nitrogen atmosphere.[9] The absolute quantity of inorganic carbon dioxide evolved is controlled by the pyrolysis temperature. As shown in Figure 7.1, carbonate decomposition becomes significant above about 500°C (cf. Figure 3.6). This is some 250°C lower than would occur were the carbonate minerals in their pure state.

Table 7.1 Typical composition of the inorganic matter in Green River and Devonian oil shales[2]

Mineral	Chemical Formula	Green River Mahogany Zone	Devonian Black Shale
		Mass %	
Dolomite	$(Mg,Fe)Ca(CO_3)_2$*	32	—
Calcite	$CaCO_3$	16	—
Quartz	SiO_2	15	28
Illite	†	19	40‡
Albite	$NaAlSi_3O_8$	10	—
Feldspar	$KAlSi_3O_8$	6	12
Pyrite, Marcasite	FeS_2	1	14
Analcime	§	1	—
Other	—	—	6
Total		100	100

* Contains about 6 percent iron.
† Potassium aluminum silicates.
‡ Includes kaolinite (hydrous aluminum silicate) and muscovite (potassium aluminum silicate).
§ Sodium aluminum silicates.

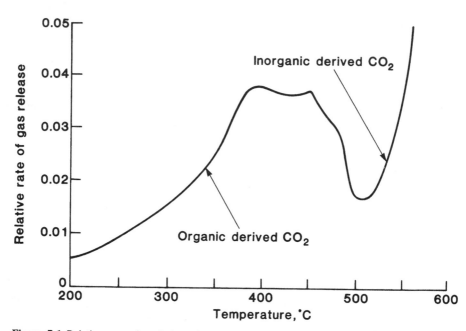

Figure 7.1 Relative rate of evolution of carbon dioxide from the organic matter and from inorganic carbonates. (The gas rate is measured relative to a constant background flow of argon.)[9]

Two minerals not shown in Table 7.1, but that are associated with specific deposits of Green River oil shale, are nahcolite and dawsonite. These are both carbonate containing minerals with decomposition temperatures considerably lower than for calcite and dolomite. Evolution of carbon dioxide occurs at around 150°C for nahcolite and 350°C for dawsonite. The name nahcolite is derived from the formula of its chemical constituent, sodium bicarbonate ($NaHCO_3$). A possible commercial market for nahcolite is as a raw material for production of industrial-grade soda ash (Na_2CO_3), while a potential market that may develop is as a chemical reactant for sulfur dioxide removal from flue gases. Dawsonite is a sodium aluminum carbonate ($NaAl(OH)_2CO_3$), which can be treated to produce both alumina and soda ash. In one area in the Piceance Creek basin in Colorado, the oil shale contains quantities of both these minerals sufficient to make their recovery along with the oil an attractive consideration. The Superior Oil Company has in fact proposed a combined process for recovery of both oil and chemicals.[2] An estimated material balance for the proposed plant suggests that yields would be in the mass ratio of 2.3 parts nahcolite, 0.5 parts soda ash, and 0.3 parts alumina for each unit of oil produced. Production of 8×10^6 L/d (50 000 bbl/day) of oil in such a multimineral plant would produce sufficient sodium salts to satisfy the current U.S. market of 8×10^6 t/yr.

A typical inorganic composition for a Devonian oil shale is included in Table 7.1. Two important differences affecting pyrolysis are the virtual absence of carbonate minerals and the high concentration of inorganic sulfur in the Devonian shales. Carbonate decomposition and the associated heat consumption is clearly not of concern with these shales. The high sulfur content, on the other hand, can lead to processing difficulties. Any sulfur evolved as hydrogen sulfide consumes hydrogen that would otherwise be available for oil formation. Published gas compositions for Devonian shales are limited, but data given in one study[4] show that the gas may contain from 20 to 30 percent hydrogen sulfide, which accounts for some 10 percent of the hydrogen in the kerogen. The high gas and low oil yields reported for eastern shales may consequently be attributed, at least in part, to the large amount of hydrogen sulfide formed.

Pyrolysis

The conversion of the organic matter in oil shale to liquid oil is accompanied by a change in chemical composition. Ultimate analyses for a Green River oil shale and shale oil are shown in Table 7.2. The most significant compositional change is the decrease in oxygen concentration which occurs as a result of the evolution of carbon dioxide, with associated carbon loss, and the formation of water, with associated hydrogen loss. Some hydrogen is also consumed in the reduction of nitrogen and sulfur to ammonia and hydrogen sulfide, which must subsequently be removed from the gas stream. Additional hydrogen is removed with the gas stream in the form of methane and molecular hydrogen. The distribution of hydrogen and other elements to the oil, gas, and residue as determined by elemental material balance is shown in Table 7.3. This balance was completed by assigning unaccounted material to the carbonaceous residue. The resulting residue composition is in good agreement with published data for pyrolyzed shales.[2]

The material balance shows that while 23 percent of the hydrogen ends up with the gas and residue, the amount of carbon "lost" to these secondary products is even higher at 27 percent. There is, consequently, a small increase in the H/C ratio of the shale oil relative to the original organic material, as is shown in Table 7.2. In fact, the H/C atom ratio of 1.63 for the shale oil compares favorably with values for the

Table 7.2 Typical composition of crude shale oil in relation to the organic matter in the oil shale and two petroleum crudes

| Mass % | Green River Oil Shale[2] | | Petroleum Crudes* | |
	Organic Matter	Shale Oil†	Heavy Asphaltic	Light Paraffinic
Carbon	80.5	84.6	86.8	83.9
Hydrogen	10.3	11.5	11.4	14.0
Nitrogen	2.4	2.0	1.7	0.1
Sulfur	1.0	0.6	0.1	2.0
Oxygen	5.8	1.0	up to 0.5	
Aromatic carbon	15–25	<25‡	20–35	
H/C atom ratio	1.54	1.63	1.58	2.00

* See Table 6.23.

† Fischer assay product.

‡ See text.

Table 7.3 Elemental material balance for Fischer assay of organic material in Green River oil shale

Element	Organic Material in Oil Shale,* kg	Shale Oil,* kg	Gas,† kg	Organic Material in Residue,‡	
				kg	mass %
Carbon	80.5	58.5	3.8	18.2	86.6
Hydrogen	10.3	7.9	0.7	1.7	8.1
Nitrogen	2.4	1.4	0	1.0	4.8
Sulfur	1.0	0.4	0.5	0.1	0.5
Oxygen	5.8	0.8	5.0	0	0
Total§	100.0	69.0	10.0	21.0	100.0

* Composition as in Table 7.2.
† Composition as in Table 3.2.
‡ By difference.
§ Based on 100 kg organic material and distribution according to Eq. (7.3).

petroleum crudes included in the table for comparison. We recall that H/C ratios for primary coal liquids were generally less than about 1.3.

The relatively high H/C ratio for crude shale oil suggests that the amount of carbon bound in aromatic rings is low. The increase in the H/C ratio on going from kerogen to shale oil further suggests that the aromaticity has decreased during pyrolysis. For this reason, the aromatic carbon content of the shale oil is shown in Table 7.2 to be less than the 25 percent maximum given for the original organic matter. The low aromatic character of the oil is confirmed by compositions determined using recently established analytical procedures for characterizing oil shale and shale oil.[6,8] Analyses based on ASTM procedures intended for natural crude oils indicate a much higher aromatic character, with up to 40 percent aromatics and a further 30 percent "polar aromatics" being reported.[10] Polar aromatics include ring compounds containing oxygen, nitrogen, and sulfur, either within the ring as heteroatoms or attached to the ring as functional groups. Such high aromaticity is not in accord with the established H/C ratio, nor with the often quoted "waxy" physical characteristic of shale oils. Part of the confusion on aromaticity may be due to terminology. For example, a compound consisting of a benzene ring with a six-membered paraffinic side chain (H/C ratio = 1.5) might be classed as aromatic, although its aromaticity is in fact 50 percent. In addition, many of the polar compounds in oil shale can be aliphatic, so the class of non-neutral substances should be labelled "polar compounds" rather than "polar aromatics." Grouping into neutral oils (saturates, aromatics, and olefins, or SAO), resins (polar compounds), and asphaltenes, as is sometimes used for natural petroleum, avoids this confusion.

Pyrolysis Variables

Some control may be exercised over the quantity and quality of oil produced by selection of the pyrolysis operating conditions. The parameters that can be varied in oil shale pyrolysis are pressure, temperature, the type of atmosphere, and, to some extent, the heating rate. The mode of heating, that is whether the heat is supplied directly by combustion of some of the organic matter, or indirectly from an external heat source, is another important variable that will be discussed subsequently.

Pressure has a significant effect on the relative yields and on the quality of the liquids and gases produced. The net effect of increasing the pressure depends both on the type of shale and on the pyrolysis atmosphere. Results reported for Green River oil shales in an inert atmosphere such as nitrogen or helium show a sharp decline in the quantity of liquid obtained with increasing pressure. In one series of tests, yields dropped to about half the Fischer assay value on increasing the pressure to 3.5 MPa. This was accompanied by an increase in the quantity of gaseous hydrocarbons produced and in the H/C ratio of the smaller quantity of oil.[2] The increase in gas yield can be explained in terms of increased cracking reactions, while increased coking would explain both the lower liquid yield and the higher H/C ratio. Increasing the pressure with a hydrogen atmosphere results in moderate increases in liquid yields from Green River shales, and substantial increases of up to a factor of 2 with Devonian shales. The presence of hydrogen inhibits coking reactions, and, especially in the case of the hydrogen-deficient Devonian shales, enhances hydrogenation reactions.

The *temperature* of pyrolysis is restricted to a relatively narrow range by the rate dependency of the decomposition reactions. We recall from Section 3.1 that liquid evolution is most rapid at about 425°C, and evolution of hydrogen and methane peaks at about 460°C. Primary decomposition to liquids and gases is virtually complete by 470°C. At temperatures above about 500°C there is a secondary decomposition of char, but this is mainly to gases and does not increase liquid yields. In view of the problem of carbonate decomposition, there is little incentive to pyrolyze Green River oil shales at temperatures above 500°C. There does, however, seem to be some justification in operating close to this temperature limit. High temperatures, for example, significantly enhance the rate of pyrolysis as is shown by Eq. (3.10). Oil in the liquid phase is subject to coking reactions which reduce yields. Operating at high temperature has the effect of producing more oil in the vapor phase and therefore increases recovery.[9] Although a decrease in coke formation by operation at higher temperature does result in a slightly lower quality oil, maximizing yields is normally of prime concern in oil shale processing.

The moderate increase in oil yield often obtained on increasing the *heating rate* can also be attributed to a decrease in coking reactions. Higher heating rates result in the oil being evolved at a higher average temperature, so more oil is in the vapor phase and coking is consequently reduced.[9] In hydrogen atmospheres where coking is not a significant factor, increased heating rates result in a moderate decrease in organic carbon recovery.[11] As explained in Section 3.1, practical heating rates for oil shale are restricted to the realm of slow pyrolysis, because of the large quantity of inert material intimately associated with the organic material.

The *pyrolysis atmosphere* can affect both the quantity and quality of the oil produced. As discussed under pressure considerations, the presence of hydrogen increases organic carbon recovery, especially with Devonian shales. It also increases the H/C ratio and decreases the sulfur and oxygen concentrations in the liquid product. Steam may be used as a less-expensive hydrogen substitute, and can result in a slightly higher oil yield of improved quality, as well as a higher hydrogen content in the gas. These improvements result in part from increases in the carbon-steam and gas shift reactions. As can be expected, the improvements are much less than obtained with pure hydrogen, and with some shales there may be little increase or even a reduction in liquid yield.[4] The most important role of steam in oil shale pyrolysis seems to be one of temperature moderation, due to its high heat capacity. Carbon dioxide also has a high heat capacity but does not affect the organic carbon conversion relative to nitrogen atmospheres, however, it does reduce carbonate decomposition. It is of interest that steam atmospheres have been found to promote carbonate decomposition.[9] Use of carbon dioxide is not economically attractive, as its separation and recovery from the product gases is more costly than for steam. The use of inert gases is of major interest in directly heated processes, as they provide a means of both limiting combustion and moderating temperatures.

Retorting

When pyrolysis is carried out in a vessel called a "retort," the process is termed "retorting." A Fischer assay unit is an example of a retort, although the word "retorting" is most generally applied to the commercial-scale recovery of shale oil. As with coal gasification, oil shale may be mined and retorted on the surface, or it may be retorted *in situ* and the released oil collected and pumped to the surface. Commercial-scale oil shale retorts are generally either moving packed beds or solids mixers. An *in situ* retort is in effect a moving bed reactor, but with the retorting zone moving through the stationary shale. Entrained flow reactors are generally not considered for oil shale processing.

Oil shale retorts, like coal gasifiers, are classified according to whether they are directly or indirectly heated. In *directly heated* processes, heat is supplied by burning a fuel, which may be recycled retort off gas, with air (or oxygen) within the bed of shale. Depending on the flow conditions some portion of either the coke residue or the unretorted organic matter may be burned as well. In many designs, most or even all the heat is provided by combustion of the kerogen or coke residue. In the *indirectly heated* processes, a separate furnace is used to raise the temperature of a heat transfer medium that is then injected into the retort to provide the heat. Two subclasses of indirectly heated retorts arise according to whether a solid or gaseous heat transfer medium is used. In the case of a gas, the shale is heated by *gas-to-solid* heat exchange as is always the case in directly heated retorts. If a solid is used, heating is by *solid-to-solid* exchange. The furnace may be fired with retort off gas or crude shale oil. If retorted shale is used as the heat carrier, then the energy content of the coke residue is partly recovered by its combustion in the furnace. The three heating methods are illustrated in Figure 7.2.

Much of the heat required for retorting is for heating the large quantity of inert material and for vaporization of the liquids evolved, with the heat of the pyrolysis reactions themselves being relatively small. A total heat requirement for raising the temperature of the feed material to about 500°C and producing the oil and gases in the vapor state at this temperature is about 0.65 to 0.75 MJ/kg shale.[2] This value may be compared to the calorific value of the shale which, according to Eq. (1.5), ranges from 5 to 8 MJ/kg for the Green River shale grades of commercial interest.

The calorific values of the shale and pyrolysis products may be used to construct a thermal balance for a retorting process. This has been done in Table 7.4, which is based on the pyrolysis of 100 kg of Green River oil shale by the modified Fischer assay procedure. These data and the heat of retorting value given above may be used

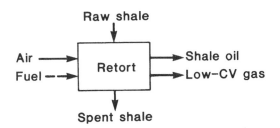

(a) Directly heated retort

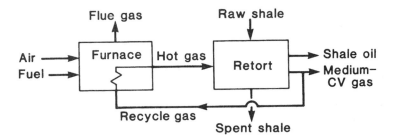

(b) Indirectly heated retort, gas–to–solid heat exchange

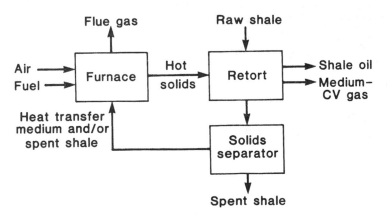

(c) Indirectly heated retort, solid–to–solid heat exchange

Figure 7.2 The three heating methods for oil shale retorting.

Table 7.4 Thermal balance for pyrolysis of a 120 L/t Green River oil shale by modified Fischer assay*

	Mass, kg	Heat Content, MJ/kg	Heat, MJ	Heat, % of Total
IN				
Organic matter	16.0	41.4†	662	109
Inorganic matter	84.0	−0.6	−52‡	−9
Total	100.0	6.1§	610	100
OUT				
Oil	11.0¶	43.0**	473	77
Gas	1.6¶	25.3**	40	7
Water (steam)	2.0††	2.7‡‡	5	1
Retorted shale	85.4§§	0.7**	60	10
Unaccounted	—	—	32§§	5
Total	100.0		610	100

* The heat of retorting is not included; see text and Eq. (7.7).
† Based on Dulong formula, Eq. (1.1).
‡ Endothermic heat of carbonate decomposition, by difference.
§ Mean value for total shale from Eq. (1.5).
¶ From Eq. (7.3).
** Based on data in Ref. 2.
†† Taken as 2 percent of the shale.
‡‡ Based on latent heat.
§§ By difference.

to calculate a thermal efficiency for a retort. Following Eq. (2.24), we define the thermal efficiency as

$$\eta = \frac{\text{GCV of oil + gas}}{\text{GCV of shale + Heat of retorting}} \qquad (7.7)$$

Assuming no heat recovery, we obtain as a lower limit for the efficiency of an idealized Fischer assay process

$$\eta = \frac{473 + 40}{610 + 70} \times 100 = 75\%$$

This efficiency value can be used to gauge the performance of the commercial retorts to be discussed. Note that in practice higher values can be obtained by sensible heat recovery, by utilizing the energy associated with the coke in the retorted shale, and by adopting retorting conditions that result in greater than Fischer assay yields.

Surface Retorting

All surface processing operations involve *mining, crushing*, and then *retorting*. The liquid product of retorting is too high in nitrogen and sulfur to be used directly as a synthetic crude for refining, and requires *upgrading* by coking and/or hydrotreating. The spent shale remaining after retorting amounts to 80 to 85 percent by mass of the mined shale, so *solid waste disposal* is a major activity. These operations are shown in Figure 7.3. In this section the emphasis is on retorting, while the following section discusses upgrading operations. Shale disposal is discussed in Chapter 9.

Some surface oil shale processes that are either well developed, or that are believed to have considerable potential, are listed in Table 7.5. An excellent review of these and other processes is given in Refs. 2 and 5. Apart from the mode of heating (Figure 7.2), many of the processes are superficially similar. As we have seen, the temperature of retorting is constrained to a narrow range, and apart from the case of retorting in a hydrogen atmosphere, there is no incentive to retort at pressures much above atmospheric. Differences among the processes are largely confined to the method of introducing and removing the shale, and the relative directions that the shale, oil, and gases move through the retort. As seen in Table 7.5, in most of the processes the shale is heated by a hot gas, although in two of the indirectly heated processes a hot solid is used to transfer the heat. For our purposes it is sufficient to describe in detail only two of the processes in the table. The Paraho process was selected as an example of a directly heated retort, and the TOSCO II system as an example of an indirectly heated retort. Relevant characteristics of the other processes will, however, be noted where appropriate.

Paraho Directly Heated Retort

The Paraho retort was developed from the U.S. Bureau of Mines' Gas Combustion retort, which in turn evolved from the Nevada-Texas-Utah (NTU) batch retort that was tested in the late 1940s. The word ''Paraho'' is derived from the Portuguese ''para homem'' meaning ''for mankind.'' The Paraho retort can operate in both the direct

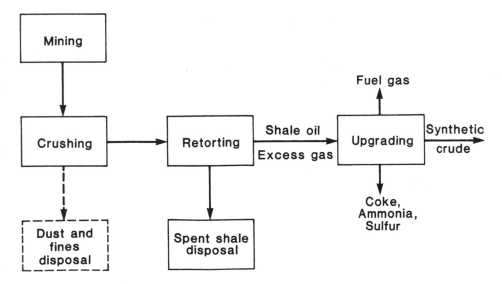

Figure 7.3 Flow diagram for surface processing of oil shale.

Table 7.5 Oil shale technologies for surface retorting

Name	Heating Mode	Heat Transfer	Stage of Development
Union Oil	Direct	Gas-to-solid	Demonstrated (discontinued)
Gas Combustion	Direct	Gas-to-solid	Discontinued
Paraho	Direct	Gas-to solid	Developed
Union Oil	Indirect	Gas-to-solid	Developed
Paraho	Indirect	Gas-to-solid	Developed
Petrosix	Indirect	Gas-to-solid	Developed
IGT Hydroretorting	Indirect	Gas-to-solid	Developing
TOSCO II	Indirect	Solid-to-solid	Developed
Lurgi-Ruhrgas	Indirect	Solid-to-solid	Commercial*

* Commercial plants for hydrocarbon cracking and coal devolatilization.

and indirect heating mode; the flow scheme in Figure 7.4 is for the retort operating in the directly heated mode. The retort, a vertical refractory-lined kiln, employs countercurrent flow of gas and solids, with the shale moving continuously downward by gravity. Recycle gas is introduced with air as oxidant towards the bottom of the retort, so that the flow system closely resembles the Lurgi coal gasifier shown in Figure 4.2.

Raw shale, crushed to between 5 and 75 mm, is fed into a hopper at the top of the retort and spread evenly onto the shale bed by a rotating distributor system. The shale is heated by the rising hot gases as it descends from the preheating zone into the retorting zone, where the kerogen is pyrolyzed to produce the oil and gas products. The rising gases carry the products out of the retort through the gas withdrawal manifold. Because the gases are cooled by the incoming shale, much of the oil product condenses and leaves the retort as a mist.

The retorted shale, which still contains 20 to 25 percent of the original organic matter as coke residue, next descends into the combustion zone, where the coke is ignited in the oxygen-containing, air/recycle-gas mixture. About half the coke residue is burned, and this provides most of the heat required for pyrolysis. Little recycle gas is burned during combustion of the coke; it serves mainly as a diluent for temperature control to limit carbonate decomposition, shale ash sintering, and clinker formation. The gas distributors are cooled with circulating water.

The oil is recovered by an oil mist separator and an electrostatic precipitator. The oil-free retort off gas has a low calorific value in the vicinity of 4.3 MJ/m³. We recall that low-CV gas is typical of directly heated air-blown processes because of the dilution by nitrogen, and by carbon dioxide formed during combustion. About two-thirds of the gas is recycled to the retort, with the excess drawn off for use as a plant fuel. The spent shale is removed from the retort through a hydraulically operated grate which can be adjusted to control the downward velocity of the shale. A shale with an initial organic content of some 16 percent (120 L/t) would typically contain about 2 percent coke residue on leaving the retort.

The countercurrent gas-solids flow scheme allows for excellent sensible heat

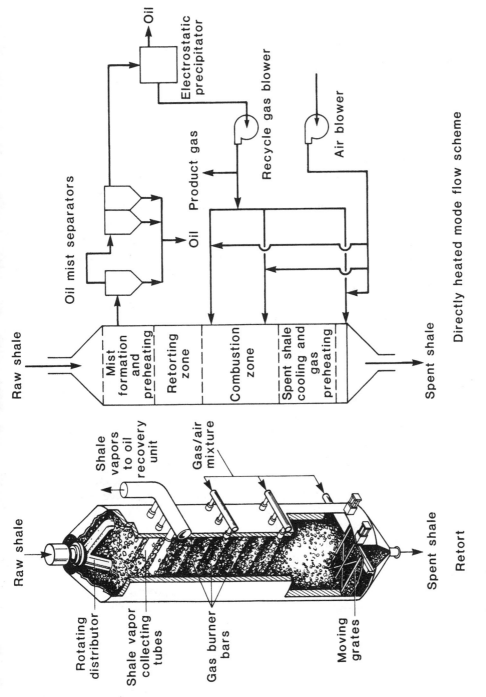

Figure 7.4 Paraho retorting process.

335

recovery. The processed shale leaves the retort at approximately 200°C, while the retort gases are removed at about 65°C. A lower exit-gas temperature is not desirable, as it would result in condensation of significant quantities of water vapor in contact with the oil. Combustion of the coke residue further enhances the thermal efficiency of the process, although it is estimated that some 22 percent of the carbonates are decomposed. Oil yields are about 90 percent of Fischer assay, while gas yields seem to be significantly greater than Fischer assay, probably due to secondary decomposition of the coke in the combustion zone. The calorific value of the gas amounts to about 15 percent of the calorific value of the combined oil-plus-gas product. A thermal efficiency of 86 percent has been estimated for the retort.[12]

Two retorts, a pilot-scale and a semiworks unit, have been operated by Paraho at Anvil Points, Colorado. The pilot retort had a 0.76 m inside diameter, was 18 m tall, and processed shale at about 1 kg/(m²·s), or about 40 t/d. The semiworks unit shown in Figure 7.5 has a 2.6 m inside diameter, is 23 m high, and throughputs have reached 360 t/d. These data suggest shale residence times of from 4 to 6 hours. A full-scale commercial unit is designed to have 20 times the planned capacity of the semiworks unit. It would be about 32 m tall, have an outside diameter of 12.8 m, and handle 10 500 t/d of shale. Production when using 120 L/t grade shale is projected to be about 1.2×10^6 L/d, so a commercial plant operating 6 retorts would have a crude shale oil output of 7.2×10^6 L/d.

The principal differences between the Paraho retort and its predecessor, the Gas Combustion retort, are in the feed distribution and shale discharge mechanisms. These modifications and the ability to more closely regulate temperatures were introduced into the Paraho retort to avoid clinkering, which was a continual problem in the earlier Gas Combustion tests. In the retorts of the Union Oil Co., an example of which is shown in Figure 7.6, the solids-handling problem is approached in a more direct way. Shale is charged into the lower and smaller end of a truncated cone and is pushed upward by a piston referred to as a "rock pump." Gas flow is countercurrent from the top. In the directly heated mode, air without recycle gas is used, and nearly all the energy of the residual carbon is recovered by combustion. Cooling water is not required. Thermal efficiencies of 83 percent have been reported.[2]

Both the Paraho and the Union retorts can be operated in the indirect heating mode. The Union B retort, which is the retort proposed for commercialization, is intended for operation in the indirect mode. In this mode, recycle gas is heated in a furnace before being recycled to the retort. Combustion does not occur within the retort. The Petrosix retort shown in Figure 7.7 was developed for the Brazilian government and was built specifically for operation in the indirectly heated mode. The name Petrosix is derived in part from Petrobras, the Brazilian national oil company. Its design was also based on the Gas Combustion retort and it is consequently very similar to the Paraho system. One difference is that the spent shale in the Petrosix system is discharged into a water bath, as is the case for the Union retort. In the Paraho retort, the shale is discharged dry. A 5.5 m diameter demonstration unit designed to handle 2000 t/d of shale has been operated in Brazil since mid-1972. Construction of a 6.8×10^6 L/d commercial facility based on twenty 11-m-diameter Petrosix retorts is currently being considered by the Brazilian government.

Figure 7.5 Paraho semiworks retort at Anvil Points, Colorado; the retort is 23 m high and has a 3.2 m outside diameter.

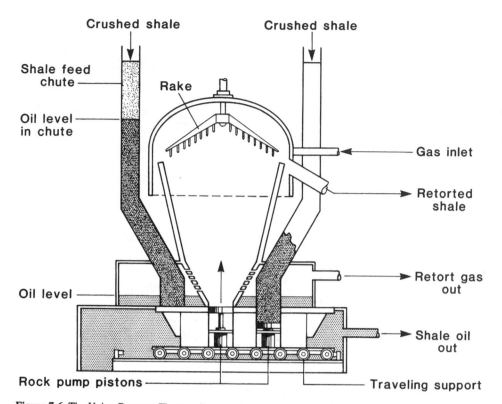

Figure 7.6 The Union B retort. The traveling support system moves each rock pump from its supply chute to the retort feed cone.

TOSCO II (Indirectly Heated) Retort

All the retorts mentioned so far, whether operated in the direct or indirect mode, heat shale by contact with a hot gas. As shown in Figure 7.2, in indirectly heated retorts this gas-to-solid heat transfer may be replaced with solid-to-solid heat transfer by mixing a preheated solid with the shale. The hot solid may be recycled spent shale, or may be a special-purpose material introduced into the system specifically for heating the shale. An example of this latter case is the TOSCO II process shown in Figure 7.8, in which 13-mm-diameter ceramic balls are used as the heat transfer medium.

TOSCO (previously The Oil Shale Company) purchased rights to the Aspeco retort, that had been patented by the Swedish inventor Aspegren in 1964. Further development work carried out by the Denver Research Institute led to the process shown schematically in Figure 7.8. Crushed shale of less than 13 mm in size is preheated by pneumatically conveying the shale upward through a vertical pipe cocurrently with hot flue gases from the ball heater. The flue gas is cooled during this process and the cooled gas is passed through a venturi wet scrubber to remove shale dust before venting to the atmosphere, at a temperature of about 50 to 55°C.

The ceramic balls are heated in a vertical furnace and then fed along with the preheated shale, which has been separated from the flue gas in settling chambers and cyclones, into a horizontal rotating kiln where the pressure is slightly above atmospheric. The mixture of balls and shale flows through the kiln, bringing the shale to a retorting temperature of about 510°C through conductive and radiative heat exchange with the balls. The resulting hydrocarbon and water vapors are drawn off and fractionated, leaving behind a mixture of balls and processed shale.

The ceramic balls are separated by size from the fine-powdered spent shale by passage through a trommel, a heavy-duty rotating cylinder with many small holes punched in its shell. Warm flue gas is used to remove residual dust from the ball circulation system. The dust is removed from the flue gas with a venturi wet scrubber.

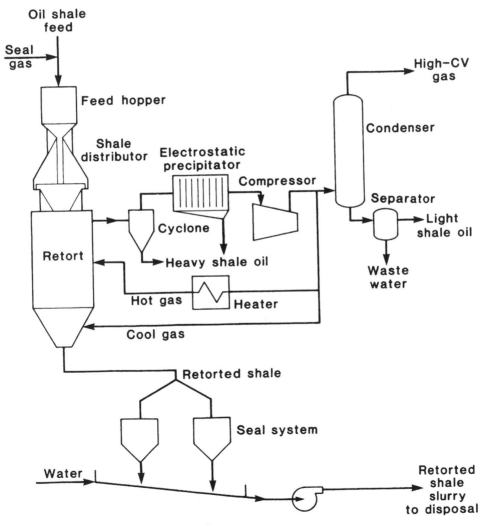

Figure 7.7 Petrosix indirectly heated retorting process.

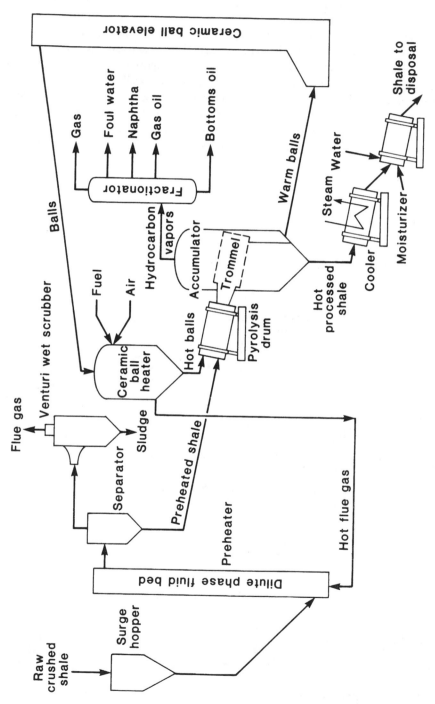

Figure 7.8 TOSCO II retorting process.

With a bucket elevator, the balls are then circulated back to the ball heater for reheating by burning some of the product gas.

The processed shale is cooled in a rotating drum steam generator, then moistened to approximately 14 percent moisture content in a rotating drum, after which it is transported by a conveyor belt for disposal. The steam and processed shale dust produced in the moisturizing process are passed through another venturi wet scrubber to remove the dust before discharge to the atmosphere. The importance of the moisturizing is that addition of the water to the TOSCO-type processed shale, at a predetermined shale temperature, appears to lead to cementation of the shale after proper compaction.

In 1964 TOSCO, the Standard Oil Co. of Ohio (SOHIO), and Cleveland Cliffs Iron Co., formed the Colony Development Company and subsequently operated a 900 t/d semiworks plant near Grand Valley in Colorado. A commercial-scale unit would be 10 times larger. *Oil* yields obtained were 93 percent of Fischer assay, with approximately 0.66 kg oil and 0.12 kg high-CV gas produced per unit mass of organic matter in 158 L/t shale. About 20 percent of the original organic matter remains in the spent shale, which contains approximately 4.5 percent organic carbon on a dry basis. As is typical for indirectly heated retorts, the gas is undiluted by combustion products, and has a relatively high calorific value of about 28 MJ/m^3. A disadvantage of the indirectly heated processes is that there is a significant loss of heat with the flue gases, and thermal efficiencies can consequently be expected to be lower than for directly heated processes. Efficiencies of 76 percent and 73 percent have been estimated for the TOSCO II and Paraho indirectly heated retorts respectively.[12]

A further disadvantage of the indirectly heated retorts discussed so far is that the energy associated with the residual coke is not recovered. In a proposed modification of the TOSCO process, called TOSCO V, this deficiency is overcome by replacing the ceramic balls with spent shale. A similar process is that of Lurgi-Ruhrgas, which has been developed commercially for the devolatilization of subbituminous coal and the cracking of liquid hydrocarbons. Units suitable for the retorting of 4500 t/d of oil shale are presently in operation for cracking naphtha. In the oil shale retorting version of the process, as shown in Figure 7.9, spent shale is mixed with the incoming raw shale in a sealed screw-conveyor, which acts as the retort. All of the material is discharged into a surge bin. The vapor stream is drawn off to a condenser to produce the gas and liquid fractions. Some spent shale is withdrawn from the surge bin as waste and the remaining solids are transferred to the end of a lift pipe. Here they are heated by combustion of the carbonaceous residue on the spent shale (and by burning supplemental fuel if needed). The hot solids are then used once again for retorting.

Hydroretorting

A very different concept of retorting than hitherto described is that of pyrolysis in a hydrogen atmosphere at elevated pressure. Hydroretorting is of particular interest for processing the low-hydrogen, high-aromatic Devonian oil shales of the eastern United States. One study has estimated that the known recoverable resources in these deposits are sufficient to support over 1000 commercial-size plants with a potential of yielding

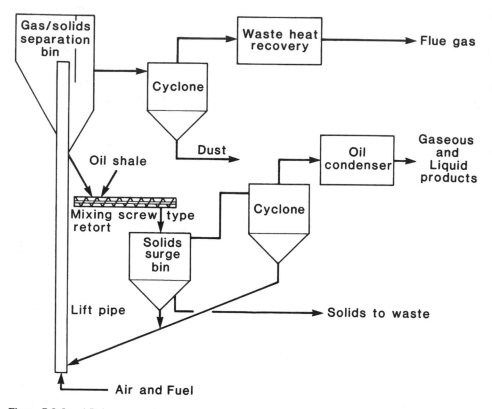

Figure 7.9 Lurgi-Ruhrgas retorting process.

oil at a rate of some 8×10^6 L/d, or 52×10^{12} liters of oil over the 20-year life span of the plants.[13] These oil yields are based on nearly 90 percent organic conversion, and are justified on the basis of preliminary hydroretorting tests conducted by the Institute of Gas Technology (IGT).[11,13] Several schemes for processing the eastern shales have been investigated by IGT. Initially, gasification processes were studied with the object of producing a high-CV gas. More recent tests indicate that either a middle distillate oil or a high-CV gas can be selectively produced by appropriate control of the operating conditions.

A bench-scale unit with a nominal capacity of 1 t/d, and a process development unit with about 10 times this capacity have been operated by IGT with hydrogen pressures of up to 3.5 MPa and more. Both Green River and Devonian oil shales have been processed, with organic conversions of 95 percent in the former case, and up to 90 percent in the latter case, being achieved. In a commercial plant producing a synthetic crude, hydrogen for retorting and upgrading could be made from steam reforming the byproduct retort gas. If SNG were the desired product, hydrogen could be obtained in part by gasification of the crude shale oil produced. As can be expected for a high pressure system, the retort is more complex than units described previously. In addition, the overall process is more involved due to the need to produce hydrogen.

Nevertheless, it is stated that overall costs for hydroretorting eastern shales are comparable to those for conventional processing of western shales.[13]

In Situ Retorting

In situ retorting offers the possibility of eliminating the problems associated with the disposal of the large quantities of spent shale that occur with surface retorting. In addition, mining costs are largely reduced or eliminated, so that recovery of lower-grade shales down to about 40 L/t becomes a possibility. A major problem arises with *in situ* retorting if the oil shale deposit is associated with groundwater. Aquifers are found both above and below shale zones in some oil shale formations. In such cases the deposit to be retorted *in situ* must first be dewatered, the water disposed of, and a means provided to ensure that the groundwaters are not contaminated by contact with the organic residue remaining after retorting is completed.

True in situ (TIS) retorting[2,5] involves fracturing the shale in place, igniting the fractured shale, feeding in air to sustain the combustion for pyrolysis, and pumping out the product oil and gas. The process is superficially similar to underground coal gasification discussed in Section 4.5. Oil shale is not porous and generally does not lie in permeable formations, so adequate flow paths are difficult to create. The result is that the gas flow is very difficult to control and the oil yields are low.

To overcome this difficulty, an alternative approach known as *modified in situ* (MIS) has been developed.[2,5] In this procedure, some portion of the shale is mined out using conventional techniques, and the remaining shale is "rubblized" by exploding it into the mined void volume. The resulting oil shale rubble constitutes the retort.

Occidental Oil Shale, Inc., a subsidiary of Occidental Petroleum Corporation, has been particularly active in developing MIS retorting towards commercialization. Their "Oxy" process has evolved from tests initiated in 1972 at Logan Wash near Debeque, Colorado, where the shale grade is about 60 L/t. Commercial-scale retorting is currently being planned for Federal lease tract C-b in Colorado, where the average shale grade is substantially higher, at about 110 L/t.

Each commercial-size retort in the Oxy system may measure 60 m × 60 m × 95 m high. About 25 percent of the shale in this block is mined out according to a pattern designed to give a uniform voidage distribution on explosion. After the retort is formed, connections are made to the surface for withdrawing the oil and gas, and for pumping down air and steam. The completed retort is shown in Figure 7.10. Retorting is started at the top of the rubble pile, which is ignited by injecting air, and burning fuel gas. Heat from the burning zone is carried downward and pyrolyzes the shale in lower layers. The combustion zone advances vertically down the retort, preceded by the pyrolysis zone. The fuel gas is required only to initiate combustion; thereafter, the residual carbon remaining in the pyrolyzed shale is sufficient to supply the required heat. The oil vapors and gases are swept down by the overall gas flow to the lower regions of the retort. The oil and some water vapor condense out in the cooler lower regions and then percolate down to the sump at the bottom of the retort, from where they are pumped to the surface. The underground process operates as a packed bed, directly heated batch retort.

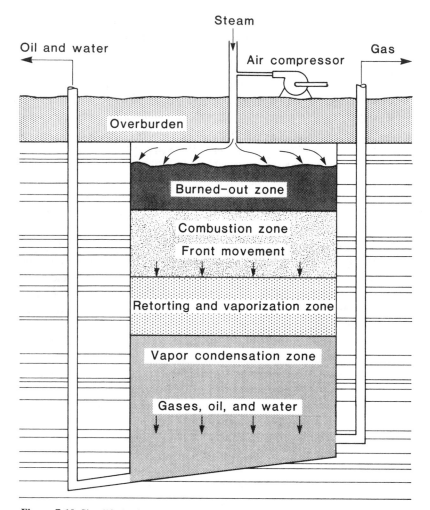

Figure 7.10 Simplified schematic of Oxy MIS retort.

Oil losses can occur by combustion due to injection of excessive oxygen, by thermal degradation to coke and gases due to excessive temperatures, and by partial or total bypassing of wall or other areas due to non-uniform flow. Two conflicting requirements are the need to limit oxygen supply and the need for a high gas flow. The high gas flow serves as a mass transport medium to carry off the product vapors, as a heat transport medium to convect the heat of combustion downstream to the pyrolysis zone, and is also necessary to achieve a flat velocity profile across the retort, with a uniformly advancing combustion/pyrolysis front.

The use of a diluent gas with a high heat capacity has been shown in model studies[14] and retorting tests to significantly increase oil yields. Recycled retort off gas was earlier used as diluent, but after startup its fuel value is not needed and this practice was considered wasteful of a potentially valuable byproduct. Carbon dioxide and steam are both suitable diluent gases with relatively high heat capacities. As discussed previously, steam is cheaper to use and can result in a slightly improved oil quality. Steam, at a pressure of about 1.7 MPa to achieve uniform distribution, has been selected for the Oxy process.

To obtain a uniform retorting pattern and to minimize bypassing, it is essential that the rubblized shale have a uniform voidage and acceptable particle size distribution. Considerable effort has been expended in determining the optimum mining patterns and explosion sequencing to achieve the desired uniformity. A secondary concern in mining out the shale necessary to create the voids is to selectively remove shale of above-average grade. As one part of shale is removed for every three parts retorted underground, surface retorting of the mined shale will probably be practiced in conjunction with *in situ* operations to maximize resource recovery. Selection of higher-grade shale for the surface operation effectively reduces the cost of mining and retorting per unit of oil produced.

Feed and product rates. The projected mass flow rates to an Oxy MIS retort are about 7.2 parts air and 2 parts steam for each part of oil produced. These rates can be used to make an approximate estimate of retorting conditions. It may be assumed that the oxygen is completely consumed according to

$$C + O_2 \rightarrow CO_2 \qquad \Delta H^\circ_{298} = -393.5 \text{ kJ/mol} \qquad (7.8)$$
$$\text{kg: } 12 \quad 32 \qquad 44$$

As air contains 23.2 percent oxygen by mass, the 7.2 kg air will consequently consume $7.2 \times 0.232 \times 12/32 = 0.63$ kg carbon.

If we assume that the oil yield is 100 percent of Fischer assay, then Eq. (7.3) can be used to show that each 1 kg of oil would derive from $1/0.69 = 1.45$ kg organic matter and leave $0.21 \times 1.45 = 0.30$ kg residue. According to Table 7.3 the residue is 86.6 percent carbon, so that about 0.26 kg carbon remains for combustion for each kg oil produced. As shown, the air supplied is sufficient to consume 0.63 kg carbon. This means that our assumption of 100 percent Fischer assay oil yield is wrong. The above calculational procedure can be used to solve for the oil yield directly, and shows that oil production is about 63 percent of Fischer assay. This is in good agreement with the 60 percent projected for the commercial retorts. The difference between the projected and calculated Fischer assay yields is due to uncombusted carbon that remains on the spent shale. A detailed material balance in which consumption of oxygen to form water is included, shows that about 17 percent of the original organic carbon remains behind on the spent shale, and that about 48 percent of the carbon in inorganic carbonate is released as carbon dioxide.

The heat of reaction for Eq. (7.8) amounts to 20.5 MJ/kg of oil for the quoted air-to-oil mass flow ratio. For a 60 percent Fischer assay yield and assuming a 15 percent organic content for the 110 L/t oil shale, the heat produced by combustion translates to about 1.3 MJ/kg of shale. This is nearly double the 0.65 to 0.75 MJ/kg range for the heat of retorting specified earlier. To this must be added the heat to generate the injected steam, which is some 5.4 MJ/kg of oil produced, or 0.3 MJ/kg of shale retorted, as well as the compression costs for the large air requirement.

Clearly, energy consumption per unit of oil produced is relatively high for *in situ* retorting, which may be ascribed in part to the large heat capacity of the retort and to heat losses through the "walls" of the retort. This high energy consumption must be weighed against the advantages of reduced mining, spent shale disposal, and capital costs for *in situ* processing.

The mass flow rate of the dry retort off gas is about 9.5 times the oil rate. Due to the large air feed rate and the production of carbon dioxide by combustion and carbonate decomposition, the dry gas contains about 60 percent nitrogen and 25 percent carbon dioxide by volume. As can be expected, the calorific value is very low, at about 2.8 MJ/m³. On an energy basis the crude shale oil represents approximately 68 percent and the gas 32 percent of the total production. In spite of the low calorific value of the gas, it is anticipated that the excess gas remaining after satisfying plant thermal requirements can be used for generation of electricity either in a gas turbine or a combined cycle system.

The steam entering the retort, together with the water formed by reaction and the free and combined water released from the shale, must leave the retort with the oil and gases. Some fraction of the total water vapor will condense out with the oil in the retort, and will be later recovered as a very dirty water. Much of the remainder that leaves with the gases will be condensed out during gas purification to form a less dirty water stream. The water that condenses in the retort is called a "retort water," while the water from the gas stream is called "gas condensate." The retort water rate will initially amount to about half the oil rate, and gradually diminish as the *in situ* retort heats up. The wastewater rates are considerably higher than arise with surface retorting, as is the source water requirement for steam production. There is consequently some interest in using the untreated wastewater directly for raising a low-quality steam suitable for retorting purposes.

The *in situ* Oxy retorts at the commercial site at tract C-b will be arranged in groups called "clusters," and the clusters will in turn be grouped to form "panels." Current plans call for 8 retorts per cluster, and 32 clusters per panel. A total of 15 panels may be retorted in sequence. Based on a production rate of 11 million L/d of crude shale oil, a panel would last about 4 years, and plant life would be about 50 years with a total oil yield of nearly 0.2×10^{12} liters.

7.2 INTEGRATED OIL SHALE PLANTS

Crude shale oil is the major product produced in conventional retorting of Green River oil shales. Characteristics of these crude oils typical of the various retorting processes discussed in the previous section are listed in Table 7.6. In the table, an oil produced by a Union B retort has been selected as being representative of indirectly heated retorts, and is compared with oil from a directly heated Paraho retort. Although a MIS retort is also directly heated, processing conditions are significantly different, and an oil typical of the Oxy process has therefore been included in the table. One difference between the oils produced by the two heating modes is the oxygen concentration. A higher oxygen content is typical of published compositions for directly heated retorts,

Table 7.6 Typical properties of Colorado Green River crude shale oils[10,15-17]

	Surface Retorts		
	Indirectly Heated (Union B)	Directly Heated (Paraho)	MIS Retort (Oxy)
Elemental analysis (mass %)			
Carbon	85.3	84.5	84.3
Hydrogen	11.2	11.3	11.9
Nitrogen	1.77	1.96	1.45
Sulfur	0.61	0.64	0.58
Oxygen	1.12	1.60	1.77
Total	100.00	100.00	100.00
Hydrocarbon type (mass %)			
Asphaltenes	4	4	—
Resins*	24	26	—
Oils	72	70	—
Total	100	100	
Trace elements (μg/g)			
Arsenic	52	28	18
Iron	60	100	27
Nickel	2.5	4	7
Mean molar mass	—	297	279
H/C atom ratio	1.56	1.60	1.68
Gravity, °API	21.2	21.4	25.0
Specific gravity	0.927	0.925	0.904
GCV, MJ/kg	42.7	—	—

* Polar compounds.

but it is not clear whether the additional oxygen is organically bound, or whether it results from a higher content of (emulsified) water in samples from directly heated processes.[15]

Apart from oxygen concentration, differences between the oils from directly and indirectly heated surface retorts appear minor and probably are as much due to differences in the raw shales retorted as to differences in retorting mode. Although not apparent from the tabulated compositions, it is sometimes stated that oil from indirectly heated retorts is lighter and more amenable to hydrotreating. A moderate improvement in quality is apparent from the data given here for the MIS oil relative to that from both the surface-retorted products. The H/C ratio and API gravity are higher, while nitrogen and arsenic contents are lower in the MIS case.

Distillation curves for the representative shale oils of Table 7.6 are compared in Figure 7.11, and further demonstrate the difference in quality between the products of MIS and surface retorting. The curves for the Union B and Paraho oils very nearly coincide and have been drawn as a single line for clarity. The MIS oil is seen to have a flatter distillation curve with less naphtha as well as less heavy ends than the Union B and Paraho oils. The differences are attributable both to the slower retorting rate

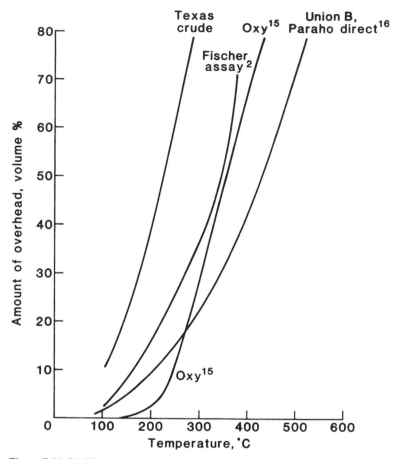

Figure 7.11 Distillation curves for Colorado shale oils and a petroleum crude.

of the large size shale rubble, and the higher partial pressure of steam in the MIS retort. The Oxy oil reportedly undergoes mild coking during retorting, and some combustion of the heavy ends occurs as well.[15,16]

As can be expected from their higher H/C ratio, crude shale oils more closely resemble petroleum crudes than do primary coal liquids. A comparison of the two classes of synthetic crudes can be made from Figures 6.19 and 7.11 using the curve for the high-quality Texas crude as a guide. While it is clear that shale oils require less hydrogenation than do primary coal liquids, the two fuels may be viewed as being complements rather than alternatives for supplying the broad-based liquid fuel market.[5] As we saw in Section 6.5, mild upgrading of coal liquids produces some naphtha suitable as a gasoline feedstock, with the remainder being useful mainly as a heavy boiler fuel. Shale oils, on the other hand, lie mainly in the mid-distillate range and appear to be ideal for conversion to diesel and jet fuels.

In reviewing the possible uses for shale oils, we again begin by considering the possibility of using the crude product directly as a boiler fuel. The API gravity range of 20 to 25 shown for the shale oils in Table 7.6 puts these oils in the class of No. 4 to No. 5 fuel oils. However, sulfur and nitrogen levels are as high as in solvent refined coal, and the same provisos regarding the emission of sulfur and nitrogen oxides as discussed in Section 6.5 for SRC combustion apply. Distribution of crude shale oil presents a further problem. In spite of their relatively high H/C ratios, crude shale oils are, in general, not pumpable, probably due to their high nitrogen levels.[18] Addition of chemical viscosity depressants has in some cases been successful in improving transportation properties. Conventional petroleum crudes are also available in the Colorado area, and are being considered as blending agents to decrease viscosity. This may, however, be a wasteful use of the petroleum.

At the other end of the spectrum, shale oils may be viewed as feedstock for the production of high-grade transportation fuels. As with coal liquids, the shale oils would require initial hydrotreatment before being subjected to "conventional" refining. Complete refining to high-grade products is not considered an option for an integrated oil shale plant. However, while some developers might export the raw shale oil with no more than, say, addition of a viscosity depressant, others have indicated that a certain degree of upgrading will be carried out to produce a synthetic crude. The options available to the developers are illustrated in Figure 7.12.

Upgrading

Probably the most detrimental characteristic of shale oils in relation to refining is the *nitrogen* concentration of around 1.7 percent. This value is higher than the 1 percent typical of primary coal liquids, and of course much greater than the 0.3 percent typical of conventional refinery feedstock. Reference 18 is a detailed review of nitrogen compounds in shale oils and upgraded products. We recall that high nitrogen concentrations cause rapid poisoning of cracking catalysts and are associated with poor product stability. Therefore, as with coal liquids, denitrogenation is an essential upgrading requirement. Hydrodenitrogenation is again an important processing step, although in the case of shale oil, coking may also denitrogenate because much of the nitrogen is present in high-molar-mass compounds. As shown in Figure 7.13, nitrogen concentrations increase from about 1 percent in the naphtha to more than 2 percent in the heavy fractions.

Sulfur concentrations are not significantly higher than in many petroleum crudes, and (as shown in Figure 7.13) are distributed uniformly over the boiling fractions. Hydrodesulfurization occurs more readily than hydrodenitrogenation, so that control of nitrogen to satisfactory levels is expected to result in acceptable sulfur removal. Although *oxygen* concentrations are generally higher than in petroleum crudes, removal of oxygen is not considered a potential refining problem. With reference to oxygen combined as water, it has been found that water emulsions are readily broken on heat treatment and that the water may then be removed by settling.[15]

Of the trace elements shown in Table 7.6, oil soluble *arsenic* is the most troublesome, as it rapidly deactivates certain refinery catalysts. While arsenic may possibly be removed by alkali or thermal treatment, many published upgrading schemes include a catalyst guard bed to protect the downstream catalyst.

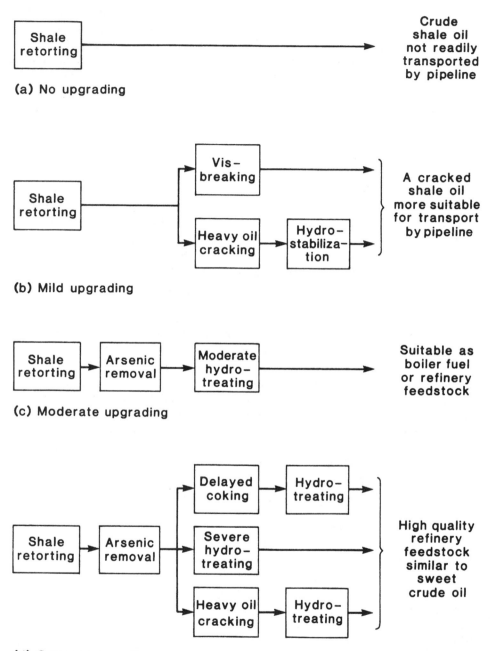

(a) No upgrading

(b) Mild upgrading

(c) Moderate upgrading

(d) Severe upgrading

Figure 7.12 Processing alternatives for shale liquids according to the degree of upgrading required.[16]

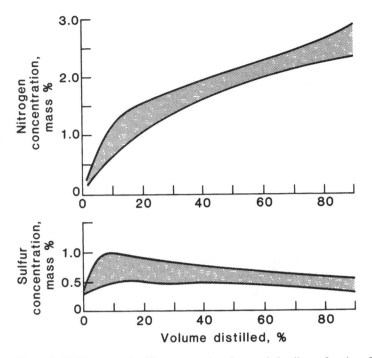

Figure 7.13 Nitrogen and sulfur concentrations in raw shale oil as a function of distillation range.[10]

Visbreaking has been shown in Figure 7.12 as a mild upgrading step to produce a pumpable oil. The name derives from the process of cracking, or breaking, large molecules to reduce the viscosity of the product. The crude shale oil is heated to between 420°C and 480°C and maintained in this temperature range for from several seconds to several minutes depending on the desired severity of treatment. The decrease in viscosity is accompanied by some loss of product in the form of a gas released during cracking. There is little change in nitrogen, sulfur, and oxygen concentrations. The process is regarded as being low in capital costs but having a relatively high energy consumption. The gas formation represents an energy loss, while sensible heat requirements depend on the degree of heat exchange effected between feed and product.

In the case of more severe hydrotreating typified by Figure 7.12d, the first step may be distillation into three or four fractions, followed by hydrotreatment, cracking, or coking. A simplified flow scheme of one possible upgrading process is shown in Figure 7.14. Based on the shale oil distillation curves shown earlier, the split of liquid products can be expected to be about 10 percent naphtha, 50 to 60 percent middle distillates, with the remainder as bottoms. Of interest is that the naphtha fraction typically has a H/C atom ratio greater than 1.8, while that in the bottoms is greater than about 1.4.[10] Even the heavy fraction contains relatively more hydrogen than a typical primary coal liquid.

The relative quantity of naphtha produced can be increased by introducing a

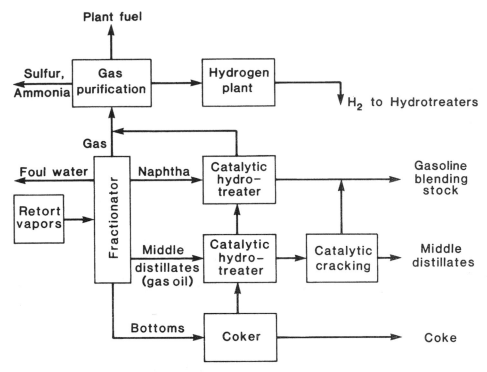

Figure 7.14 Simplified shale oil upgrading process.

catalytic hydrotreatment or coking step prior to fractionation. Hydrogenation of the whole oil is advantageous, as the more difficult heavy fraction is hydrogenated along with the lighter ends and little of the low-value residuum remains. A first stage hydrotreatment may be used to increase the yield of gasoline blending stock.

Coking may be used as a first step to increase the yield of diesel and heating fuels, and has the advantage of simultaneously removing solids and ash present in the oil. Coking is in general less costly than hydrotreatment, but not as energy efficient for production of gasoline. While coking represents an oil loss in the form of coke, hydrotreatment represents an energy loss for production of the required hydrogen.

It is stated in Section 6.5 that coking is not appropriate to coal liquids, as it produces only a coke and a gas. The higher H/C ratio in shale oil results in coking which produces mainly a lighter liquid product called a coker distillate. Coking tests with a Paraho shale oil, for example, show that yields are about 71 percent liquid, 18 percent coke, and 11 percent gas. The liquid yield obtained with a MIS Oxy oil was higher, at nearly 89 percent, and can be attributed in part to some coking having previously occurred in the underground retort.[15] Most of the iron was removed, arsenic was reduced to about 6 μg/g, and some reduction in nitrogen, oxygen, and sulfur occurred during coking.

Like visbreaking, coking is a thermal process and does not require a catalyst. Temperatures are higher and residence times longer than in visbreaking, allowing polymerization reactions to proceed, resulting in the deposition of coke and so increasing the H/C ratio in the remaining oil. Delayed coking, fluid coking, and Flexicoking are the main procedures used. *Delayed coking* is the oldest and most developed process. The oil is preheated to 480°C in a furnace and then charged to one of two coke drums which may be 3 m in diameter and 30 m tall. The pressure in the drums is sufficient to prevent vaporization, and the coking reactions proceed in the liquid state. Coke is allowed to deposit until the drum is nearly full, when the feed is switched to the second drum and the accumulated coke is removed from the first. A typical cycle time is about 8 hours. The degree of coking is determined by the temperature and residence time in the drum, as well as by the amount of oil recycled.

The coke particles are fluidized in the *fluid coking* process. Steam is used as the fluidizing medium and serves also to strip the distillate oil from the coke particles and carry the product off overhead as a vapor. The coke particles are continuously withdrawn from the coker and fed to a second fluidized bed called the "burner." Here they are fluidized in air and some of the coke is burned, thereby heating the remaining coke to about 620°C. The hot coke is returned to the coker, where it heats the oil feed which vaporizes to yield distillate oil, leaving more coke on the coke particles. The amount of coke burned to provide the coker heat requirement is only about 5 percent of the coke produced, so coke is continuously withdrawn to maintain a constant inventory in the system. The advantages of fluid coking are that it is a continuous, more rapid process, and the shorter pyrolysis time results in improved oil yields over delayed coking.

In the *Flexicoking* process, patented by Exxon, conventional fluid coking is combined with gasification of the coke product. This is essentially a mechanism for recovery of the energy normally lost in the coke, in the form of increased gas yield. The gas may in turn be used to produce hydrogen for hydrotreatment. The gasifier is fed with coke from the burner, and the hot ash and ungasified coke at about 980°C are returned to the burner to recover some of their sensible heat. The burner is required as an intermediate vessel to provide a means of controlling temperatures between the high temperature gasifier and the coker.

Hydrotreatment and cracking processes were discussed previously in connection with upgrading coal liquids. In this section we estimate the hydrogen requirements for hydrotreatment of shale oil. The Paraho product shown in Table 7.6 may be represented by the approximate molar formula

$$CH_{1.59}O_{0.014}N_{0.02}S_{0.003}$$

Based on the reactions with hydrogen to form $CH_{1.8}$, water, ammonia, and hydrogen sulfide (see for example Eqs. 3.53 to 3.56), we can show that 2.1 kg or 24 m^3 of hydrogen are required per 100 kg of crude oil. This is in good agreement with values of 28 to 30 m^3/100 kg of oil reported for hydrotreatment of a Paraho oil to a synthetic crude having a H/C atom ratio of 1.8 and containing about 0.3 percent nitrogen, 0.5 percent oxygen, and less than 0.002 percent sulfur.[17] In practice, hydrogen consumption might be as high as 40 m^3/100 kg of crude oil treated.[16] The difference between the estimated and actual hydrogen requirements is due to the hydrogen consumed in the formation of gases. In the Paraho tests, more than 90 percent of the hydrotreater feed was recovered as C_{6+} liquids, with about 3 percent removed as ammonia, water, and hydrogen sulfide, and the remainder as hydrocarbon gases.

The yields quoted for the hydrotreatment of the Paraho oil were determined in pilot-scale and full-scale refinery equipment. In total, some 14×10^6 liters of crude shale oil were hydrotreated and then further refined to military-specification diesel and jet fuels. The light naphtha fraction obtained on fractionation of the hydrotreated product was very paraffinic and had a low octane rating. Pilot-scale reforming to increase the octane rating was only partially successful due to catalyst deactivation and poor stability of the product. In the refinery run, the naphtha fraction was "discarded" by blending with feed to another hydrotreater. The bottoms fraction was in general equivalent to a good quality No. 6 fuel oil with a relatively high hydrogen content but low viscosity. The major middle distillate fraction produced specification jet and diesel fuels using conventional refining techniques followed by acid treatment that was required to produce a stable product.[17]

These and other tests[19] have shown that refining of oil shale is not entirely without problems. Catalyst deactivation, high hydrogen consumption, and product stability are some of the areas of concern. Nevertheless, both coking and hydrotreatment of whole shale oil have been adequately demonstrated, and these processes are reportedly available for commercial-scale processing.[16] It is concluded that shale oil is amenable to conversion to transportation fuels in existing refineries, but that costs will be higher than for petroleum refining.

Retort gases

Depending on whether the retort is directly or indirectly heated, either a low- or a medium-CV gas is produced along with the oil. Typical gas compositions and the relative flow rates of gas, oil, and combustion air are shown in Table 7.7 for an indirectly heated and a directly heated surface retort, and a MIS retort. In terms of calorific value, the gas produced amounts to 16, 18, and 47 percent respectively of the oil output. Although the 47 percent in the case of the MIS retort is relatively high, the gas is of too low a calorific value for even local distribution. Of interest in this case is the relatively high hydrogen production, which can be attributed to the injected steam. The calorific value in the case of the directly heated surface retort is not much higher, so these gases will invariably be used within the plant boundaries. The excess gas remaining after satisfying plant heating requirements will most likely be used for electricity generation.

While the medium-CV gas from an indirectly heated retort is suitable for SNG production, the amount of gas produced is relatively small. For example, a nominal standard-size oil shale plant will produce gas at a rate of only 10 to 15 percent of a standard-size SNG plant. An important use for the gas, apart from supplying process heat, is as a source of hydrogen for refining. It is of some interest that were the total gas product in the Union B case converted to hydrogen, assuming ideal conversion by methane reforming and gas shift, it would yield about 2.3 kg H_2/100 kg oil. This is close to the theoretical hydrogen requirement of 2.1 kg/100 kg of oil estimated above.

Regardless of the use of the gas, means must be provided for control of the pollutants present. Ammonia and hydrogen sulfide must either be removed directly from the retort off gas, or alternatively if the gas is burned, sulfur and nitrogen oxides may be removed from the flue gases. These options and processes are examined in Sections 5.1 and 5.5, and are also discussed in Section 9.2. As mentioned, spent shale is an effective SO_2 absorbent and provides an additional means of pollution control specific to oil shale plants.

Mining

On average, some 85 000 t/d of shale must be mined to supply a plant producing 10 million liters per day of crude shale oil. This is some 70 percent more than the copper ore mined in the San Manuel mine in Arizona, currently the largest underground mine in the U.S. It has been estimated that the initial investment for mine equipment will be 10 to 15 percent of the plant capital cost, and that mining costs might be as high as 30 to 40 percent of total operating costs.[20] Clearly, mining represents an important facet of an oil shale operation. However, based on the experience gained in development operations, mining more than 3.5 million tons of shale from the Green River formation, it appears that mining is not likely to prove a constraint in oil shale commercialization.

Detailed reviews of techniques and past experience in oil shale mining have recently appeared.[20,21] Both surface and underground mining techniques are applicable

Table 7.7 Typical properties of retort off gases

		Surface Retorts		
	M	Indirectly Heated (Union B)	Directly Heated (Paraho)	MIS Retort (Oxy)
Constituent, mol %, dry				
H_2	2	25	5.5	9.4
CO	28	5	2.9	1.8
CO_2	44	16	24.2	25.0
CH_4	16	24	2.4	1.6
C_2H_4	28	⎰ 10	0.7	0.12
C_2H_6	30	⎱	0.6	0.26
C_{3+}	~50	16	1.8	0.49
NH_3	17	—	0.6	0.57
H_2S	34	4	0.3	0.16
N_2	28	—	61.0	60.6
Total		100	100.0	100.00
Mean molar mass		25.0	30.5	29.4
GCV, MJ/kg		29.1	3.2	2.1
MJ/m³		32.5	4.3	2.8
Flow rates, kg/kg				
Retort off gas/oil		0.23	2.3	9.5
Combustion air/oil		0	1.8	7.2

to oil shale. The mining method selected will depend on site-specific factors including overburden ratios, oil shale grade, groundwater conditions, and rock strength. *Surface mining* may be considered if the overburden ratio is less than 2.5:1, although ratios of less than about 1:1 are more realistic. Surface mining is less costly, and may consequently be used to recover lower-grade shales than would be economical with underground mining. Open pit rather than strip mining is the procedure proposed for oil shale. Although surface mining can provide high resource recovery, the slope of the pit walls will have to be about 45° to ensure a stable mine, so much of the resource would be left behind in small-scale surface mining operations.

Environmental considerations are also of concern with surface mining. To minimize environmental impact and maximize recovery, it has been proposed that large-scale open pit mining of the rich Green River deposits of the Piceance Creek basin owned by the federal government, be undertaken. Overburden ratios in much of this area are less than 0.5:1. It would be possible to establish a pit 300 m deep by 1500 m wide, which if advancing at 400 m per year would yield up to 800 million liters of oil per day. Spent shale and waste would eventually be returned to the pit and the land restored to its present use.[22]

Several *underground mining* techniques have been assessed and it appears that the room-and-pillar system will in general be the economically preferred method. In this procedure, 40 to 60 percent of the resource is recovered, with the remainder being left in place as pillars. Costs are higher than for surface mining, so underground mining would normally be undertaken only for richer deposits associated with high overburden ratios.

While mining is a capital intensive operation, it is of interest to estimate its energy requirements. Both diesel fuel and electrical energy are required for running the mining equipment, and blasting represents an additional energy consumption. Diesel fuel will generally be used for drilling, crushing, and hauling, while most of the electrical energy in an underground mine will be for ventilation. Data from development plans suggest that an underground operation will consume diesel fuel and electrical energy in a ratio of about 2:1. Assuming a 33 percent heat-to-electricity conversion efficiency, the total thermal energy consumed in mining comes to about 140 MJ/t of shale mined. For a 140 L/t grade shale and a 100 percent Fischer assay recovery, this translates to 1 MJ/L of crude oil produced. As the calorific value of the oil is about 45 MJ/L, this estimate suggests mining energy is 1 MJ/45 MJ of oil, or over 2 percent of the crude oil production. A higher proportion of energy will, of course, be required for lower-grade shales, and also if some of the mined shale is unsuitable for retorting, and if oil yields are lower than Fischer assay. These considerations may escalate mining energy requirements to 4 to 5 percent of the oil yield. For the same grade of shale, less energy is required for combined surface mining–MIS operations. It is stressed that the values given here are not based on actual plant experience, and should be considered as rough estimates that are dependent on numerous site-specific factors.

Integrated Oil Shale Plants

All the important steps for the production of an upgraded shale oil have now been discussed, and are summarized in Table 7.8. Based on the assessments made of these

Table 7.8 Important steps in preparing a synthetic crude oil in integrated oil shale plants*

	Surface Retorts		MIS Retort
	Indirectly Heated	Directly Heated	
Mining and crushing	+	+	−
Retorting	+	+	+
Steam/air supply to retort	−	×	+
Gas purification	−	×	+
Electricity generation	†	×	×
Water treatment‡	−	−	+
Spent shale disposal	+	+	−
Upgrading	See Figure 7.12		

* Indicated by +, ×, −, depending on whether a major, intermediate, or minor operation.

† Electricity will probably be imported and the high quality retort gas used for hydrogen production.

‡ Water treatment requirements will increase if the mine or underground retort is associated with aquifers.

processes, it is possible to determine their relative energy requirements. This has been done for the major process steps and the results are summarized in Table 7.9. In calculating these energy data it has been assumed that the total energy for upgrading is for the hydrogen consumed only, and that the energy content of the synthetic crude produced is equal to that of the total raw shale oil retorted. Depending on the degree of upgrading, and the relative amount of hydrogenation and coking, more or less energy may be consumed in this step.

According to the data in Table 7.9, the directly heated surface retorting procedure has the lowest energy requirement per unit of oil produced. This is due to the heat of

Table 7.9 Illustrative energy requirements, as a percentage of the synthetic crude calorific value, for major processing steps in integrated oil shale plants*

	Surface Retorts		MIS (Oxy) Retort
	Indirectly Heated	Directly Heated	
Mining and crushing†	3	4	2
Heat of retorting	13‡	0	0
Air/recycle gas to retort§	3	3	24
Steam to retort¶	0	0	13
Upgrading**	12	12	9
Total	31	19	48
Energy in retort gas	16	18	47

* Based on estimates made in this and the previous section; see text.

† Underground mining.

‡ External energy input can be reduced if spent shale is burned.

§ Heat-to-work conversion efficiency for compression taken as 30 percent. Pressure ratios of 1.5 for surface and 17.0 for MIS assumed. Compressor efficiency is 80 percent.

¶ Heat to raise steam is 2.7 MJ/kg.

** Based on 130 MJ/kg of hydrogen (Section 5.4) and a hydrogen requirement per 100 kg of crude oil of 3.5 kg for surface and 2.5 kg for MIS.

retorting being derived entirely from combustion of the coke residue on the spent shale with no additional energy input being assumed. The lower oil yield obtained with directly heated retorts is reflected in the higher mining energy requirement, and there is, in addition, compression energy consumed in supplying air to the retort. These requirements are, however, relatively small. The heat of retorting shown for indirectly heated retorts assumes that 20 percent of the total heat required is provided by sensible heat recovery. Additional energy savings may be effected if spent shale is burned to supply some of this heat.

The most outstanding energy requirements in the table are those for air and steam supply to the MIS retort. In calculating the air compression requirements for this case it was assumed that the air is compressed to the same 1.7 MPa pressure as the steam. Energy for withdrawing the products from the underground retort has been neglected. It was shown that while oil yields in MIS retorting are about 60 percent of Fischer assay, the amount of energy in the gas produced is relatively high. In fact, as seen from the table, the energy in the MIS retort gas is equivalent to the process energy requirements.

It is again stressed that the data in Table 7.9 are estimates, and are not based on plant operating experience. In addition, the relative quantities will vary depending on actual retorting procedures, shale grade, and other site-specific parameters. We note further that energy requirements for gas purification, water treatment, and shale disposal have not been included. With these caveats, we may use the estimated energy values to calculate representative thermal efficiencies for the three types of processes considered, using

$$\eta = \frac{\text{GCV of oil} + \text{gas}}{\text{GCV of shale} + \text{Heat consumed (Table 7.9)}} \times 100\% \qquad (7.7a)$$

For the case of 100 percent Fischer assay yield, we find from Table 7.4 that the heat rate of the raw shale feed is 610/473, or about 130 MJ per 100 MJ of oil produced. If oil yields are lower than Fischer assay, then the raw shale feed rate must be higher for the same product oil rate. Assuming that an indirectly heated retort operates at 95 percent of Fischer assay, then the raw shale heat rate is 130/0.95 = 137 MJ per 100 MJ of oil. This value and the data in Table 7.9 may be substituted in Eq. (7.7a) to obtain a representative thermal efficiency for an indirectly heated retort

$$\eta = \frac{100 + 16}{137 + 31} \times 100 = 69\%$$

For a directly heated process in which the oil yield is 90 percent of Fischer assay, the raw shale GCV is 130/0.9 = 144 MJ and the thermal efficiency becomes $\eta_{DH} = 72$ percent. Assuming a 60 percent Fischer assay oil yield for MIS, we get $\eta_{MIS} = 56$ percent.

In comparing these thermal efficiencies we must bear in mind that MIS is not considered to be a competitive process with surface retorting. Rather MIS should be viewed as a means of processing a shale resource that is not amenable to recovery by conventional mining/surface retorting procedures. In addition, preliminary estimates of capital costs suggest that the investment in a MIS plant would be about 60 percent of that for a surface process of equivalent capacity. Although detailed cost estimates have not been published, relative costs for a Lurgi-Ruhrgas process in Australia are given in Table 7.10 to provide some indication of the distribution of costs in an oil shale plant. It is of interest that total costs are fairly evenly distributed amongst mining, retorting, and upgrading.

7.3 TAR SANDS RECOVERY

Two options for the recovery of oil from tar sands are of importance: mining of the tar sands followed by above-ground bitumen extraction and upgrading, and *in situ* extraction in which the bitumen is released underground by thermal and/or chemical means and then brought to the surface for processing and upgrading.[2,23-26] Although the approaches are superficially similar to the options discussed in the last two sections for the recovery of oil from shale, there are important differences.

Because tar sand deposits are generally unconsolidated (Section 1.3), they are not structurally competent, and conventional underground mining methods, such as the room-and-pillar technique used for oil shale recovery, are precluded. As a consequence, tar sands mining is carried out almost entirely above ground, although some underground hydraulic fracturing techniques to form a slurry which is then pumped to the surface have been investigated. *In situ* tar sand techniques also differ from their oil shale counterparts because tar sand deposits are permeable or can be made permeable to fluid flow, in contrast to oil shale formations which are impermeable consolidated deposits. This leads to *in situ* procedures for the separation and recovery of the bitumen from the sand which may be quite different from the above-ground procedures. On the other hand, *in situ* methods for the recovery of shale oil are really similar to their surface counterparts, except that the retort is formed underground.

The economic feasibility of surface mining tar sands, as with surface mining coal, is dictated mainly by the overburden that must be stripped away to expose the tar sands for mining, and the thickness of the tar sand zone. Where the overburden is soft and unconsolidated, as in the Canadian deposits, surface mining is considered practical when the overburden thickness is 45 m or less and the ratio of overburden to tar sand zone thickness is less than 1. In the surface mining operations at the Suncor facility

Table 7.10 Capital and operating costs as a percent of total costs* for a proposed Lurgi-Ruhrgas process in Australia[22]

	Capital	Operating	Total
Mining	24	39	33
Retorting	58	21	36
Upgrading	18	40†	31
Total	100	100	100

* Onsite–offsite costs distributed proportionately.

† Reflects a higher degree of upgrading than in Table 7.9.

(formerly called the Great Canadian Oil Sands facility) this ratio averages somewhat greater than 0.3, with the overburden thickness averaging about 16 m and the tar sand zone thickness about 50 m.[26] The Syncrude operation is located immediately adjacent to the Suncor lease, so deposit characteristics may be assumed to be similar. It is important to note that the characteristic thicknesses required for economical surface mining in Utah would be different, since most of the deposits are located beneath consolidated, rock-like material that must be drilled and blasted in the recovery operation.

As noted in Section 1.4, no more than 10 percent of the Canadian and U.S. tar sands reserves are recoverable by surface mining. Recovery of the bitumen from the deeper tar sands deposits requires the use of *in situ* methods. These methods are generally considered appropriate where the overburden depth is greater than 200 to 300 m. In the recovery operation to drive out the bitumen, a minimum depth is necessary to contain the pressurized fluids injected, as for example high pressure steam. As for surface mining, the tar sand zone should also be of a minimum thickness in order that the recovery operation be economical, one aspect of which is to minimize any heat losses to the surrounding overburden and underburden from any hot fluids injected, or from combusted tar sands (cf. Section 4.5). A thin tar sand zone may be considered to be one less than about 15 m thick.

An important factor in defining whether the recovery is economical, either for surface mining or for an *in situ* operation, is the bitumen saturation of the tar sands. For present surface mining methods, a lower limit of about 10 mass percent is considered necessary for economical recovery. This value could be somewhat lower for *in situ* methods. At the time of writing, the commercial Suncor and Syncrude operations in Canada both use surface mining. However, a commercial *in situ* operation has been proposed for the recovery of bitumen from Alberta tar sands.

Another important parameter defining the recoverability of tar sands bitumen is its viscosity. At reservoir temperatures the bitumen is immobile and has the consistency of tar. However, as with other members of the petroleum family, its viscosity decreases with increasing temperature. For reservoir conditions of 10°C, bitumen viscosities are typically several orders of magnitude higher than a petroleum crude. This is illustrated in Figure 7.15 which compares the viscosity-temperature behaviors for several tar sand bitumens and a high-quality petroleum crude.[25,27,28] The ordinate of Figure 7.15 is a highly compressed scale representing the logarithm of the logarithm of the viscosity, and in this coordinate system petroleum viscosity-temperature characteristics generally plot as a straight line. It can be seen that Utah bitumens are considerably more viscous than Canadian bitumens, a fact which will make their separation and recovery more difficult with both above-ground and *in situ* techniques. In this regard, we may note that the *in situ* recovery of the bitumen from the porous sand matrix without moving the solid necessitates lowering the viscosity of the bitumen, usually by heating so that it will flow as a liquid and can be collected in wells and removed to the surface. Above-ground techniques generally also involve separating the bitumen by thermal means to facilitate the extraction. The Cold Lake bitumen, whose viscosity at reservoir conditions is shown in Figure 7.15 to be 10 times lower than that of the Athabasca bitumen, is from the Alberta deposit that has been proposed for an *in situ* commercial operation.

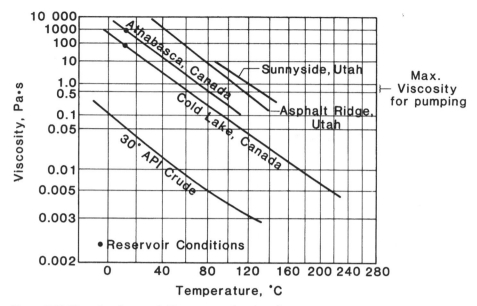

Figure 7.15 Viscosity of tar sands bitumens as a function of temperature.

The recovery of tar sand bitumens is also affected by the fact that they are considerably heavier than crude oils and have specific gravities of about 1 (10°API) compared with about 0.8 to 0.9 for most crudes (45 to 25°API). As discussed in Section 6.5, this is a manifestation of the lower hydrogen content of the bitumens.

A most important characteristic affecting the extraction procedure, brought out in Section 1.3, is whether the tar sand is water wet (hydrophilic), so the oil in the pores is not in direct contact with the sand grains, or whether the grains are not water wet (hydrophobic). Because Canadian tar sands are hydrophilic, this makes the separation of the bitumen from the sand grains by water displacement relatively easy since the displacing water swells the water film and the bitumen is floated away. On the other hand, the Utah deposits are hydrophobic and without the water film there are strong interfacial forces between the bitumen and the sand, which make the removal of the bitumen by water-washing much more difficult.

A final characteristic to be noted in relation to tar sands processing methods is that although the sand grains in Alberta tar sands have sizes roughly in the range of 0.1 to 0.3 mm, the sands can contain a significant fraction of small fines which tend to form an emulsion in water, which then requires an additional separation step. The quantity of fines in the mineral fraction increases almost linearly with decreasing bitumen content, amounting to only a few percent for bitumen contents from 14 to 18 percent. Utah sands have a smaller fraction of fines.[25]

After the bitumen has been extracted by either above-ground or *in situ* methods, it generally will require upgrading to provide a higher-quality fuel. This will include sulfur removal, increasing the hydrogen-to-carbon ratio, and lowering the molar mass. Usually this is done by coking followed by hydrotreating. Below we consider the various above-ground and *in situ* extraction methods and discuss upgrading in connection with integrated plants in the following section.

Surface Recovery

If the tar sands solids are mined, a number of methods are available for recovering the bitumen including hot water and cold water extraction, wherein the bitumen which is not soluble in water is gravity separated from the sand. Solvent extraction either alone or in combination with water extraction can also be applied. Finally, the tar sands may be coked directly, thereby converting the bitumen to a vapor and carrying out the separation from the sand in this way.

At the time of writing, two full-scale commercial tar sands plants are in operation and both use surface mining and hot water extraction. The plants are located near Fort McMurray in the Athabasca tar sands area of Alberta. One, the Suncor facility, was built in the late 1960s with an initial design capacity of 7.2×10^6 L/d (45 000 bbl/day) of synthetic crude oil and was expanded recently to 9.2×10^6 L/d (58 000 bbl/day). The larger Syncrude facility was built in the middle to late 1970s and has a capacity of some 20×10^6 L/d (125 000 bbl/day) of synthetic crude oil.

Hot water extraction. This technique is illustrated in the simplified flow diagram of Figure 7.16. In the "conditioning" step, the tar sand feed is first heated and mixed with water to form a pulp of 60 to 85 percent solids at about 90°C. Sodium hydroxide and chemical conditioners are added to control the pH and facilitate the subsequent separation of the bitumen from the sand. The sodium hydroxide is thought to react with the carboxylic acids in the bitumen to form detergents. The conditioning is done in a horizontal rotating drum heated by steam. The layers of tar sand disintegrate in this process from the heat and mechanical action, the warmed layers sloughing off and

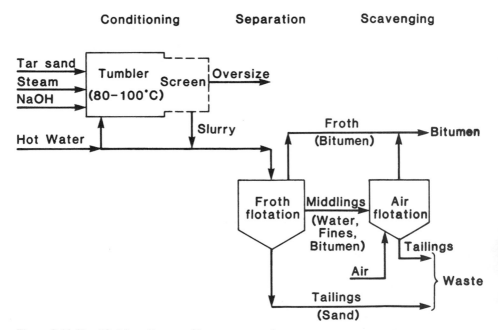

Figure 7.16 Simplified flow diagram of hot water extraction process.

exposing the cooler inner layers. The pulp from the conditioning drum is screened to remove oversize material and lumps of tar sand. The screened pulp is then mixed with more hot water to give the proper consistency and enable subsequent gravity separation of the bitumen.

Large quantities of hot water and low pressure (0.35 MPa) process steam are required in the conditioning step. In the Syncrude operation, approximately 7 kg of hot water and 0.5 kg of steam must be supplied for every kg of raw bitumen recovered.[2] Assuming the water to be heated 80°C at an 85 percent boiler efficiency, this would require 0.39 MJ/kg of water heated. The heat requirement to raise process steam at the same boiler efficiency would be about 3.2 MJ/kg of steam raised. The gross calorific value of the extracted bitumen is about 41 MJ/kg from which we may estimate that some 10 percent of the energy in the extracted bitumen is required to heat the water and raise the steam for the conditioning step. When measured as a fraction of the calorific value in the product synthetic crude, which is about 77 percent of that in the extracted bitumen, the heat requirement goes up to about 13 percent of the calorific value of the synthetic crude output.

The conditioned pulp is pumped to banks of separation cells. The separation cells are froth flotation units in which the sand settles to the bottom and the bitumen froths to the surface where it is skimmed off. Middlings from the separation cell which consist mostly of water, but with suspended fine material, may be recycled or sent to another recovery or "scavenging" step. The scavenging unit is generally an air flotation device into which low pressure air is pumped. The bitumen particles adhere to the rising air bubbles, which come to the surface as a froth. In the Syncrude plant, approximately 93 percent of the bitumen originally in the tar sands is recovered in the froth from the separation cells and scavenging units. At this stage the extracted bitumen froth contains air, minerals, and water impurities which must be separated out in subsequent upgrading. The bitumen recovery in this step is about 98 percent.

Cold water extraction. Because of the large energy requirement for heating in the hot water extraction process, a cold water extraction process would be desirable. In this procedure, the tar sand is broken up, kerosene is added in equal mass to the bitumen, and then water is added in an amount 2 to 3 times the mass of bitumen. After the addition of chemicals for pH control and a wetting agent to enhance separation, the mixture is agitated and the sand is separated from the bitumen by the shearing action that takes place. The mixture is subsequently classified. Several variants on the cold water process have been tested at pilot scale, although none appear to have been incorporated in commercial plants. It is unlikely that such a procedure would be appropriate for hydrophobic U.S. tar sands.

Solvent extraction. In this procedure, the bitumen is dissolved by mixing a light hydrocarbon solvent with the tar sands and then draining the resulting mixture from the mineral sands. The mixture is then pumped to a vessel where the solvent is recovered by distillation and recycled. The principal limitation of the method is that the solvent has a higher value than the bitumen so that a large fraction of it must be recovered if the process is to be economical. Separation of the solvent from the bitumen by thermal stripping is straightforward, although it can be energy intensive. The recovery of the residual solvent physically entrained in the waste sand interstitial void volume is considerably more difficult. The solvent that is not drained from the sand is usually a volatile hydrocarbon and can be driven off and recovered by heating the sand. This results in a substantial energy loss because of the need to heat the waste sand itself, from which it is not possible to recover all of the heat.

Solvent extraction processes have been tested on small pilot plant scale on both Canadian and U.S. tar sands. However, because of the high solvent recovery requirements, it is unlikely that a pure solvent extraction process is economically viable. It is probable that the inability to recover the solvent from the waste sand is the main limitation to the economic viability of the method. Nevertheless, in relation to the extraction of Utah tar sands, it does have the advantage that it is a ''dry'' procedure in that it does not require the use of large quantities of water—which is severely limited in the areas where the largest U.S. tar sands deposits are located. It may also prove somewhat more successful with Utah tar sands because, as noted earlier, these sands have less fines than Canadian sands, and it is the separation of the solvent from the fines that presents the greatest difficulty.

Direct coking. A method for recovering the bitumen from tar sands that is most similar to the retorting procedures discussed for shale oil recovery is that of direct coking. The method has the advantage that in a single, relatively simple pyrolysis operation an upgraded synthetic crude product is produced directly. One procedure employing indirect heating, that has been tested in Canada, is the fluidized bed coking technique shown schematically in Figure 7.17. In this method, the tar sand is pyrolyzed in the coker at 480°C with the heat supplied by clean sand from which the coke has been burned off. In the coker the volatile matter is distilled from the tar sand with coke being deposited on the sand grains by thermal cracking. The coked solids are then fluidized with air and transferred to the burner, which operates at about 760°C. Part of the hot sand (20 to 40 percent) is rejected, with the remainder transferred to the coker and fluidized by the pyrolysis off gases. Table 7.11 compares the properties

Table 7.11 Properties of synthetic crude oil from direct fluidized bed coking of Athabasca tar sands[29]

Mass %	Original Bitumen	Synthetic Crude Oil
Carbon	80.8	85.9
Hydrogen	9.9	11.8
Nitrogen	1.1	0.4
Sulfur	5.1	1.6
Oxygen	3.1	0.3
H/C atom ratio	1.47	1.65
Molar mass	532	238

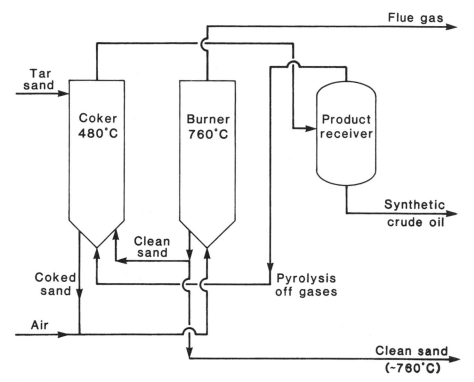

Figure 7.17 Direct fluid coking of tar sands (after Ref. 23).

of the synthetic crude product from the direct fluidized bed coking of Athabasca tar sand with the original bitumen properties. The synthetic crude is seen to be comparable with a heavy petroleum crude (see Table 7.2).

The technology is similar to a number of coal and shale pyrolysis processes discussed earlier, as well as to fluid coking techniques for petroleum upgrading (Section 7.2). The principal disadvantage is the need to handle a large quantity of solids in relation to the bitumen pyrolyzed. A mass ratio of sand to oil in the coker is reported to be about 35.[23] This value results from the need to have a large recirculation rate of hot sand to the coker to raise the coker temperature. For comparison, typical catalyst-to-oil mass ratios in petroleum refinery fluidized bed catalytic crackers are about 1. This need to recycle large quantities of hot sand would seem to be a main limitation of the method and a development goal might be to limit the solids transfer without reducing the coker temperature. One means of accomplishing this would be to raise the burner temperature. A problem with Athabasca sands might be fines carryover, although this should not be a major one.

It may be noted that the clean sand leaving the burner does so at a temperature of about 760°C. The heat capacity of the sand is approximately 0.84 kJ/(kg·K), which corresponds to a sensible heat of 0.63 MJ/kg of clean reject sand. If S_b is the bitumen fraction originally in the tar sand, sometimes referred to as the bitumen saturation, then since the amount of sand leaving equals that entering, it follows that the sensible heat in the reject sand as a fraction of the gross calorific value in the bitumen is given by

$$\frac{\text{Sensible heat in reject sand}}{\text{GCV of bitumen}} = \frac{0.63(1 - S_b)}{41S_b} \tag{7.9}$$

Here we have taken the gross calorific value of the bitumen to be 41 MJ/kg. For a bitumen fraction of 12 percent the sensible heat fraction is 11 percent. Clearly, heat recuperation techniques to recover the heat from the reject sand would have to be developed.

A pyrolysis process using indirect heating that has been suggested as being applicable to the direct coking of tar sands is the Lurgi-Ruhrgas process described in connection with oil shale retorting and mentioned in connection with coal pyrolysis. In tar sands applications the heat carrier is, as in the fluid coking process, the separated sand. Pilot plant tests have been carried out on California tar sands and the system has also been piloted on a small scale for Athabasca tar sands.

Since direct coking processes do not require water, they should be most appropriate for processing Utah tar sands which are not water wet and which are found in arid regions. One may even ask why hot water extraction followed by coking is more economical than the straightforward direct coking of the tar sands, particularly since fines handling does not entail a separate operation as it does with hot water extraction. The answer must, of course, lie with the detailed economics, which as we have suggested may be quite different for processing U.S. and Canadian tar sands.

In Situ Recovery

Because of their depth, the exploitation of much of the U.S. and Canadian tar sands reserves will undoubtedly require the development of *in situ* technologies. No commercial plant is presently in operation, although Esso Resources Canada, Ltd., is proposing to build an *in situ* plant in Alberta's Cold Lake deposit to produce 22 × 10^6 L/d (140 000 bbl/day) of synthetic crude oil. It should be noted that the bitumen in this deposit at reservoir conditions has a viscosity some 10 times lower than the Athabasca bitumen (see Figure 7.15).

In situ methods for the recovery of tar sand bitumen are for the most part similar to those employed in the enhanced recovery of conventional crude oil. Most of the current approaches involve heating and pressurizing the deposit. Heating reduces the bitumen viscosity, as discussed in connection with Figure 7.15, until at 120 to 150°C it will flow freely, so with pressurization it can be easily driven out of the deposit. The thermal methods include injecting steam into the deposit alone or in conjunction with hot water, and combustion of the tar sands. Solvent extraction may also be used to reduce the bitumen viscosity in place, although solvent losses in the deposit generally make this procedure economically unacceptable as is the case in above-ground processing, though solvent extractants may be used as aids in thermal processing. Water emulsion techniques have been tried on pilot-scale but not exploited further.

The tar sand formations have porosities of 25 to 40 percent, so they are sufficiently porous to enable fluids to flow through them under relatively small pressure gradients. At reservoir conditions, however, the permeability is greatly reduced because of the immobility of the bitumen filling the void spaces.

Steam injection. Of the various *in situ* procedures, steam injection is the most advanced. The injection can be either cyclic or continuous, and perhaps the simplest of the processes is single-well cyclic steam flooding, also termed the "huff and puff" method. In this procedure, a well is driven into the tar sands and high pressure, saturated steam is then injected into the reservoir for a period of time ranging from weeks to months. The steam spreads into the formation and both heats up the tar sands and builds up pressure. The steam is then shut off and the well is sealed in for a period of time, again measured in weeks or months, to allow the heat to penetrate a sufficiently broad zone around the well. The well is then opened and a mixture of water and the bitumen which has been rendered mobile flows back into the well and is withdrawn to the surface.

The cyclic process is repeated until it becomes uneconomical to continue or until there is breakthrough to an adjacent well. At this time, conversion to a continuous steam drive process is possible, wherein the steam is injected at one well and bitumen is removed from a producing well. Esso Canada, Ltd., which has proposed the commercial *in situ* operation mentioned above, has carried out extensive field tests on the "huff and puff" method.

In continuous steam drive schemes, injection and production wells are drilled and if permeable pathways do not exist between them, such pathways are made by fracturing the formation horizontally to connect the bases of the wells (cf. underground gasification procedures, Section 4.5). The fracturing is normally accomplished either with water or steam. Most schemes have a central injection well surrounded by four to six production wells at spacings of 60 to 150 m. In the steam injection process, the steam front moves from the vicinity of the injection well outward toward the production well. The heated bitumen ahead of the front is mobilized downward by gravity through the porous medium into the permeable path at the base, from which it is drawn to the production well.[26]

The process of steam injection into a reservoir and the subsequent displacement of the bitumen in the porous strata is a highly complex thermal-fluid-solid interaction. For example, as steam is injected, the rising temperature vaporizes the distillable components of the bitumen. The steam then carries the lighter components further into the reservoir, which then condense together with the water because of the temperature drop. The fluid-thermal mechanisms of steam flooding are illustrated in Figure 7.18. It can be seen that the process involves not only steam distillation and miscible gas drive, but hot- and cold-water drive and reduction in bitumen viscosity, together with thermal expansion of the bitumen and any water initially present. In almost all cases, there is as shown in Figure 7.18 an emulsion formed, so the fluid flow characteristics are extremely complex.[28] Moreover, because the injected fluid is more mobile, the interface between the bitumen and this fluid may be unstable, resulting in the phenomenon of "fingering." Finally, there are fluid-rock interactions not indicated in the figure, in which dissolution and precipitation of the rock may occur together with plugging of the cracks through which the bitumen-water mixture flows.

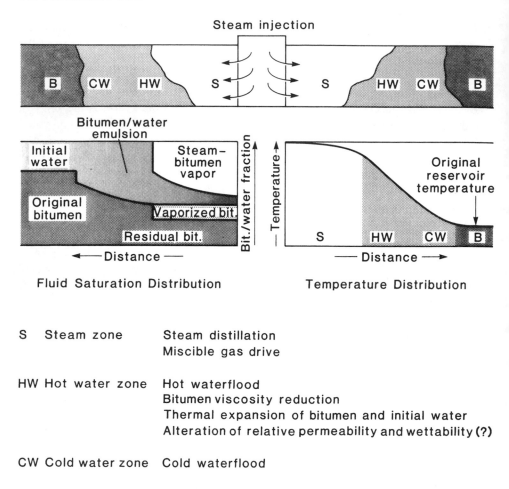

Figure 7.18 Mechanisms of steam flooding (after Ref. 28).

Two factors of importance in steam flooding are the amount of bitumen recovered from the deposit and the heat input needed to recover the bitumen. The heat input must equal the heat losses to the overburden and underburden, the heat retained by the tar sand, and the sensible heat of the produced water and bitumen. The heat loss to the surrounding burden is proportional to the tar sand zone surface area, while the heat retained is proportional to its volume. It follows that

$$\frac{\text{Heat loss}}{\text{Heat retained}} \sim \frac{1}{H} \tag{7.10}$$

where H is the tar sand zone thickness (cf. Eq. 4.18).

Neglecting the sensible heat of the produced water and bitumen in comparison with the calorific value of the bitumen, we may take the thermal output from the *in situ* process as proportional to the bitumen saturation in the tar sand. Of importance in steam flooding is the reduced bitumen saturation, S_{red}, defined as equal to the product of the average crude bitumen saturation, S_b, multiplied by the volumetric fraction of the tar sand zone swept by the steam, E_V:

$$S_{red} = S_b E_V \qquad (7.11)$$

A typical value for E_V, which is called the "volumetric sweep efficiency," is 0.5 for cyclic steam injection operation in the Alberta deposits.

Assuming that over the operating time the heat loss is large in comparison with the heat retained, then from Eq. (7.10) the ratio of the heat input to the calorific value of the recovered bitumen is given by the proportionality

$$\theta = \frac{GCV\ bitumen}{Heat\ input} \sim HS_{red} \qquad (7.12a)$$

The quantity is called the "thermal ratio," which is the inverse of a thermal efficiency defined in terms of the wellhead product. Detailed calculations by Outtrim and Evans[30] of the thermal ratio as a function of the tar sand zone thickness and saturation, support the relation given above, at least approximately. These calculations also take into account the efficiency of generating the steam. From their results we may write, approximately,

$$\theta \approx \frac{H(meters)S_{red}(\%)}{36} \qquad (7.12b)$$

with the units as indicated. This approximation is generally valid for $\theta > 3$, which is considered to be about the minimum value necessary for economical recovery.[30] For a 12 percent saturation, a 30-m thickness, and $E_V = 0.5$, the thermal ratio is 5. From the same relation, it can be seen that for tar sand zone thicknesses less than about 15 m it is unlikely that steam flooding would prove economical. This is of considerable importance in Utah, where tar sand zone thicknesses are generally less than 15 m. These simplified estimates are only for illustration and not intended to replace detailed calculations.

The other parameter of importance is the bitumen recovery from the deposit.[30] Assuming that the residual bitumen saturation in the deposit after recovery is 3 percent, then the recoverable fraction of the in-place crude bitumen is

$$Recoverable\ crude\ fraction = \frac{(S_b - 3)E_V}{S_b} \qquad (7.13)$$

On the other hand the net production as a fraction of the recoverable crude is

$$Net\ production = 1 - \frac{1}{\theta} \qquad (7.14)$$

From these relations we may write that

$$Deliverable\ fraction = Net\ production \times Recoverable\ crude\ fraction \qquad (7.15)$$

For the example given above, the deliverable fraction is 0.375×0.8, or 30 percent, of the bitumen in the deposit. This figure is reduced further, of course, when the efficiency of upgrading to a synthetic crude ($\sim$75 percent) is factored in.

Combustion. An alternative to steam flooding to heat the bitumen is in-place combustion. The procedures used are in many respects similar to those employed in underground gasification (Section 4.5). As with underground gasification and steam flooding, the process depends on a sufficiently high strata permeability and uniformity to prevent channeling of the combustion. If the permeability is inadequate, then fracturing of the tar sands zone prior to ignition is necessary.

In the simplest embodiment of combustion recovery, two wells connected by a permeable pathway are drilled into the deposit. Through one well, high pressure air is injected, and through the other the product is recovered in the form of a vapor. In the forward combustion mode of operation, the reservoir is heated by igniting the bitumen at the base of the injection well and by sustaining the combustion through continued injection of air that moves in the same direction as the combustion front. The region immediately ahead of the moving combustion front becomes strongly heated, typically to 350 to 450°C, and the bitumen there is thermally cracked, while ahead of the cracking region the bitumen is vaporized. The cracked oils, gases, and vapors are driven toward the production well by the pressurized injected air moving in the same direction as the combustion front. The fuel which supports the combustion is a coke-like material laid down by the cracking and coking which takes place.

An advantage of this *in situ* recovery procedure is that the wellhead product is a synthetic crude. The product quality depends, however, not just on the original bitumen characteristics but on the tar sand zone characteristics, the bitumen saturation, the air flux, and the system temperature.[31]

The main energy requirement and operational cost of the *in situ* combustion process is for compressing air. The quantity of air required depends on the amount of fuel available and therefore relates to the extent of the thermal cracking and coking reactions. The large amount of energy needed to compress the air can be appreciated form the reported air requirements of 2.3 to 3.6 m^3/L oil recovered.[2,32] This refers to air at standard conditions which will be compressed typically to 3.5 to 7 MPa. From Eq. (14) in Table 2.16, we can compute the work required to compress the air which, assuming three stages and a compressor efficiency of 80 percent, ranges from 0.57 to 0.71 MJ/m^3 air. For a 30 percent conversion efficiency of heat to work, the corresponding thermal energy requirement is 1.9 to 2.4 MJ/m^3 air, or 4.4 to 8.6 MJ/L oil recovered. The calorific value of the crude is about 41 MJ/L, so the thermal energy required to compress the air ranges from about 10 to 20 percent of the calorific value in the recovered oil. This represents a substantial fraction of the recovered energy and we may conclude that the reduction in the air needed for combustion must represent an important effort in the development of *in situ* recovery of tar sands by combustion.

Among the advantages of forward combustion are that it burns the least desirable fraction of the bitumen, leaves a clean stripped sand, and is a relatively efficient heat-generation process. A drawback of the procedure is that plugging of the strata can occur as the hydrocarbon vapors enter the colder regions and condense. To minimize this problem, reverse combustion is employed, in which combustion is started near the production well and the combustion progresses toward the injection well (cf. the analogous procedure used for linking underground coal gasification wells, Section 4.5). Because the produced liquids pass through the heated portion of the tar sands zone, the possibility of plugging is greatly reduced. The fuel for this process is an intermediate fraction of the original bitumen. Here, unlike in forward combustion, the coke remains in the strata after combustion and the recovery is correspondingly less. However, the recovered oil is a product of both vaporization and thermal cracking, and laboratory tests indicate that a highly upgraded light synthetic crude with a gravity as high as 25°API (s.g. = 0.904) could be produced.

An interesting procedure developed by Amoco Canada is a combined forward combustion and fluid injection process.[32] The process has been termed COFCAW, an acronym for "combination of forward combustion and waterflood" and proceeds in three steps: a heat-up phase, a blow-down phase, and a displacement phase. In the heat-up phase, air is injected at or somewhat above the fracturing pressure of the formation and forward combustion is initiated, with the aim of heating up the reservoir. Oil withdrawal from the production wells is controlled to maximize the volumetric heating of the formation. In the blow-down phase the air injection is stopped or reduced, and continued production lowers the formation pressure. In the final displacement phase, air and water are injected into the deposit. The air keeps the combustion going, and the water which vaporizes both acts as a steam drive and serves to dissipate the heat, so that a much larger proportion of the reservoir is uniformly heated to a temperature sufficient to cause the bitumen to flow.

In the original tests of the procedure, carried out in the Athabasca deposit, a square well pattern 45 m on edge was used with four production wells at the corners and a central injection well. The heat-up period lasted about 8 months and the blow-down period about 4 months, during the last month of which the injection well was shut in. The final water and air injection phase continued for about 6 months. The overall air/oil ratio for these tests was 2.5 m³/L oil recovered. A little over 7 percent of the in-place bitumen was burned, while some 60 percent was heated to 65°C or higher. A 50 percent recovery is estimated as feasible for this process. The procedure's success can be expected to be quite dependent on the site. Moreover, because of the complex thermal-fluid-solid interactions, characterization of the process must necessarily be heavily dependent on field test data.

Because U.S. tar sands formations are not as thick as those in Alberta and because most of the overburden is consolidated, it is likely that *in situ* combustion processes or some form of them will be required for recovery.

7.4 INTEGRATED TAR SANDS PLANTS

Commercial Canadian tar sands plants, either operating or proposed, all have as their final product an upgraded high-quality synthetic crude oil. This is in part a consequence of the location of the deposits in relation to pipelines, refineries, and markets. In order that we may compare and utilize the Canadian data, we consider the product from an integrated tar sands plant to be a high-quality synthetic crude similar to that from the Suncor and Syncrude operations. However, the data are presented in a manner that enables estimates to be made of both the oil quality and process efficiencies, should the product be pipelined with no upgrading or only a mild upgrading. This is the main option being considered at present in the United States for shale oil and is a possible course for oil recovered from tar sands, since the largest deposits are in an arid region removed from existing refineries.

The synthetic crude oil from the Suncor and Syncrude operations is of a higher quality than conventional crude oil in that it is lower in sulfur. Table 7.12 compares typical properties of an Athabasca bitumen and a refined synthetic crude from the Suncor pilot plant. Comparison of the synthetic crude properties with those given in Table 7.2 for conventional petroleum crudes illustrates the high quality of the product.

There are a number of routes by which the bitumen can be upgraded to a crude of the quality shown in Table 7.12. The extent of the upgrading that is needed depends on the extraction process. For example, if direct coking is used in surface recovery or if combustion is used to recover *in situ*, then the extracted product is already partially upgraded. With surface hot water extraction or *in situ* steam flooding, the product is essentially crude bitumen. In selecting the procedures to upgrade the bitumen from Alberta tar sands, an important consideration is that straight thermal treatment is not sufficient because of the high levels of organically bound sulfur. This sulfur can only be removed in the presence of hydrogen with or without the aid of a catalyst. On the other hand, with low-sulfur U.S. bitumen this would not necessarily be true.

Both of the operating Canadian plants and most of the proposed ones employ

Table 7.12 Typical properties of an Athabasca bitumen and refined synthetic crude oil from Suncor pilot plant[29]

	Original Bitumen	Synthetic Crude Oil
Elemental analysis (mass %)		
Carbon	83.4	86.3
Hydrogen	10.4	13.4
Nitrogen	0.5	0.02
Sulfur	4.5	0.03
Oxygen	1.2	—
Hydrocarbon type (mass %)		
Asphaltenes	19	0
Resins*	32	0
Oils (aromatics)	30	21
Oils (aliphatics)	19	79
Trace metals (μg/g)		
Vanadium	250	0.01
Nickel	90	0.01
Iron	75	—
Copper	5	0.02
H/C atom ratio	1.50	1.86
Gravity, °API	6.5	37.6
Specific gravity	1.025	0.837

* Synonymous with polar compounds (see Section 7.1).

Table 7.13 Major steps in preparing a refined synthetic crude oil in integrated tar sands plants

	Mining	Extraction	Coking or Extraction/ Coking	Hydrotreating
Surface				
Hot water	×	×	×	×
Direct Coking	×	−	×	×
In situ				
Steam Flooding	−	×	×	×
Combustion	−	−	×	×

coking followed by hydrotreating to upgrade the separated bitumen. The surface plants all use hot water extraction, and the one proposed *in situ* plant uses steam flooding. An alternative to coking would be to use a mild thermal cracking process, such as vacuum distillation, although it would then be necessary to deal with high yields of residuum. In one commercial proposal it has been suggested that the residuum be upgraded by solvent deasphalting followed by hydrocracking. This procedure of distillation followed by hydrotreating of the light product and deasphalting and hydrocracking of the residuum could potentially yield a higher liquid output than coking followed by hydrotreating. It has been pilot tested but has so far not been used commercially. For our purposes, the precise procedures for upgrading are less important than approximate estimates of the overall and individual energy and capital costs of the major processing steps. We assume therefore that the upgrading procedure is coking followed by hydrotreating. When the extraction involves coking then this step is assumed to be absent from the upgrading. This is illustrated in Table 7.13, which lists the major steps in preparing a refined synthetic crude oil in an integrated tar sands plant. Shown in the table are the two major surface recovery procedures and the two major *in situ* recovery procedures discussed in the last section.

Mining, coking, and hydrotreating. Before considering overall material and thermal balances for integrated tar sands plants, we briefly discuss the energy requirements and some of the characteristics of the mining, coking, and hydrotreating operations.

The surface mining operations being carried out at the Suncor and Syncrude plants are extremely large and consume a considerable amount of energy. In both cases the tar sands are extracted by open-pit mining. Bucket wheel excavators are used at Suncor and electric draglines at Syncrude. The magnitude of the operations may be appreciated by the fact that for a bitumen concentration of 12 mass percent, approximately 1600 kg of tar sands must be processed to produce 135 kg of synthetic crude oil. This assumes about 7 percent of the bitumen is lost in the extraction tailings and a 75 percent thermal efficiency for conversion. Stated differently, 1.6 tons of tar sands must be mined to yield 160 liters of synthetic crude oil, which is about 1 barrel. For a ratio of overburden to tar sand zone thickness of around 1, taking into account that the overburden must be removed and that the separated sand must be disposed, this means almost 5 tons of material has to be handled for each barrel of oil produced. For this reason, the Canadian tar sands operations presently constitute the largest earth-moving operations in the world.

It has been estimated[33] that approximately 10 percent of the product-fuel calorific value is required for the surface mining in an integrated tar sands plant. In specifying this value we have assumed that conversion of the synthetic crude calorific value to electricity or work is done at about a 30 percent efficiency.

The main coking processes are delayed coking, in use at Suncor; fluidized bed coking, in use at Syncrude; and Flexicoking, proposed for the Cold Lake *in situ* plant. These processes have been described previously in connection with shale oil upgrading and the EDS coal liquefaction process. An appreciation of the magnitude of the tar sands coking operation can be seen in Figure 7.19, which is a photograph of one of the two fluidized bed cokers at Syncrude.

All of the coking operations on tar sand bitumen are relatively low temperature ones carried out in the range 480 to 815°C. As discussed previously in the book, the specific product distributions depend principally on the reaction time and temperature. A representative product yield from the delayed coking of bitumen at the Suncor plant is given in Table 7.14.[29] More generally, the gases may range from 7 to 12 mass percent of the product and the coke from 17 to 23 percent. The very high sulfur content of the products is evident, particularly in the coke, where the sulfur can be as much as 8 percent by mass. Under existing environmental regulations in Canada, this coke cannot be burned without controlling the sulfur oxide emissions by flue gas desulfurization (Section 9.2). This has led to stockpiling of the coke at Syncrude. Since such a large fraction of the product is coke, its utilization within the plant is important in any integrated design.

Early estimates of the energy requirements for coking[29] and estimates we have made based on reported material balances about the coker section of the Suncor and Syncrude plants indicate that approximately 5 percent of the calorific value of the bitumen fed to the cokers is required for the coking operation. This represents about 7 percent of the calorific value of the synthetic crude oil product.

Table 7.14 Representative product yield from delayed coking of Athabasca bitumen at Suncor plant

	Component, Mass %	Sulfur	
		Mass % of Component	Mass % of Total
Gases (C_1–C_4; H_2S)	7.9	6.3*	0.50*
Naphtha (C_{5+})	12.7	2.2	0.28
Kerosene	15.0	2.7	0.41
Gas oil	36.2	3.8	1.38
Fuel oil	6.0	~5.0†	0.30
Coke	22.2	6.0	1.33
Total	100.0		4.20

* By difference as H_2S; for a 4.2% sulfur content in bitumen.
† Estimate.

The last major step is hydrotreating for sulfur and nitrogen removal, and general liquid quality improvement. Following hydrotreatment, the different liquid hydrocarbons are blended to produce the synthetic crude. For reasons of optimization, the upgrading is generally carried out in three separate hydrotreaters which handle different boiling ranges of liquid product: naphtha, kerosene, and gas oil. From the material balance for the Syncrude design[2] the hydrogen requirement is 23 g/kg syncrude.

Figure 7.19 Photograph of one of two fluidized bed cokers at 20×10^6 L/d (125 000 bbl/day) Syncrude tar sands plant. Vessel on left is reactor and is 63 m high, with an inside diameter of 9.0 m. Vessel on right is burner with inside diameter of 16.3 m. *(Courtesy of Syncrude Canada, Ltd.)*

Assuming that the hydrogen is produced at a 65 to 70 percent thermal efficiency (Section 5.4), we estimate that about 10 percent of the calorific value of the product is needed to manufacture the hydrogen. This will be the major energy requirement for the hydrotreatment operations and it may be assumed to be approximately the same for all the processes.

Canadian integrated plants. Most of the information presented on tar sands plant operations and energy requirements has been derived from commercial and pilot plant experiences in Alberta. Although some of the data may be applicable to proposed United States operations, care must be exercised in its transfer. For example, the large hydrogen requirements might not be necessary for processing a Utah bitumen because of the much lower sulfur content. On the other hand, surface-mining energy requirements will be higher because of the consolidated nature of the overburden, and the energy to extract the bitumen could also be higher because of its higher viscosity.

In Table 7.15 are collated illustrative energy requirements, as a percent of the calorific value of the synthetic crude output, for the major steps in preparing a refined synthetic crude oil in integrated plants. The numbers are drawn from this and the preceding section and are based almost exclusively on Canadian experiences. Most of the values given may be found in the text. The direct-coking extraction requirement assumes that 75 percent of the loss is sensible heat loss in the sand, and the steam flooding extraction estimate assumes a thermal ratio of 5. The extraction energy range for *in situ* combustion is based on the previously calculated range of 10 to 20 percent of the calorific value in the extracted oil and the assumption that 80 percent of this calorific value is recovered in the synthetic crude product.

The totals shown in Table 7.15 do not represent all the plant energy needs but only those for the major operations, so overall energy requirements can be expected to be somewhat higher. The result that the hot water and steam-flooding energy requirements are higher than those for direct coking and *in situ* combustion should be viewed with caution. The important point is that the hot water extraction process is commercial and the steam-flooding process is on the verge of becoming commercial.

Table 7.15 Illustrative energy requirements, as a percent of synthetic crude calorific value, for major operations in integrated tar sands plants, as derived mainly from Canadian experiences

	Mining	Extraction	Coking or Extraction/ Coking	Hydrotreating*	Total
Surface					
Hot Water	10	13	7	10	40
Direct Coking	10	—	15	10	35
In situ					
Steam Flooding	—	26	7	10	43
Combustion	—	—	13–25	10	23–35

* Hydrogen production only.

On the other hand, the direct coking and combustion processes have not advanced beyond the pilot plant stage and any projections for them would undoubtedly rise in their passage to commercialization. Despite this caveat, these procedures do hold considerable promise and are, moreover, of particular importance to U.S. tar sands development since some form of direct coking or *in situ* combustion is probably dictated by the nature of the resource.

In Figure 7.20 we have shown a material balance and a simplified flow diagram for the Syncrude plant drawn from Ref. 2 and scaled to the eventual projected synthetic crude output of 20×10^6 L/d (125 000 bbl/day). The balance shows clearly the large quantities of tar sands and hot water processed in a commercial-scale plant. Table 7.16 is the corresponding overall thermal balance, and unlike the other thermal balances in the book, this one is based on net calorific value, as these were the values reported. The thermal efficiency for conversion of the bitumen to synthetic crude is seen to be 68 percent. This is about 5 percent lower than a comparably calculated efficiency using the energy requirements in Table 7.15, which is to be expected since only the major energy losses are included in that listing.

A capital cost breakdown has been reported for the proposed Alsands hot water extraction plant to produce 22×10^6 L/d (140 000 bbl/day) from the Athabasca

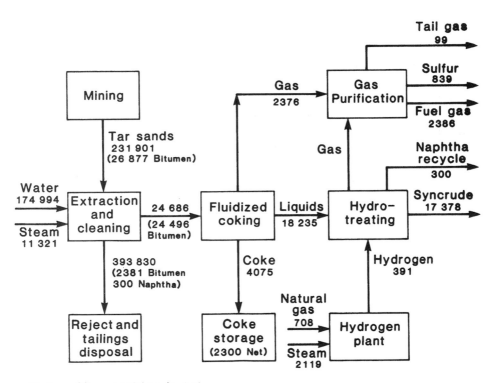

Note: All quantities in t/d

Figure 7.20 Material balance and simplified flow diagram for Syncrude plant producing 20×10^6 L/d (125 000 bbl/day) of synthetic crude oil from Athabasca tar sands by hot water extraction.

Table 7.16 Thermal balance for Syncrude plant producing 20 × 10⁶ L/d (125 000 bbl/day) of synthetic crude oil from Athabasca tar sands by hot water extraction process

	Mass Flow, kg/s	Heat Content, MJ/kg	Heat Flow, MJ/s	Heat Flow, % of Total
IN				
Bitumen mined	311.1	38.96	12 120	95.9
Natural gas	—	—	501	4.0
Electric power	—	—	15	0.1
Total			12 636	100.0
OUT				
Bitumen in tailings	27.6	38.96	1 075	8.5
Naphtha loss in tailings	3.5	43.85	153	1.2
Heat loss (by difference)	—	—	1 930	15.3
Net coke	26.6	31.63	841	6.7
Sulfur	9.7	9.28	90	0.7
Synthetic crude	201.1	42.50	8 547	67.6
Total			12 636	100.0

deposit.[25] The project is largely modeled on Syncrude. In the first column of Table 7.17 is given a breakdown of the initial capital costs expressed as a percent of the total cost.

In the second column the "onsite—offsite" costs have been distributed proportionately among the other capital costs. The utility costs have been combined with the extraction costs since the principal utility costs will be the boilers for heating the hot water and raising the steam needed for the extraction. In the last column is shown the percentage of energy lost in each of the categories indicated; the values are based on Table 7.15. The figures show a consistency between the capital and energy costs.

Table 7.17 Capital and energy costs as a percent of total costs for hot water extraction plant producing a synthetic crude oil from Athabasca tar sands
Capital costs based on proposed Alsands project.

	Initial Capital Cost*	Reduced Capital Cost†	Lost Energy‡
Mining	24.7	31.6	25.0
Extraction	9.2	25.4	32.5
Utilities	10.7		
Upgrading	33.7	43.0	42.5
Onsite–offsite	21.7	—	—
Total	100.0	100.0	100.0

* Indirect plus preproduction costs approximately equal to capital costs.
† Onsite–offsite costs distributed.
‡ From Table 7.15.

They also show that the combined cost (capital plus energy) breakdown is about 30 percent for mining, 30 percent for extraction, and 40 percent for upgrading.

As we have emphasized throughout, care must be exercised in extrapolating Canadian data to projected U.S. tar sands plants. Estimates which have been made indicate that the costs for producing oil from U.S. tar sands would be 2 to 3 times that from Athabasca tar sands.[25] Moreover, as we have previously indicated, the recovery of the U.S. resources would probably be by *in situ* combustion or surface direct coking, both of which require development, in contrast to the hot water and steam-flooding methods used or projected for use in Canada.

REFERENCES

1. Baker, J. D., "World Oil Shale Resources and Development History," in *Symposium Papers: Synthetic Fuels from Oil Shale, Atlanta, Georgia, December 3-6, 1979*, pp. 1-16. Institute of Gas Technology, Chicago, Ill., 1980.

2. Baughman, G. L., *Synthetic Fuels Data Handbook*. 2nd Edition, Cameron Engineers, Inc., Denver, Colorado, 1978.

3. Schächter, Y., "Oil Shale; Burn It Already," *ChemTech* **9**, 568-570, 1979.

4. Duvall, J. J., and Mason, K. K., "Effect of Oil Shale Type and Retorting Atmosphere on the Products from Retorting Various Oil Shales by the Controlled-State Retort," Report No. LETC/RI 80/2, Laramie Energy Technology Center, U.S. Department of Energy, Laramie, Wyoming, 1980.

5. Office of Technology Assessment, "An Assessment of Oil Shale Technologies," Stock No. 052-003-00759-2., U.S. Government Printing Office, Washington, D.C., June, 1980.

6. Tucker, W. F., *et al.*, "Thermal Decomposition of Kerogen: Retorting, Separation and Characterization of Shale Oil," *Proc. 13th Oil Shale Symposium* (J. H. Gary, ed.), pp. 122-148. Colorado School of Mines Press, Golden, Colo., 1980.

7. Schmidt-Collerus, J. J., *et al.*, "Investigations on Colorado Oil Shale of the Green River Formation," Center for Fundamental Oil Shale Research, First Annual Report, Denver Research Institute, Denver, Colo., 1966.

8. Miknis, F. P., "Current Methods of Oil Shale Analysis and New Developments (NMR)," in *Symposium Papers: Synthetic Fuels from Oil Shale, Atlanta, Georgia, December 3-6, 1979*, pp. 245-270. Institute of Gas Technology, Chicago, Ill., 1980, and personal communication, December, 1980.

9. Burnham, A. K., "Studies of Oil Shale Reaction Chemistry at LLL," in *Symposium Papers: Synthetic Fuels from Oil Shale, Atlanta, Georgia, December 3-6, 1979*, pp. 323-352. Institute of Gas Technology, Chicago, Ill., 1980.

10. Lovell, P. F., "Production of Utah Shale Oils by the Paraho DH and Union 'B' Retorting Processes," in *Proc. 11th Oil Shale Symposium* (J. H. Gary, ed.), pp. 184-192. Colorado School of Mines Press, Golden, Colo., 1978.

11. Weil, S. A., "Oil Shale Hydrotreating Laboratory Studies at IGT," in *Symposium Papers: Synthetic Fuels from Oil Shale, Atlanta, Georgia, December 3-6, 1979*, pp. 353-376. Institute of Gas Technology, Chicago, Ill., 1980.

12. Probstein, R. F., and Gold, H., *Water in Synthetic Fuel Production*. MIT Press, Cambridge, Mass., 1978.

13. Feldkirchner, H. L., "The HYTORT Process," in *Symposium Papers: Synthetic Fuels from Oil Shale, Atlanta, Georgia, December 3-6, 1979*, pp. 489-524 (see also pp. 21-116), Institute of Gas Technology, Chicago, Ill., 1980.

14. Braun, R. L., and Chin, R. C. Y., "Computer Model for *In Situ* Oil Shale Retorting: Effects of Gas Introduced into the Retort," in *Proc. 10th Oil Shale Symposium* (J. H. Gary, ed.), pp. 166-179. Colorado School of Mines Press, Golden, Colo., 1977.

15. Sullivan, R. F., and Stangeland, B. E., "Catalytic Hydroprocessing of Shale Oil to Produce Distillate Fuels," in *Refining of Synthetic Crudes* (B. M. Harvey and M. L. Gobarty, eds.). ACS Symposium Series No. 179, American Chemical Society, Washington, D.C., 1979.

16. Energy Engineering Board, Assembly of Engineering, "Refining Synthetic Liquids from Coal and Shale," National Research Council, National Academy Press, Washington, D.C., 1980.

17. Robinson, E. T., "Refining of Paraho Shale Oil into Military Specification Fuels," in *Proc. 12th Oil Shale Symposium* (J. H. Gary, ed.), pp. 195-210. Colorado School of Mines Press, Golden, Colo., 1979.

18. Poulson, R. E., "Nitrogen and Sulfur in Raw and Refined Shale Oils," *Am. Chem. Soc., Div. of Fuel Chem., Preprints* **20(2)**, 183-197, 1975.

19. Lanning, W. C., "The Refining of Shale Oil," Report No. BERC/IC-77/3, Bartlesville Energy Research Center, U.S. Department of Energy, Bartlesville, Oklahoma, May 1978.

20. Crookston, R. B., and Weiss, D. A., "Oil Shale Mining—Plans and Practices," in *Symposium Papers: Synthetic Fuels from Oil Shale, Atlanta, Georgia, December 3-6, 1979*, pp. 117-163. Institute of Gas Technology, Chicago, Ill., 1980.

21. Brady, R. A., and Rutledge, P. A., "Oil Shale Mining: Technologies and Environmental Consequences," in *Symposium Papers: Synthetic Fuels from Oil Shale, Atlanta, Georgia, December 3-6, 1979*, pp. 179-197. Institute of Gas Technology, Chicago, Ill., 1980.

22. Lewis, A. E., "Oil from Shale: The Potential, the Problems, and a Plan for Development," *Energy* **5**, 373-387, 1980.

23. Camp, F. W., "Tar Sands," in *Kirk-Othmer, Encyclopedia of Chemical Technology*, Vol. 19, pp. 682-732. 2nd Edition, Interscience Publishers, New York, 1969.

24. Redford, D. A., and Winestock, A. G. (eds.), *The Oil Sands of Canada-Venezuela 1977*. CIM Special Vol. 17, The Canadian Institute of Mining and Metallurgy, Montreal, 1977.

25. de Nevers, N., Glenne, B., and Bryner, C., "Analysis of the Environmental Control Technology for Tar Sand Development," Report No. COO-4043-2, U.S. Department of Energy, Washington, D. C., June 1979.

26. Mossop, G. D., "Geology of the Athabasca Oil Sands," *Science* **207**, 145-152, 1980.

27. Bunger, J. W., Cogswell, D. E., and Oblad, A. G., "Thermal Processing of a Utah Tar Sand Bitumen," in *The Oil Sands of Canada-Venezuela 1977* (D. A. Redford and A. G. Winestock, eds.), pp. 178-182. CIM Special Vol. 17, The Canadian Institute of Mining and Metallurgy, Montreal, 1977.

28. Flock, D. L., and Lee, J., "An Experimental Investigation of Steam Displacement of a Medium Gravity Crude Oil," in *The Oil Sands of Canada-Venezuela 1977* (D. A. Redford and A. G. Winestock, eds.), pp. 386-394. CIM Special Vol. 17, The Canadian Institute of Mining and Metallurgy, Montreal, 1977.

29. Speight, J. G., "Thermal Cracking of Athabasca Bitumen," in *Bitumens, Asphalts and Tar Sands* (G. V. Chilingarian and T. F. Yen, eds.), pp. 124-154. Developments in Petroleum Science, 7, Elsevier, New York, 1978.

30. Outtrim, C. P., and Evans, R. G., "Alberta's Oil Sands Reserves and Their Evaluation," in *The Oil Sands of Canada-Venezuela 1977* (D. A. Redford and A. G. Winestock, eds.), pp. 36-66. CIM Special Vol. 17, The Canadian Institute of Mining and Metallurgy, Montreal, 1977.

31. Hayashitani, M., *et al.*, "Thermal Cracking of Athabasca Bitumen," in *The Oil Sands of Canada-Venezuela 1977* (D. A. Redford and A. G. Winestock, eds.), pp. 233-247. CIM Special Vol. 17, The Canadian Institute of Mining and Metallurgy, Montreal, 1977.

32. Burger, J. G., "In-Situ Recovery of Oil from Oil Sands," in *Bitumens, Asphalts and Tar Sands* (G. V. Chilingarian and T. F. Yen, eds.), pp. 191-212. Developments in Petroleum Science, 7, Elsevier, New York, 1978.

33. Puttagunta, V. R., Sochaski, R. O., and Robertson, R. F. S., "A Role for Nuclear Energy in the Recovery of Oil from the Tar Sands of Alberta," in *The Oil Sands of Canada-Venezuela 1977* (D. A. Redford and A. G. Winestock, eds.), pp. 498-519. CIM Special Vol. 17, The Canadian Institute of Mining and Metallurgy, Montreal, 1977.

EIGHT

BIOMASS CONVERSION

8.1 RESOURCES

Biomass is any material that is directly or indirectly derived from plant life and that is renewable in time periods of less than about 100 years. More conventional energy resources such as petroleum and coal, as well as kerogen and tar sands bitumen, are, of course, also derived from plant life, but are not considered renewable. Typical biomass resources are energy crops, farm and agricultural wastes, and municipal wastes. Animal wastes are also biomass materials in that they are derived, either directly or via the food chain, from plants which have been consumed as food.

About one-quarter of the carbohydrate formed by photosynthesis is later oxidized in the reverse process of respiration to provide the energy for plant growth. The excess As with conventional fuels, the energy in biomass is the chemical energy associated with the carbon and hydrogen atoms contained in oxidizable organic molecules. The source of the carbon and hydrogen is carbon dioxide and water. Both of these starting materials are in fact products of combustion and not sources of energy in the conventional sense. The conversion by plants of carbon dioxide and water to a combustible organic form occurs by the process of photosynthesis. Two essential ingredients for the conversion process are solar energy and chlorophyll. The chlorophyll, present in the cells of green plants, absorbs solar energy and makes it available for the photosynthesis, which may be represented by the overall chemical reaction

$$nCO_2 + mH_2O \xrightarrow[\text{chlorophyll}]{\text{sunlight}} C_n(H_2O)_m + nO_2 \qquad \Delta H° = +470 \text{ kJ/mol} \qquad (8.1)$$

$C_n(H_2O)_m$ is used here to represent the class of organic compounds called "carbohydrates," several of which are made in the course of the reaction. We recall from Table 2.11 that both sugars and cellulose are carbohydrates.

About one-quarter of the carbohydrate formed by photosynthesis is later oxidized in the reverse process of respiration to provide the energy for plant growth. The excess carbohydrate is stored. The plant typically contains from between 0.1 to 3.0 percent of the original incident solar energy, which is a measure of the maximum energy recoverable from the plant if converted into a synthetic fuel. Some of this energy may, however, be degraded in the formation of intermediate products, and there will be additional losses in converting the biomass material into a conventional fuel form.

The carbon dioxide which provides the carbon for biomass formation is present in the atmosphere (0.03 percent by volume), and in water either as the dissolved gas or as carbonate species. It is estimated that the amount of readily available carbon dioxide in the water/air reservoir passes through the life cycle every 350 years, and that more than two-thirds of this occurs through the flora of the oceans. Carbon dioxide that is formed by respiration, biological degradation, or combustion is thus eventually reconverted to oxidizable organic molecules by photosynthesis. Provided an equivalent quantity of vegetation is replanted, the use of biomass as a fuel will consequently not result in a net change in atmospheric carbon dioxide levels. However, the carbon dioxide equivalent to the carbon in fossil fuels which has been removed from circulation for millions of years, is now being returned to the atmosphere in ever-increasing amounts. The resulting increase in carbon dioxide levels is of some concern in relation to possible climatic effects, as is discussed in the next chapter.

One of the reasons for the great interest in biomass as a fuel source is that it does not affect atmospheric carbon dioxide concentrations. More important, perhaps, is that this energy source is renewable. In addition, biomass fuels are clean burning in that sulfur and nitrogen concentrations are low, and because the hydrogen-to-carbon ratio is generally high. It is nevertheless not expected that biomass will make a major contribution to overall energy requirements in the near future. One study estimates that in the United States, the energy produced from biomass sources by the year 2010 will be less than 10 percent of the total energy consumed.[1] The results of another study are summarized in Table 8.1 and show that the biomass contribution could be as much as 19 percent, assuming maximum development of this resource. This upper value is dependent also on available land not being required for food production, as well as on the availability of adequate water supplies.

The limitations of extensive development of biomass resources are its high land

Table 8.1 Estimated gross* energy potential from biomass in the United States for the year 2000[2]

Source	Biomass	Percent of Total Energy†
Commercial forestland and mill wastes	Wood	5–10
Cropland used for intensive agriculture	Crop residues	0.8–1.2
Available land‡	Grasses and/or grain and sugar crops	0–5
Animal husbandry	Animal wastes	0.1–0.3
Food/cotton processing	Agricultural wastes	0.1
Other	Aquatic plants	0.1
Municipal and solid wastes[1,3]	—	1.6–1.9
Total		8–19

* Does not include deductions for cultivation and harvest energy, or losses.
† Assuming annual energy consumption is 100×10^{18} J (100×10^{15} Btu, or 100 "quads").
‡ The amount of land available for energy crops is strongly dependent on future food requirements.

and water requirements (see Chapter 9), and the competition with food production. From an economic standpoint, large-scale production of energy from biomass faces competition with the synthetic fuels discussed previously in this book. In particular, there is currently not a large energy market for ethanol and methanol, the two liquid fuels most readily produced from biomass. A market for ethanol as a gasohol blend could, however, be relatively rapidly developed, while a market for methanol as a transportation fuel is dependent on future trends in internal combustion engines.

Properties and Conversion Processes

The oxidizable organic materials that are produced by photosynthesis, and that determine the properties of the plant matter of relevance to biomass energy utilization, are carbohydrates and lignin. All the carbohydrates present are "saccharides," that is, they are either sugars or polymers of sugars. A well-known sugar is sucrose, $C_{12}H_{22}O_{11}$, which is the main sugar in the sap of plants, and occurs in high concentrations in sugar cane and beet. Glucose, $C_6H_{12}O_6$, is the main sugar in corn and grapes and is found in many fruits. It is known also as blood sugar. The level of glucose in human blood is held constant at about 0.8 g/L under the control of the hormone insulin which converts excess sugar to fat. In muscle tissue, glucose is degraded with release of energy.

As with other organic compounds, there are many isomeric forms in which the atoms in sugar molecules can be arranged. For example, $C_6H_{12}O_6$ represents several sugars including glucose, mannose, and fructose.

CHO	CHO	CH₂OH
|	|	|
HCOH	HOCH	C=O
|	|	|
HOCH	HOCH	HOCH
|	|	|
HCOH	HCOH	HCOH
|	|	|
HCOH	HCOH	HCOH
|	|	|
CH₂OH	CH₂OH	CH₂OH
D-glucose (dextrose)	D-mannose	D-fructose

D-glucose and D-mannose differ only in the orientation of the hydrogen and hydroxyl groups attached to the carbon atom and are termed "stereoisomers." The mirror image of a sugar structure represents a special stereoisomer called an "enantiomorph." Enantiomorphs are differentiated by the prefixes D- and L-, which originate from abbreviations for dextrorotatory and levorotatory, as determined by the direction of rotation of polarized light by solutions of reference sugars. Glucose and mannose contain the aldehyde functional group CHO, and are aldoses, which differ chemically from ketoses such as fructose, which contain the ketone group C=O.

Probably the most important property of sugars relevant to biomass conversion is whether or not they are fermentable. The fermentable sugars are D-glucose, D-mannose, D-fructose, D-galactose, and maltose. Other sugars can be converted to fermentable sugars by hydrolysis, usually in the presence of an acid or an enzyme. For example, sucrose, which can be thought of simply as a combination of two sugars, glucose and fructose (and is called a "disaccharide"), is hydrolyzed by the enzyme invertase present in yeast:

$$\underset{\text{Sucrose}}{C_{12}H_{22}O_{11}} + H_2O \xrightarrow{\text{invertase}} \underset{\text{Glucose}}{C_6H_{12}O_6} + \underset{\text{Fructose}}{C_6H_{12}O_6} \qquad (8.2)$$

Once hydrolyzed, the resultant sugars can be fermented to produce ethanol.

The other carbohydrates present in plant matter are cellulose, starch, and hemicellulose. These are all polymers of sugars and are called "polysaccharides." The base sugars and polymers that are of interest in biomass processing are listed in Table 8.2. *Cellulose* is a fibrous polysaccharide that is the main constituent of the cell walls of land plants. It is the most abundant naturally occurring organic substance. Cotton fiber is virtually pure cellulose, and cellulose is a major constituent in wood, hemp, and straw. The important chemical properties of cellulose that affect its conversion in biomass processes are its extreme insolubility and chemical inertness, in particular its resistance to acidic and enzymatic hydrolysis.

Starch is a granular polysaccharide which accumulates in the storage organs of

Table 8.2 Selected saccharide carbohydrates and lignin

	Chemical Formula	Molar Mass
Monosaccharides		
Xylose*	$C_5H_{10}O_5$	150
D-glucose† (corn/grape sugar)	$C_6H_{12}O_6$	180
Fructose (fruit sugar)	$C_6H_{12}O_6$	180
Mannose*	$C_6H_{12}O_6$	180
Disaccharides		
Sucrose (cane/beet sugar)	$C_{12}H_{22}O_{11}$	342
Maltose (malt sugar)	$C_{12}H_{22}O_{11}$	342
Lactose (milk sugar)	$C_{12}H_{22}O_{11}$	342

	Approximate Representation	Monomer Building Block	Molar Mass
Polysaccharides			
Cellulose	$(-C_6H_{10}O_5-)_n$	D-glucose	>100 000
Starch	$(-C_6H_{10}O_5-)_n$	D-glucose	35 000–90 000
Hemicellulose	—	Various sugars*	10 000–35 000
Wood	$C_{38}H_{54}O_{23}$	Lignocellulose	—
Lignin	—	Hydroxyphenylpropane‡ $(C_3H_7 . C_6H_4 . OH)$	5 000–10 000

* Xylose and mannose are the principal sugar constituents of wood hemicellulose.
† Also called dextrose.
‡ A compound related to phenol, C_6H_5OH.

plants such as seeds, tubers, roots, and stem pith. It is an important constituent of corn, potato, rice, and tapioca. Starch consists of 10 to 20 percent α-amylose, which is water soluble, and 80 to 90 percent amylopectin, which is insoluble. Both the constituents of starch are polymers of D-glucose, with amylose linked in chain structures, while amylopectin is a highly branched structure. Starch is not as chemically resistant as cellulose, and can be readily hydrolyzed by dilute acids and enzymes to fermentable sugars.

Hemicelluloses are polysaccharides that occur in association with cellulose. They are chemically different from cellulose, are amorphous, and have much lower molar masses (Table 8.2). While cellulose is built from the single sugar D-glucose, most hemicelluloses contain two to four different sugars as building blocks. D-glucose is a component of some hemicelluloses, although xylose is a dominant sugar in hardwood hemicellulose, and mannose is important in softwood hemicellulose. Unlike the other sugars described so far, xylose contains only 5 carbon atoms and is a pentose.

$$CHO$$
$$|$$
$$HCOH$$
$$|$$
$$HOCH$$
$$|$$
$$HCOH$$
$$|$$
$$CH_2OH$$

D-xylose

The fraction of the cellulose containing xylose polymers is often referred to as "pentosan." Hemicellulose is more soluble than cellulose, is dissolved by dilute alkaline solutions, and can be relatively readily hydrolyzed to fermentable sugars. Cellulose and hemicellulose together are known as "holocellulose."

Lignin is the final major constituent of plant material important to biomass processing. It is not a carbohydrate, but a polymer of single benzene rings linked with aliphatic chains. The phenolic compound p-hydroxyphenylpropane is an important monomer group in lignin.[3]

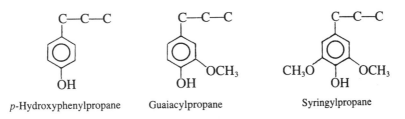

p-Hydroxyphenylpropane   Guaiacylpropane   Syringylpropane

Lignin is a characteristic constituent of the walls of woody cells, and acts as a natural glue providing the plant with added mechanical strength. As can be inferred from its name, lignin is an important constituent of wood. Like hemicellulose, it is amorphous and more soluble than cellulose. It may be removed from wood by steaming or by dissolving in hot alkali or bisulfite solutions. Lignin totally resists hydrolysis, and is resistant to microbial degradation.

In discussing the properties of specific biomass materials, we do so in relation to the possible processes for their conversion to synthetic fuels. The potential processes considered here are biochemical conversion by fermentation and anaerobic digestion, and the thermal processes of pyrolysis and gasification. Fermentation produces mainly liquids, pyrolysis results in both liquid and gaseous products, while gasification and anaerobic digestion produce gaseous fuels. Most biomass materials can be gasified using processes similar to those discussed in Chapter 4, and the resulting gas may be used for synthesis of liquid fuels or SNG. Direct combustion of biomass is always an option, and in some instances may be the only viable approach. The conversion processes of interest are summarized in Figure 8.1.

The plants in which the sugars and lignocellulosic compounds discussed above occur and that are important for biomass energy are trees, grasses and legumes, grain and sugar crops, and aquatic plants. *Wood* from commercial forestland is the largest potential source of biomass in the United States.[2] Most of this potential lies in wood processing byproducts such as sawdust and spent paper-pulping liquor, and forest management byproducts such as thinning and logging residues. One study for the U.S.

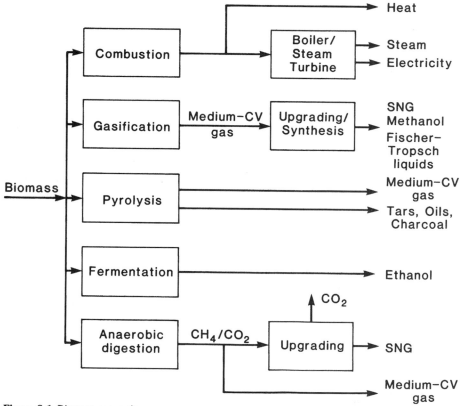

Figure 8.1 Biomass conversion processes.

estimated that in 1974 the GCV of unused wood residues was at least half that of the total fossil fuels consumed industrially.[4]

Wood is about two-thirds holocellulose and one-quarter lignin, with the remainder being extraneous materials such as resins, gums, tannins, and waxes. The amounts of the various constituents present are different for hardwoods and softwoods and vary also among the many tree species. Nevertheless, the composition of many common species falls within the range shown in Table 8.3. The hemicellulose and about one-quarter of the cellulose in wood are readily hydrolyzable to fermentable sugars,[2] but the remaining cellulose and lignin resist acidic or enzymatic attack. Removal of the lignin by steaming or organic solvent extraction can make the cellulose more available for biochemical conversion, and the removed lignin may be burnt to meet plant energy requirements. Nevertheless, delignification increases process costs and currently does not appear to be a viable option. As mentioned, lignin is not digestible, and further seems to interfere with the conversion of the holocellulose. Biochemical processes consequently have limited application to wood biomass.

Thermal processes offer a more effective means for the recovery or conversion of the energy content of wood and other lignocellulosic biomass. Wood contains 80 percent or more of volatile organic matter that may be recovered as a gas and tar on pyrolysis. Pyrolysis of wood was at one time used for obtaining creosote oils as well as acetic acid (wood vinegar) and some methanol (wood spirits). The gases evolved, essentially hydrogen, methane, carbon monoxide, and carbon dioxide, had no value for illuminating, and were used to supply the heat for the pyrolysis. Gasification of wood may be used to produce either a low- or a medium-CV gas if a gaseous fuel is desired, or a synthesis gas that can be converted to liquid fuels using one of the indirect liquefaction processes. Gasification followed by a synthesis step is expected to yield a higher-quality liquid than can be obtained by pyrolysis.

Table 8.3 Selected properties of representative biomass materials[2,3]

Mass %, dry	Wood	Grain*	Municipal Solid Waste†	Animal Wastes (Manure)
Carbon	50–53	45.0	47.6	35.1
Hydrogen	5.8–7.0	5.8	6.0	5.3
Nitrogen	0–0.3	2.4	1.2	2.5
Sulfur	0–0.1	0	0.3	0.4
Oxygen	38–44	42.5	32.9	38.7
Volatile matter	77–87	~80	77	76.5
Fixed carbon	13–21	—	11	0
Ash	0.1–2.0	4.0	12	23.5
H/C atom ratio	1.4–1.6	1.5	1.5	1.8
GCV, MJ/kg (dry)	19.8–21.0	16.8	19	13.4
Moisture, %	25–60	16	20	7–35‡

* "Red corn cob;" corn stover contains about 25% cellulose, 10% lignin, and 15% moisture.

† Combustible portion; may contain 9% metals and 12% glass/ceramics on an as-received basis.

‡ Air dried.

Combustion offers the most direct route for energy recovery, and is an effective means of utilizing the total energy content of whole wood and other biomass. Technology for small-scale combustion is certainly well developed. Larger-scale operation for electricity generation at the 5 to 50 MW level should be feasible, although it is not expected to be economically competitive with petroleum- or natural-gas-fired systems. Due to the large land area over which wood must be harvested, the seasonal nature of the supply, and the large volume of material to be transported and stored, the reliable provision of wood to a large plant can present considerable problems. It is in fact stated[2] that wood fueled systems above 50 MW present logistical problems of "overwhelming proportions." A 60 MW plant would, for example, require 1000 t/d of dry wood, supplied from 400 to 800 square kilometers of intensely managed forestland.

The large volume of biomass that has to be handled is in part a result of the open pore structure or unconsolidated nature of these materials. For instance, dry wood has a specific gravity of about 0.6, while the specific gravity of many coals falls in the range 1.2 to 1.4. The calorific value per unit volume of wood is about half that of lignite and one-quarter that of bituminous coal. Another problem with biomass materials is their relatively high moisture content. As shown in Table 8.3, raw woods may have moisture contents of up to 60 percent by mass. Air drying is a time-consuming process but can reduce this down to 15 to 20 percent, at which stage the heat required to evaporate the moisture remaining is equivalent to about 3 percent of the GCV of the dry wood.

Grain crops include corn, wheat, rice, barley, and other cereals. The seeds of these plants are typified by their high starch content that can be hydrolyzed to fermentable sugars for ethanol production. The *sugar crops,* including sugar cane, beet, and sweet sorghum, are preferable to the starch crops to the extent that sucrose is more readily hydrolyzed to fermentable sugars. However, sugar cane production is restricted to warm climates, and requires both high-quality land and irrigation. Sugar beets produce only 40 percent as much fermentable sugar as does sugar cane per unit growing area, and reportedly have a higher value as cattle feed than as an energy feedstock. Sweet sorghum is currently a minor crop in the United States, but has considerable advantages over sugar cane as an energy crop. It can grow in a variety of soils and climatic conditions and its production costs have been estimated to be just over half those for sugar cane.[2]

In the United States, the cropland not currently in use can provide grain and sugar crops for the production of between 4×10^9 and 11×10^9 L/yr of ethanol, which is about 1 to 3 percent of U.S. gasoline consumption. Corn, which is the largest field crop in the United States, might be considered particularly attractive as it has a high photosynthetic efficiency. Nevertheless, much of the potential croplands are poorly suited to corn, and switching to *grasses, legumes*, and short rotation trees may be more beneficial from an energy standpoint. These latter crops are lignocellulosic, and therefore less suitable for fermentation to ethanol. If the versatility of a liquid fuel is desired, they may be gasified and the gas used for methanol synthesis. Otherwise they may be burned directly for heat or for electricity generation.

A most important factor that arises in considering the use of cropland for energy is the competition with the food market. It has been concluded[2] that there is no assurance that cropland will in the future be available in the U.S. for energy uses, and the cultivation of specific energy crops may have to be limited to avoid inflation of food prices. The same study found that the use of crop residues, the material left in the field after harvesting, could supply up to 1 percent of U.S. energy requirements. This figure is based on the recovery of 20 percent of the crop residue, with the remaining 80 percent required for soil conditioning. The crop residues do not, of course, have the high starch or sugar content of the crop itself, but are essentially lignocellulosic materials. As in the case of grasses, crop residues are probably more suited to gasification or direct combustion. Processes for the conversion of cellulosic biomass to ethanol are, however, being investigated and may prove viable.

The crop residue attainable per unit area annually is, based on data for corn, less than 350 t/km². Based on an estimated conversion rate of 6.1 kg per liter of ethanol,[2] a farm measuring 25 km by 25 km would be required to supply an ethanol plant rated at 0.1×10^6 L/d. In addition, some 200 000 t of crop residue would have to be stored to assure a steady supply of raw materials from this annual resource. It is noted also that typical residence times in a fermentor are two days, while residence times in coal liquefaction reactors are of the order of an hour. On taking all these factors into account, it would seem that plant sizes much larger than the quoted 0.1×10^6 L/d are not realistic. We recall that a nominal synthetic fuel plant producing liquids from coal or oil shale was rated at 10×10^6 L/d of oil or gasoline. In fact, the ethanol production from the total useable crop residue of 35×10^6 to 55×10^6 t/yr in the whole United States,[2] is at most equivalent to the energy output of two nominal-sized coal liquefaction plants.

Animal wastes, or manure, as a source of biomass has the advantage that it is not competitive with other uses for this material. However, it must of necessity come from confined operations such as dairy farms and cattle feedlots, and has a potential limited to less than one-third of 1 percent of total U.S. energy consumption. It is further estimated that about half this resource is from farms having less than 100 animals, so that the scale of operation is small. Small-scale conversion on farms appears most appropriate, and could result in animal husbandry operations being self-sufficient in energy. Anaerobic digestion to produce methane is the most applicable technology.

Agricultural wastes are defined as the waste material remaining after processing crops to a marketable material. Common wastes are sugar cane bagasse and cotton gin trash. Many agricultural wastes are sold as animal feed, for which they realize a better price than if used for energy. Materials unsuitable as feed are typically burned on site and some processing plants are reportedly self-sufficient in energy. The total potential for agricultural wastes is less than one-tenth of 1 percent of U.S. energy requirements.[2]

Aquatic plants include kelp from the ocean, and freshwater plants such as algae, water hyacinth, and duckweed. Production of freshwater plants is associated with high capital costs and water requirements, while harvesting difficulties and the particularly high water content of this form of biomass are added disadvantages. Enhancing the growth rate of these plants by increased nutrient supply, for example from carbon dioxide in flue gases, or growing suitable plants in conjunction with municipal wastewater treatment facilities, has received some attention. While quantitative data and costs for this resource are limited, full development of the aquatic biomass energy potential is not expected to approach 1 percent of total U.S. energy requirements.[2]

Municipal solid waste is a source of energy probably with as much potential as wood. Non-organic constituents such as metals and glasses must be removed, and may be sold for their recycle value to at least partly compensate for the cost of the separation step. The remaining refuse contains a high proportion of cellulosic material (paper) and other components not readily amenable to biochemical conversion. The characteristics of municipal solid waste can vary significantly with the time of the year, which further detracts from its suitability for biochemical conversion. Thermal processes are consequently of interest. Direct combustion for district heating is an obvious application practiced in some countries, while gasification is another possibility. Of particular interest is a scheme for mixing coal, municipal solid waste, and dewatered sewage sludge to produce briquettes suitable for gasification. Apart from supplementing the coal energy, the waste material renders the coals non-caking. Appropriation of these waste materials for energy is not competitive with other uses, and in fact reduces their disposal problem. An additional advantage over other biomass resources is that collection or "harvesting" procedures are both established and financed.

In principal, biomass resources can be converted using any of the conversion processes shown in Figure 8.1. However, depending on their composition and other properties discussed above, some processes can be expected to be more effective than others in recovering energy from a specific resource. The process/biomass combinations believed to offer the greatest potential are summarized in Table 8.4. In cases where more than one process is suitable, the end product will be an important consideration in selecting among conversion routes. Liquid products are more versatile than gas and heat. Methanol is not as well suited as ethanol for blending with gasoline to form gasohol, but may prove to be an acceptable stand-alone fuel. Electricity is readily produced in conjunction with direct combustion, but may be inefficient relative to other processes if used for heating.

8.2 BIOCHEMICAL CONVERSION

Fermentation and anaerobic digestion are the two biochemical processes of interest. In its broad sense, fermentation refers to any chemical change of organic material that is accompanied by effervescence, normally without the participation of oxygen. According to this description, both the biochemical processes to be considered here could be classed as fermentation processes. We choose to reserve the term *fermentation* for the anaerobic decomposition of carbohydrates to alcohols in the presence of enzymes. Ethanol is the principal product of the fermentation processes appropriate to biomass conversion, although other alcohols, as well as organic acids, ketones, and aldehydes may be produced either as main products or as byproducts. *Anaerobic digestion* is the decomposition of any organic material by the metabolic action of

Table 8.4 Biomass conversion processes

Process	Principal Biomass	Scale*	Product
Combustion	Wood, municipal solid waste, grasses, crop residue	Small Large	Heat Steam, electricity
Air or oxygen gasification	Wood, municipal solid waste (grasses, crop residue)	Large	Low-CV or synthesis gas, methanol
Pyrolysis	Wood, sewage sludge	Large	Medium-CV gas, tars
Fermentation	Grain and sugar crops	Small, large	Ethanol
Anaerobic digestion	Animal wastes, aquatic biomass	Small	Methane

* Small implies domestic or farm application; large is industrial-scale processing of up to 1000 t/d of biomass.

bacteria without the participation of atmospheric oxygen. Methane and carbon dioxide are the main products of the decomposition. The source of the oxygen in the carbon dioxide is the combined oxygen in the organic molecules and in the water.

The important differences between fermentation and anaerobic digestion are the nature of the product produced and the character of the biological contribution. Fermentation produces a liquid product in the presence of enzymes, while anaerobic digestion yields a gaseous product as a result of the metabolic activity of bacteria. The distinction between the two processes is, however, somewhat artificial. Bacterial digestion is in effect accomplished by enzymes. Further, certain bacteria produce acids and alcohols as the principal degradation products. In some cases, it is not clear whether the degradation proceeds as a result of bacterial metabolism, or whether it can be achieved by non-growing cells.[5] Nevertheless, the distinction between the two processes is convenient for presentation purposes, and should not cause confusion in classifying the important biochemical processes currently in contention.

The substances called "enzymes" that are stated to be necessary for fermentation can be considered to be the catalysts of life processes. They are complex proteinaceous compounds produced by living cells, and they catalyze hydrolysis and oxidation reactions. Glycolysis is a special kind of fermentation that occurs in muscle tissue and converts glucose to lactic acid with the release of energy. Biomass fermentation to produce ethanol is very similar to glycolysis, but the use of different enzymes results in different end products. Respiration, like glycolysis, releases energy for life processes, but is an aerobic process requiring a source of oxygen. The biological process of respiration may be likened to the thermal process of combustion. This loose analogy may be extended to relate the biochemical processes of fermentation and anaerobic digestion to the thermal process of pyrolysis. In other words, instead of providing heat to enhance decomposition reactions, in biochemical processes a biological means is used to induce decomposition at near-ambient temperatures. The chemical decomposition reactions in the two cases are not the same, so that the end products are generally different. In addition, the biochemical reaction rates are from 1 to 2 orders of magnitude slower than the thermally induced reactions.

Fermentation

A simplified flow diagram for a process to convert biomass to ethanol by fermentation is shown in Figure 8.2. If the biomass is a sugar crop, then the initial hydrolysis steps are not required. Although the fermentation process is otherwise similar for all sources of biomass, there are significant differences in some of the initial stages which are introduced to maximize yields for a specific feedstock.

Our description of fermentation is based on the process for fermenting corn, a starch crop, as discussed by Kelm.[6] Traditionally the corn is dry-milled and then suspended in water which is boiled. The purpose of the milling and boiling is to free the starch molecules for subsequent reaction. However, the non-convertible constituents cannot be separated at this stage and are carried on through the entire process before being recovered from the distillate bottoms (called "stillage") as a byproduct for

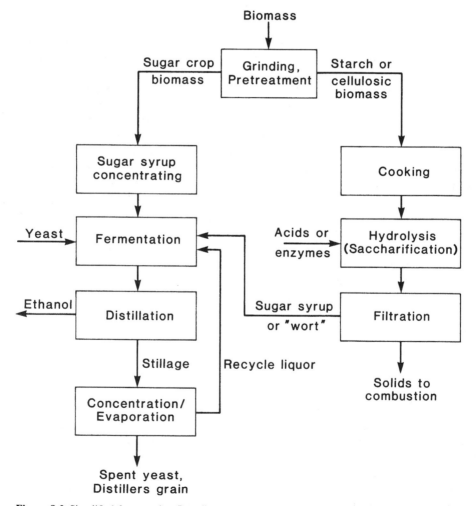

Figure 8.2 Simplified fermentation flow diagram.

animal feed. In a modified procedure, the soluble protein and carbohydrates are first leached out by steeping in a warm sulfurous acid solution for 25 to 40 hours. The undissolved solids are then coarsely milled in the wet state, following which it is possible to separate out the germ in a hydrocyclone for subsequent corn oil recovery. The reported power required for the wet-milling operation is equivalent to about 1.2 MJ/L of ethanol produced. The advantages of the steeping and wet-milling procedures are firstly that more valuable byproducts are recovered, and secondly that the absence of solubles in the stillage allows more water to be recycled. Previously much of this water had to be evaporated to recover the solubles as "distillers grain" used for animal feed.

The milled corn-water slurry called the "mash" is then boiled in cookers. This cooking step, which lasts about an hour, can be improved by the addition of the enzyme α-amylase, which prevents the gelatinized starch from regressing or "setting back" and so becoming unavailable for hydrolysis. Enzymes are also added to catalyze the subsequent hydrolysis step. "Malt," which is traditionally used, is barley grain that has been allowed to germinate, so producing a mixture of the enzymes α- and β-amylase, sometimes called "diastase." Glucoamylase is a pure enzyme that is now available and that can be used to advantage. Both diastase and glucoamylase hydrolyze the starch to the disaccharide maltose. The hydrolysis is carried out at 50 to 60°C and, for fuel production, is sufficiently complete after an hour or so. The mixture leaving the hydrolysis reactors is called a "wort."

While enzymatic hydrolysis seems to be most economical for starch, acids are generally used for hydrolysis of the more resistant cellulosic biomass resources. Dilute acid hydrolysis processes generally use 0.5 percent sulfuric acid at a temperature of 140°C to 190°C. In concentrated acid processes, 70 to 80 percent sulfuric acid or 35 to 42 percent hydrochloric acid is used at about 45°C or lower. When using concentrated acids, the acid must be recovered for reuse. Hemicellulose may be separated out in an initial dilute acid treatment and fermented separately. The lignin is of course not converted. Processes using enzymes (cellulase) and bacteria[5] are being investigated for cellulose conversion.

Following hydrolysis, the wort is cooled to below 30°C and its pH is adjusted to about 5. Yeast is added and the fermentation proceeds until the sugars are consumed, which in a batch process may take 2 to 3 days. Yeast is a microscopic single-cell plant that supplies the enzymes required to catalyze the formation of ethanol. In particular, maltase catalyzes the degradation of maltose to glucose, and zymase catalyses the fermentation of the glucose to ethanol. Yeast also produces invertase which, as we have seen (Eq. 8.2), is required to "invert" any sucrose present to fermentable sugars.

The fermentation reaction is strongly exothermic, and the heat of the reaction must be removed to maintain the temperature below 30°C to assure the viability of the yeast. At this low temperature, the heat released is "unavailable" for recovery or interchange and has to be dissipated by cooling. Based on a heat of reaction of 105 kJ/mol of ethanol formed, the heat dissipated amounts to about 8 percent of the GCV of the product alcohol.

Fermentation proceeds in dilute aqueous solutions. The sugar concentration in the feed is generally in the range of 14 to 18 percent by mass, and the reaction continues until the alcohol concentration approaches about 14 percent. At higher concentrations the process becomes self-inhibitory and the metabolic activity is arrested. Consequently the alcohol is obtained as a very dilute solution and has to be concentrated, for example, by distillation. Energy requirements for distilling off the alcohol depend on the required purity of the product alcohol and the extent of heat recovery practiced in distillation. Published estimates[2,6] for producing 99.5 percent ethanol suitable for gasohol blending suggest a steam requirement of 7 to 11 MJ/L of product ethanol. This is equivalent to 30 to 45 percent of the GCV of the product.

Material and Energy Balance

The amount of alcohol produced per unit of starch feed can be estimated from the appropriate chemical reactions.

(a) Saccharification

$$2(-C_6H_{10}O_5-) + H_2O \xrightarrow{\text{diastase}} C_{12}H_{22}O_{11} \qquad (8.3)$$
$$\text{Starch} \qquad\qquad\qquad \text{Maltose}$$

(b) Fermentation

$$C_{12}H_{22}O_{11} + H_2O \xrightarrow{\text{maltase}} 2C_6H_{12}O_6 \qquad (8.4)$$
$$\text{Maltose} \qquad\qquad\qquad \text{Glucose}$$

and

$$2C_6H_{12}O_6 \xrightarrow{\text{zymase}} 4C_2H_5OH + 4CO_2 \qquad (8.5)$$
$$\text{Glucose} \qquad\qquad \text{Ethanol}$$

The equations can be used to show that 184 kg of ethanol is derived from 324 kg of starch. In practice, about 10 percent of the starch is consumed in the formation of byproducts including higher alcohols (fusel oils), glycerine, and ethers. Correcting for byproduct formation and assuming the corn is 61 percent starch, leads to a value of 3.2 kilogram corn per kilogram of ethanol, or 2.6 kilogram corn per liter of ethanol produced.

These conversions have been used together with published energy requirements to construct the material and thermal balance given in Table 8.5. As shown, the thermal efficiency with respect to the ethanol product is 46 percent, while the process-energy input in the form of steam and electricity is equivalent to some 65 percent of the ethanol product. These values do not include the energy required for cultivating and harvesting the corn crop, which is estimated to be equivalent to some 42 percent of the ethanol product.[2] The gross energy consumed in corn fermentation is consequently about 7 percent more than is produced. If the production of alcohol is not to be a net consumer of premium fuels, coal or crop residues must be used to supply at least the process-energy requirements. In this context it is stated that the octane-boosting effect of alcohol in gasohol is equivalent to a saving of 0.4 L of gasoline in refinery energy for each liter of ethanol used. On the assumption that non-premium fuels are used for processing and that the ethanol product is used for gasohol blending, it has been

Table 8.5 Material and thermal balance for fermentation of corn to ethanol*

	Mass, kg	GCV, MJ/kg	Heat, MJ	Heat, % of Total
IN				
Corn (61% starch)	100	14.1	1410	70.4
Dilution water†	336	—	—	—
Power for milling (as heat)	—	—	9	0.4
Process steam (cooking, etc.)	—	—	179	8.9
Steam for distillation	—	—	268	13.4
Byproduct recovery, etc.	—	—	138	6.9
Total	436		2004	100.0
OUT				
Ethanol (39 L)	31	29.7	921	46.0
Water	329	—	—	—
Byproduct and losses (by difference)	76	—	1083	54.0
Total	436		2004	100.0

Thermal efficiency* $= \dfrac{921}{2004} \times 100 = 46\%$

* Energy requirements based on data in Ref. 2.
† Calculated for 16 percent glucose solution to fermentor.
‡ Does not include cultivation and harvesting energy; see text.

concluded that corn fermentation has the potential for reducing gasoline consumption by nearly a liter for each liter of ethanol produced.[2]

The large energy consumption in fermentation has led to proposals for modifications of the process, in particular to reduce the distillation costs. Membrane separation techniques (reverse osmosis) may, for example, be used to concentrate the alcohol.[7] In another proposed scheme, the fermentation is carried out under vacuum to remove the alcohol as it is formed. These systems have yet to be demonstrated on a commercial scale.

Anaerobic Digestion

The anaerobic decomposition of complex organic molecules to methane and carbon dioxide can be perceived as occurring in three stages.[2] As in fermentation, the first stage is one of hydrolysis in which the insoluble carbohydrates, proteins, and oils are converted to soluble substances including sugars and alcohols. Next, the soluble materials are converted to fatty acids, esters, carbon dioxide, and hydrogen. In the final stage, called "methanogenesis," the intermediate compounds are converted to methane. The hydrogen formed in the initial stages is consumed while still in solution, for the synthesis of methane, and is not evolved as a product. The mixture of methane and carbon dioxide that is evolved from solution is referred to as "biogas," although the constituent gases are no different from those produced, say, from the reaction of carbon with steam (Eq. 2.47).

The reaction sequence for anaerobic digestion shown in Figure 8.3 shows that four different bacterial groups are involved. Ideally, each step in the reaction sequence should be conducted in a separate reactor so as to accommodate the specific reaction rates and preferred ''operating'' conditions associated with each group of bacteria. In practice, the process is generally carried out in a single reactor vessel, the domestic septic tank being a typical example. In using less readily digestible resources such as lignocellulosic crop residue biomass, an initial pretreatment hydrolysis step may be incorporated. This may be accomplished by either acid or enzymatic hydrolysis as described for fermentation. While such a pretreatment does result in greater gas yields, it is estimated that the improved production may not offset the increased capital and operating costs.[2]

A simplified flow scheme for an anaerobic digestion process is shown in Figure 8.4. It is assumed that biomass resources not as readily converted will be subjected to an initial hydrolysis step as is done in the case of fermentation (Figure 8.2). In comparing the two processes, it can be seen that an equalization step is introduced in the case of anaerobic digestion. This recognizes that the bacteria are sensitive to changes in their environment, and perform best under invariant conditions. The equalization pond is therefore designed to even out any fluctuations in feedstock composition and should have a residence time in accordance with anticipated temporal variations in the composition of the biomass resource. Apart from equalizing composition, the pond can serve to equalize flows by storing feedstock provided from a non-steady or periodic source.

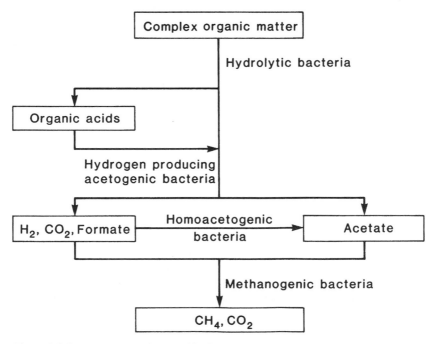

Figure 8.3 Successive stages in anaerobic digestion.[8]

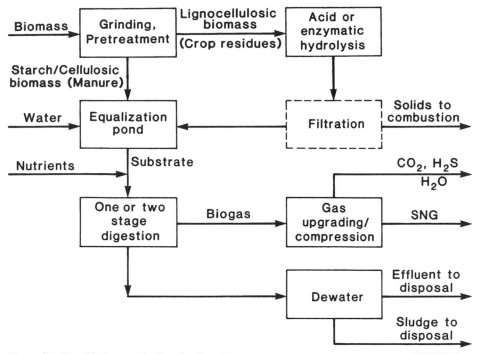

Figure 8.4 Simplified anaerobic digestion flow diagram.

The heart of the process is the digestor itself. The purpose of this vessel is to provide an environment for the bacteria to propagate, to ensure adequate contact between the bacteria and the digestible organics, and to hold the mixture for sufficient time for the nearly complete consumption of the degradable organics. In addition, the vessel must exclude the access of oxygen, and provide a means of collecting and delivering the biogas. This latter function is not a design criterion for municipal sewage treatment plants, where the important function is to produce a stabilized sludge. In fact, municipal systems are now largely aerobic processes that oxidize the organics to carbon dioxide and water, and do not produce methane. Aerobic bacteria tend to be more tolerant to system changes, and require shorter residence times to complete the decomposition.

The mixture of biomass material, water, and any added nutrients is called the "substrate." Bacteria are generally present in the biomass material itself, and if the required conditions are provided, will propagate. Those bacteria best able to metabolize the feedstock will, by a process of natural selection, soon dominate. This process of producing a thriving microorganism population is known as "acclimation." The acclimated population will be maintained in a continuous process by recycling some of the sludge, whereas in a batch process some sludge from a previous batch may be used to "seed" subsequent batches. Important variables requiring careful control in acclimating and maintaining a viable microorganism population include substrate concentration, temperature, and pH, as well as the presence of nutrients and absence of toxicants. As discussed in Section 9.1, neutral solutions have a pH of 7.0, with lower values indicating acidic solutions and higher values indicating basic, or alkaline, solutions.

The *concentration* of biomass in the substrate is limited by the toxicity of the components present. As in fermentation, the substrate has to be relatively dilute, equivalent to a solids concentration of 7 to 12 percent of the original biomass material. Again, therefore, a considerable supply of water is required to provide the necessary dilution. Acetate concentrations above 2 g/L and ammonia concentrations in excess of 3 g/L are toxic and will decrease or arrest the methane production rate.[8] Oxygen is highly toxic to methanogenic bacteria, while trace metals and antibiotics may be troublesome.

Unlike fermentation, the digestion process is only mildly exothermic, and heat may have to be supplied to the substrate to achieve the optimum operating *temperature*. Apart from their decompositional activity, bacteria can be grouped according to the temperature at which they thrive. Temperature ranges are 0 to 20°C for psychophilic bacteria, 20 to 45°C for mesophilic bacteria, and 45 to 65°C for thermophilic bacteria. While the temperature range is fairly broad for each group, an established population requires much narrower confines of temperature for optimum performance. The temperature of thermophilic processes should be controlled to within 5°C, and to within 2°C for mesophilic processes. The selection of an operating temperature not only determines the group of bacteria that will dominate, but also affects the degree of conversion, residence time, and cost.

Thermophilic processes generally have the highest decomposition rates and hence have reduced residence times. The associated reduction in capital costs must, however, be weighed against the increased operating costs incurred in heating the substrate. While mesophilic decomposition rates are lower, there is evidence that these bacteria achieve a greater degree of conversion with some feedstocks. In fact, feedstocks such as ocean kelp are not effectively converted by thermophilic bacteria and must, therefore, be processed under mesophilic conditions. Decomposition rates at psychophilic temperatures are too slow for practical application. Depending on the biomass, residence times may range from 5 to 10 days for thermophilic processes, to 10 to 15 days or longer for mesophilic processes. Small- or farm-scale processes will probably operate under mesophilic conditions, while with larger-scale processes it may prove economical to incorporate more extensive heat recovery systems for operation at thermophilic temperatures. In all designs, a tradeoff must be made between residence time and the degree of conversion to product gas.

The *pH* of the substrate should be maintained between 6.6 and 7.6, and preferably between 7.0 and 7.2.[8] Unconventional feedstocks such as ocean kelp may, however, require different pH conditions. *Nutrients* such as nitrogen, phosphorus, and alkali metals must be added if not present in adequate concentrations in the biomass material. Animal manures usually contain all the required nutrients. However, the high ammonia concentrations in some animal wastes may lead to ammonia toxicity problems. In this respect, the *carbon-to-nitrogen ratio* (C/N) is thought to affect the digestion process.[2,8] The rate of carbon consumption is typically about 30 times that for nitrogen, whereas in animal wastes C/N is closer to 15. It is found that wastes such as dairy manure with higher than average C/N ratios can be processed at relatively high solids concentrations and require relatively short retention times. Poultry manure, which has a high nitrogen content, requires lower solid concentrations and longer residence times. In all cases, the concentration of nitrogen in the solids remaining after decomposition is higher than in the original feedstock.

Both batch and continuous digestors are commercially available. While batch systems may be suitable for some applications, they are labor intensive, and in developed countries are probably not appropriate to the steady production of gas on any significant scale. Mixing may be used in both batch and continuous digestors to enhance contact between the bacteria and substrate. It has, however, been argued that while mixing may increase conversion rates, it is not energy efficient. In two stage systems such as shown in Figure 8.5, only the first stage is mixed. Multistage systems provide optimum conditions for the successive digestive steps, and further development may prove them to be economical for some biomass materials.

Apart from methane and carbon dioxide, the raw biogas will contain ammonia, hydrogen sulfide, and water vapor. A range of compositions covering typical biomass resources and operating conditions is shown in Table 8.6. If the desired product is a SNG, then the carbon dioxide, trace gases, and moisture must be removed using the upgrading processes described in Section 5.1. For farm use the gas may be burned directly, although it will generally be desirable to first remove the toxic and corrosive hydrogen sulfide. This may be done on a small scale by absorption into iron oxide to form ferric sulfide. If the gas is not required for immediate consumption, it will have to be compressed for storage. Even after compression to 3.5 MPa, the volume required to store pure methane is 24 times that required to store an equal amount of energy as gasoline. The danger of leaks and explosions are an important consideration in the handling and compression of biogas, especially in small-scale and batch processes.

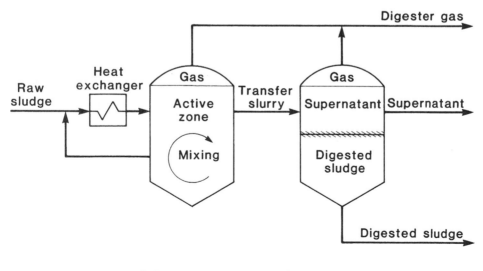

Figure 8.5 Two stage anaerobic digester.[2]

Table 8.6 Typical biogas compositions[2,8]

Constituent	Mol %, dry*	Element	Mass %, dry
Methane	50–65	C	40–46
Carbon dioxide	50–35	H	7–10
Ammonia	1	O	53–44
Hydrogen sulfide	0.1–1.0		
Mean molar mass	30–26		
GCV, MJ/m³	20–26		
MJ/kg	15–22		

* Moisture content, if saturated at atmospheric pressure, will range from 5 percent at 35°C to 20 percent at 60°C.

The final stage in the process is the disposal of the wastes. Essentially all the water entering the system will be present in the effluent. Direct recycle of the water after separation from the sludge cannot be practiced, as this will result in a buildup of toxic substances and eventual failure of the system. It is possible that some fraction of the water may be safely recycled, although data on this aspect have not been seen. Treatment of this large volume of water to remove the toxics is costly and is likely to be uneconomical other than in large-scale operations. It is normally assumed that the water, which is considered to have some nutritional value, will be used for irrigation. Due consideration will have to be given to the effect of trace metals and other toxic materials present, and the final disposition of the irrigated crops. The sludge, too, may be used for land application and, as stated, will generally have a higher nitrogen content than the original material. On an absolute basis there is, of course, less nitrogen due to losses as ammonia, and there may be additional losses on land application, as the nitrogen in the sludge is in a more volatile form. Nevertheless, the sludge resulting from the anaerobic digestion of animal manure is considered to have improved fertilizer value over the original material.

Mass and Energy Balance

The overall conversion process may be represented in terms of cellulose by

$$(-C_6H_{10}O_5-) + H_2O \xrightarrow{\text{bacteria}} 3CO_2 + 3CH_4 \qquad (8.6)$$

kg: 162   18   132   48

Theoretically, therefore, 0.3 kg methane or 0.4 m³ methane can be obtained per kg of biomass. In practice, the amount of biomass converted ranges from 25 to 50 percent depending on the feedstock and process conditions. Typical methane yields are consequently in the range of 0.07 to 0.15 kg/kg, or 0.1 to 0.2 m³/kg of biomass.

Most of the energy required by the process is for heating the substrate. The gross heat required for this function in a thermophilic process operating at 55°C with an ambient temperature of 10°C has been estimated to be equivalent to nearly 40 percent of the product.[8] Heat recovery is therefore essential, but due to the relatively small temperature differences, large and costly heat exchangers will be needed. The energy requirements for the anaerobic digestion of cattle manure, assuming a 50 percent heat recovery from the effluent, are summarized in Table 8.7 and indicate a process thermal efficiency of 53 percent. Assuming that the effluent water is not treated, the process energy input is equivalent to about 30 percent of the methane output. This is less than half the 65 percent value required for ethanol fermentation, and reflects the fact that the biogas is spontaneously separated from the substrate and does not require an energy intensive (e.g. distillation) recovery step. To the extent that biogas processes are restricted to processing waste materials, the energy associated with production and harvesting of the feedstock is not as significant as for corn growing.

8.3 THERMAL CONVERSION

Thermal processes include pyrolysis, gasification, and combustion. A graphic description of these processes and their products is given in the triangular diagram of Figure 8.6. Note that the apices in Figure 8.6 represent carbon, atomic hydrogen, and atomic oxygen, whereas the coordinate system used in Section 3.2 is based on the reacting species carbon, steam, and molecular oxygen. The common factor in all the thermal processes portrayed in Figure 8.6 is, of course, heat. Pyrolysis normally proceeds at

Table 8.7 Material and thermal balance for anaerobic digestion of cattle manure*

	Mass, kg	Heat Content, MJ/kg	Heat, MJ	Heat, % of Total
IN				
Cattle manure (dry)	100†	13.4	1340	84.7
Dilution water (10% solids)	900	—	—	—
Substrate heating‡	—	—	157	9.9
Mixing energy	—	—	55	3.5
Gas scrubbing	—	—	21	1.3
Methane compression (to 1 MPa)	—	—	9	0.6
Total	1000		1582	100.0
OUT				
Methane (21 m³)	15	55.6	834	52.7
Carbon dioxide	41	—	—	—
Moisture in gas§	6	—	—	—
Effluent water	888	—	—	—
Sludge (dry) and losses (by difference)	50	—	748	47.3
Total	1000		1582	100.0

Thermal efficiency $= \dfrac{834}{1582} \times 100 = 53\%$

* Energy requirements based on data in Ref. 8.
† About 22 bovine-days worth.
‡ Assuming 50 percent heat recovered from the effluent.
§ Assuming gas saturated at 55°C and 101 kPa.

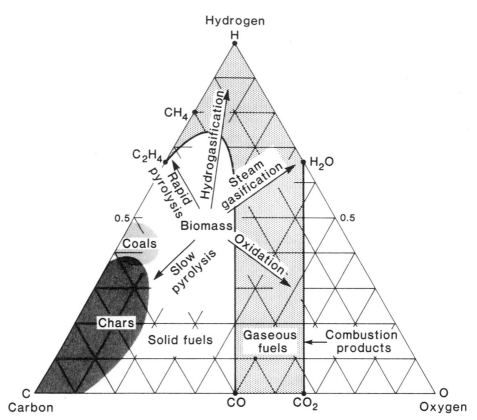

Figure 8.6 Thermal conversion of biomass.[3]

temperatures below about 600°C, while in gasification processes temperatures range from 800 to 1100°C or even higher in slagging systems, and combustion temperatures are 1500°C or more.

Pyrolysis occurs in the absence of oxygen, and other chemical reactants are usually not used. The products of pyrolysis are gases, liquids, and a carbonaceous char, in relative amounts dependent on the properties of the biomass, the rate of heating, and the final temperature attained. As shown in Figure 8.6, slow pyrolysis produces mainly char, while with rapid pyrolysis the liquid and gaseous yields are increased. Pyrolysis in a hydrogen atmosphere (hydropyrolysis) also increases the gaseous yield relative to char, and generally produces a high quantity and quality of liquid product.

In the case of gasification one or more reactants, such as oxygen, steam, or hydrogen, are introduced into the system. These chemical reactants combine with solid carbon at the higher gasification temperatures, so increasing the gas yield while consuming char. The amount of char byproduct remaining on gasification is in fact essentially zero with biomass materials, while the small quantity of tars and oils evolved may be recycled to extinction. The introduced reactants also enter into gas phase reactions which, together with the shift in equilibrium and the change in relative reaction rates at the higher temperatures, results in a significantly better-quality gas than that obtained on pyrolysis. Important distinctions between pyrolysis and gasification are therefore the improved gas yield and the elimination of solid and liquid byproducts. These advantages are gained at the cost of a more complex process.

Gasification results in only a partial oxidation of the carbon constituent. Much of the calorific value of the original fuel leaves the gasifier in the form of chemical energy in carbon monoxide, methane, and hydrogen. In combustion, air containing sufficient oxygen to completely oxidize the hydrocarbon is used. The resulting gas (flue gas) is essentially all carbon dioxide, water vapor, and diluent nitrogen, and has no calorific value. The energy of the original fuel leaves the system as sensible heat that can be used for steam raising and, if required, electricity generation. Combustion offers a practical means for recovering the energy in biomass materials, using currently available technology. Further, it is well suited to small-scale utilization, a factor that seems to be of importance on account of the dispersed nature of the resource. Strictly speaking, however, combustion does not produce a synthetic fuel, and so in keeping with the constraints of this text, is not discussed here.

Pyrolysis and gasification fundamentals are presented in Chapter 3, while practical processes are described in Chapters 4 and 6. Although the emphasis in these earlier chapters is on fossil fuel conversion, much of the subject matter is directly applicable to biomass. In this section, we therefore confine our remarks to those aspects in which biomass conversion differs from the conversion of conventional fuels, while also giving an indication of the properties and merits of the synthetic fuels produced. The reactor types under development fall into the categories described previously in this book, so that there is little new to be learned in presenting descriptions of these systems. The interested reader is referred to Ref. 3 which contains an up-to-date summary of the important gasification/pyrolysis equipment and processes being developed.

Properties

The properties of biomass that have a significant bearing on its thermal conversion are its relatively high moisture, oxygen, hydrogen, and volatile matter content, and low energy density. The high oxygen and hydrogen contents account for the high proportion of volatile matter and consequent high yields of gases and liquids on pyrolysis. The low yield of char relative to that obtained on pyrolysis of lignite and coals is depicted in Figure 3.2. Also shown in this figure is the relatively high water yield which results in turn from the high oxygen concentration in biomass, and which consumes considerable hydrogen. Consequently the advantages of the high H/C ratio associated with biomass (Table 8.3) are not reflected in the products to the extent that might be expected. In fact, pyrolysis gases can be deficient in pure hydrogen (Figure 3.3), and pyrolysis liquids are highly oxygenated, viscous tars.

An additional and significant source of water vapor in biomass gases is the high moisture content of the source materials. In countercurrent flow schemes such as the Lurgi moving bed gasifier, this water is evolved in the relatively low temperature drying and pyrolysis zones and does not partake in gas phase or carbon-steam gasification reactions. On the other hand, in fluidized bed systems the moisture is evolved in the high temperature well-mixed reaction zone and therefore does participate in the reactions. If the system is directly heated and air blown, the additional heat required to evaporate the water will result in more nitrogen being introduced, and more carbon dioxide being produced, so reducing the calorific value of the product gas. As the gas from air-blown processes is, in any case, a low-CV product, this factor is probably of little consequence other than with very wet feedstock. In oxygen-blown systems, however, the additional pure oxygen required and higher carbon dioxide content of the medium-CV off gas may be of sufficient impact to dictate some degree of drying as a pretreatment.

Apart from drying, additional beneficiation may be undertaken to yield a resource of higher energy density. These operations will normally be undertaken at the source, so transport and subsequent storage costs may be reduced as well. Beneficiation steps include size reduction and densification. Waste heat, if available, may be used for drying, while size reduction and compression to form pellets or briquettes is estimated to require less than 2 percent of the energy in the dry biomass.[3] Nevertheless, these operations are time consuming, and can be either labor or capital intensive.

Some advantages of biomass over conventional fossil fuels, not so far discussed, are their low-sulfur content and their relatively highly reactive chars. In addition, biomass materials do not cake and can therefore be easily handled in both fluidized and moving bed reactors. Finally, catalyst poisons are not present in biomass in significant concentrations. This can be important for the initial thermal processing as well as for subsequent upgrading operations.

The most suitable biomass resources for thermal conversion are wood and the organic portion of municipal solid waste. Crop residues and grasses are of intermediate value, although thermal conversion might prove more effective than fermentation. If the versatility of a liquid fuel is desired, gasification may be combined with methanol or Fischer-Tropsch synthesis. Very wet resources such as aquatic biomass and animal wastes are not suited to thermal conversion, and are best reserved for anaerobic digestion.

Pyrolysis

The relative quantity of char, tars, and gases evolved on pyrolysis is strongly dependent both on the rate of heating and the final temperature attained. Slow-heating, low temperature pyrolysis favors char yields. In theory, cellulose can be carbonized to yield carbon and water only:

$$(-C_6H_{10}O_5-) \xrightarrow[200-280°C]{\text{slow pyrolysis}} 6C + 5H_2O(g) \qquad \Delta H° = -463 \text{ kJ/mol} \qquad (8.7)$$

In practice, the initial decomposition product of cellulose pyrolysis is levoglucosan, $C_6H_{10}O_5$, an anhydride of glucose. Subsequent coking and cracking reactions result in a char that contains oxygen and hydrogen in addition to the carbon. The char remaining on pyrolysis of wood contains about 80 percent carbon, 17 percent oxygen, and 3 percent hydrogen.

Hemicellulose is the most reactive fraction of the lignocellulosic portion of wood, being decomposed slightly more rapidly than cellulose. Cellulose, however, produces the least char, while lignin, which decomposes slowly, produces the most char and is also responsible for the aromatic content of the liquid product. As a whole, wood is thermally decomposed more rapidly, more completely, and over a narrower temperature range than is coal.

Apart from char and water, some tars and gases are always produced, as is shown in Figure 3.2. The watery distillate evolved from wood at 160 to 175°C is called "pyroligneous acid" and contains some 5 to 10 percent acetic acid and 1.5 to 3 percent methanol. This used to be the source of methanol when it was produced by the destructive distillation (pyrolysis) of wood. Clearly, methanol is only a minor byproduct of wood pyrolysis. The heavier tar fraction produced at higher temperatures is more important as a source of liquid fuels. It contains aromatic hydrocarbons and "creosote oil" evolved from the lignin, as well as aliphatics. The creosote oil, which consists of high-molar-mass phenols, used to be considered the most valuable constituent of the tar fraction, in part due to its potential for resin manufacture. As a whole, the tar fraction is a viscous, highly oxygenated oil that is unstable and corrosive.

The composition of a typical wood pyrolysis oil is given in Table 8.8. These data show the pyrolysis oil to be low in sulfur, to have a moderate nitrogen content of about 1 percent, and to have a relatively high H/C atom ratio of 1.6. The high oxygen content of 33 percent is, however, a dominant feature of the composition. We recall that primary coal liquids and shale oils have oxygen contents of less than 2 percent, while oxygen concentrations below 0.5 percent are typical of petroleum crudes. The high oxygen content of biomass pyrolysis oil accounts for its low calorific value of about half that of coal liquids on a mass basis, and probably also for its corrosiveness and low stability. Further, the oxygen can be expected to result in excessive hydrogen consumption in upgrading operations. It appears, therefore, that pyrolysis tars are not desirable liquid fuel precursors, and are probably best consumed on site to meet process energy requirements.[3]

Both liquid and gas yields are enhanced by increased heating rates and by higher pyrolysis temperatures. Reported gas compositions vary from essentially carbon dioxide

Table 8.8 Typical properties of wood pyrolysis oil[2]

Constituent	Mass %
Carbon	57.0
Hydrogen	7.7
Nitrogen	1.1
Sulfur	0.2
Chlorine	0.3
Oxygen	33.2
Ash	0.5
Total	100.0
H/C atom ratio	1.6
s.g.	1.30
GCV, MJ/kg	24.7
Pour point, °C	32

and water vapor only, to gases containing mainly carbon monoxide but with significant concentrations of hydrogen, methane, and higher hydrocarbons. The heating rate, maximum temperature reached, residence time at the maximum temperature, method of withdrawing and quenching the gas, and the extent of moisture and oxygen present in the system all appear to influence the final gas composition. However, the effect of these variables and the specific reaction schemes involved are not yet fully understood.

While recognizing that gas formation mechanisms have not been well defined, and that operating conditions are not always fully specified for the reported gas analyses, some trends in composition are evident. At pyrolysis conditions favoring char formation, the evolved gases contain a high proportion of carbon dioxide, some carbon monoxide, but little hydrogen or other gases. As operating conditions are changed to favor gas yields, carbon monoxide becomes the major component present, while the concentrations of hydrogen, methane, and other hydrocarbon gases increase significantly. At very rapid heating rates (flash pyrolysis), the yield of ethylene and higher olefins becomes important, and the char yield is correspondingly reduced.[3] The reduced char yield associated with increased temperatures and higher carbon monoxide and hydrogen yields suggests the influence of the gasification reactions

$$C + CO_2 \rightleftharpoons 2CO \tag{8.8}$$

$$C + H_2O \rightleftharpoons CO + H_2 \tag{8.9}$$

The equilibrium of both these reactions shifts to the right with increasing temperature.

The composition of a gas obtained on pyrolysis of dried sawdust in an indirectly heated fluidized bed system at 825°C is given in Table 8.9. The gas consists mainly of carbon monoxide and hydrogen, with smaller amounts of carbon dioxide and methane. The reported gas yield contains 95 percent of the carbon, and also represents about 95 percent of the calorific value of the original sawdust. While the heat of retorting based on Eq. (8.7) is equivalent to a 15 percent exothermic loss as sensible heat when char is the main product, decomposition to gases is approximately thermally neutral. The maximum heat loss in the sawdust pyrolysis is therefore only the heat

Table 8.9 Gas evolved on rapid pyrolysis of dry sawdust at 825°C[3]

Constituent	Mole %	Mass %
CO	43.8	58.4
CO_2	12.5	26.2
H_2	29.3	2.8
CH_4	11.0	8.4
C_2H_6	0.3	0.4
C_2H_2	3.1	3.8
Total	100.0	100.0
Mean molar mass	21.0	
H_2/CO ratio	0.67	17.0 MJ/kg
GCV	15.9 MJ/m³	

required to raise its temperature to pyrolysis conditions. Based on a heat capacity of 2.5 kJ/(kg·K), this amounts to about 10 percent of its calorific value. Assuming some heat recovery is practiced, a large portion of the heat requirements could be met by the small amount of char and liquid byproducts produced. In other words, the reported gas yield indicates a process thermal efficiency, to a desirable medium-CV gas product, of 90 percent or more.

Several factors must be considered when assessing the general utility of biomass pyrolysis in relation to the above yields. Most important is that the rate of heating possible with the dry, small-particle-size sawdust in a fluidized bed operation is relatively high, which in turn accounts for the high gas yields. Other data for the same operating conditions showed gas yields based on calorific content of about 70 percent for municipal solid waste and lower for other materials. All materials were predried and comminuted; lower yields can be expected without such pretreatment. Reduced gas yields are not so much a problem as are the increased yields of undesirable solid and liquid byproducts which cannot be consumed on site.

It appears therefore that there are limited resources such as sawdust that are well suited to pyrolysis, while other biomass materials may remain undesirable feedstock even after extensive pretreatment. The vast municipal solid waste resource is amenable to pyrolysis, although increased gas yields can be expected in gasification systems.

Gasification

The primary objective of biomass gasification is to convert as much of the feed material as possible to a gaseous product. A secondary objective might be to increase the H_2/CO ratio of the gas over that obtained in pyrolysis so as to facilitate downstream upgrading. As is shown in Section 5.5, however, this latter objective can be effectively achieved using a subsequent gas shift step. Apart from upgrading to SNG, liquid products can be produced by using Mobil M or Fischer-Tropsch synthesis.

Gasification with air results in a low-CV gas that can be used for combined cycle electricity generation. We recall that the increased efficiencies attained with combined cycle systems gasifying conventional fossil fuels arise in part from the more effective pollution control techniques that may be used. In the case of biomass, sulfur concentrations are very low, so its direct combustion does not require an attendant energy-consuming flue gas desulfurization step. Other factors including NO_x emissions, the inherent greater efficiency of gas turbines, and the limited scale of operations, will determine whether or not combined cycle generation with biomass will be more viable than conventional steam-electric generation.

If heat or steam is the desired product, then again air gasification appears to offer little advantage over direct combustion as far as pollution control is concerned. It is possible, however, that air gasification may prove useful as a means of providing a substitute fuel for existing gas-fired boilers. A gasifier supplying fuel to an adjacent boiler is termed a "close-coupled" system. It has been argued that a close-coupled system can have a higher efficiency than is achievable with direct combustion of moist fuels such as biomass. High-moisture feedstocks have lower flame temperatures, which can result in particulate and tar emissions and lower boiler efficiencies. Once gasified, however, the moisture is in the vapor state and improved combustion efficiencies can be achieved.[2] Nevertheless, high-moisture-content fuel gases may have undesirable characteristics for existing equipment, so some degree of predrying may in any case be required. The final selection between close coupling a gasifier with an existing boiler, or replacing or modifying the boiler to handle solid fuels, will depend on capital costs, environmental effects, and space availability, as well as overall combustion efficiency.

In relation to pyrolysis, gasification will normally require the provision of a steam supply, while if a medium-CV or synthesis gas is desired, a pure oxygen supply may be required as well. For gasification to be viable, the cost of the oxygen and steam must be compensated for by increased gas yields and the elimination of undesirable solid and liquid byproducts. We note here that the situation is different with coal, which has a relatively low-volatile-matter content, so gas yields are indeed considerably larger than achievable by pyrolysis. Biomass also differs from coal in that the chars are much more reactive. The combined factors of high volatile content and high-char reactivity suggest that steam and oxygen requirements are lower than for coal.

In Union Carbide Corporation's Purox process, shown in Figure 8.7, municipal solid waste (MSW) is gasified using oxygen but no steam. Apart from shredding and magnetic separation of ferrous metals, no other pretreatment is provided prior to gasification at slagging temperatures. The oxygen rate shown in the figure, assuming the raw MSW has a 60 percent organic content with properties as summarized in Table

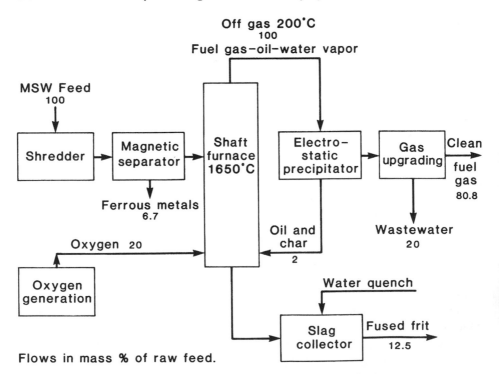

Figure 8.7 Union Carbide Corporation's Purox municipal solid waste gasification process.[3]

8.3, converts to about 20 g/MJ of product gas. This is similar to that for the Lurgi slagger, while Texaco coal gasification consumes some 50 percent more oxygen. The final gas, with composition as shown in Table 8.10, may be distributed locally as a medium-CV fuel or upgraded to SNG or liquid products. The reader may verify that the cold gas thermal efficiency based on the flows shown in Figure 8.7 is about 91 percent. Some further energy is expended in the shredding and magnetic separation stages as well as in gas purification, but the overall process nevertheless appears very attractive. A demonstration unit sized to process 180 t/d of MSW has been installed in South Charleston, West Virginia. We state again that MSW is probably the only biomass material that is available in quantity, that is not otherwise desirable, and for which an infrastructure for collection and concentration is currently operative.

There is considerable interest in using catalysts to enhance the performance of both pyrolysis and gasification systems. As biomass contains low concentrations of sulfur and other catalyst poisons, the rapid deactivation of some catalysts that occurs with fossil fuels is not a problem. Alkali metal carbonates have been shown to increase the rate and extent of low temperature wood gasification. The use of these catalysts is discussed in Section 4.4 in connection with coal gasification. As mentioned, biomass materials have a high-volatile-matter content, and the small amount of char remaining on rapid pyrolysis is highly reactive relative to coal chars. The role of a catalyst would therefore seem to be more one of controlling gas phase reactions rather than promoting gas yields.

As discussed in Section 4.4, alkali metal catalysts promote methane formation. On the other hand, steam gasification of wood in the presence of nickel and silica-alumina results in a synthesis gas containing mainly hydrogen and carbon monoxide.[9] In preliminary tests with these catalysts at 750 to 850°C, the off gas had a H_2/CO ratio of about 2. The silica-alumina function is to crack the higher hydrocarbons thus reducing tar yields, while the nickel reforms methane to hydrogen and carbon monoxide. The resultant gas is ideal for synthesis to methanol or Fischer-Tropsch liquids. The

Table 8.10 Gas composition on oxygen gasification of municipal solid waste[3]

Constituent	Mole %	Mass %
CO	44	59
CO_2	13	28
H_2	31	3
CH_4	4	3
C_2H_4	1	1
N_2	1	1
H_2O	6	5
Total	100	100
Mean molar mass	21	
H_2/CO ratio	0.7	
GCV	11.7 MJ/m³	12.5 MJ/kg
Gas rate,* per t MSW	868 m³	808 kg

* Estimated.

gas yields obtained in the preliminary tests are reported to have the potential of producing methanol in net quantities of about 60 percent by mass of the dry wood. This converts to a thermal efficiency in excess of 60 percent. Estimates for methanol via non-catalytic wood gasification suggest thermal efficiencies in the 33 to 40 percent range.[2,10]

REFERENCES

1. National Academy of Sciences, *Energy in Transition 1985-2010*, Final Report, Committee on Nuclear and Alternative Energy Systems, National Research Council, 1979. W. H. Freeman and Co., San Francisco, 1980.
2. Office of Technology Assessment, *Energy from Biological Processes*, Vols. I, II, III A-C. U.S. Government Printing Office, Washington, D.C., 1980.
3. Solar Energy Research Institute, "A Survey of Biomass Gasification," Vols. I-III, Report No. SERI/ 7R-33-239, Golden, Colorado, July 1980.
4. Berry, R. I., "An Ancient Fuel Provides Energy for Modern Times," *Chem. Eng'g.* **87**(8), 73-76, 1980.
5. Wang, D. I. C., *et al.*, "Anaerobic Biomass Degradation to Produce Sugars, Fuels and Chemicals," *Proc. Second Annual Symp. on Fuels from Biomass* (W. W. Shuster, ed.). Rensselaer Polytechnic Institute, Troy, N.Y., 1978.
6. Kelm, C. R., "Fuels and Chemical Feedstocks from Renewable Sources," *Ind. Eng. Chem. Product Res. Dev.* **19**, 483-489, 1980.
7. Gregor, H. P., and Jeffries, T. W., "Ethanolic Fuels from Renewable Resources in the Solar Age," *Annals N.Y. Acad. Sci.* **326**, 273-287, 1979.
8. Chen, Y. R., Varel, V. H., and Hashimoto, A. G., "Chemicals from Cellulosic Materials," *Ind. Eng. Chem. Product Res. Dev.* **19**, 471-477, 1980.
9. Mitchell, D. H., *et al.*, "Methane/Methanol: By Catalytic Gasification of Biomass," *Chem. Eng. Prog.* **76**(9), 53-57, 1980.
10. Cheremisinoff, N. P., *Gasohol for Energy Production*. Ann Arbor Science, Ann Arbor, Mich., 1979.

NINE

ENVIRONMENTAL ASPECTS

9.1 ENVIRONMENTAL EFFECTS AND THEIR MEASURE

All synthetic fuel plants will utilize land and water resources in addition to the material being converted, and generate gaseous, liquid, and solid wastes that will be discharged or disposed. In view of the scale of these plants, it is necessary that controls be placed on their utilization of limited resources, to minimize ecosystem disruption, and on their discharge of pollutants, to prevent damage to health and the ecosystem. Standards for quantity and quality parameters used to measure ecosystem disruption and damage, or health hazards, are still matters of considerable controversy. For example, a limited discharge of dissolved organics in water can provide food necessary to sustain aquatic life. On the other hand, aerobic biological decomposition of organics consumes free oxygen, and if the oxygen demand of the discharged material represents a significant fraction of the available supply of oxygen dissolved in the water, this can be fatal to aquatic life.

The materials brought into the plant consist principally of the carbonaceous matter to be converted, water, air, and inorganic mineral matter. These materials must leave either in a product, byproduct, or waste stream. Some of the largest emissions may have little environmental impact, while some small emissions may be hazardous and require strict control. By far the largest gaseous waste streams leaving a synthetic fuel plant are water vapor from cooling towers and carbon dioxide. Since all synthetic fuel plants are net consumers of water, wastewaters are generally not discharged but are recycled and reused within the plant boundaries. This reflects the fact that returning water to a source is not economical when to meet discharge criteria the water must be cleaned to a quality equal to or better than the source water. The largest quantity of solid wastes leaving a fossil fuel conversion plant is the inorganic mineral fraction of the coal, oil shale, or tar sands. In bioconversion, a solid cellular material generally will be produced as a waste.

Gaseous Emissions

Water used for cooling passes through a heat exchanger where the heat to be dissipated is transferred to the water. The means which will most often be used in synthetic fuel plants for removing the heat from the water before it is recirculated is a wet or evaporative cooling tower. In this type of tower, the water leaving the heat exchanger is pumped to the top and allowed to fall through the tower in direct contact with the air, which removes the heat by evaporation and convection. The water that is evaporated and the heat that is lost are absorbed by the air moving through the tower by natural draft or forced draft. The water vapor leaving the tower is a gaseous emission which is for the most part innocuous, although small amounts of inorganic salts may be carried out. The large quantities of water required for evaporative cooling are perhaps of greater environmental significance. The amount of water vapor passing through such towers and emitted to the atmosphere may perhaps best be appreciated by considering the size of the natural draft (hyperbolic) cooling towers shown in the overview photograph of the Sasol indirect liquefaction plant in Figure 6.3.

The other major gas emission from any synthetic fuel plant will be carbon dioxide, and the relative amount emitted will be larger the lower is the thermal efficiency. Much concern has been expressed recently over the "greenhouse effect," wherein increased concentrations of CO_2 in the atmosphere from the burning of fossil fuels are conjectured to lead to global increases in air temperature and long-term climatic changes. The possibility that changes in atmospheric CO_2 levels could lead to changes in world climate is not new and has been suggested since at least the end of the nineteenth century.[1] In the simple picture of this effect,[2] the CO_2 and water vapor in the atmosphere absorb part of the longwave (infrared) radiation from the earth's surface, while the atmosphere transmits the incident shortwave (visible) radiation from the sun. The greater the CO_2 concentration, the greater the absorption. The subsequent re-radiation of the absorbed energy causes the earth's surface to reach a higher equilibrium temperature, resulting in an increase in mean air temperature. This picture of the raising of the mean temperature of the earth's surface is called the greenhouse effect, because the differential transmissivity of the atmosphere plays the same role as the glass in a horticultural greenhouse. However, the name is considered by some to be inappropriate because the actual global picture is considerably more complex than the simple equilibrium model given here would indicate. Many other important physical processes and climatic feedback mechanisms must be considered, and the extent of the coupling of the effects is still sufficiently in doubt that it is not known with certainty whether the earth's surface temperature will increase or decrease.[1]

It has been argued that the burning of hydrocarbon fuels with the highest hydrogen-to-carbon ratio would minimize the CO_2 problem in that it would produce the least CO_2 per unit of heat released. This is correct for the burning of natural fuels, but is only part of the picture for synthetic fuels, where the amount of CO_2 generated in their manufacture must also be taken into account. In Table 9.1 are shown estimates of the amount of CO_2 produced per unit of synthetic fuel calorific value in the manufacture of the product from carbon, and in its burning. The total values shown in the table are taken from the results given in Table 3.4 for the carbon contained in the fuel plus the carbon consumed in producing the fuel by the procedure noted. Our assumption

Table 9.1 Comparison of CO_2 produced in idealized synthetic fuel manufacturing processes and in the burning of the fuel

Fuel	mol CO_2/MJ of Product (NCV)		
	Manufacture	Burning	Total
C	—	2.54	2.54
CH_2	1.36*	1.61	2.97
CH_4	2.17†	1.25	3.42
CH_3OH	1.92	1.57	3.49

* Case (b), Table 3.4.
† Case (g), Table 3.4.

here is that all of the carbon must be burned to CO_2, either in the manufacture or in the burning, and that one mole of CO_2 will be produced for each mole of carbon entering the synthetic fuel plant. The values for burning shown in the table are obtained by taking one mole of CO_2 to be produced from burning each mole of carbon in the fuel. The values for manufacturing are then obtained by difference between the total values and the values for burning. For comparison purposes we have also shown the CO_2 produced by the burning of coal (carbon) itself. Consistent with our earlier remarks on the comparative efficiency of burning coal, it produces the least amount of CO_2. The manufacture of gasoline and its burning produces some 17 percent more CO_2 than does burning coal, while the manufacture and burning of methane or methanol leads to CO_2 production levels some 35 percent more. This manifests the inefficiencies attendant on the manufacture of the fuel.

In shale oil production, there is an additional release of carbon dioxide due to the decomposition (calcining) of the carbonates that make up the shale (see Sections 2.1 and 7.1).[3] The decomposition only becomes pronounced above 500°C and is therefore not expected to be a large factor with indirectly heated processes. However, it can become appreciable with directly heated retorts and MIS operations where temperatures of 700 to 1100°C are reached in the combustion zones. From data given earlier (Section 1.3 and Table 7.1), calcium carbonate constitutes about 50 mass percent of the inorganic matter in Green River shale. For a high-grade oil shale of 125 L/t the organic matter is 16.5 percent (Eq. 1.4), and the GCV from Eq. (1.5) is 6.3 MJ/kg of raw shale. For conversion to a synthetic crude product at 70 percent thermal efficiency, which is representative of a directly heated retort (Section 7.2), this means there is produced 4.4 MJ product/kg of raw shale. In such a directly heated retort, about 25 percent of the carbonate is calcined to CO_2 and CaO (Eq. 2.18) from which we may write

$$\frac{1 \text{ kg raw shale}}{4.4 \text{ MJ product}} \times \frac{1 \text{ kmol } CO_2}{100 \text{ kg } CaCO_3 \text{ calcined}}$$

$$\times \frac{0.25 \times 0.5(1 - 0.165) \text{ kg } CaCO_3 \text{ calcined}}{1 \text{ kg raw shale}} = 0.24 \frac{\text{mol } CO_2}{\text{MJ product}}$$

This is not a large value. However, for a MIS retort the efficiency is about 55 percent and some 80 percent of the $CaCO_3$ is calcined, so for the same shale grade, 0.96 mol CO_2/MJ product is produced, which is appreciable. For lower-grade shales the amount is even larger and for an 85 L/t grade we calculate 1.55 mol CO_2/MJ product.

In accordance with the discussion in Section 8.1, the minimization of the CO_2 problem would argue in favor of biomass conversion, so long as an equivalent amount of vegetation is planted to that converted.

Although CO_2 is potentially a pollutant of global concern, its emission is not controlled by present air pollution regulations and it is not treated as a pollutant. Pollution regulations currently are focused on small-scale events and the effects of trace concentrations of materials.[4]

In the United States, the Clean Air Act of 1970 and the 1977 amendment to it provide the basis for the regulatory constraints imposed on air emissions from synthetic fuel plants. The major provisions included promulgation by the U.S. Environmental Protection Agency of National Ambient Air Quality Standards (NAAQS). The specific regulated pollutants are presently particulates, sulfur dioxide, photochemical oxidants, hydrocarbons, carbon monoxide, nitrogen oxides, and lead. Also included are provisions for establishment by the EPA of National Emission Standards for hazardous pollutants and standards of performance for new sources. Finally, there is a requirement for the establishment of permissible incremental increases in ambient concentrations of pollutants from new sources for the "prevention of significant deterioration" in air quality where it is better than the National Ambient Air Quality Standards.

To give the reader a measure of allowed pollution levels, we have shown in Table 9.2 the National Ambient Air Quality Standards. The primary standards are based on health considerations and the secondary ones on environmental considerations. It

Table 9.2 National Ambient Air Quality Standards ($\mu g/m^3$)

Pollutant	Averaging Time	Primary Std.*	Secondary Std.*
Sulfur oxides	Annual	80	—
	24 h	365	—
	3 h	—	1 300
Particulates	Annual	75	60
	24 h	260	150
Carbon monoxide	8 h	10 000	10 000
	1 h	40 000	—
Ozone	1 h	240	240
Hydrocarbons	3 h	160	160
Nitrogen oxides	Annual	100	100
Lead	3 mo	1.5	1.5
Photochemical oxidants	1 h	160	160

* Not specified where no value is indicated.

should be emphasized that state standards often exceed the federal ones. For example, for sulfur dioxide and nitrogen oxides about half of the states have stricter standards than imposed by NAAQS.

At the time of writing, emission standards have not been formulated for synthetic fuel plants, although they have been established for similar or related sources including coal mining, iron and steel manufacturing, petroleum refining and steam-electric power generation. The standards currently in effect are based on the "best practicable treatment currently available (BPT)," with stricter standards being developed based on the "best available treatment economically achievable (BAT)."

The requirement for the "prevention of significant deterioration" of air quality as presently formulated could affect the development of synthetic fuel plants. In areas designated as Class I, which include national parks and wilderness areas, the air quality must remain virtually unchanged. In Class II areas, some additional air pollution corresponding to moderate industrial growth is allowed. In already industrialized areas, designated as Class III, significant deterioration can be avoided by reducing emissions from existing sources to offset new emissions. In the western part of the United States, where synthetic fuel development is likely to take place, there are no established industrialized areas. The requirement to prevent significant deterioration could impose a severe constraint on development. Even if efforts are made to minimize the emissions from the first few plants built in a given area, the sum of their emissions might approach the allowable emission standards and make further expansion difficult, even with improved pollution control technology.

Wastewater Discharge

The capacity of a receiving water to accept an industrial wastewater without significant damage depends largely on the relative flow rates of the effluent and the receiving stream, although other factors such as stream depth and temperature are also of importance. In the United States, under the provisions of the Federal Water Pollution Control Act and the National Pollutant Discharge Elimination System (NPDES), the EPA has established in-stream water quality standards and industrial effluent discharge standards in a parallel manner to that described above for industrial air pollution control. The in-stream water quality standards serve as a measure of minimum acceptable receiving water quality, while the effluent discharge standards are based on best practicable and economically achievable wastewater treatments.

Water pollutants may be broadly classed as:

Suspended matter including colloids
Undissolved oils and greases (including emulsions)
Dissolved gases (ammonia, carbon dioxide, and hydrogen sulfide)
Dissolved inorganic salts including heavy metals
Dissolved organics

These pollutants have traditionally been characterized by a number of water quality parameters, with the unit mg/L being the most common. The quantity of suspended matter is given by the parameter total suspended solids (TSS), while oil and grease are reported as such. Parameters characterizing the dissolved gases include total dissolved gases, total ammonia (gas plus ions), and as a measure of carbon dioxide, total alkalinity. Acidity and basicity are measured by the pH of the solution, defined as the negative logarithm of the hydrogen ion concentration. For pure water, the hydrogen ion concentration is 10^{-7} mol/L so that the neutral pH is 7. When an acid is added to water the hydrogen ion concentration increases, resulting in a lower pH, and when a base is added it decreases, raising the pH. The pH scale from 0 to 7 is the acid range, and from 7 to 14 the base range. Dissolved inorganic salts are most often given in terms of total dissolved solids (TDS).

Dissolved organics are measured by a number of parameters, the most common one of which is the biochemical oxygen demand (BOD), which is used to evaluate the effect of wastewaters on the oxygen resources of the receiving waters. The biochemical oxygen demand is a bioassay measuring the oxygen consumed by microorganisms utilizing the organic matter (Eq. 2.19). It is usually reported in mg/L. The most common period of observation is 5 days and the resulting measurement is denoted by BOD_5. For reference, the BOD_5 of drinking water is less than 2 mg/L, which is about the lower limit of detectability; a raw municipal sewage may have a BOD_5 of 300 mg/L, while the value of a phenolic wastewater condensate from a gasifier might be 10 000 mg/L. The BOD value will not include organics that are biologically refractory. In industrial wastes the BOD may be considerably lower than the chemical oxygen demand (COD), which is the amount of oxygen required for chemical oxidation of the organic matter to carbon dioxide and water by strong oxidants. No uniform relationship exists between BOD and COD except that COD must always be greater than BOD by at least the extent to which nonbiodegradable matter is represented in the organic content. Synthetic fuel process condensates can have COD/BOD ratios as high as 1.5 to 2. Another common measure of the organic matter in wastewaters is the total organic carbon (TOC), determined by measuring the amount of carbon dioxide evolved when vaporizing water in a stream of oxygen at high temperature. The COD/TOC mass ratio normally lies between 2.7, the value for carbon, and 5.3, the value for methane. The presence of organic matter is also indicated by the amount of dissolved oxygen in the water, which at saturation under standard conditions is around 8 mg/L. The most common measure of the potential presence of disease producing organisms in wastewater is taken to be the total coliform organism count in a fixed volume of water.

The water quality parameters we have described may not be sufficient to adequately characterize the pollutants in a wastewater, as they do not necessarily reflect important hazardous contaminants such as heavy metals and toxic organics, which may have to be specified as concentrations of specific compounds. In fact, at present the EPA is required to regulate 129 toxic organic and inorganic pollutants where they are found in significant amounts.[5] Many of these pollutants will be found in condensate waters from synthetic fuel plants. For example, mercury is among the more volatile of the heavy metals which are present in trace amounts in coal. Therefore it will tend to be concentrated, for example, in coal gasification wastewater condensates. Other toxic compounds and toxic organics may be similarly found. Their removal prior to discharge may require costly treatment techniques.

To illustrate relative water qualities in terms of the parameters described, we show in Table 9.3 federal water quality standards for drinking water, together with the quality of the foul condensate water recovered from a dry ash Lurgi gasifier after gasification of a Wyoming subbituminous coal. The particular process condensate shown is among the dirtiest of wastewaters generated in synthetic fuel processing. The quality of the process condensates is dependent on the gasifier type, operating conditions, and the coal. It is clear from the table that treatment for discharge to a clean source water is not economical when compared to recycling the water within the plant after any necessary treatment to enable its use. For example, neither cooling nor solids disposal require a discharge-quality source water. Wastewater reuse is made even more necessary economically by the fact that effluent discharge criteria are based on achievable wastewater treatments, which can therefore reflect a water quality that is actually better than the receiving stream. For example, BAT effluent standards for the petroleum refining industry are 2 mg/L BOD, which as we noted above approaches the lower limit of detectability. In many cases this would mean that the effluent was of better quality than the receiving or intake stream with respect to BOD. For these reasons, synthetic fuel plants will embody the design concept of "zero wastewater discharge." An added incentive for maximizing water recycle and reuse, and mini-

Table 9.3 Comparison of federal water quality standards for drinking water and quality of condensate water recovered from dry ash Lurgi gasifier after gasification of Wyoming subbituminous coal

	Drinking Water*	Lurgi Condensate[6]
Parameter (mg/L)		
TDS	250	2 480
Ammonia	—	7 610
Carbon dioxide	—	13 600
Hydrogen sulfide	—	55
Oil and grease	Virtually free	150
BOD$_5$	(<2)†	10 600
COD	—	22 800
TOC	—	7 640
Phenols	0.001	5 000
Trace elements (μg/L)		
Arsenic	50	400
Barium	1	<10
Cadmium	10	260
Lead	50	310
Mercury	2	150
Selenium	10	130
Silver	50	40
Other		
Total coliform bacteria	1 count/100 ml	—
pH	5–9	8.2

* Standard not specified where no value is indicated.

† Not specified; below detectable limit.

mizing water consumption, is the relative scarcity of water in many of the western regions where synthetic fuel plants are expected to be built.

Solid Wastes

The quantities of solid wastes from fossil fuel conversion plants may be very large and the disposal of the coal ash, spent shale, or tar sand mineral matter can constitute an important solids handling, reclamation, or disposal problem. Under current regulations, these materials are not considered hazardous, although trace amounts of toxic compounds are present. Organic fossil fuel residues such as coke and unprocessed fines are produced by some processes, but represent only a small fraction of the residues. Other solid wastes include spent catalysts and ancillary residues such as those from air and water pollution treatments, including flue gas scrubber sludges, evaporator brines, and raw-water softening sludges. These wastes are not large and their degree of toxicity is much less than that of spent catalysts.

The primary federal regulations governing solid wastes disposal derive from the Resource Conservation and Recovery Act.[7,8] Although this law is aimed at the proper disposal of solid waste, its major emphasis and intent is the control of hazardous materials. The criteria for identifying a waste as hazardous are based on ignitability, corrosivity, reactivity, and toxicity. The measure of toxicity is of most significance for synthetic fuel plant solid wastes. It consists of a 24-hour leaching test in a pH 5 solution. In order that the waste not be considered hazardous, the leachate must not contain any dissolved matter that exceeds 100 times drinking water standards, for a specified list of toxic pollutants. For example, arsenic, lead, and silver must not exceed 5 mg/L, while mercury must not exceed 0.2 mg/L. Consideration is being given to reducing these concentrations to a level 10 times that of drinking water standards. Currently no fossil fuel residues have been classed as hazardous, though it is likely that catalyst wastes will be. It appears, therefore, that at present the principal problems associated with solid waste disposal from synthetic fuel plants are the disposal procedures themselves and associated reclamation practices.

Land and Water Resources

Land use has already been indicated as one measure of environmental impact. In Table 9.4 are shown the land requirements to produce raw fuel sufficient for a standard-size synthetic fuel plant producing 10^7 L/d of liquid fuels or 10^7 m³/d of pipeline gas (see Table 1.4). The land requirements for the conversion facilities, exclusive of any disposal sites, are relatively small and in the range of 1 to 5 km². The figures for surface mining are based on strip mining of western coal and assume that reclamation of the land takes from 2 to 5 years. Surface mining requirements are seen to be quite low and manifest the high energy density of the material being recovered. Since most shale will be mined underground, the principal land use is for disposal of the spent shale, consisting mainly of inorganic mineral matter. Disposal area requirements have been estimated for a number of proposed commercial plants, and we have assumed in the table that this area would be required over the life of the plant, typically 20

Table 9.4 Land requirements for synthetic fuel plants producing 10^7 L/d of liquid fuels or 10^7 m^3/d of pipeline gas*

	km^2/yr	
·	Min.	Max.
Surface coal mining†	1	10
Shale disposal area‡	4	10
Logging residues§	2 600	4 400
Algae ponds	2 900	4 400
Silviculture	3 500	10 000
Agriculture (corn)	13 000	19 000
Crop residues§	25 000	38 000
Forestry¶	32 000	126 000

* Biomass data derived from Ref. 9, assuming 50 percent thermal efficiency in conversion.

† Data from Ref. 10, assuming 2 to 5 year reclamation time.

‡ Data from Ref. 11, assuming total disposal area is disturbed over life of plant.

§ Land required for collection; already in use for forestry or agriculture.

¶ Only 2 to 5 percent of indicated areas harvested annually.

years. This may be considered conservative in that it represents a maximum land use figure, since spent shale revegetation will undoubtedly begin earlier.

The biomass land area requirements are seen to be from 2 to 5 orders of magnitude larger than that for coal recovery. As noted in Section 8.1, the scale is so large that it is unlikely that plants of the output indicated would be built using biomass as the feedstock. However, the figures do serve to indicate the relative land requirements for a fixed energy output. The land areas given are based on an efficiency for conversion of the raw feedstock of 50 percent. Fermentation efficiencies would, for example, be lower, while biogasification efficiencies may be somewhat higher. The areas listed can, however, be scaled up for a lower efficiency or down for a higher one.

In the case of residues, it may be properly argued that the land is already used for the purpose noted, and as pointed out in the table the size is only indicative of the collection task. For forests, the land would likely have a concurrent use including wildlife support, providing recreational areas, etc. Since the harvesting of forests would be done on a renewable basis with harvesting, say, every 20 to 50 years, this would mean that only 2 to 5 percent of the areas shown would be harvested annually. Despite these caveats, it is clear that the land use for biomass production on a large scale would be very extensive and probably constitute a limiting factor, although small-scale development need not be precluded by this argument. On the other hand, what is clear is that land requirements for coal and shale production are really minimal and a relatively small disturbed area could serve to support a very large number of plants that could provide a substantial synthetic fuel output, measured as a fraction of U.S. energy requirements.

The other major non-fuel resource required for any synthetic fuel plant is water. In the United States this is an important consideration for fossil fuel conversion since much of the easily mined coal, and all of the high-grade oil shale and tar sands are in the arid West. This also poses a problem for biomass cultivation, which can have large water requirements for irrigation since much of the available land would also be in western areas where water is limited.

In the conversion of fossil fuels, the principal consumptive requirements for water are for cooling, as a source of hydrogen, and for mining and residuals disposal. Cooling is generally the major consumptive use in any synthetic fuel plant, and although various degrees of dry (air) cooling may be employed to reduce this consumption, it can only be done at a higher cost. In biomass conversion, the water requirements are generally for the same purposes, although there may be no requirement for hydrogen in bio-conversion. The hydrogen requirement is in any case generally small in comparison with that for cooling and disposal. On the other hand, as we have noted, water may be needed for irrigation and this can constitute a very large quantity even in comparison with cooling needs.

In order to keep plant water requirements to a minimum, and for the economic reasons related to environmental constraints on discharge that were discussed earlier, water is reused and recycled to extinction within the plant. The actual consumption will, of course, depend on the uses to which the water must be put, the cooling required, and the means of treating the water. However, as a measure of water needs we note that for fossil fuel conversions about 1 to 4 volumes of water are required per volume of liquid product or its equivalent. For biochemical conversion of biomass, because of lower thermal efficiencies, the consumption, excluding any irrigation requirements, can be somewhat higher.

9.2 AIR POLLUTION CONTROL

Air emissions from a synthetic fuel facility depend not only on the efficiency of the control placed on each pollutant but also on the process, and the quantities of the compounds making up the pollutants that are present in the raw material being converted. In fossil fuel conversion facilities, which we shall emphasize here, the largest air emissions, apart from carbon dioxide and water vapor, are particulates, sulfur dioxide, and nitrogen oxides. There are other pollutants emitted including hydrocarbons, ammonia, and trace metals, but the quantities are small, though not necessarily unimportant environmentally. The air emissions from fossil fuel conversion facilities may be broadly classed as those originating from the major processing steps—pretreatment, conversion, and upgrading—and those originating from ancillary steps. In Figure 9.1 we diagram the generic emission sources and indicate the major pollutants that will be emitted to the atmosphere in most plants. Below we discuss how these pollutants arise, and from crude estimates of their amount and the degree of pollution control that can be applied, indicate the order of magnitude of air emissions that can be expected. The crudeness of the estimates results from the fact that the mass rate of air emissions is very small in comparison with the mass flow of raw material brought into the plant; the pollutant fractions are measured in hundredths or thousandths of a percent. As a consequence, generalized estimates based on mass balances are subject to error, since small differences in plant configuration and operating mode result in large percentage differences in specific pollutant emission rates. Despite this caveat, however, the order of the absolute emissions can be developed.

We consider particulates first and note that a major generic source of air polluting emissions from any synthetic fuel plant is in the preparation of the raw feedstock, specifically the crushing, screening, and storage operations, and in its subsequent feeding to the process reactors. In this category we include any drying of coal or preheating of shale, as in the TOSCO II process. For coal conversion processes, another major source of particulates is in ancillary combustion operations where fly ash is produced when coal, or ash-containing carbonaceous materials, is burned to generate process heat, raise steam, or produce electric power. For oil shale processes, it is usual to burn pyrolysis gases for these purposes, so particulate generation is negligible.

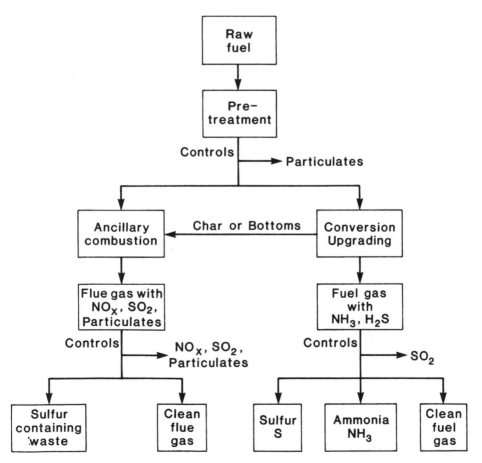

Figure 9.1 Generic emission sources and major air pollutant emissions in synthetic fuel plants.

Based on data from a number of proposed commercial coal and oil shale conversion facilities, we have estimated the quantity of particulate matter produced in the pretreatment operations. The particulates produced are found to range in amount from 0.3 to 0.8 mass percent of the raw fuel fed to the plant. For purposes of estimation, we may assume an average value of 0.5 percent for the particulate fraction generated. In synthetic fuel facilities, it is likely that control of 99 mass percent of the emitted particulates would be required, although 98 percent control is more typical of what currently is achieved in commercial facilities. The particulate emissions to the atmosphere from the pretreatment steps in coal or oil shale plants can, for estimation purposes, therefore, be taken to be 0.005 percent by mass of the raw fuel fed to the plant.

In coal conversion plants, the amount of particulates generated in the ancillary operations of raising process heat, steam, or electricity can be estimated in relation to the plant thermal efficiency. We may assume that for every kg of coal fed to the plant, $(1 - \eta)(1 - \beta)$ kg of coal, char, or ash-containing carbonaceous material is burned for ancillary needs. Here, η is the fractional thermal efficiency and β the fraction of heat supplied by burning fuel gas or deashed liquid or solid products. Generally, $\beta \gtrsim 0.5$, although this need not always be the case as, for example, in the Sasol plants, where coal is burned to meet all the ancillary needs, so $\beta = 0$. The assumption that the fuel combusted corresponds to the heat dissipation fraction $(1 - \eta)$ is also not quite correct, since there is some heat recovery from the combustion processes. However, the approximation is sufficient for our purposes here.

In modern pulverized-coal furnaces, 20 percent of the ash leaves as bottom ash and 80 percent as fly ash. For a 99 percent particulate control, the rate of fly ash emission to the atmosphere may therefore be expressed as

$$\text{Ash emission rate} = 0.8(1 - 0.99)(1 - \eta)(1 - \beta) \times \text{coal rate} \times \text{ash fraction} \quad (9.1a)$$

$$= 0.004(1 - \eta) \times \text{coal rate} \times \text{ash fraction} \quad (9.1b)$$

Here we have set $\beta = 0.5$, so Eq. (9.1b) is representative of the upper range of emissions.

In the ancillary combustion operations, sulfur dioxide and nitrogen oxides are produced as combustion products. For reasons described in Sections 5.1 and 5.5, including the more proven nature of gas purification processes, it is most likely that any process generated fuel gases would be cleaned of ammonia and hydrogen sulfide prior to burning. Sulfur dioxide is therefore not emitted by the burning of clean fuel gases, although nitrogen oxides are, as the result of the oxidation of atmospheric nitrogen. The amount of nitrogen oxides so formed may be less than that produced by the oxidation of fuel nitrogen but for purposes of this estimate, we shall not distinguish between the sources.

Control of nitrogen oxides is achieved at present mainly through modification of the combustion process, although other methods, including flue gas treatments, are being developed. A reasonable estimate for current combustion-modified, coal-fired utility boilers is that approximately 0.22 kg/10^9 J of coal fed will be emitted as oxides of nitrogen (0.5 lb/10^6 Btu). Flue gas control techniques may achieve significantly higher control levels. Under the assumption that the emissions of nitrogen oxides will be the same per unit of heating value for combustion of either high-or low-nitrogen fuels, we may write

$$\text{NO}_x \text{ emission rate} = \text{calorific output} \times \frac{(1 - \eta)}{\eta} \times \frac{0.22 \text{ kg}}{10^9 \text{ J}} \qquad (9.2)$$

The sulfur dioxide emitted from burners and furnaces in ancillary operations may be estimated by taking the sulfur fraction of the sulfur-containing fuels to be half that in the original coal. If raw coal is the furnace fuel, this is approximately correct since about half of the sulfur is inorganic and much of this sulfur is separated out in washing operations or during pulverizing. If char or bottoms is the fuel, the sulfur fraction has generally been reduced by about half by hydrogenation of much of the organic sulfur and some of the inorganic sulfur to hydrogen sulfide. Essentially all the sulfur in the fuel will be oxidized to sulfur dioxide. Current flue gas desulfurization techniques are capable of removing about 90 percent of this sulfur dioxide, so that we may write for the emission rate

$(\text{SO}_2 \text{ emission rate})_{\text{ancillary}}$

$$= 0.5(1 - 0.9)(1 - \eta)(1 - \beta) \frac{64}{32} \times \text{coal rate} \times \text{sulfur fraction} \qquad (9.3)$$

Here, β represents the fraction of the fuel which is a clean-burning product from which the sulfur has been removed. We take β to have the same value as in Eq. (9.1).

In the conversion and upgrading steps, most of the organic sulfur in the coal or oil shale will be converted to hydrogen sulfide, as will some of the inorganic sulfur. We may estimate that for coal conversion plants about half of the total sulfur in the coal fed to the reactors will be converted to hydrogen sulfide and for oil shale all of the organic sulfur in the kerogen will be converted. In oil shale facilities, essentially all the raw shale entering the plant will be retorted, while in coal conversion facilities all the coal entering the plant may not necessarily pass through the reactors. The reason for this is indicated in Figure 9.1, where it can be seen that part of the coal may be used directly for ancillary purposes. We shall assume here that all the coal entering the plant passes through the reactors. As discussed in Section 5.1, after removal of the hydrogen sulfide from the process generated gas, it is usually converted to elemental sulfur by the Claus process. Typically about 92 percent of the sulfur is recovered, and in one common procedure the remaining 8 percent is incinerated to sulfur dioxide. For our purposes we shall assume this to be the procedure followed. If 90 percent of the sulfur dioxide generated is controlled by flue gas desulfurization, this means that about 0.4 percent of the total sulfur originally in the coal is emitted as sulfur dioxide. An expression for the sulfur dioxide emission rate is then given by

$$(\text{SO}_2 \text{ emission rate})_{\text{process}} = 0.004 \frac{64}{32} \times \text{coal rate} \times \text{sulfur fraction} \qquad (9.4)$$

Any conversion of the fuel nitrogen or atmospheric nitrogen in the fuel conversion step itself or in a subsequent upgrading step is to ammonia and not nitrogen oxides. As we noted in Section 5.1, ammonia is very soluble in water and so is readily removed by washing. Ammonia-containing water is toxic and cannot be discharged. The treatment of this wastewater will be discussed in connection with water management.

By way of example consider a process for the direct liquefaction of an Illinois coal (Table 1.8) to produce 10^7 L/d of liquid fuels (4×10^8 MJ/d) at a 65 percent thermal efficiency. The calorific value of the coal is 24.9 MJ/kg, from which the total coal feed rate to the plant is 2.47×10^7 kg/d. In Table 9.5 are given the emission levels found by applying the estimating rules discussed. The mass emissions per unit of product calorific value are also presented in British units, which are the units in which EPA regulations are specified for New Source Performance Standards. We would emphasize, however, that at the time of writing, such standards have not been established for synthetic fuel plants. The estimates, by the nature of how they were made, tend to represent upper limits. They show that with the achievable control levels specified, the emissions are not all that large from a single full-scale (63 000 bbl/day) plant, although they indicate that control of nitrogen oxides may be the problem of greatest concern.

The characteristics of specific technologies that might be employed to control the levels of pollutant emissions have so far not been considered and are discussed below. Particulate, SO_2, and NO_x controls are emphasized. A good description of the various air pollution control technologies can be found in Ref. 12. A later summary of current and projected technologies is given in Ref. 4, from which we have drawn a number of descriptions in part.

Particulates

Current particulate standards are generic in their requirements and not specific to chemical composition or particulate size. However, increasing emphasis is being placed on inhalable particulates less than 15 μm in size, and especially on fine particulates less than 2 to 3 μm.

Among the conventional control technologies are cyclones, scrubbers, electrostatic precipitators, and fabric filters. In Section 5.1, brief descriptions are given of cyclone separators and wet scrubbers. Fabric filtration, which is used for fine particulate removal, is accomplished in a so-called "baghouse" in which are hung a number of

Table 9.5 Order of magnitude estimates of major air pollutant emissions in tons per year from a 10^7 L/d direct coal liquefaction plant converting an Illinois No. 6 coal*

	Particulates	Sulfur Dioxide (SO_2)	Nitrogen Oxides (NO_x)
Preparation	450	—	—
Processing	—	2090	—
Burning	930	4580	17 300
Total	1380	6670	17 300
kg/10^{12} J product	9.5	45.7	118
lb/10^6 Btu product	0.022	0.106	0.276

* Significant figures given are not indicative of accuracy.

filter bags through which the particle-laden gases are forced. The particles are generally removed from the bags by gravity. Electrostatic precipitators (ESP), which are also best suited for fine particulate removal, operate on the principle of charging the particles with ions and then collecting the ionized particles on a surface from which they are subsequently removed. In one simple embodiment, a high voltage is established between a central wire electrode and a grounded outer tube which is the collecting surface. The ions in the gap between the wire and the grounded surface charge any particles entering the space, following which they migrate to the collecting surface under the action of the electric field. When particles reach the collecting surface, they lose their charge and adhere to the surface. The particles are physically dislodged by rapping, washing, or other means. Electrostatic precipitators are in common use for fly ash removal fiom power plant flue gases.

In Figure 9.2 is shown the effectiveness of various conventional devices for particulate control, as measured by the mass fraction of particles collected, a quantity termed the "collection efficiency." The curves indicate that the technologies are least efficient in removing particles in the range of 0.1 to 1 μm. For wet scrubbers and fabric filters, this is explained by the fact that the very small particles (<0.1 μm) are removed mainly under the action of diffusional forces (Brownian diffusion), which are relatively greater the smaller the particles; whereas larger particles (>1 μm) are collected principally by inertial forces (impaction), which are greater the larger the particles. The minimum in the fractional efficiency curves is in the transition range between removal by diffusional and inertial forces. There is an analogous explanation for the minimum in the same size range for the electrostatic precipitator curves; for particles ≤0.3 μm, diffusional charging dominates, while field charging dominates for particles >1 μm.

All the conventional devices have certain advantages and disadvantages. Among the advantages of cyclone separators are that they can operate at high pressures and temperatures, are relatively simple, and handle large particles and high dust loadings. However, as is evident from Figure 9.2, they have a relatively low collection efficiency for particles <5 μm. Wet scrubbers are efficient removers of fine particulates but not submicrometer-size particles. For synthetic fuel plants they have the advantage that they are able to clean and cool high temperature gases at the same time that particulates are removed. However, they are consumptive of water, are limited by scaling and fouling, and are fairly energy intcnsive.

Because fabric filters are dry collectors, have extremely high collection efficiencies, and are relatively easy to operate and maintain, there is a trend toward their use. Among their disadvantages in application in synthetic fuel plants is that high temperature gases above about 230°C must be cooled, and even then the fabric must be fiberglass with special graphite coatings. In addition, the bags are susceptible to sulfuric acid and other chemical attack. Even trace concentrations of sulfur trioxide in gases can condense out at temperatures as high as 170°C to form a relatively concentrated sulfuric acid on the filter fabric and lead to rapid bag failure. Woven fabric made from ceramic fibers that resist acid attack and that can be used at temperatures of 815°C and higher are currently being evaluated for commercial application.

High particle-removal efficiencies for micrometer- and submicrometer-size par-

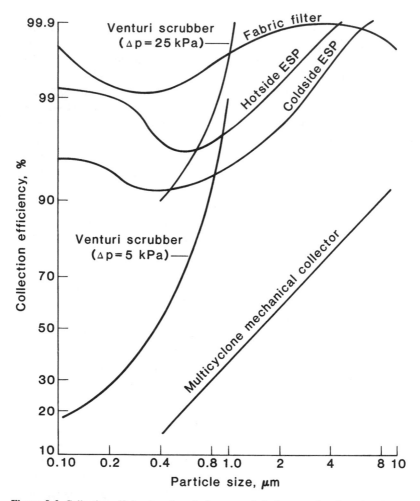

Figure 9.2 Collection efficiencies of particulate control devices as a function of particle size.[4]

ticles are also obtainable with electrostatic precipitators. Of importance in synthetic fuel plants is that the particles can be collected wet or dry and that the units can be operated at high temperatures up to 300 to 450°C. It is of interest that electrostatic precipitators are one method considered for separation of the oil vapors from the off gas leaving the Paraho directly heated retorts. However, ESP units have a relatively high capital cost and are sensitive to the resistivity of the particulate matter. Dust of either high or low resistivity is not easily collected. Another disadvantage of ESP units is that nonuniformities in the flow can lead to particle re-entrainment. One procedure currently under development to enhance particulate collection is wet electrostatic scrubbing in which the particles and/or the scrubbing liquor droplets are charged so that the same scrubber removal efficiency can be obtained at lower-power inputs.

Flue Gas Desulfurization

Sulfur dioxide represents 98 percent of the sulfur oxide pollutants generated by combustion. The removal of the SO_2 from the combustion gases before they are released to the stack is, as we have noted, called "flue gas desulfurization," to which the acronym FGD is applied. A large number of removal procedures exist or are under development. They may be classified as wet or dry, depending on whether a water mixture is used to absorb the SO_2 or whether the acceptor is dry. The procedures are further classified as non-regenerative or regenerative, depending on whether the chemical used to remove the SO_2 is disposed of or regenerated and reused. Most commercial installations currently are wet non-regenerative lime/limestone procedures in which a calcium sulfite or calcium sulfate sludge is produced that is disposed. In wet regenerative processes either elemental sulfur or sulfuric acid is normally recovered.

Wet scrubbing processes contact the flue gas with a solution or slurry into which the SO_2 is absorbed. As noted, the most common procedures are the non-regenerative systems that utilize limestone ($CaCO_3$), hydrated or slaked lime ($Ca(OH)_2$), or both. In the wet limestone process, the flue gas contacts a wet limestone slurry in the scrubber and the SO_2 is removed following the reaction

$$CaCO_3 + SO_2 \rightarrow CaSO_3 + CO_2 \tag{9.5}$$

with

$$CaSO_3 + \tfrac{1}{2}O_2 \rightarrow CaSO_4 \tag{9.6}$$

The scrubber effluent generates a spent slurry following the reactions

$$CaSO_3 + \tfrac{1}{2}H_2O \rightarrow CaSO_3 \cdot \tfrac{1}{2}H_2O \downarrow \tag{9.7}$$

$$CaSO_4 + 2H_2O \rightarrow CaSO_4 \cdot 2H_2O \downarrow \tag{9.8}$$

In the wet lime scrubbing system, the gas containing the SO_2 is reacted with a wet lime slurry following the reaction

$$Ca(OH)_2 + SO_2 \rightarrow CaSO_3 + H_2O \tag{9.9}$$

Reactions (9.6) to (9.8) follow.

In the so-called double alkali, non-regenerative procedure, the flue gas is scrubbed with a soluble alkali such as sodium sulfite, which is subsequently regenerated with lime to form insoluble calcium sulfite. Typical reactions in the scrubber and reaction tank are, respectively,

$$Na_2SO_3 + SO_2 + H_2O \rightarrow 2NaHSO_3 \tag{9.10}$$

$$2NaHSO_3 + Ca(OH)_2 \rightarrow Na_2SO_3 + 2H_2O + CaSO_3 \tag{9.11}$$

The spent calcium sulfite slurry must be disposed of, presenting similar problems to the lime/limestone procedures. The Wellman-Lord process is an example of a wet regenerative procedure in which absorption is by sodium sulfite, as represented by reaction (9.10). The spent absorbent, sodium bisulfite, is regenerated thermally following the reaction

$$2NaHSO_3 \xrightarrow{\text{heat}} Na_2SO_3 + SO_2 + H_2O \tag{9.12}$$

The water vapor is condensed and the active sodium sulfite dissolved in it for recycling to the scrubber. The concentrated SO_2 stream can be processed to elemental sulfur or sulfuric acid in a suitable plant.

In all these procedures, water leaves the system as vapor in the flue gas. The amount may be estimated by assuming the gas is saturated with water vapor at the temperature and pressure at which it leaves the scrubber, in the final absorber and before any reheating.[10] Reheating is usually necessary to provide sufficient buoyancy to the stack gases cooled down in the scrubbing procedures and to prevent corrosion when the gases enter any downstream equipment.

In either the lime/limestone or sodium/lime scrubbing processes, there are two major solid wastes, $CaSO_3 \cdot \frac{1}{2}H_2O$ or $CaSO_4 \cdot 2H_2O$. In addition, unreacted limestone, $CaCO_3$, and slaked lime, $Ca(OH)_2$, are also wasted. From data on the composition of lime and limestone sludges, the mass of solids per unit mass of sulfur has been calculated for the lime and limestone processes and is given in Table 9.6. Also shown in the table are the comparable figures for the crystalline forms of calcium sulfate and calcium sulfite. For a mixed lime/limestone process, an average of 5.9 kg solid/kg sulfur is appropriate. For the sodium/lime process represented by Eq. (9.11), a fractional mass of about 5 kg solid/kg sulfur should provide a satisfactory estimate. In addition, slurry water in about equal mass is disposed with the solids. Clearly, this represents a large solids-disposal problem.

The large solids-waste handling and disposal problem has fostered efforts to develop alternative dry scrubbing procedures for the removal of SO_2. Dry systems have the additional advantage of reducing the pumping requirements attendant on the high liquid-to-gas ratios in wet systems, and may save reheat energy. The most promising direction currently is to use dry lime scrubbing systems just upstream of fabric filters or electrostatic precipitators. In this way a dry, solid-waste product is produced that can be handled by conventional fly ash removal systems. Other dry processes include the Shell Copper Oxide process, which also has the advantage of being simultaneously able to remove nitrogen oxides. In this process, SO_2 is absorbed by supported copper oxide at temperatures around 400°C. The copper oxide acceptor is then regenerated at about 400°C with a mixture of hydrogen and steam. The off gas from the regeneration step contains from 4 to 7 percent SO_2, with the remainder mainly steam. For subsequent processing, the SO_2 concentration is then increased to about 90 percent by removing the water in an absorption/stripping system.

**Table 9.6 Mass of solids per unit mass
of sulfur in lime and limestone sludges,
and for crystalline forms of calcium
sulfate and sulfite[10]**

Crystal or Process	kg Solid/kg Sulfur
$CaSO_3 \cdot \frac{1}{2}H_2O$	4.0
$CaSO_4 \cdot 2H_2O$	5.4
Lime	5.2
Limestone	6.6

Nitrogen Oxides

Nitrogen oxides are formed during burning by the thermal oxidation at high flame temperatures of the nitrogen in the combustion air and by the oxidation of the organic nitrogen in fuel, which can take place at lower temperatures. This has given rise to the terminology "thermal NO_x" and "fuel NO_x." In synthetic fuel plants, NO_x emissions are produced principally by the burning of high-nitrogen fuels including coal, char, and bottoms. The NO_x that is produced is therefore both thermal and fuel, but the fuel NO_x predominates. The formation of thermal NO_x depends strongly on the temperature and can be controlled by reducing combustion temperatures. The formation of fuel NO_x depends primarily on the amount of oxygen available in the flame. In Table 9.7 are listed the principal combustion modification techniques employed to reduce thermal NO_x and fuel NO_x emissions. The procedures are listed under the category in which the major reduction is accomplished.

Low excess air firing is among the simplest of the combustion modifications. Lowered oxygen availability in the flame zone reduces the fuel NO_x, and the lesser amount of nitrogen also lowers the thermal NO_x. Staged combustion has also been successful in reducing fuel NO_x, wherein combustion is first carried out in a fuel-rich zone, and subsequently in a second lower temperature, fuel-lean zone. Flue gas recirculation in which a portion of the flue gas is recirculated to the furnace lowers the gas temperature and thereby lowers the thermal NO_x formation. Moreover, the oxygen concentration is also reduced, which therefore tends to reduce the fuel NO_x as well. Reduced firing rates lower the gas temperature by cutting back on the volumetric heat release rate which then reduces the thermal NO_x formation.

Newer coal-fired boilers are being built with burners designed to reduce NO_x formation. This is accomplished by, among other things, cutting the level of the flame turbulence in order to lower flame temperatures and in this way reduce thermal NO_x formation. These burners also delay fuel/air mixing in order to establish fuel-rich combustion zones so that the reduced oxygen level then lowers the fuel NO_x production.

Among the more recent and interesting procedures, noted in Section 5.5, are those which employ ammonia to reduce the nitrogen oxides to nitrogen and ammonia. In one non-catalytic technique,[4] ammonia is injected into the post combustion zone in the boiler in the presence of oxygen. The nitric oxide, for example, is reduced to nitrogen and ammonia in a series of gas phase reactions, the overall reaction for which may be written

$$4NO + 4NH_3 + O_2 \rightarrow 4N_2 + 6H_2O \qquad (9.13)$$

Table 9.7 Principal combustion modification techniques for reducing NO_x emissions

Fuel NO_x	Thermal NO_x
Low excess air	Flue gas recirculation
Staged combustion	Reduced firing rates
Low NO_x burners	Low NO_x burners

Optimum reaction temperatures are quite high, in the range 925 to 975°C. Catalytic procedures using iron oxide, vanadium oxide/aluminum oxide, and iron/chromium systems have also been reported.[13] Very high nitric oxide reductions by ammonia in the presence of oxygen have been reported at temperatures from 300 to 400°C.

9.3 WATER MANAGEMENT

In the United States much of the easily mined coal and almost all the high-grade oil shale and tar sands deposits are found in the arid western areas of the country. Increased biomass production, which could also require large supplies of irrigation water, is also envisaged for some of the same regions. Water availability could therefore be an important resource constraint to synthetic fuel production in the West, so proper water management would be necessary to minimize consumption. In water-sufficient areas such as the eastern coal regions, effluent discharge regulations will dictate the degree of water conservation, recycle, and reuse. But even in these humid areas, away from major rivers, surface water supplies are less reliable and the same incentives to minimize consumption may exist as in the West.

Water Requirements

The major consumptive uses of water in most synthetic fuel plants are for cooling, hydrogen, and solids handling and dust control. The use of wet FGD could also constitute a large consumptive use. Biochemical conversion processes do not require water for hydrogen production although it may be needed for irrigation. In defining the water requirements for an integrated synthetic fuel plant, it is assumed that all effluent streams are recycled or reused within the plant or mine after any necessary treatment. These streams include the organically contaminated waters generated in the conversion and the highly saline blowdown waters from evaporative cooling systems. Water only leaves the plant as vapor, as bonded hydrogen, or as occluded water in solid residues. As discussed in Section 9.1, dirty water is cleaned, but only for reuse and not for returning it to a receiving water. In the discussion which follows, we shall for simplicity of presentation confine our attention to surface processing plants.

Cooling water. For surface coal conversion and most biomass conversion plants, the largest quantity of water consumed will be in wet evaporative cooling for the purpose of transferring unrecoverable thermal energy to the environment. The quantity of water consumed in evaporative cooling depends principally upon the plant thermal efficiency. It is important, however, to recognize that not all the unrecovered heat must be dissipated by cooling. Even if this were desirable, it would not be possible, since an appreciable fraction of this heat will be lost directly to the atmosphere up flue gas stacks, in heat radiated from boilers, and from other sources over which one has little or no control. Disposing of all the remaining unrecovered heat by evaporating water is generally not economical; some of the heat should be transferred to the atmosphere by forced air cooling. But even in a wet evaporative cooling tower, the air will also carry away heat, and although 2.3 MJ will evaporate a kilogram of water, the actual amount of heat carried away is generally at least 3.0 MJ for every kilogram of water evaporated. The extent to which water is evaporated to remove the unrecovered heat that must be dissipated by cooling in a properly engineered design, should depend upon whether the plant is located in a region where water is plentiful or scarce, coupled with the true cost of water. In water-rich areas, typically only 35 to 50 percent of the total unrecovered heat should be dissipated by evaporating water, with 40 percent being a reasonable average. In arid regions or where water is expensive, no more than 15 to 30 percent of the unrecovered heat should go to evaporating water, with 20 percent being a reasonable average.[10,14] With η the plant efficiency, we may therefore write

$$\text{Cooling water evaporated, L/MJ product} = \frac{(1 - \eta) f}{\eta} \frac{f}{3} \qquad (9.14a)$$

where

$$f = 0.2 \qquad \text{arid regions} \qquad\qquad (9.14b)$$

$$f = 0.4 \qquad \text{humid regions}$$

Hydrogen. Water is consumed as a source of hydrogen. Excess steam may be introduced for temperature control or to drive a reaction, but this water can be assumed to be recovered and treated for reuse or recycle. In principle, therefore, there is no net consumption of process water beyond that needed to supply hydrogen. This does not, of course, include water required for cleaning, purification, or other ancillary operations, some of which may be lost through evaporation. Hydrogen requirements have been discussed extensively throughout the book and we recall that hydrogen is used not only to increase the hydrogen-to-carbon ratio of the fuel but also to reduce the fuel sulfur, nitrogen, and oxygen. Any water produced from the fuel oxygen is considered recoverable, so water is consumed mainly in raising the hydrogen content of the feed.

The amount of water needed for a synthetic fuel conversion process can be estimated from the overall stoichiometric relation

$$(4 + y)CH_x + 2(y - x)H_2O \rightarrow (4 + x)CH_y + (y - x)CO_2 \qquad (9.15)$$

The water-equivalent hydrogen requirement per unit of product calorific value follows as

Water for hydrogen, L/MJ product
$$= \frac{2(y - x) \text{ kmol } H_2O}{(4 + x) \text{ kmol } CH_y} \cdot \frac{18 \text{ kg/kmol } H_2O}{\text{GCV in MJ/kmol } CH_y} \qquad (9.16)$$

Calorific values for typical products are given in Table 2.12. For example, to produce methane ($y = 4$) from a typical coal ($x = 0.8$) the water needed is 0.027 L/MJ product. Hydrogen requirements for both oil shale and tar sands conversion are relatively low ($x \sim 1.5$), since a synthetic crude of little different hydrogen-to-carbon ratio is normally the product. In oil shale and tar sand processes, as in refinery operations, a significant proportion of the hydrogen consumed is used for heteroatom hydrogenation. This consumption of hydrogen is not included in Eq. (9.16). For estimation purposes, Eq. (9.16) is most appropriate for coal conversion, the error being lower the larger the amount of hydrogen that is added.

Solids handling and dust control.[10,14] Water is required for a variety of solids handling operations including mining, crushing, and feeding of the raw material, and for the disposal of residuals. In mining, crushing, and feeding, the main use of water is for dust control. In disposal, water may be required for quenching, dust control, compaction, and reclamation. The amount of water for these purposes is dependent principally on the mass of material handled.

A reasonable average for the water required for raw shale handling is about 5 percent by mass of the shale. The figure for underground coal mining may be half again as much, although with water conservation or with surface mining, the value of 5 percent can be used. The more significant difference, however, lies in the very much greater mass of material that must be handled for shale conversion than for coal conversion, with the factor anywhere from 2.5 to 5 times for the same product calorific value.

In the surface processing of shale, the largest quantity of water can be required for the disposal and compaction of the spent shale. The spent shale generally constitutes 80 to 90 percent by mass of the raw shale depending on the shale grade, and consists mainly of inorganic mineral matter. The amount of water needed for this purpose is still a matter of some debate, although a figure of approximately 15 percent by mass of the material is considered to be optimum for compaction and cementation and is the value used in the TOSCO II commercial design. In this design, an additional 15 percent is needed for evaporation losses in moisturizing the spent shale and for revegetation requirements. For a 10^7 L/d plant processing a high-grade shale of 150 L/t, water equal in mass to 15 percent of the spent shale would amount to about 0.03 L/MJ of product. According to Eq. (9.14) this is about the same amount of water that would be used for cooling in an arid region for a plant with a 70 percent thermal efficiency, which is a value appropriate for an integrated oil shale plant.

Ash disposal only requires water in an amount about 10 percent by mass of the ash. Since the ash typically represents only 10 percent by mass of the coal, the water requirement is only about 1 percent by mass of the feed coal, which is small in comparison with the other requirements considered. Although we have not discussed tar sands water needs, it can be seen from Figure 7.20 that the amount of water currently used for disposal in hot water processing is even larger than that used for spent shale disposal. Although this is by no means a minimum or conservative usage, nevertheless it is unlikely that the process would be employed in the arid regions of the United States where the largest tar sands deposits are located, because of the large water needs, among other reasons.

Other requirements such as water for flue gas desulfurization are much more process specific, and simple generalizations are somewhat difficult to make. However, these other requirements generally do not represent a large fraction of the total consumption. In Figure 9.3 is shown a histogram of the average net water consumption for coal and shale conversion processes based on detailed estimates from conceptual designs. The coal conversion values represent averages for a number of different processes and different coals, with the limits indicated for both high and minimum practical wet cooling options. For oil shale processing an intermediate wet cooling option was used, with the difference in water consumption between processes ascribable mainly to the amount assumed to be needed for shale disposal and revegetation. The high water consumption is based on the TOSCO II disposal option with a water requirement for disposal and revegetation of about 30 percent by mass of the spent shale. The low water consumption is based on the Paraho disposal option with a water use for disposal and revegetation about one-quarter that for the TOSCO II design. The reader may verify that the estimating rules given for water usage would result in values comparable to those shown in the figure, for both the coal and shale conversion processes.

For a product GCV of 40 MJ/L, it follows from Figure 9.3 that about 1 to 4 volumes of water are required per volume of liquid fuel product or its calorific equivalent. Shale processing is the highest because of the large disposal requirement, while refining to a solvent refined coal is lowest because it has the highest thermal efficiency and lowest hydrogen requirement. In any case, the actual water requirements are seen to be relatively small with water recycle and reuse practiced, and with the use of partial dry cooling where economically appropriate. It may be noted that the irrigation water requirements for production of biomass by silviculture or agriculture, measured per unit calorific value of converted product, have a typical range from 100 to 1000 times the water requirements shown in Figure 9.3.[9]

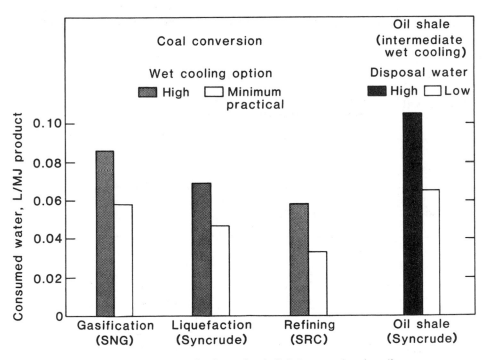

Figure 9.3 Average net water consumption for coal and oil shale conversion plants.[10]

Water Treatment

In our estimates of water requirements it was assumed that dirty waters produced in any synthetic fuel facility would be cleaned and treated for recycle or reuse, and that no water streams leave the facility except as vapor, hydrogen in the product, or occluded water in solid residues. The water management scheme for any plant and the treatment procedures used depend not only on the quantity and quality of the process waters generated but also on the quantity and quality of water available to the plant.[10] In any synthetic fuel plant there are three main uses of water, each requiring a particular quality: high quality for the process, medium quality for cooling, and low quality for disposal and mine uses. Figure 9.4 diagrams one simplified water reuse scheme. The scheme assumes that the water source is a relatively clean surface water and that the effluent from the process is of low quality and insufficient to meet all the plant's cooling needs.

Figure 9.5 is an amplification of Figure 9.4 and represents a general water treatment scheme for a coal conversion plant generating a foul process water contaminated with phenol. The scheme is not unique but does contain the main components of any water treatment plant: boiler feed water preparation, process water cleanup, and cooling water. The three main streams are shown with heavy lines. Illustrated in the figure is the treatment of the raw water to boiler feed quality, the treatment of the foul process condensate into makeup for the cooling tower plus water for dust control and, if needed, makeup for flue gas desulfurization. The use of raw water in the cooling tower is also shown. The dashed lines for the products from the foul condensate treatment indicate alternative schemes. In one, the phenol may be recovered by solvent extraction followed by ammonia distillation. Alternatively, the ammonia separation may precede

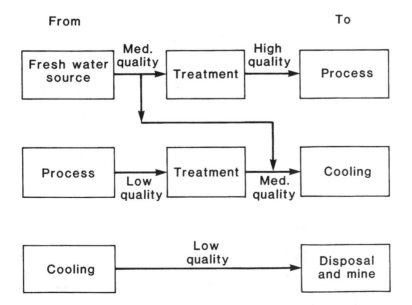

Figure 9.4 Simplified water reuse scheme in a synthetic fuel plant.[10]

biological oxidation in which the organic matter is oxidized and produces a waste sludge. All three approaches may also be used. The treated foul condensate may in part be used for ash disposal, or the cooling tower blowdown by itself may be used to meet this requirement.

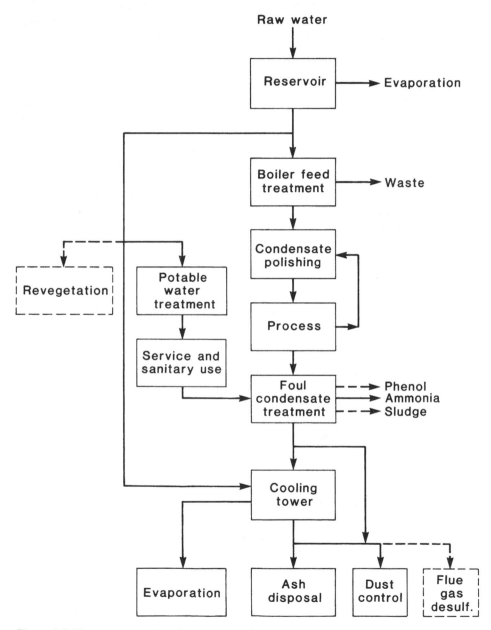

Figure 9.5 Water treatment flow diagram for coal conversion plant generating dirty process water. (Requirements shown dashed are not necessary for every plant.)[10]

Boiler feed water. Water fed to boilers to generate high pressure steam must be of very high purity, or the boiler tubes will become corroded or scaled with mineral matter. The treatment of boiler feed water to remove dissolved inorganic salts is for the most part accomplished by ion exchange, in which cations and anions in solution are exchanged with hydrogen and hydroxyl ions in ion exchange resins that are capable of producing a water virtually free of common salts. The resins are regenerated with relatively strong acid and alkali solutions and the regenerant wastes have to be controlled. Boiler feed water treatment is, however, not unique to synthetic fuel plants and is for the most part well-understood engineering practice.

Process condensate. Treatment of the foul process condensates recovered from coal liquefaction, and low and intermediate temperature gasification plants is one of the more difficult water treatment problems in synthetic fuel plants because the waters have a high chemical oxygen demand and a high biochemical oxygen demand. They are heavily contaminated with dissolved organic matter, ammonia, and acid gases (see Table 9.3).

To remove or destroy phenol and other organic matter, many possible procedures exist including solvent extraction, adsorption on activated carbon or synthetic polymers, biological or chemical oxidation, and others. These procedures may be used alone or in combination. Ammonia separation and recovery may also be accomplished in a number of ways, although distillation, relying on the volatility of ammonia, is usually a main feature of any method used for the concentrations present in synthetic fuel facility wastewaters. Several procedures are in use, and likely candidates for water treatment plants of synthetic fuel complexes include the procedures suggested for the water treatment scheme of Figure 9.5; namely ammonia separation by distillation, phenol and organic matter recovery by solvent extraction, or destruction by biological oxidation.

Being volatile, ammonia and the acid gases carbon dioxide and hydrogen sulfide are not difficult to separate from water by distillation, as noted in Section 5.1. When the water containing the dissolved gases is boiled, the vapor is more concentrated in the gases than was the liquid. If the vapor is condensed and reboiled, the vapor is still more concentrated. This is the principle of fractional distillation which was discussed in connection with oxygen production. However, separation of the gases from the water is not enough. The gases must be separated into ammonia and acid gases. In one process for carrying this out, proprietary to Lurgi Gesellschaft, use is made of the fact that the vapor pressure of carbon dioxide and hydrogen sulfide over the process condensate is higher than the vapor pressure of ammonia. If clean water is poured down the distillation column, called "refluxing," it is possible to operate so that the carbon dioxide and hydrogen sulfide go overhead as a vapor with very little ammonia and water, while water, ammonia, and some small amount of carbon dioxide leave as a liquid at the bottom. Ammonia is separated from water in a second distillation column.

When biological oxidation is used to destroy the organic matter in process condensates, ammonia separation will precede it because high concentrations of ammonia are toxic to the bacteria that biologically degrade the organic matter. Aerobic biological decomposition of organic matter in wastewater to a settleable cellular sludge, carbon dioxide, and water requires the presence of aerobic bacteria and dissolved oxygen. For most wastewaters, including process condensates, the concentration of bacteria yielded by the decomposition of the organic matter in a once-through reaction tank, into which air or oxygen is mixed, is not high enough to obtain an adequate rate of reaction. Usually the sludge is separated from the slurry leaving the oxidation tank and most is returned to the tank, maintaining a high bacterial concentration in the reactor at all times. The balance of the sludge is disposed. This procedure is called the "activated sludge process" and is one of the more likely procedures for the reduction of organic matter in organically contaminated process condensates.

In contrast to biological oxidation, the extraction of phenol from the process condensate has the advantage of recovering the phenol, at least for use as a fuel, or for a product with an even higher value. Moreover, extraction may remove the necessity for biological treatment or at least minimize its costs and problems. To extract phenol from the dirty process water, the water must be contacted either with a solid that absorbs the phenol on its surface or an immiscible liquid solvent that dissolves phenol preferentially over water. Liquid solvents appropriate for process condensate treatment include kerosene and ethers. The usual procedure is to separate the solvent and phenol by distillation. The solvent extraction process is fairly energy intensive, with the energy consumed mostly to distill the solvent.

Cooling water. The last major component in a water treatment plant for a synthetic fuel facility is that part dealing with cooling water treatment. A wet cooling tower is an evaporator, and salts dissolved in the makeup water concentrate, often to the point of precipitation. The precipitates tend to adhere to heat transfer surfaces, forming a hard scale. As in boilers, this must be prevented. Moreover, circulating cooling water is warm and well oxygenated and is an ideal habitat for microbial growth. Untreated cooling systems are subject to fungal rot of the wooden parts of the tower, bacterial corrosion of iron, growth of algae, etc. To control biological growth, biocidal chemicals must be added, while scale and corrosion is customarily prevented by "blowing down" the cooling water to avoid its concentration. Disposal of this highly saline blowdown water must be controlled. Alternatively, chemical treatments are used to prevent scaling, including lime "softening" to precipitate salts, and acid addition. As with boiler feed water treatment, this falls within the category of engineering practice not specific to synthetic fuel plants.

Although control technologies for treating synthetic fuel plant wastewaters are available, the costs and energy requirements are much less well defined. However, it can be said that the energy required for the water treatment plants is very much controlled by the amount needed for ammonia separation, and this amount is directly proportional to the rate of production of the foul process condensate. Most estimates made indicate that the water treatment costs should not exceed 5 percent of the cost of the product fuel.

9.4 SOLID WASTES DISPOSAL

Every synthetic fuel plant generates solid wastes that may be broadly classed as fossil fuel residues, ancillary process residues, and spent catalysts. The fossil fuel residues consist mainly of the mineral fraction of the coal, oil shale, or tar sands, and for some processes may also include small amounts of unprocessed fines and coke. After burning or converting coal, the mineral content remains as coal ash that is disposed of together with any occluded water from quenching, dust control, or transport. Taking the coal residue to be proportional to the mass fraction of ash in the raw coal, then per unit of product calorific value we may write

Coal residue, kg/MJ product

$$= \frac{(1 + \text{occluded } H_2O \text{ fraction})(\text{coal ash fraction})}{\eta \cdot (\text{coal GCV, MJ/kg})} \tag{9.17a}$$

$$\approx \frac{1.1}{\eta} \cdot \frac{\text{coal ash fraction}}{\text{coal GCV, MJ/kg}} \tag{9.17b}$$

where, as before, η is the process thermal efficiency. Here, the occluded water mass fraction has been set equal to 0.1, consistent with the amount of water noted in the last section as being typically required for ash disposal. For the direct liquefaction plant example of Table 9.5, converting an Illinois coal with an ash content of 7.4 percent and a calorific value of 24.9 MJ/kg at a 65 percent thermal efficiency, from Eq. (9.17b) the disposed wet residue for our nominal standard-size plant is a little over 2000 t/d.

A relation analogous to Eq. (9.17) can be written for the amount of spent shale residue. Since essentially all the shale calorific value is contained in the organic fraction, the equation takes the form

Shale residue, kg/MJ product

$$= \frac{(1 + \text{occluded } H_2O \text{ fraction})(\text{shale mineral fraction})}{\eta \cdot (\text{organic GCV, MJ/kg})(\text{shale organic fraction})} \tag{9.18}$$

where the sum of the shale mineral fraction plus organic fraction is equal to 1. Normally the shale grade is specified by the Fischer assay yield. By writing the mineral fraction as the complement of the organic fraction and using Eq. (1.4), which relates the organic fraction to the yield, Eq. (9.18) can be put into the form

$$\text{Shale residue, kg/MJ product} \approx \frac{0.028}{\eta} \left(\frac{811 - \text{yield, L/t}}{\text{yield, L/t} + 11} \right) \tag{9.19}$$

Here, we have used the GCV value of 41.1 MJ/kg appropriate to the organic component of Green River oil shale (Section 1.3), and have assumed that the spent shale is compacted and cemented with 15 percent occluded water. For a plant with a 70 percent thermal efficiency feeding a high-grade shale of 150 L/t, the residue to be disposed according to Eq. (9.19) is 0.164 kg/MJ of product. This translates into 66 000 t/d for our nominal standard-size plant, or some 30 times more material than that from a direct coal liquefaction plant producing the same product. This is, of course, a consequence of the very high mineral content of the shale, averaging some 85 percent, compared with an average ash content of 10 percent for U.S. coals. Tar sands residues, consisting mainly of quartz grains, would be similarly large.

Although fossil fuel residues contain trace amounts of toxic materials, they are believed to be non-hazardous, and as noted in Section 9.1, are treated as such under current regulations. Nevertheless, they do present a problem of management in their disposal, particularly with the shale wastes because of the relatively large quantities involved.

Ancillary processes including air and water pollution control technologies also generate solid wastes. Estimates of the quantity of these wastes are strongly process- and design-dependent. Where wet flue gas desulfurization is used, the sludges generated can sometimes represent a sizeable fraction of the material to be disposed of from a coal conversion plant. An order of magnitude estimate of the amount of FGD sludge can be made using the results for SO_2 emissions, obtained from the procedure outlined in Section 9.2. Consistent with the assumption made there of 90 percent control, this means there are 9 kg of SO_2 removed for each 1 kg emitted or alternatively, 4.5 kg of sulfur removed for every kg of SO_2 emitted. It is also noted in Section 9.2 (Table 9.6) that for a mixed lime/limestone process, about 5.9 kg of solids are wasted in addition to slurry water. If the scrubber sludge is to be disposed as landfill, for physical stability it may be dewatered to as high as 65 to 70 percent solids. This means that about 7.8 kg of wet solids are wasted for each kg of sulfur removed.

By way of example we use the results from Table 9.5, which were calculated for a coal liquefaction plant of 65 percent thermal efficiency burning an Illinois coal. Taking the SO_2 emission estimate and multiplying it by (7.8×4.5) kg S/kg SO_2 emitted gives an FGD sludge rate of 640 t/d. Using the results derived above of 2000 t/d of coal residues shows the FGD sludge in this case to represent about 24 mass percent of the total solids disposed. This value may be considered generally to be at the upper end of the range expected, since the SO_2 emission values that were used are correspondingly at the high end of their range. Other ancillary sludges generated include raw water softening sludges, sludges from the activated sludge treatment of contaminated wastewaters, and evaporator brines. However, the levels of these sludges will generally represent only a small fraction of the total solid wastes and are in any case mostly of low toxicity, as is FGD sludge.

Solid wastes that will most probably require special handling are spent catalysts. Although their volumes are very small, far less than 1 percent of the wastes disposed, they are considered hazardous due to their high-heavy-metals content. However, because their quantity is relatively small, it may be that their quantity as well as toxicity will be considered in setting environmental rules for their disposal.

In summary, we may estimate that for coal conversion processes, typically the fossil fuel residues constitute about 85 percent of the solid wastes with 15 percent a representative average for the ancillary process residues. In oil shale and tar sands processing, the solid wastes are mostly made up of the inorganic residues from the processed fuel.

Solid Waste Management

Solid waste management techniques can be broadly classed as disposal procedures or recycle and reuse procedures. Disposal procedures include conventional landfill, se-cured landfill, mine disposal, and ponding. Recycle and reuse procedures include catalyst regeneration and product recovery.

In conventional landfill, layers of solid wastes are compacted and covered with layers of soil. This is most commonly used for non-hazardous industrial wastes and is a likely choice for the non-hazardous wastes we have discussed. The co-disposal of fossil fuel residues and ancillary treatment sludges by conventional landfilling techniques could be environmentally acceptable under proper conditions. For example, the landfill sites should exhibit a relatively impervious sub-soil layer, should be contoured to control runoff and infiltration, and should be monitored to detect possible aquifer contamination.

Hazardous wastes can only be disposed of in secured landfills. A secured landfill is defined to be located in an area having a sufficiently high atmospheric evaporation rate and to be bounded on its sides and base by material of very low permeability. The specific evaporation rate and permeability are set by regulation. An alternative hazardous waste management procedure is to render the waste non-hazardous by chemical fixation or encapsulation. The relatively high costs of these methods make them most suitable for low-volume wastes such as spent catalysts.

Return of the solid wastes to the mine site is likely to be used where the mine is above ground. The disposal procedure would be similar to landfilling with overburden used for fill material, and would be subject to restrictions similar to those applied to conventional and secured landfills. This is the disposal method used in the Athabasca tar sands operations. Underground mine disposal is considerably more difficult and involves among other problems a holding period in which the wastes cannot be disposed in the mine because it is unavailable for this purpose.

In connection with either landfill or surface mine disposal, it may be noted that under the right conditions, both spent shale and coal ash cement up when wetted with the proper amount of water, and this property might be used to seal the landfill. This has been suggested in a commercial design for an oil shale plant built around the Paraho process. A proposed commercial design for the TOSCO II process envisages fully cementing the wastes by the addition of dirty process water at an appropriately high temperature, followed by compaction of the shale. The cemented shale appears to permanently ''freeze in'' the moisture that is added and to become effectively impermeable to percolation and leaching.[10]

Ponds can also be used for the temporary or permanent disposal of non-hazardous wastes. The ponds are usually earthen reservoirs which serve as particle settling basins and to enhance evaporation. When the pond is filled it may be either abandoned or dredged, with any dredge material generally disposed by landfill.

Recycle or reuse procedures will tend to minimize environmental effects. Moreover, the wastes may have economic value. For example, the cementation properties of ash or spent shale could make these materials useful as a replacement for cement. However, this is not necessarily obvious since all economic factors must be taken into account, including supply/demand considerations and proximity to markets. Regeneration of spent catalysts is another possibility, but this must be weighed against the costs of the alternative disposal procedures discussed. Although various options exist for using biological treatment sludges—including composting, land application, and incineration—the relatively small quantities produced in coal or oil shale plants will probably not make this worthwhile economically. On the other hand, the uses mentioned would undoubtedly be important for the wastes from synthetic fuel plants based on biochemical conversion methods, as discussed in Section 8.2.

REFERENCES

1. Schneider, S. H., and Chen, R. S., "Carbon Dioxide Warming and Coastline Flooding: Physical Factors and Climatic Impact," in *Annual Review of Energy, Vol. 5, 1980* (J. M. Hollander, M. K. Simmons, and D. O. Wood, eds.), pp. 107-140. Annual Reviews, Inc., Palo Alto, Calif., 1980.
2. Stern, A. C., *et al.*, *Fundamentals of Air Pollution*. Academic Press, New York, 1973.
3. Sundquist, E. T., and Miller, G. A., "Oil Shales and Carbon Dioxide," *Science* **208**, 740-741, 1980.
4. Buonicore, A. J., "Air Pollution Control," *Chem. Eng.* **87**(14), 81-101, 1980.
5. Robertson, J. H., Cowen, W. F., and Longfield, J. Y., "Water Pollution Control," *Chem. Eng.* **87**(14), 102-119, 1980.
6. U.S. Environmental Protection Agency, "Pollution Control Guidance Document for Indirect Coal Liquefaction," Coal Gasification and Indirect Liquefaction Working Group. Industrial Environmental Research Laboratory, U.S. Environmental Protection Agency, Research Triangle Park, N.C., 1981 (to appear).
7. Lynch, J. W., "The New Hazardous-Waste Regulations—Part I," *Chem. Eng.* **87**(15), 55-59, 1980.
8. Gradet, A., and Short, W. L., "Managing Wastes Under RCRA—Part II," *Chem. Eng.* **87**(15), 60-68, 1980.
9. Holdren, J. P., Morris, G., and Mintzer, I., "Environmental Aspects of Renewable Energy Sources," in *Annual Review of Energy, Vol. 5, 1980* (J. M. Hollander, M. K. Simmons, and D. O. Wood, eds.), pp. 241-291. Annual Reviews, Inc., Palo Alto, Calif., 1980.
10. Probstein, R. F., and Gold, H., *Water in Synthetic Fuel Production*. MIT Press, Cambridge, Mass., 1978.
11. National Research Council, "Surface Mining of Non-Coal Minerals, Appendix II: Mining and Processing of Oil Shale and Tar Sands," National Academy of Sciences, Washington, D.C., 1980.
12. Stern, A. C. (ed.), *Air Pollution, Vol. IV, Engineering Control of Air Pollution*. 3rd Edition, Academic Press, New York, 1977.
13. Naruse, Y., *et al.*, "Properties and Performances of Various Iron Oxide Catalysts for NO Reduction with NH_3," *Ind. Eng. Chem. Product Res. Dev.* **19**, 57-61, 1980.
14. Gold, H., and Goldstein, D. J., "Water-Related Environmental Effects in Fuel Conversion," Report Nos. EPA-600/7-78-197a and b, U.S. Environmental Protection Agency, Washington, D.C., October, 1978. [Also Reports No. FE-2445 (V.1) & (V.2), U.S. Department of Energy.]

CHAPTER TEN

ECONOMICS AND PERSPECTIVE

10.1 ECONOMIC CONSIDERATIONS

Rational methodologies have been presented in this book that enable estimates to be made of synthetic fuel plant operating characteristics, efficiencies, product slates, and emissions without the need for recourse to full-scale design studies. An analogous simplified methodology for the estimation of product costs for a given process would be difficult to derive at the present stage of synthetic fuels development, although cost trends with changes in some variables can be represented analytically.

Many detailed and extensive studies have been made of the costs of producing synthetic fuels, with wide divergencies among estimates made at the same time by different groups and with even wider divergencies among the costs estimated at earlier and later times; the costs being invariably higher the more recent the estimate. The reason is simply that any economic estimate shows costs perceived at a given time defined by a given set of technological, economic, and regulatory factors. Any one or all of the factors together with the cost estimating procedures, may change with time because of different assumptions employed.[1-3]

Probably the most important change responsible for past uncertainties in estimated costs relates to the technologies themselves. Synthetic fuel plants have not been reduced to standardized designs and have not been operated on commercial scale in the United States, so process performance, reliability, and operability cannot be defined with certainty. Changes in feedstock fuel costs, the bases for plant investment, and the economic analyses—including inflationary factors, financing methods, tax credits, return on investment, and product marketability—have also contributed importantly to wide variations in synthetic fuel cost estimates. As pointed out in Chapter 9, current and future environmental requirements have not as yet been defined for synthetic fuel plants, and this also has had an important effect on estimations that have been made. Lastly, we mention one current suggestion, not yet grounded in economic theory, that large-size projects like synthetic fuel plants may incur additional costs related to their protracted startup period, size, and complexity, contrary to the usual engineering economics concept of "economy of scale."

The differences in cost patterns as a consequence of economic assumptions may be seen in the differences reported for the allocations of estimated product costs. In coal conversion plants, cost fractions are generally reported as capital, operating, and fuel, although most mines are likely to be integrated with the plant. The product cost breakdown for coal conversion plants is given in Ref. 1 roughly as 50 percent capital, 25 percent operating, and 25 percent fuel. The breakdown from values reported in Ref. 4 averages roughly 25 percent capital, 25 percent operating, and 50 percent fuel. These very marked differences are probably ascribable to differences in assumptions related to interest charges, tax factors, return on investment, depreciation, and other economic factors.

Synthetic fuel plants involve many technical and economic uncertainties, require large investments of capital that is more costly and less readily obtainable, and therefore may be expected to command a larger return on investment than would be the case for new utilities, or new standard chemical or petroleum plants. One way this is manifested is in the assumed rate at which the invested capital is to be recovered. The product cost for a coal conversion plant may be expressed in the form

$$\text{Product cost (\$/MJ)} = \frac{F(\$/\text{yr}) + O(\$/\text{yr}) + k(1/\text{yr})C(\$)}{\text{GCV produced (MJ/yr)}} \qquad (10.1)$$

where F is the annual fuel cost, O the annual operating cost, C the capital investment, and k the so-called capital recovery factor. Recent guidelines suggest that a capital recovery factor of about 24 percent per year may be appropriate for new synthetic fuel plants. This value is likely to be representative of that used in Ref. 1, where a 50 percent capital cost fraction was reported. In Ref. 4, a value for k averaging 10 percent per year was used, and this may explain the much lower capital cost fraction, of 25 percent, that was reported there. For comparison, it may be noted that capital recovery factors of 11 percent per year are used for electric utilities, which are considered mature industries with relatively easy access to capital markets, while values as low as 7 percent per year are presently used for natural gas utilities.

For an oil shale plant with an integrated mine, where the fuel costs are not calculated separately, it is reported[1] that capital represents 70 percent of the product cost and operating expenses 30 percent. This is consistent with the breakdown based on the capital investment and operating costs reported for a proposed commercial hot water extraction tar sands plant in Canada,[5] if a capital recovery factor of about 20 percent per year is assumed. The higher capital cost fraction for both oil shale and tar sands plants is related to the capital investments in the mining operations, which include the costs of the reserve acquisition and mine development.

It is evident that the wide range in cost fractions makes the presentation of any absolute costs difficult and beyond the scope of the book. However, as noted above, a most important factor is the technology itself, and the better defined the process the better will be the estimate of the cost to build any plant. It is therefore appropriate to consider in relative terms how the technology factor enters the final product cost.

During the development of any new process, such as the ones being considered for synthetic fuels manufacture, as the process gets closer to commercialization there is a tendency for a lowering of yield and an increase in complexity, with concomitant higher costs. However, parallel with this cost increase there is a decrease in the technical uncertainties involved and hence a decrease in the uncertainties as to the cost of the first plant. This is illustrated in Figure 10.1, where the cost uncertainties during the development phase are represented by the shaded area which narrows up to the time of commercial inception. From experience with the commercialization of new pioneer petroleum and chemical processes it has been estimated[1] that at the very early stages of development, when only initial engineering studies are available, the contingencies necessary to bring the estimated investment cost up to the probable level for the first commercial plant can be equal to the unadjusted cost estimate.

After the first plant is built, the investment costs for subsequent plants tend to fall as a result of knowledge and experience gained with the first plant. This is the so-called "learning curve" effect, and for new pioneer petroleum and chemical processes, the drop in plant costs in "constant dollars" may be as much as 25 to 40 percent over a period of 10 years.[1] In this regard, it may be noted that the reliability of plant cost estimates for producing liquid fuels by indirect liquefaction using Sasol Fischer-Tropsch technology could be expected to be quite high, since a complete commercial plant is built and operating. The cost estimates for direct coal liquefaction are probably least accurate, since none of the candidate processes has as yet been demonstrated on large scale. The reliability of cost estimates for SNG and oil shale processes would fall somewhere in between.

Although the above remarks indicate how investment costs for a developing process change with time, they do not indicate the extent to which improvements in technology can lower cost. For synthetic fuel plants, which are largely dominated by heat and mass transfer operations, probably the most important single measure of improved technology is improved thermal efficiency. But for any heat or mass transfer process, increasing the transfer area will decrease the driving potential needed to

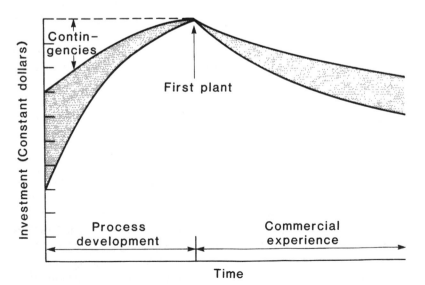

Figure 10.1 Effect of stage of commercialization on synthetic fuel plant investment cost.[1]

transfer a given quantity of heat or mass, which results in smaller system losses and higher thermal efficiencies. In economic terms, this translates into the fact that it is usually possible to increase thermal efficiency by increasing the investment cost in equipment.

In an energy-intensive process, we may consider the product cost to be divided into capital costs and energy (operating) costs. With increasing transfer area, the energy cost per unit of product will decrease, while the capital cost per unit product will increase. This behavior of energy and capital costs is illustrated in Figure 10.2, which supposes a plant with a fixed total product rate. The precise shapes of the curves and hence the optimum product cost shown will depend, naturally, on the details of the specific process. The absolute minimum energy cost indicated in the figure corresponds to the minimum thermodynamic energy necessary to carry out the conversion, and is equal to the difference in free energy between the product and the feed. From the discussion in Section 2.3, this minimum energy is small compared to the energy that is dissipated. The concepts embodied in the curves of Figure 10.2 have been termed[6] "thermoeconomics," which is a statement of the fact that thermodynamics can fix a minimum requirement but not an optimum one that will give the lowest possible product cost for the prevailing economic and technological constraints.

If it is assumed that heat and mass transfer conditions play a critical role in setting the energy consumed and hence the thermal efficiency of a synthetic fuel plant, then we may reduce the plant to a "black box" in which a generalized transfer process takes place through a transfer surface having some given resistance. The capital equipment necessary for the process is that which supplies the necessary generalized

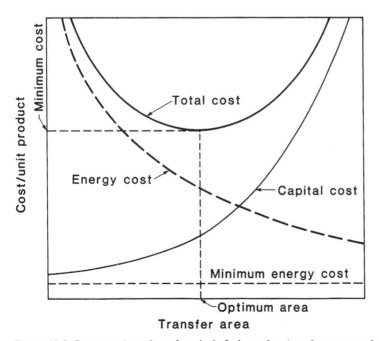

Figure 10.2 Cost per unit product of synthetic fuel as a function of system transfer area.

transfer conditions. The capital equipment is assumed to be assignable a cost in a manner analogous to the way an energy cost is assignable.

Following the above model, we may represent the costs of producing a specified quantity of product fuel by any synthetic fuel process as made up of three components. The first component is a capital factor proportional to the transfer area, the second is an energy factor proportional to the heat (or work) input to accomplish the transfer, and the third component is composed of those auxiliary costs that are independent of the flux. The product cost can therefore be written

$$C = k_c A + k_e q + k_{aux} \qquad (10.2)$$

where C is the product fuel cost per unit mass, \$/kg, A is the transfer area per unit mass of product produced per unit time, $m^2 \cdot yr/kg$, q is the energy input, MJ/kg, k_c is the capital recovery factor per unit area of transfer surface, $\$/(m^2 \cdot yr)$, k_e is the energy cost, \$/MJ, and k_{aux} is the cost per unit mass of product from sources independent of flux, \$/kg. The cost expression given above was first set out in this form by Simpson and Silver[7] in modeling desalination plant costs.

By definition, the energy input is inversely proportional to the thermal efficiency η, and consistent with a Fourier or Fickian law of heat or mass transfer, is directly proportional to the transfer resistance and inversely proportional to the transfer area. We may therefore write

$$C = k_c A + \frac{k_e \rho}{A \eta} + k_{aux} \qquad (10.3)$$

where ρ is the transfer resistance in appropriate units. Optimizing the cost with respect to the transfer area by setting $dC/dA = 0$, we find that

$$A_{opt} = \left(\frac{k_e}{k_c} \frac{\rho}{\eta} \right)^{1/2} \qquad (10.4)$$

When the above relation is substituted into Eq. (10.2), it is seen that the transfer area costs and energy costs are equal to each other and the minimum total cost is given by

$$C_{opt} = 2 \left(\frac{k_c k_e \rho}{\eta} \right)^{1/2} + k_{aux} \qquad (10.5)$$

This problem of balancing capital costs against operating costs to obtain an optimum product cost is known as "Kelvin's problem," and was first solved by Lord Kelvin in connection with defining the optimum cross section of a conductor.[8]

In Figure 10.3 are shown results of detailed economic studies for the Exxon Donor Solvent coal liquefaction process. Shown are changes in product cost resulting from changes in thermal efficiency and capital investments. It is assumed that the higher efficiencies above the base level could be achieved with development of the process. The plus and minus signs on the constant percent lines represent increased or decreased capital investment, respectively, relative to the base case indicated. A 10 percent increase in thermal efficiency is seen to reduce the product cost by typically 15 percent for a fixed capital cost. It is interesting to note that from Eq. (10.5)

$$\Delta C_{opt} = \frac{\Delta \eta}{\eta^{3/2}} \tag{10.6}$$

For a change from 60 to 70 percent efficiency ($\eta_{avg} = 0.65$), the model result would indicate about a 19 percent decrease in product cost, which is consistent with the results of Figure 10.3. The trend of the other behaviors is similarly indicated by the model, although absolute comparisons cannot be made.

Any synthetic fuel plant is made up of a number of process sections, some of which are more critical to the plant performance than others. This has led Shinnar[9] to introduce the concept of differential economics for any integrated synthetic fuel plant. The approach is based on the idea that only one or two critical process sections are important in determining relative differences in process efficiency, and that to compare processes, one need only compare the efficiencies of those process sections which are radically different. Moreover, we would suggest that optimized costs for these individual sections must generally follow Eq. (10.5) despite the crudeness of the model. This would indicate that any cost reduction cannot be achieved solely by lowering the resistance, but must also reduce the capital cost parameter or at least leave it unchanged. This generally argues against increased complexity for the purpose of efficiency improvement or reduction in resistance.

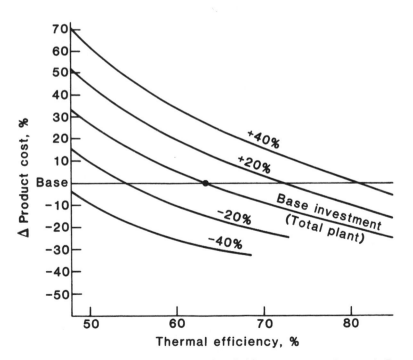

Figure 10.3 Effect of thermal efficiency and capital investment on product cost in Exxon Donor Solvent process.[1]

10.2 RESOURCE, PROCESS, AND PRODUCT CONSIDERATIONS

Synthetic fuels were introduced with the observation that their manufacture may be regarded essentially as a process of hydrogenation. By using carbon as a model feed material, it was shown that the formation of any of the desired hydrocarbon fuel species could in theory be described in terms of relatively simple chemical equations and some thermodynamic concepts including chemical equilibrium and heats of reaction. In further developing the basic theory, it became evident that the processes are in fact more complex. Feed materials are never pure carbon, but rather any of several broad categories of materials each with their own classification of types or grade, hydrogen-to-carbon ratio, concentration of the heteroatoms nitrogen, sulfur, and oxygen, as well as mineral matter, different reactivities, etc. While the reactions of the other constituents present may not significantly alter the overall chemistry, their removal and disposal can significantly affect the conversion process and its costs. Further, the overall reaction scheme is normally governed by kinetics rather than chemical equilibrium. Modeling the conversion process is consequently complicated by the plethora of intermediate compounds that may be formed, each with controlling formation and decomposition reaction rates, and for which data are invariably limited. There are, moreover, the many problems related to the handling of the feedstock and products. It is not surprising, therefore, that a host of conversion processes have arisen, each with its own advantages and disadvantages. To select among them it is necessary that they be evaluated in relation to the resource to be converted and the end product desired.

The resources considered for synthetic fuels manufacture are coal, oil shale, tar sands, and biomass. Their important properties are summarized in Table 10.1. *Coal* has the highest concentration of carbon but requires extensive hydrogenation due to its relatively low hydrogen and high heteroatom content. Currently demonstrated reserves are adequate for at least 200 nominal-size plants (Table 1.4) which would

Table 10.1 Summary of important properties of synthetic fuel resources*

	Coal	Oil Shale	Tar Sands	Biomass
Carbon, mass % dry	65	12	12	45
Heteroatoms,† mass % dry	15	1	1	40
Volatile carbon, % of total C	50	90	90	90
Mineral matter, mass % dry	14	75	75	0
H/C atom ratio	0.85	1.5	1.5	1.6
Moisture, mass % wet	20	2	4	25
GCV, MJ/kg dry	27	6	6	18
Reserves (U.S.)	Adequate	Extensive	Small	Renewable‡
Major products	Gas	Diesel	Typical	Gas
	Boiler fuel	Jet fuel	petroleum	Methanol
	Methanol	Gasoline	liquids	Ethanol
	Gasoline			

* Representative values; actual values may vary widely depending on rank or type.

† Nitrogen, sulfur, and oxygen.

‡ Up to approximately 10 percent of total energy consumption.

satisfy some 30 percent of U.S. energy needs over a 30 year period. *Biomass* also has a relatively high carbon concentration and its hydrogen-to-carbon ratio is in the range of that of liquid fuels. The high-heteroatom content (essentially all oxygen) can represent a large hydrogen demand in thermal processes, but is not a significant factor in biochemical conversion. Biomass might provide up to 10 percent of U.S. energy needs, although this figure is very dependent on a number of factors including the demand for food. Due to the dispersed nature of biomass, "large" biomass plants can be expected to be about one-fortieth the size of nominal-size coal conversion plants. *Oil shale* and *tar sands* are very similar in ultimate composition (but are very different chemically). Both are high in mineral content, but once the organic matter is separated, it may relatively readily be converted to a synthetic crude oil. Tar sands are more desirable in this respect, as they more closely represent natural petroleum. However, recoverable U.S. tar sands reserves are limited. Oil shale deposits are extensive and their contribution to the liquid fuel market is considered limited only by environmental and institutional constraints.

The conversion processes themselves may be classed as direct hydrogenation, the thermal processes of pyrolysis and gasification, and the biochemical processes appropriate to some biomass materials. In *direct hydrogenation,* a source of free hydrogen is required, and this is made to combine with the raw fuel, typically under conditions of high pressure and temperature. Coal is the only resource requiring extensive hydrogenation. Direct hydrogenation of coal to refined solid or liquid fuels appears more attractive than gasification, as less hydrogen is required and the coal undergoes less structural change. In practice, the need for a free hydrogen supply, unfavorable reaction kinetics, and problems in sulfur and nitrogen control have resulted in the more rapid development of gasification technologies.

Pyrolysis is generally the most readily accomplished process, requiring only the provision of heat. Because the temperatures required are relatively low and the reactions are approximately thermoneutral, the amount of heat required is small. In pyrolysis, hydrogenation is effected by removal of carbon, which is recovered as a char byproduct. While the char may be burned as a fuel, the synthetic fuel products are derived entirely from the volatile matter evolved. In the case of oil shale and tar sands, essentially all the organic content is recoverable by pyrolysis. Biomass also has a high volatile content, but here the liquid products are highly oxygenated and unstable. Gases produced are of medium calorific value, so pyrolysis of biomass can be attractive provided liquid yields can be kept small. As coal has a relatively low-volatile-matter content, pyrolysis is not an attractive conversion procedure unless there is a market for the large char byproduct.

Gasification of coal and biomass involves virtually complete degradation to carbon oxides, hydrogen, and methane. Temperatures are higher than required for pyrolysis and considerable heat must be provided for the strongly endothermic reactions involved. If this heat is provided by combustion within the gasifier, and if dilution of the product gas by nitrogen is not desired, a source of pure oxygen is required. The cost of the oxygen supply is then a significant portion of overall costs. Gasification produces the most highly hydrogenated of the synthetic fuel products. The source of the hydrogen is steam (Eq. 9.16), which is cheaper to provide than pure hydrogen. Furthermore, carbon reacts more readily with steam at conditions typical of slagging gasifiers than with hydrogen at conditions typical of direct hydrogenation reactors (see, e.g., Table 3.3)

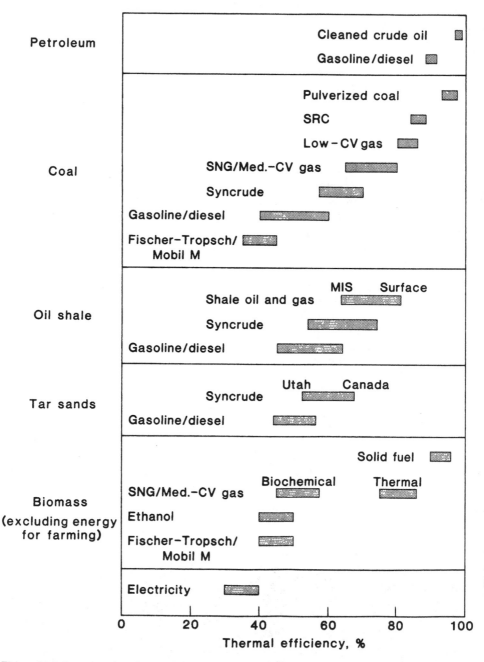

Figure 10.4 Representative thermal efficiencies of selected synthetic fuel processes (modified from Ref. 10).

An advantage of gasification is that the product gas is relatively easily separated from undesirable components such as nitrogen and sulfur compounds. Moreover, manipulation of the gas to form various final products, including liquid fuels, is feasible. A disadvantage of the gasification or indirect route to liquid fuels is that two processes are involved, each with attendant inefficiencies. The fuel source must first be completely broken down, and then put back together again. However, the final product is a clean-burning premium product requiring little if any further refining.

In the technical assessments of the synthetic fuel processes considered in the book, we have used the thermal efficiency as an implied measure of process viability. The thermal efficiencies associated with the conversion to selected synthetic fuels of the four resources under consideration are summarized in Figure 10.4. By way of comparison, efficiencies for processing petroleum crude, as well as for the production of electricity by combustion or combined cycle generation, have been included. Thermal efficiency is not by any means the only measure of the efficacy of a process. Other factors that must be considered are the economic ones discussed in Section 10.1, along with the personnel needed to design, build, and operate the plant, and the availability of required materials and equipment. Land, water, and raw fuel resources must, of course, also be available in the desired quantity and within the cost structure dictated by market conditions. Less easy to quantify are social and institutional constraints, many of which are peculiar to the United States. These range from the provision of ancillary services, development of markets for non-conventional fuels, conforming to evolving environmental regulations, to establishing public acceptance of a specific conversion process. Many of these factors are of a non-technical nature and therefore beyond the scope of this text.

While the thermal efficiency is a useful technical measure for comparing the various process/product combinations, it can be misleading. This is well demonstrated by the example of ethanol production by fermentation of grain. As shown in Figure 10.5, if premium fuels are used to supply process heat and farming energy, then alcohol production from biomass will result in a net consumption of premium fuels. A net gain in premium fuels will result, however, if non-premium fuels such as biomass residue or coal are used where feasible. As suggested in the figure, fertilizer currently produced mostly from natural gas can to some extent be replaced by ammonium sulfate byproduct from coal gasification or oil shale plants. Note that secondary effects such as octane boosting by alcohol in gasohol, and the displacement of some farming activities by the byproduct foodstuff of the fermentation process, have not been considered in Figure 10.5. Nevertheless, the figure clearly demonstrates that the net fuel production is not necessarily reflected by the process thermal efficiency.

The foregoing example leads to the perhaps obvious conclusion that in order to maximize net fuel yields, processes must be designed to utilize non-premium fuels wherever feasible. This may, however, lower overall efficiency and increase costs. For example, hydrogen for refining has customarily been produced by steam reforming light naphtha. This may be replaced by the less efficient procedure of gasification of residual asphaltenes or pitch, so leaving more naphtha for gasoline production. As a further measure, coal may be used as the principal refinery fuel.[12]

The evaluation of synthetic fuel processes is further complicated when comparing processes producing different end products. It is possible to ascribe relative values to

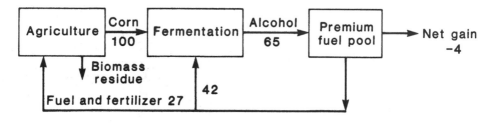

(a) Conventional use of premium fuels for process and agriculture.

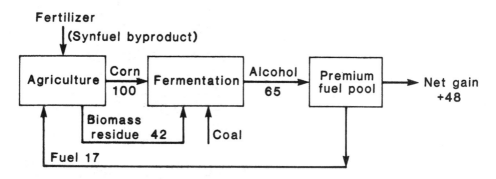

(b) Premium fuel use restricted to farming machinery.

Flows in energy units (GCV)

Figure 10.5 Simplified energy balance for ethanol production by corn fermentation (modified from Ref. 11).

fuels based on purely technical specifications. The energy density of a fuel as given in Table 10.2 may, for example, be used to rank its value. Other technical considerations include handling and storage properties, efficiency and polluting potential on combustion, safety, etc. A dominating consideration, and one which is largely non-technical in nature, is the demand for the fuel as dictated by market conditions and the prevailing energy infrastructure. Value factors for fuels which include market conditions have been published.[4] To some extent these factors are time dependent and may vary with "extraneous" issues, such as whether or not a fuel is subject to government price regulations. It is nevertheless possible to make some broad generalizations regarding the utility of specific fuels.

Gaseous fuels generally burn cleanly and with high efficiency. Production of SNG to supplement natural gas supplies for residential heating is an obvious option for countries with existing pipeline distribution networks. The creation of such a pipeline network for distribution of a synthetic fuel is probably not a viable route. The least expensive gaseous fuel to produce is low-CV gas. Because of its low energy density, it is expensive to transport even locally, so more expensive medium-CV gas may prove more economical. Both fuels would compete with the direct combustion of the solid fuel by the consumer. Advantages of purchasing a gasified product would lie in

Table 10.2 Energy content of fuels on mass and volume basis*

Gases	MJ/kg	MJ/m³
Hydrogen	142	12.7
Methane	55.6	39.8
Med.-CV gas	14.0	12.7

Liquids and solids	MJ/kg	MJ/L
Diesel fuel	48	36
Gasoline	47	32
Petroleum	44	38
Heavy fuel oil	42	43
Carbon	32.8	74.1
Ethanol	29.7	23.4
Methanol	22.7	18.0
Coal† (as-received)	22	30
Wood† (air-dried)	16	10

* Based on GCV.

† Typical values; actual values will vary widely depending on rank or type.

compatibility with existing equipment, ease of handling, no consumer storage problems, and the elimination or significant reduction in on-site pollution control equipment.

A further attraction of medium-CV or synthesis gas is that it may be relatively easily converted to *methanol*. Although development of a market for methanol as a fuel faces problems as a result of its low energy density, affinity for water, toxicity, and other problems, it has considerable potential for becoming a desirable, clean-burning fuel. It may be used for gasohol blending, but in relatively low concentrations of about 5 percent. Further upgrading of methanol to conventional transportation fuels using, say, the Mobil M gasoline process will increase its cost, but require no special accommodation by the prevailing consumer market. The more direct route to gasoline from synthesis gas via a Fischer-Tropsch process may prove less efficient in view of the numerous byproducts resulting with this route. As mentioned, one factor leading to the low efficiency associated with Fischer-Tropsch is due to the need to suppress methane formation at existing facilities. The co-production of methane and gasoline by this process may result in improved efficiencies.

Ethanol currently appears more desirable than methanol as a fuel, as it can readily be accommodated into the existing market as gasohol. While gasohol is restricted to about a 10 percent ethanol content, unlike methanol, this represents a major use of ethanol production potential. In fact, converting all gasoline to gasohol would account for more than the total U.S. grain production.

The more direct routes to liquid transportation fuels should theoretically be achievable at greater efficiency than indirect routes via gasification or by fermentation. This is certainly true of tar sands and oil shale pyrolysis, although the primary coal liquids resulting from direct hydrogenation of coal require considerable extra hydrogenation and refining. Here again, rather than attempting to maximize yields of a specific product, there would be considerable benefit in producing those fuels most suited to the resource. Coal, for instance, might be hydrogenated to heavy fuel oil; oil shale is suited to diesel and jet transportation fuels as well as light fuel oil; and petroleum crudes may thereby be reserved for gasoline production.

The choice of synthetic fuels is thus seen to be determined by infrastructure demand as much as by factors such as cost, efficiency, and pollution potential. Any crash program to produce synthetic fuels must of necessity closely mirror existing market conditions. In time, characteristics of fuels and combustion equipment can be expected to undergo mutual accommodations.[13] Introduction of fuels that might well be cheaper or cleaner than those currently in use will not have a ready market until compatible equipment becomes available.

Electricity is an energy source that, while not a synthetic fuel in the true sense, may be generated either by direct combustion of the raw fuel resource, or by the gasification/combined cycle process. It is readily transportable, non-polluting, and, in general, is recoverable at very high efficiency. As shown in Figure 10.4, the efficiency of electricity generation is at the low end of all the processes considered. If used for heating, this low efficiency will be reflected by high costs. Electricity should be regarded as "work" rather than "heat." It is most appropriately used for driving machinery, including mass transit systems. Development of the electric automobile is hampered by the large mass of storage batteries required to provide even a modest range. The range obtainable with gasoline is some 10 times that obtainable from an equivalent mass of lead-acid batteries. Increasing the mass of batteries to increase range requires carrying around more dead weight and consequently means increased power consumption. It is estimated that the energy efficiency of an electric car with a range of 45 to 72 km will be as low as that of a conventional gasoline powered vehicle.[14] Considerable improvement in battery capabilities will clearly be required before gasoline is displaced.

Research and development needs specific to synthetic fuel production have been mentioned in context throughout the book. One of the more important requirements is for improved and cheaper catalysts. Catalysts are applied in virtually all sectors of synthetic fuel manufacture, including direct hydrogenation, gasification, and refining. Although considerable industrial experience has been gained in oxygen manufacture, it is still a significant cost factor in many gasification routes. From a thermodynamic viewpoint, improvements are clearly possible, so it would seem that oxygen separation remains a cost effective field for research. The supply of pure hydrogen is likewise expensive. Research here may well be directed towards adapting engines and combustion equipment to burn lower-hydrogen fuels.[13]

As far as actual processes go, catalytic gasification, supercritical gas extraction, and the Kölbel Fischer-Tropsch reactor have been singled out in the text as having considerable potential. New *in situ* technologies will have to be developed if Utah tar sands are to be economically exploited. Improvements in process control, safety, and reliability are key factors that are to a large extent tied in with process complexity. Reductions in complexity factors, including the need for excessively high temperatures and pressures and the use of involved multistage reactors, are considered important.

Most of the processes discussed in the book either have been or are being used to supply synthetic fuels on a commercial basis. There is, therefore, little question as to the feasibility of these processes. In most cases, however, these ventures have proved economically unattractive in the face of abundant supplies of cheap natural gas and petroleum. As supplies dwindle and prices escalate, specific processes can be expected to become marginally attractive. In the United States, probably the most competitive of the synthetic fuels are shale oil, and low- and medium-CV gas. The more complex routes to liquid transportation fuels from coal are expected to be more costly. As discussed in the previous section, however, in all cases a reduction in costs will occur as experience is gained from initial plants. Coal, and eventually oil shale reserves, will, however, also become depleted. Because biomass can probably make only a limited contribution to total U.S. energy demand, other sources of energy will have to be harnessed. The development of the synthetic fuels discussed in this text is seen as very necessary to buy the time needed for the evolution of such alternative energy sources.

REFERENCES

1. Swabb, L. E., Jr., Siegel, H. M., and Vick, G. K., "Industrial Perspective on Synthetic Fuels," 109th AIME Annual Meeting, Las Vegas, Nevada, February, 1980.
2. Swabb, L. E., Jr., "Comments on Paper 'Coal Liquefaction' by G. R. Hill," *The Energy Journal* **1**(1), 105-106, 1980.
3. Sinor, J. E., and Culberson, S. F., "Overview of Synthetic Fuels Potential to 1990," in *Synthetic Fuels*, Report by Subcommittee on Synthetic Fuels, Committee on the Budget, U.S. Senate, U.S. Government Printing Office, Washington, D.C., September, 1979.
4. Rogers, K. A., and Hill, R. F., "Coal Conversion Comparisons," Report No. FE-2468-51, U.S. Department of Energy, Washington, D.C., July 1979.
5. de Nevers, N., Glenne, B., and Bryner, C., "Analysis of the Environmental Control Technology for Tar Sand Development," Report No. COO-4043-2, U.S. Department of Energy, Washington, D.C., June 1979.
6. Evans, R. B., Crellin, G. L., and Tribus, M., "Thermoeconomic Considerations of Sea Water Demineralization," in *Principles of Desalination* (K. S. Spiegler and A. D. K. Laird, eds.), Part A, pp. 1-54. 2nd Edition, Academic Press, New York, 1980.
7. Simpson, H. C., and Silver, R. S., "Technology of Sea Water Desalination," in *Desalination Research Conference, 19 June-14 July 1961*, pp. 387-413. Publication 942, National Academy of Sciences—National Research Council Press, Washington, D.C., 1963.
8. Tribus, M., and Pezier, J. P., "Concerning the Economic Value of Experimentation in the Design of Desalting Plants," *Desalination* **8**, 311-349, 1970.
9. Shinnar, R., "Gasoline from Coal: A Differential Economic Analysis," *ChemTech* **8**, 686-693, 1978.
10. Byrne, J., and Huffman, H. C., "Transportation Fuels for the Next 20 Years," *Chem. Eng. Prog.* **76**(11), 24-32, 1980.
11. Weisz, P. B., "The Science of the Possible. Not All BTU's Are Equal," *ChemTech* **10**, 653-654, 1980.
12. Anon., "Changing Priorities May Reshape Oil Refineries," *Chem. and Eng'g. News* **58**(25), 48-50, 1980.
13. Longwell, J. P., "Fuel Science and Technology," in *Science & Technology: A Five-Year Outlook* (National Academy of Sciences), Chapter 14. 2nd Edition, W. H. Freeman and Co., San Francisco, 1981 (to appear).
14. Rosen, C. L., "Challenge. What's Holding Back Electric Cars," *ChemTech* **10**, 658-660, 1980.

CONSTANTS AND CONVERSION FACTORS

PHYSICAL CONSTANTS

Quantity	Symbol	Value	SI Unit
Gas constant	R	8314	J/(kmol · K)
Mean molar mass of air	$\overline{M}_{air}$	28.97	kg/kmol
Normal pressure*	p_0	101.3	kPa
Normal temperature	T_0	273.15	K
Volume of ideal gas at NTP	V_0	22.41	m³/kmol

* Standard atmosphere.

UNITS IN USE WITH SI

Name	Symbol	Value in SI Units
minute	min	1 min = 60 s
hour	h	1 hr = 60 min = 3600 s
day	d	1 d = 24 h = 86 400 s
year	yr	1 yr ≈ 365 d
liter	L	1 L = 10^{-3} m³
ton	t	1 t = 10^3 kg
standard atmosphere	atm	1 atm = 101.3 kPa

SI PREFIXES

Factor	Prefix	Symbol
10^9	giga	G
10^6	mega	M
10^3	kilo	k
10^{-2}	centi	c
10^{-3}	milli	m
10^{-6}	micro	μ

CONVERSION FACTORS

The conversion factors to SI have in part been abstracted from the paper by J. Y. Oldshue, "AIChE Goes Metric," *Chem. Eng. Prog.* **73**(8), 135-138, 1977. An asterisk indicates that the conversion factor is exact and that all subsequent digits are zero.

To convert from	To	Multiply by
barrel (petroleum, 42 U.S. gal)	liter (L)	1.590 E + 02
British thermal unit (Btu, International Table)	joule (J)	1.055 E + 03
Btu/foot^2-hour	watt/meter^2 (W/m^2)	3.155 E + 00
Btu/hour	watt (W)	2.931 E − 01
Btu/pound-mass-degree Fahrenheit	joule/kilogram-kelvin (J/kg·K)	4.187 E + 03
Btu/pound-mole	joule/mole (J/mol)	2.326* E + 00
Btu/foot^3	joule/meter^3 (J/m^3)	3.726 E + 04
calorie (International Table)	joule (J)	4.187 E + 00
calorie (Thermochemical)	joule (J)	4.184* E + 00
calorie/gram-degree Celsius	joule/kilogram-kelvin (J/kg·K)	4.187 E + 03
centimeter	meter (m)	1.000* E − 02
degree Celsius, °C (particular temperature)	kelvin (K)	Use K = °C + 273.15
degree Celsius, °C (temperature interval)	kelvin (K)	1.000* E + 00
degree Fahrenheit, °F (particular temperature)	kelvin (K)	Use K = (°F + 459.67)/1.8
degree Fahrenheit, °F (temperature interval)	kelvin (K)	1/1.8*
degree Rankine, °R (particular temperature and interval)	kelvin (K)	1/1.8*
foot	meter (m)	3.048* E − 01
foot^2	meter^2 (m^2)	9.290 E − 02
foot^3	meter^3 (m^3)	2.832 E − 02
gallon (U.S. liquid)	liter (L)	3.785 E + 00
horsepower (550 ft·lbf/s)	watt (W)	7.457 E + 02
inch	meter (m)	2.540* E − 02
kilocalorie	joule (J)	4.187 E + 03
kilowatt-hour	joule (J)	3.600* E + 06
micron	meter (m)	1.000* E − 06
mil	meter (m)	2.540* E − 05
mile	meter (m)	1.609 E + 03
poise	pascal-second (Pa·s)	1.000* E − 01
pound-force (lbf)	newton (N)	4.448 E + 00
pound-mass (lbm avoirdupois)	kilogram (kg)	4.536 E − 01
pound-mass/foot^3	kilogram/meter^3 (kg/m^3)	1.602 E + 01
pound-force/inch^2 (psi)	pascal (Pa)	6.895 E + 03
stokes	meter^2/second (m^2/s)	1.000* E − 04
ton (short, 2000 lbm)	kilogram (kg)	9.072 E + 02
watt-hour	joule (J)	3.600* E + 03
yard	meter (m)	9.144* E − 01

SYMBOLS AND ACRONYMS

SYMBOLS

All standard mathematical symbols are taken to have their usual meaning. Symbols that are only employed locally and which are not referred to again are generally not listed here. Illustrative units are given for dimensional quantities, generally SI base units or where appropriate the SI derived units, pascal or joule. Prefixes are not shown, although they may be commonly used with a particular quantity.

Extensive thermodynamic properties that depend on the size of the system, such as internal energy, entropy, or enthalpy may be written as total system properties, specific properties (per unit mass), or molar properties (per mole). Symbolically we have not distinguished among them. In the list which follows, the property is defined as specific, except in those instances where it is normally employed on a molar basis, in which case the molar definition is given. The particular usage in the book is indicated in context.

Symbol	Definition	SI Units
a	Activity	—
A	Frequency factor in Arrhenius' law, nth order reaction	$(\mathrm{mol/m^3})^{1-n}/s$
A	Transfer area per unit mass of product produced per unit time	$\mathrm{m^2 \cdot yr/kg}$
°API	Degrees API, inverse measure of specific gravity, Eq. (6.11)	—
c	Concentration	$\mathrm{mol/m^3}$
c	Dimensionless constant in pressure drop equation (2.77)	—
C	Product fuel cost per unit mass	$\$/kg$
C	Specific heat capacity	$\mathrm{J/(kg \cdot K)}$
C_p	Specific heat capacity at constant pressure	$\mathrm{J/(kg \cdot K)}$
$C_{p,i}$	Molar heat capacity at constant pressure of species i	$\mathrm{J/(mol \cdot K)}$
C_v	Specific heat capacity at constant volume	$\mathrm{J/(kg \cdot K)}$
d_e	Characteristic length associated with particle size	m
d_s	Diameter of spherical particle	m
E	Molar activation energy	J/mol
E_V	Volumetric sweep efficiency, Eq. (7.11)	—
G	Specific Gibbs free energy	J/kg

Symbol	Definition	SI Unit
$\Delta G°$	Standard molar free energy change	J/mol
$\Delta_f G°$	Standard molar free energy of formation	J/mol
h	Henry's law constant, pressure per mole fraction	Pa
H	Coal seam or tar sand zone thickness	m
H	Specific enthalpy	J/kg
ΔH	Change in specific enthalpy	J/kg
ΔH_a	Change in specific enthalpy from original state to atmospheric ground state	J/kg
$\Delta H°$	Standard molar heat of reaction	J/mol
$-\Delta H°$	Standard molar heat of combustion	J/mol
$\Delta_f H°$	Standard molar heat of formation	J/mol
k	Rate constant, nth order reaction	$(\text{mol/m}^3)^{1-n}/\text{s}$
k	Permeability of porous medium	m^2
k	Capital recovery factor	yr^{-1}
k_{aux}	Cost per unit mass of product from sources independent of flux	\$/kg
k_c	Capital recovery factor per unit area of transfer surface	$\text{\$·yr/m}^2$
k_e	Energy cost	\$/J
K	Equilibrium constant, Eq. (2.45)	—
$K_{1,2}$	Equilibrium adsorption coefficients, Eq. (2.62)	—
$K_{I,II}$	Equilibrium constants in ideal gasifier example, Eqs. (3.21), (3.22)	—
m	Mass	kg
m_c	Mass of catalyst	kg
M	Molar mass	kg/kmol
$\bar{M}$	Mean molar mass	kg/kmol
n	Order of reaction	—
n	Number of moles	—
p	Pressure	Pa*
p_i	Partial pressure of species i	Pa*
$p_{v,i}$	Vapor pressure of species i	Pa
$\Delta p/L$	Pressure gradient	Pa/m
q	Thermal energy or heat per unit mass	J/kg
Q_{GCV}	Gross calorific value	J/kg
Q_{NCV}	Net calorific value	J/kg
r	Reaction rate	mol/s
R	Universal gas constant, 8.314	J/(mol·K)
Re	Reynolds number, $d_e\rho u/\mu$	—
s	Number of stages of compression	—
s.g.	Specific gravity, ratio of density of material to density of pure water at 15.6°C	—
S	Specific entropy	J/(kg·K)
S_b	Bitumen fraction or crude bitumen saturation	—
S_{red}	Reduced bitumen saturation, Eq. (7.11)	—
ΔS	Change in specific entropy	J/(kg·K)
ΔS_a	Change in specific entropy from original state to atmospheric ground state	J/(kg·K)
$\Delta_f S°$	Standard molar entropy of formation	J/(mol·K)
t	Time	s
T	Thermodynamic (absolute) temperature	K
u	Fluid velocity	m/s
u_f	Fluid velocity at which particles in packed bed are entrained	m/s
U	Specific internal energy	J/kg

Symbol	Definition	SI Unit
ΔU	Change in specific internal energy	J/kg
V	Volume	m^3
V	Volatile matter evolved	kg
V_∞	Total volatile matter evolved	kg
w	Work per unit mass done on system	J/kg
w_u	Useful work per unit mass done on system	J/kg
x_i	Mole fraction of gas species i in solution	—
y_i	Mole fraction of species i in gas mixture	—
y_w	Mole fraction of water vapor in gas mixture	—
β	Fraction of plant heat supplied by burning fuel gas or deashed liquid or solid product	—
γ	Ratio of specific heat capacities, C_p/C_V	—
ϵ	Voidage of packed bed, Eq. (2.81)	—
η	Thermal efficiency	—
$\eta_{\text{cold gas}}$	Cold gas efficiency, Eq. (2.60a)	—
$\eta_{\text{modified cold gas}}$	Modified cold gas efficiency, Eq. (2.60b)	—
η_{I}	First law efficiency, Eq. (2.24)	—
η_{II}	Second law efficiency, Eq. (2.28)	—
θ	Thermal ratio, Eq. (7.12a)	—
λ	Friction factor, Eq. (2.78a)	—
μ	Absolute viscosity	Pa·s
ν_i	Stoichiometric coefficient of species i	—
ρ	Mass density	kg/m^3
τ	Residence time, Eq. (2.84)	s
τ'	Space time, Eq. (2.83)	s
τ_H	Characteristic heating time for pyrolysis	s
τ_R	Characteristic reaction time for pyrolysis	s
ϕ	Specific availability, Eq. (2.52)	J/kg
ϕ_0	Standard chemical availability per unit mass	J/kg
$\Delta\phi$	Change in specific availability	J/kg

Subscripts

a	Atmospheric ground state	
F	Feed condition	
i	With reference to species i	
$\sqrt{k}$	Parameter defined in terms of length $\sqrt{k}$	
opt	Optimum	
rev	Reversible	
T	Reference temperature in kelvins	
0	At normal temperature and pressure	
298	Reference temperature of 298 K	

* Unit of pressure is atm in equations for chemical equilibrium.

ACRONYMS

As with symbols, acronyms that are only employed locally are generally not included. Also not included are acronyms for specific processes. These may be found in the subject index.

Acronym	Meaning
BAT	Best available treatmeant economically achievable
BFW	Boiler feed water
BOD	Biochemical oxygen demand
BPT	Best practicable treatment currently available
BTX	Benzene, toluene, and xylene
COD	Chemical oxygen demand
CV	Calorific value
daf	Dry and ash-free
DME	Dimethylether
EPA	U.S. Environmental Protection Agency
ESP	Electrostatic precipitator
FCC	Fluid catalytic cracking
FGD	Flue gas desulfurization
GCV	Gross calorific value
HC	Hydrocarbons
IBP	Initial boiling point
MIS	Modified *in situ* retorting
MSW	Municipal solid waste
NAAQS	National Ambient Air Quality Standards
NCV	Net calorific value
NPDES	National Pollutant Discharge Elimination System
NTP	Normal temperature and pressure
SAO	Saturates, aromatics, and olefins
SNG	Substitute natural gas
SRC	Solvent refined coal
syncrude	Synthetic crude oil
TDS	Total dissolved solids
TIS	True *in situ* retorting
TOC	Total organic carbon
UCG	Underground coal gasification
VM	Volatile matter

INDEX

INDEX

A CATALOG OF SELECTED
DOVER BOOKS
IN SCIENCE AND MATHEMATICS

Engineering

DE RE METALLICA, Georgius Agricola. The famous Hoover translation of greatest treatise on technological chemistry, engineering, geology, mining of early modern times (1556). All 289 original woodcuts. 638pp. 6¾ x 11.   0-486-60006-8

FUNDAMENTALS OF ASTRODYNAMICS, Roger Bate et al. Modern approach developed by U.S. Air Force Academy. Designed as a first course. Problems, exercises. Numerous illustrations. 455pp. 5⅜ x 8½.   0-486-60061-0

DYNAMICS OF FLUIDS IN POROUS MEDIA, Jacob Bear. For advanced students of ground water hydrology, soil mechanics and physics, drainage and irrigation engineering and more. 335 illustrations. Exercises, with answers. 784pp. 6⅛ x 9¼.
0-486-65675-6

THEORY OF VISCOELASTICITY (Second Edition), Richard M. Christensen. Complete consistent description of the linear theory of the viscoelastic behavior of materials. Problem-solving techniques discussed. 1982 edition. 29 figures. xiv+364pp. 6⅛ x 9¼.   0-486-42880-X

MECHANICS, J. P. Den Hartog. A classic introductory text or refresher. Hundreds of applications and design problems illuminate fundamentals of trusses, loaded beams and cables, etc. 334 answered problems. 462pp. 5⅜ x 8½.   0-486-60754-2

MECHANICAL VIBRATIONS, J. P. Den Hartog. Classic textbook offers lucid explanations and illustrative models, applying theories of vibrations to a variety of practical industrial engineering problems. Numerous figures. 233 problems, solutions. Appendix. Index. Preface. 436pp. 5⅜ x 8½.   0-486-64785-4

STRENGTH OF MATERIALS, J. P. Den Hartog. Full, clear treatment of basic material (tension, torsion, bending, etc.) plus advanced material on engineering methods, applications. 350 answered problems. 323pp. 5⅜ x 8½.   0-486-60755-0

A HISTORY OF MECHANICS, René Dugas. Monumental study of mechanical principles from antiquity to quantum mechanics. Contributions of ancient Greeks, Galileo, Leonardo, Kepler, Lagrange, many others. 671pp. 5⅜ x 8½.   0-486-65632-2

STABILITY THEORY AND ITS APPLICATIONS TO STRUCTURAL MECHANICS, Clive L. Dym. Self-contained text focuses on Koiter postbuckling analyses, with mathematical notions of stability of motion. Basing minimum energy principles for static stability upon dynamic concepts of stability of motion, it develops asymptotic buckling and postbuckling analyses from potential energy considerations, with applications to columns, plates, and arches. 1974 ed. 208pp. 5⅜ x 8½.
0-486-42541-X

METAL FATIGUE, N. E. Frost, K. J. Marsh, and L. P. Pook. Definitive, clearly written, and well-illustrated volume addresses all aspects of the subject, from the historical development of understanding metal fatigue to vital concepts of the cyclic stress that causes a crack to grow. Includes 7 appendixes. 544pp. 5⅜ x 8½.   0-486-40927-9

ROCKETS, Robert Goddard. Two of the most significant publications in the history of rocketry and jet propulsion: "A Method of Reaching Extreme Altitudes" (1919) and "Liquid Propellant Rocket Development" (1936). 128pp. 5⅜ x 8½.   0-486-42537-1

STATISTICAL MECHANICS: PRINCIPLES AND APPLICATIONS, Terrell L. Hill. Standard text covers fundamentals of statistical mechanics, applications to fluctuation theory, imperfect gases, distribution functions, more. 448pp. 5⅜ x 8½.

0-486-65390-0

ENGINEERING AND TECHNOLOGY 1650–1750: ILLUSTRATIONS AND TEXTS FROM ORIGINAL SOURCES, Martin Jensen. Highly readable text with more than 200 contemporary drawings and detailed engravings of engineering projects dealing with surveying, leveling, materials, hand tools, lifting equipment, transport and erection, piling, bailing, water supply, hydraulic engineering, and more. Among the specific projects outlined-transporting a 50-ton stone to the Louvre, erecting an obelisk, building timber locks, and dredging canals. 207pp. 8⅜ x 11¼.

0-486-42232-1

THE VARIATIONAL PRINCIPLES OF MECHANICS, Cornelius Lanczos. Graduate level coverage of calculus of variations, equations of motion, relativistic mechanics, more. First inexpensive paperbound edition of classic treatise. Index. Bibliography. 418pp. 5⅜ x 8½.   0-486-65067-7

PROTECTION OF ELECTRONIC CIRCUITS FROM OVERVOLTAGES, Ronald B. Standler. Five-part treatment presents practical rules and strategies for circuits designed to protect electronic systems from damage by transient overvoltages. 1989 ed. xxiv+434pp. 6⅛ x 9¼.   0-486-42552-5

ROTARY WING AERODYNAMICS, W. Z. Stepniewski. Clear, concise text covers aerodynamic phenomena of the rotor and offers guidelines for helicopter performance evaluation. Originally prepared for NASA. 537 figures. 640pp. 6⅛ x 9¼.

0-486-64647-5

INTRODUCTION TO SPACE DYNAMICS, William Tyrrell Thomson. Comprehensive, classic introduction to space-flight engineering for advanced undergraduate and graduate students. Includes vector algebra, kinematics, transformation of coordinates. Bibliography. Index. 352pp. 5⅜ x 8½.   0-486-65113-4

HISTORY OF STRENGTH OF MATERIALS, Stephen P. Timoshenko. Excellent historical survey of the strength of materials with many references to the theories of elasticity and structure. 245 figures. 452pp. 5⅜ x 8½.   0-486-61187-6

ANALYTICAL FRACTURE MECHANICS, David J. Unger. Self-contained text supplements standard fracture mechanics texts by focusing on analytical methods for determining crack-tip stress and strain fields. 336pp. 6⅛ x 9¼.   0-486-41737-9

STATISTICAL MECHANICS OF ELASTICITY, J. H. Weiner. Advanced, self-contained treatment illustrates general principles and elastic behavior of solids. Part 1, based on classical mechanics, studies thermoelastic behavior of crystalline and polymeric solids. Part 2, based on quantum mechanics, focuses on interatomic force laws, behavior of solids, and thermally activated processes. For students of physics and chemistry and for polymer physicists. 1983 ed. 96 figures. 496pp. 5⅜ x 8½.

0-486-42260-7

Mathematics

FUNCTIONAL ANALYSIS (Second Corrected Edition), George Bachman and Lawrence Narici. Excellent treatment of subject geared toward students with background in linear algebra, advanced calculus, physics and engineering. Text covers introduction to inner-product spaces, normed, metric spaces, and topological spaces; complete orthonormal sets, the Hahn-Banach Theorem and its consequences, and many other related subjects. 1966 ed. 544pp. 6⅛ x 9¼. 0-486-40251-7

ASYMPTOTIC EXPANSIONS OF INTEGRALS, Norman Bleistein & Richard A. Handelsman. Best introduction to important field with applications in a variety of scientific disciplines. New preface. Problems. Diagrams. Tables. Bibliography. Index. 448pp. 5⅜ x 8½. 0-486-65082-0

VECTOR AND TENSOR ANALYSIS WITH APPLICATIONS, A. I. Borisenko and I. E. Tarapov. Concise introduction. Worked-out problems, solutions, exercises. 257pp. 5⅝ x 8¼. 0-486-63833-2

AN INTRODUCTION TO ORDINARY DIFFERENTIAL EQUATIONS, Earl A. Coddington. A thorough and systematic first course in elementary differential equations for undergraduates in mathematics and science, with many exercises and problems (with answers). Index. 304pp. 5⅜ x 8½. 0-486-65942-9

FOURIER SERIES AND ORTHOGONAL FUNCTIONS, Harry F. Davis. An incisive text combining theory and practical example to introduce Fourier series, orthogonal functions and applications of the Fourier method to boundary-value problems. 570 exercises. Answers and notes. 416pp. 5⅜ x 8½. 0-486-65973-9

COMPUTABILITY AND UNSOLVABILITY, Martin Davis. Classic graduate-level introduction to theory of computability, usually referred to as theory of recurrent functions. New preface and appendix. 288pp. 5⅜ x 8½. 0-486-61471-9

ASYMPTOTIC METHODS IN ANALYSIS, N. G. de Bruijn. An inexpensive, comprehensive guide to asymptotic methods–the pioneering work that teaches by explaining worked examples in detail. Index. 224pp. 5⅜ x 8½ 0-486-64221-6

APPLIED COMPLEX VARIABLES, John W. Dettman. Step-by-step coverage of fundamentals of analytic function theory–plus lucid exposition of five important applications: Potential Theory; Ordinary Differential Equations; Fourier Transforms; Laplace Transforms; Asymptotic Expansions. 66 figures. Exercises at chapter ends. 512pp. 5⅜ x 8½. 0-486-64670-X

INTRODUCTION TO LINEAR ALGEBRA AND DIFFERENTIAL EQUATIONS, John W. Dettman. Excellent text covers complex numbers, determinants, orthonormal bases, Laplace transforms, much more. Exercises with solutions. Undergraduate level. 416pp. 5⅜ x 8½. 0-486-65191-6

RIEMANN'S ZETA FUNCTION, H. M. Edwards. Superb, high-level study of landmark 1859 publication entitled "On the Number of Primes Less Than a Given Magnitude" traces developments in mathematical theory that it inspired. xiv+315pp. 5⅜ x 8½. 0-486-41740-9

Paperbound unless otherwise indicated. Available at your book dealer, online at **www.doverpublications.com**, or by writing to Dept. GI, Dover Publications, Inc., 31 East 2nd Street, Mineola, NY 11501. For current price information or for free catalogues (please indicate field of interest), write to Dover Publications or log on to **www.doverpublications.com** and see every Dover book in print. Dover publishes more than 500 books each year on science, elementary and advanced mathematics, biology, music, art, literary history, social sciences, and other areas.